# Animal Genetics: Epigenetics and Chromatin

# Animal Genetics: Epigenetics and Chromatin

Edited by George Gardner

SYRAWOOD
PUBLISHING HOUSE

New York

Published by Syrawood Publishing House,
750 Third Avenue, 9th Floor,
New York, NY 10017, USA
www.syrawoodpublishinghouse.com

**Animal genetics : epigenetics and chromatin**
Edited by George Gardner

International Standard Book Number: 978-1-68286-782-2 (Hardback)

**Cataloging-in-Publication Data**

Animal genetics : epigenetics and chromatin / edited by George Gardner.
    p. cm.
Includes bibliographical references and index.
ISBN 978-1-68286-782-2
1. Animal genetics. 2. Epigenetics. 3. Chromatin. I. Gardner, George.
QH432 .A55 2019
591.35--dc23

# TABLE OF CONTENTS

# PREFACE

Over the recent decade, advancements and applications have progressed exponentially. This has led to the increased interest in this field and projects are being conducted to enhance knowledge. The main objective of this book is to present some of the critical challenges and provide insights into possible solutions. This book will answer the varied questions that arise in the field and also provide an increased scope for furthering studies.

Animal genetics refers to the study of genes, variation and heredity in animals. An animal's behavior, development and genetic processes are influenced by their environment and experiences. Genetic changes occur due to mutations, natural selection and evolution. The study of the heritable changes in gene function that do not arise due to a change in the DNA sequence is under the domain of epigenetics. Histone modification and DNA methylation are few mechanisms that can produce such changes. Macromolecules found in the cells, which consist of protein, RNA and DNA form a complex, which is called the chromatin. Its fundamental function is to prevent DNA damage, allow mitosis by reinforcing the DNA macromolecule, pack DNA more densely and compactly, and control the DNA replication and gene expression. The topics covered in this extensive book deal with the core aspects of animal genetics, epigenetics and chromatin. Scientists and students actively engaged in this field will find this book full of crucial and unexplored concepts. It traces the progress of this discipline and highlights some of its key concepts and applications.

I hope that this book, with its visionary approach, will be a valuable addition and will promote interest among readers. Each of the authors has provided their extraordinary competence in their specific fields by providing different perspectives as they come from diverse nations and regions. I thank them for their contributions.

**Editor**

# Histone and DNA methylation control by H3 serine 10/threonine 11 phosphorylation in the mouse zygote

Jie Lan[1,2†], Konstantin Lepikhov[1†], Pascal Giehr[1] and Joern Walter[1*]

## Abstract

**Background:** In the mammalian zygote, epigenetic reprogramming is a tightly controlled process of coordinated alterations of histone and DNA modifications. The parental genomes of the zygote show distinct patterns of histone H3 variants and distinct patterns of DNA and histone modifications. The molecular mechanisms linking histone variant-specific modifications and DNA methylation reprogramming during the first cell cycle remain to be clarified.

**Results:** Here, we show that the degree and distribution of H3K9me2 and of DNA modifications (5mC/5hmC) are influenced by the phosphorylation status of H3S10 and H3T11. The overexpression of the mutated histone variants H3.1 and 3.2 at either serine 10 or threonine 11 causes a decrease in H3K9me2 and 5mC and a concomitant increase in 5hmC in the maternal genome. Bisulphite sequencing results indicate an increase in hemimethylated CpG positions following H3.1T10A overexpression suggesting an impact of H3S10 and H3T11 phosphorylation on DNA methylation maintenance.

**Conclusions:** Our data suggest a crosstalk between the cell-cycle-dependent control of S10 and T11 phosphorylation of histone variants H3.1 and H3.2 and the maintenance of the heterochromatic mark H3K9me2. This histone H3 "phospho-methylation switch" also influences the oxidative control of DNA methylation in the mouse zygote.

## Background

The epigenetic reprogramming in mouse zygote involves an extensive rearrangement of the epigenetic landscape, including chromatin reorganization and comprehensive changes in DNA modifications. These changes require a coordinated control of epigenetic "writers", "readers", "erasers" and "remodelers" on the level of histones and DNA after fertilization. The interplay between histone variants, chromatin modifications and DNA modifications has been studied to a great detail. Here, we analyse the synergetic dynamics of different post-translational modifications in histone H3 variants H3.1, H3.2 and H3.3 which are found in different epigenetic compartments of chromatin [1]. In mouse zygotes, these histone variants show an asymmetrical deposition into parental pronuclei: H3.3 is a predominant histone variant in the newly formed paternal pronucleus, while H3.1 and H3.2 only appear during the first replication of the paternal chromatin. In contrast, the maternal chromatin is initially enriched for H3.1/H3.2 and accumulates H3.3 at later zygotic stages [2–4]. This asymmetry in histone variant composition is accompanied by an asymmetric allocation of histone modifications in both pronuclei [2, 5, 6]. While the paternal chromatin is mainly marked by open chromatin modifications such as H3K4me3, the maternal chromatin shows a high abundance of H3K9me2 heterochromatic mark, which is slightly reduced during the first cell division [5, 7]. In pre-replicative paternal chromatin, H3K9me2 is almost absent and only becomes detectable at late replication stages. In both pronuclei, the abundance of H3K9me2 is linked to differences in DNA modifications. The H3K9me2 containing maternal pronucleus maintains 5mC as the predominant modification, and the level of DNA methylation decreases only slightly during

*Correspondence: j.walter@mx.uni-saarland.de
†Jie Lan and Konstantin Lepikhov contributed equally to this work
1 FR 8.3, Biological Sciences, Genetics/Epigenetics, University of Saarland, Campus A2.4, 66123 Saarbrücken, Germany
Full list of author information is available at the end of the article

the first DNA replication [8]. It has been shown that the presence of H3K9me2 in the maternal pronuclei protects against Tet3-mediated oxidation of 5mC to 5hmC [9]. As a consequence of this, the maternal chromosomes appear to maintain 5mC levels in contrast to the more oxidized paternal chromosomes, which are practically devoid of H3K9me2 at early stages of DNA replication and where 5mC is extensively converted to 5hmC by Tet3, reducing DNA methylations by about 50% at the end of the first cell cycle [10].

The current knowledge suggests that H3K9me2 has an important protective role for the maintenance of 5mC. Work by Nakamura et al. showed that Stella protein while present in both pronuclei only protects the maternal DNA against Tet3 oxidation due to the presence of H3K9me2 [9, 11]. However, previous data also suggest that this epigenetic control could be linked to the asymmetric distribution of the major histone variants H3.1, H3.2 and H3.3 [4]. H3S10 phosphorylation has been shown to negatively control H3K9 methylation in fruit fly [12]. In vitro biochemical assays demonstrated a protective role of H3T11 phosphorylation against H3K9me3 demethylation [13, 14]. However, interactions of H3S10phos and H3T11phos with H3K9me2 in mammalian cells have not yet been described. The vicinity of the K9, S10 and T11 residues in the N-terminus of H3 suggests a possible influence or crosstalk of modifications at these residues. This crosstalk might influence "writers" or "erasers" of individual modifications or alternatively affect the interaction with modification "readers". Phosphorylation of histone H3 fulfils multiple roles: it participates in mitotic chromosomes condensation and segregation, but also modulates gene expression in the context-dependent manner (reviewed in [15]). Our intention was to examine the potential links between the asymmetric distribution of histone variants and the various layers of epigenetic control in both pronuclei before and after replication. In particular, we analysed the dynamics of H3S10 and H3T11 phosphorylation in the three histone H3 variants during the first cell cycle and their impact on the replication-dependent control of H3K9me2 and DNA modifications in both zygotic pronuclei. Our data reveal a direct or indirect crosstalk between H3S10 and H3T11 phosphorylation, histone variant-dependent H3K9me2 methylation and DNA methylation.

## Results

### H3S10phos and H3T11phos have different dynamics and associations with histone H3 variants in the mouse zygote

We first determined the dynamics of H3S10phos and H3T11phos in the developing mouse zygotes. In line with previous reports, we observe that H3S10phos is clearly detectable in G1 (PN1/2), disappears in S (PN3) phase and reappears at late G2 (PN4/5). The mainly perinucleolar accumulation of H3S10phos is most pronounced in the paternal pronucleus at all stages [16] (Fig. 1). H3T11phos also accumulates in the perinucleolar heterochromatin but follows a different dynamic: it is absent in G1, becomes first visible during early S-phase and gradually accumulates during S-phase to remain as a strong signal up to G2 (Fig. 1). In contrast to H3S10phos which clearly shows signal intensity differences between maternal and paternal pronuclei, H3T11phos signals are equally absent or present in both pronuclei. We conclude that the neighbouring phosphorylation marks have similar nuclear patterns but different dynamics during the first cell cycle. While H3T11phos marks the replicative S and the G2 phase, H3S10phos is mainly present in the non-replicative G1 and G2 phases.

We next investigated the dynamic of H3S10 and H3T11 phosphorylation on all three histone variants. We therefore microinjected mRNAs encoding either histone wild-type H3 variants (WT) or S10A or T11A mutated forms, respectively, into early pre-replicative (2–3 h post-fertilization) mouse zygotes. The ectopically expressed wild-type or mutated forms of H3 variants were fused to GFP reporter (C-terminal fusion). This allowed us to follow their import in the pronuclei. We observe that all WT and mutant forms were readily expressed and efficiently imported into the pronuclei (Additional file 1). We assumed that the mutated non-phosphorylatable H3 variants will be incorporated into nucleosomes generating a "dominant-negative" phosphorylation effect in nucleosomes after replication at PN4/5.

Indeed, the overexpression of all three H3S10A mutated isoforms (H3.1-GFPS10A, H3.2-GFPS10A and H3.3-GFPS10A) leads to a significant decrease in H3S10phos in G2 zygotes, as visualized and measured by immunofluorescence (IF) (Fig. 2a, b). In contrast, H3T11phos was reduced when we injected and overexpressed the mutated H3.1-GFPT11A and H3.2-GFPT11A variants but remained unaffected with the H3.3-GFPT11A variant (Fig. 3a, b). Our data suggest that while all three H3 variants are equal substrates for H3S10 phosphorylation, only H3.1 and H3.2 variants are the predominant targets for H3T11 phosphorylation.

Note that the overexpression of H3WT-GFP variants in the majority of cases did not change the H3S10 and H3T11 phosphorylation pattern. However, in each experiment we observe a few examples in which overexpression leads to a reduction in the respective phosphorylation signals (Additional file 2). This effect may be caused by the time of injection or a variable amount of injected material, leading to a higher abundance of H3.1WT-GFP and H3.2WT-GFP overexpressed proteins

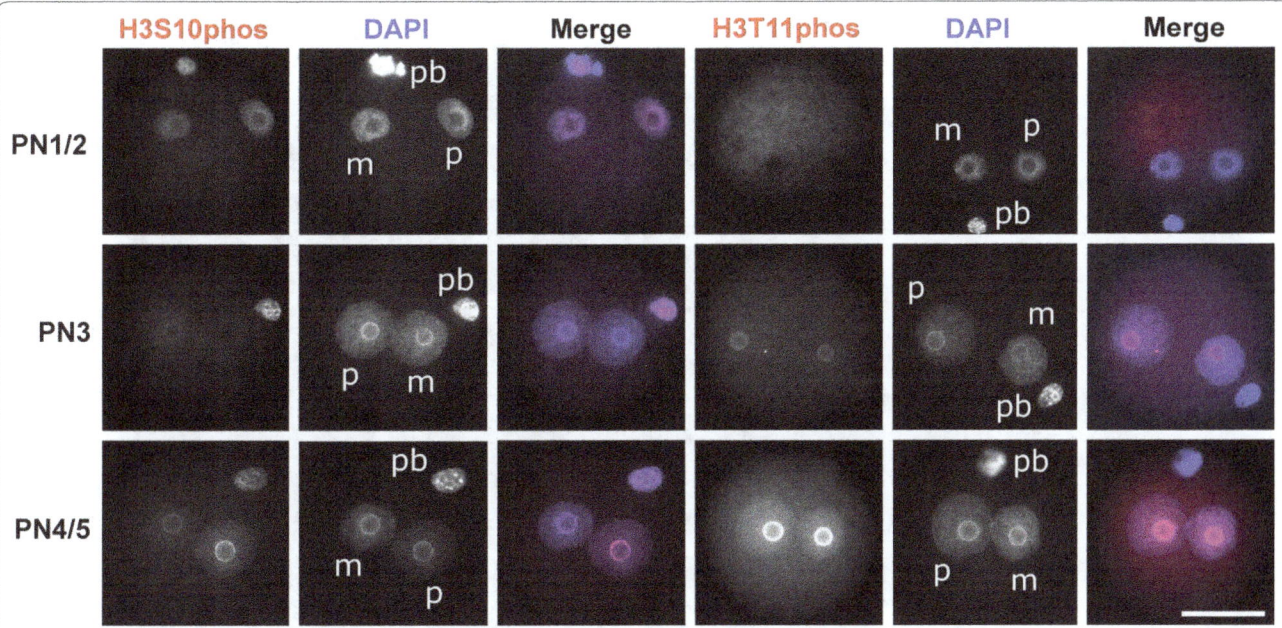

**Fig. 1** Dynamic patterns of H3S10phos and H3T11phos in mouse zygotes. Representative images of zygotes at different PN stages stained with antibodies against H3S10phos and H3T11phos, respectively. DNA is visualized by DAPI. *m* Maternal pronucleus, *p* paternal pronucleus, *pb* polar body. *Scale bar* 50 μm

competing with endogenous H3 for the kinase activity. Zygotes injected with H3.3WT-GFP mRNA did not show variation in H3T11phos signals (Additional file 2), supporting the notion that H3.3 is not a substrate for H3T11 phosphorylation-specific reaction.

### H3S10 and H3T11 phosphorylation is coupled to H3K9me2 histone methylation

The perinucleolar heterochromatic signature of H3S10 and H3T11 phosphorylation prompted us to investigate whether: (1) phosphorylation is linked to canonical heterochromatic marks such as H3K9me2 and (2) such effects are found for all histone variants. Our approach was to individually overexpress H3S10A mutants of all three histone variants and analyse the H3K9me2 status in G2 zygotes at PN4/5. We indeed find that the overexpression of all three mutant variants caused a measurable reduction in H3K9me2 at G2 phase compared to non-injected control. A reduction in about 30% signal intensity was found in maternal pronuclei of H3.1-GFPS10A and H3.2-GFPS10A injected groups, while the

H3.3-GFPS10A injected group only showed an average reduction in about 15% (Fig. 4a, b).

Next, we analysed H3K9me2 signals in zygotes overexpressing H3T11A mutants. We observe a strong and highly significant reduction in H3K9me2 signals in maternal pronuclei when overexpressing H3.1-GFPT11A and H3.2-GFPT11A (reduction in about 70 and 30%, respectively) but no change in both pronuclei when overexpressing H3.3-GFPT11A (Fig. 5a, b). Note that the IF signals remain constant in polar body nuclei (Figs. 4a, 5a). For H3.1-GFPT11A, we even find a mild but significant reduction in the low-level H3K9me2 in the paternal G2 pronuclei. We also performed a side-by-side comparison of zygotes, expressing H3.1-GFPT11A to zygotes expressing H3.1-GFPWT (Additional file 3). This comparison revealed a clear reduction in H3K9me2 levels in the mutants. The overexpression of WT histone H3.1 does not cause a significant reduction.

From all these experiments, we conclude that H3S10 and H3T11 phosphorylation particularly of variants H3.1

(See figure on next page.)
**Fig. 2** Effects of H3.1/2/3-GFPS10A expression in mouse zygotes on H3S10phos. **a** Shown are the representative images of PN4/5 stage zygotes stained with antibodies against H3S10phos. DNA is visualized by DAPI. *m* Maternal pronucleus, *p* paternal pronucleus, *pb* polar body. *Scale bar* 50 μm. **b** Quantification of H3S10phos signals, normalized against DNA signals in both parental genomes of zygotes at PN4/5. Relative signal intensities in control groups are set to 1. Statistical significance was calculated using *t* test (***P < 0.001)

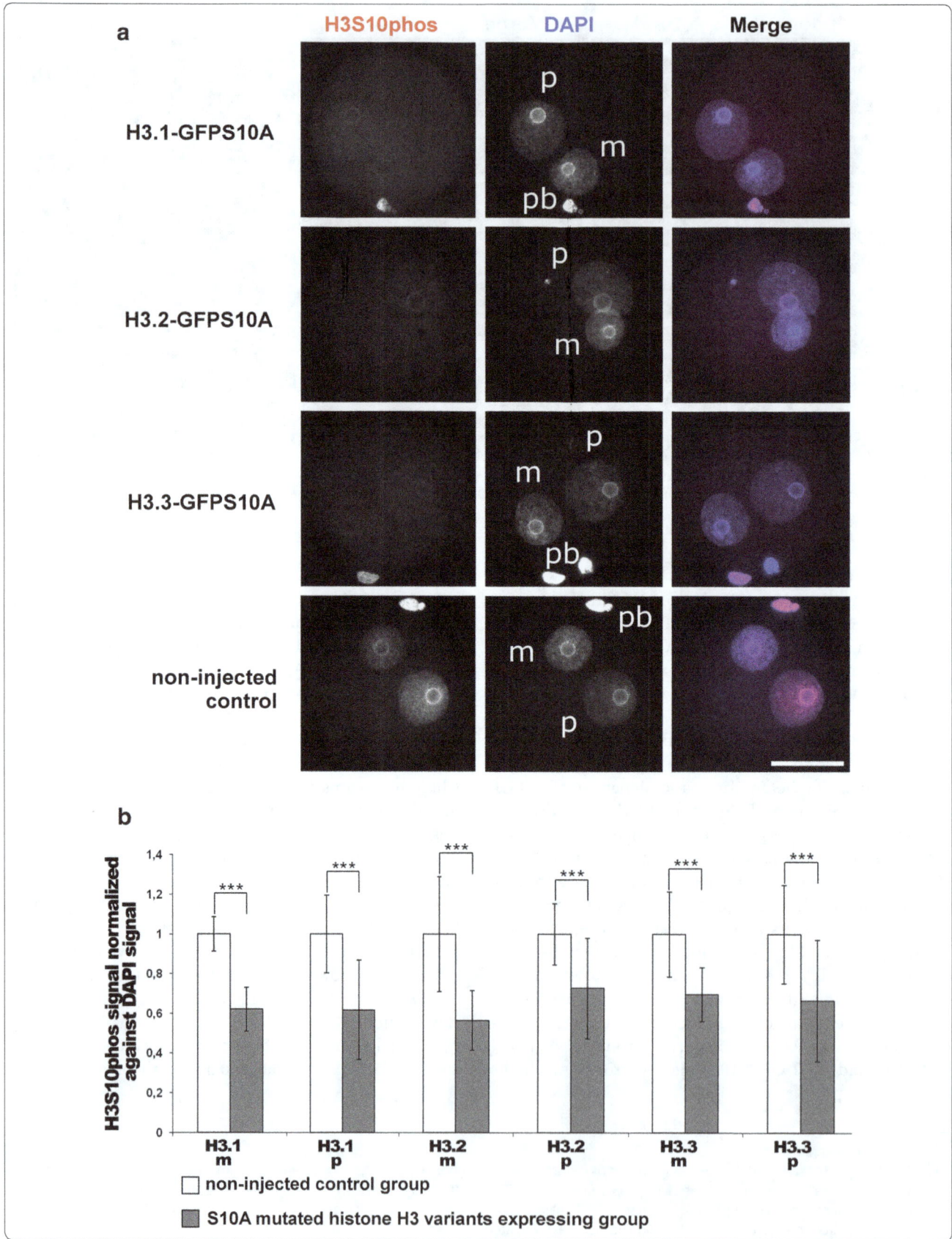

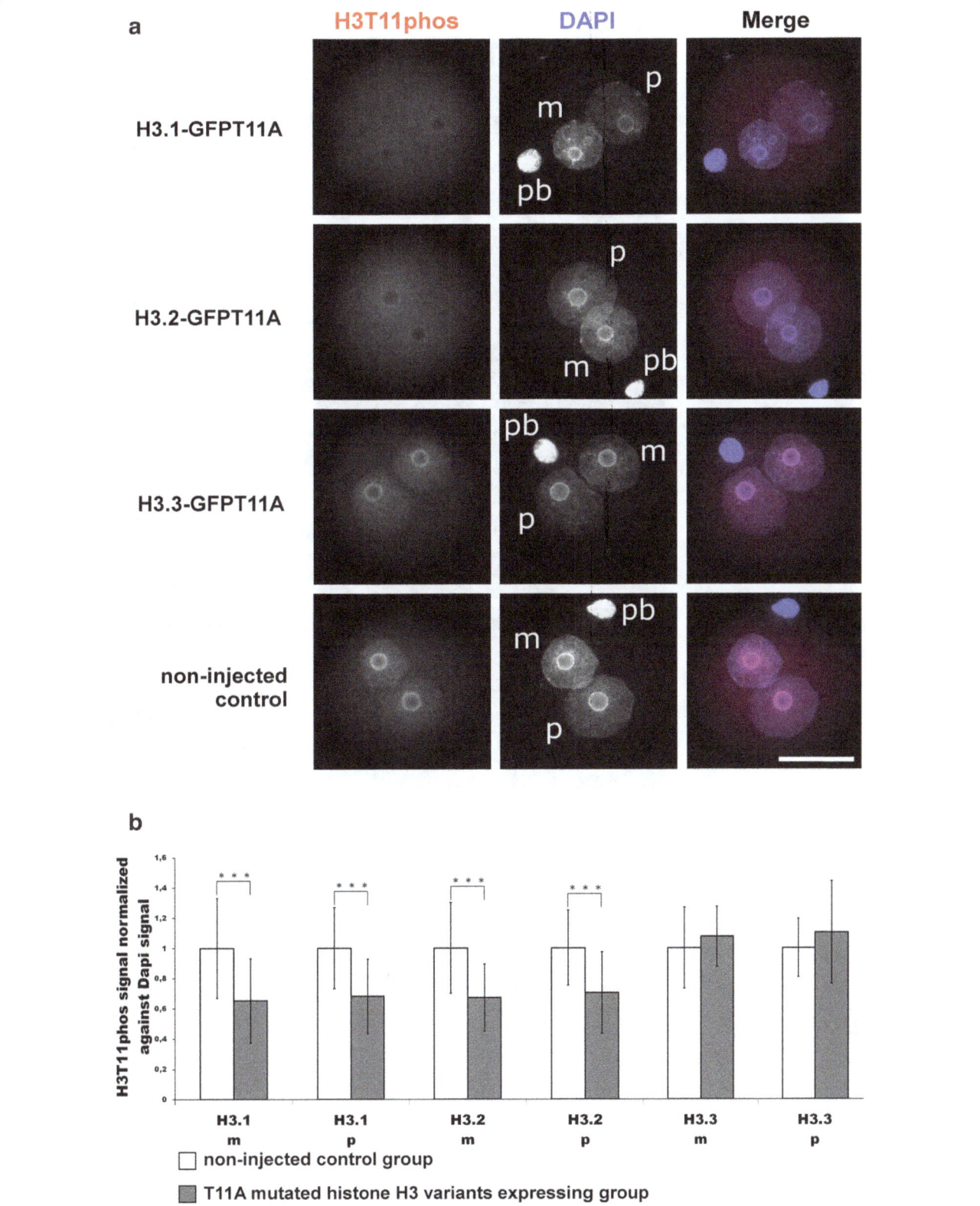

**Fig. 3** Effects of H3.1/2/3-GFPT11A expression in mouse zygotes on H3T11phos. **a** Shown are the representative images of PN4/5 stage zygotes stained with antibodies against H3T11phos. DNA is visualized by DAPI. *m* Maternal pronucleus, *p* paternal pronucleus, *pb* polar body. *Scale bar* 50 μm. **b** Quantification of H3T11phos signals, normalized against DNA signals in both parental genomes of zygotes at PN4/5. Relative signal intensities in control groups are set to 1. Statistical significance was calculated using *t* test (***P < 0.001)

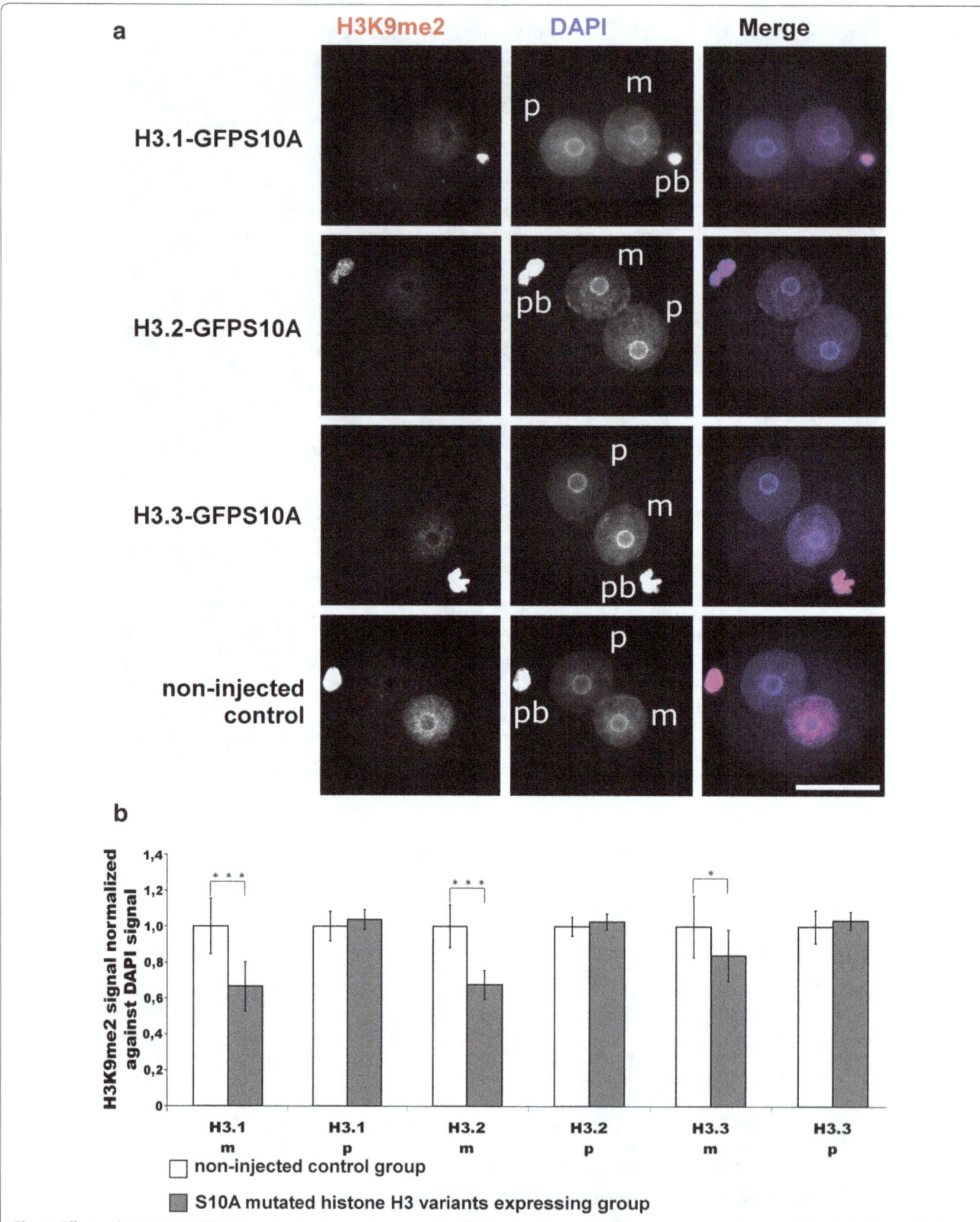

**Fig. 4** Effects of H3.1/2/3-GFPS10A expression in mouse zygotes on H3K9me2. **a** Shown are the representative images of PN4/5 stage zygotes stained with antibodies against H3K9me2. DNA is visualized by DAPI. *m* Maternal pronucleus, *p* paternal pronucleus, *pb* polar body. *Scale bar* 50 μm. **b** Quantification of H3K9me2 signals, normalized against DNA signals in both parental genomes of zygotes at PN4/5. Relative signal intensities in control groups are set to 1. Statistical significance was calculated using *t* test (***$P < 0.001$; *$P < 0.05$)

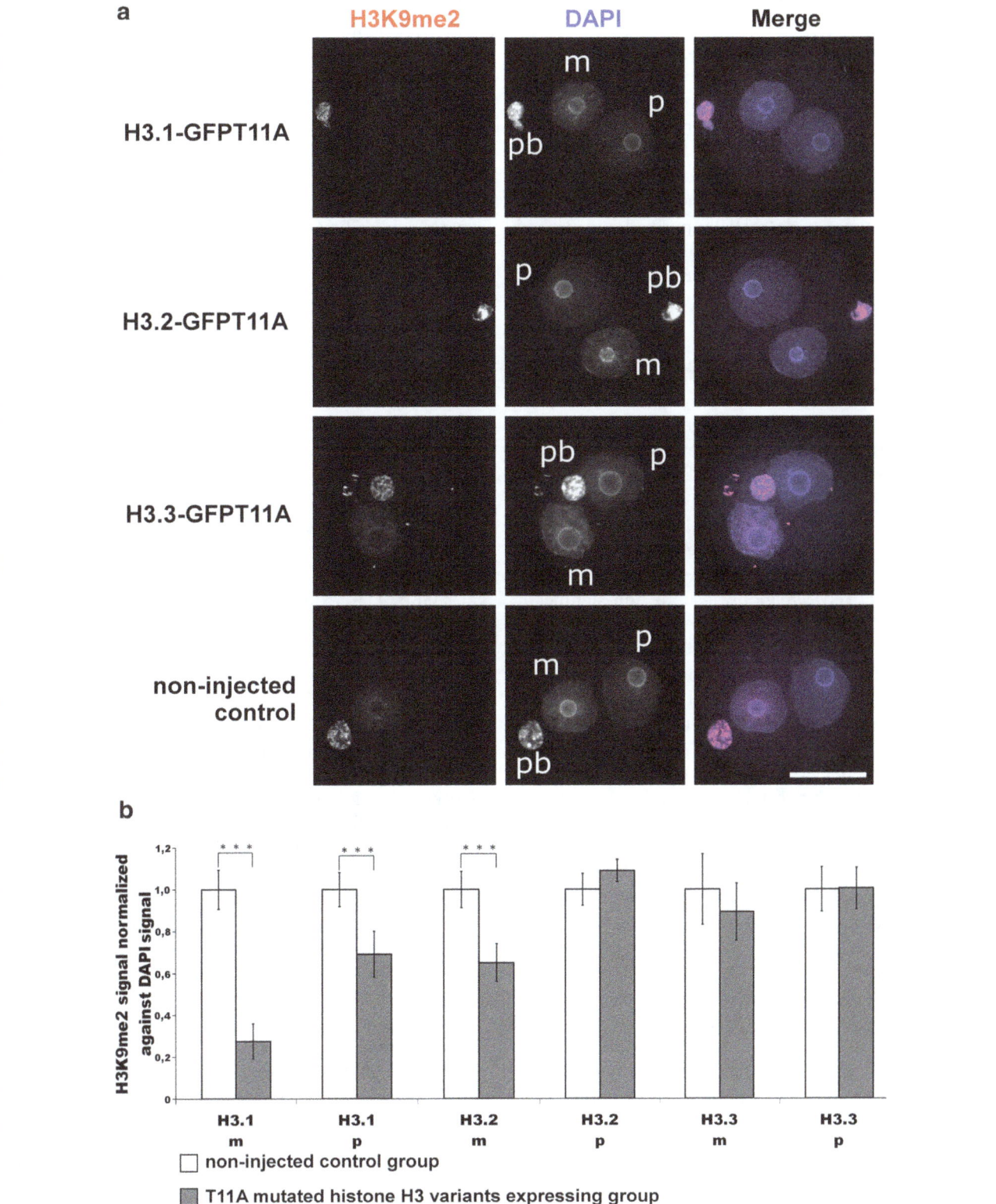

**Fig. 5** Effects of H3.1/2/3-GFPT11A expression in mouse zygotes on H3K9me2. **a** Shown are the representative images of PN4/5 stage zygotes stained with antibodies against H3K9me2. DNA is visualized by DAPI. *m* Maternal pronucleus, *p* paternal pronucleus, *pb* polar body. *Scale bar* 50 μm. **b** Quantification of H3K9me2 signals, normalized against DNA signals in both parental genomes of zygotes at PN4/5. Relative signal intensities in control groups are set to 1. Statistical significance was calculated using *t* test (***$P < 0.001$)

and H3.2 strongly influences the post-replicative levels of H3K9me2.

A simple explanation for a reduction in H3K9me2 signals following mutant overexpression is that the recognition and binding of anti-H3K9me2 antibody to its epitope are affected by the mutation. To examine this possibility, we co-expressed either H3.1-GFPT11A or H3.1-GFPWT together with the catalytical domain of G9a histone methyltransferase (G9aCat) in *E. coli* cells. The wild-type or mutated histones were partially purified and probed by Western blot using anti-H3K9me2 antibody. Indeed, the antibody clearly detects the H3K9me2 modification after co-expression on both WT and T11A mutated form (Additional file 4).

## H3S10 and H3T11 phosphorylation influences DNA modifications

H3K9me2 has been shown to be linked to the presence of 5mC in the maternal pronucleus (reviewed in [17]). We therefore investigated whether the observed reduction in maternal H3K9me2 also affected the corresponding 5mC levels. Indeed, we find that the levels of 5mC in maternal genomes are significantly reduced by about 20–15% when overexpressing H3.1-GFPS10A and H3.2-GFPS10A mutants, respectively, while only a subtle non-significant reduction is found in H3.3-GFPS10A expressing zygotes (Fig. 6a, b). Moreover, the loss of 5mC was accompanied by a gain of 5hmC. Surprisingly, the gain of 5hmC was observed for all three H3.1, H3.2 and H3.3 S10A mutants (Fig. 6a). Hence, despite only a very subtle change in the 5mC signal in H3.3-GFPS10A expressing zygotes, the maternal pronuclei show a clear increase in 5hmC (Fig. 6a). Note that due to technical obstacles (antibody compatibility) we were unable to directly quantify the 5hmC signal, normalized against DNA antibody signal (as done for 5mC quantification). We therefore adjusted the denaturation conditions allowing us to detect anti-5hmC antibody signals and DNA signals (via propidium iodide, PI) simultaneously. Using this strategy, we were able to quantify the ratio of DNA and IF signal, and we find a clear and highly significant increase in 5hmC in the maternal pronuclei (Fig. 6c). In addition, we calculated the paternal-to-maternal ratio of 5hmC and found a significant increase in the maternal 5hmC content ratio in overexpressing zygotes suggesting a relative increase in maternal 5hmC levels (Additional file 5). Both the loss of 5mC and the gain of 5hmC in maternal pronuclei are more pronounced in H3.1-GFPS10A expressing zygotes (Fig. 6 and Additional file 5). In summary, our data suggest that the incorporation of non-phosphorylatable H3S10 variants has a variant-specific influence on the maintenance of H3K9me2 and the conversion of 5mC to 5hmC.

We found that overexpression of H3-GFPT11A mutants generated a very similar (almost identical) spectrum of variant-specific DNA modification changes. 5mC was significantly reduced in both H3.1- and H3.2-GFPT11A groups (Fig. 7a, b), and 5hmC signals were strongly enhanced in both H3.1-GFPT11A and H3.2-GFPT11A expressing groups compared to the controls (Fig. 7a, c and Additional file 5). Again, no significant change of 5mC was found in H3.3-GFPT11A expressing zygotes, while 5hmC in maternal pronuclei was increased in H3.3-GFPT11A expressing group (Fig. 7a).

Next, we examined the changes in 5hmC/5mC at the molecular level using sequencing-based approaches. We concentrated our analysis on zygotes overexpressing H3.1T11A and for which we observed the most extensive effects on 5mC and 5hmC in our IF analysis. We used hairpin bisulphite sequencing to monitor the strand-specific methylation status (methylated, unmethylated and hemimethylated) at individual CpG positions after replication [18, 19]. In contrast to the significant reduction seen in IF analysis, we only observed a small reduction in the total Line1 methylation in the H3.1-GFPT11A expressing zygotes. However, we found a strong and significant increase in hemimethylated positions in the H3.1-GFPT11A expressing group in contrast to non-injected controls (Fig. 8). This indicated that the overexpression of H3.1GFPT10A affects the DNA methylation maintenance at replication [20, 21].

## Additional H3K9me2 methylation causes only subtle changes in DNA modifications

H3K9me2 is not detectable on the paternal chromatin before replication and appears weakly during late replication in the mouse zygote [5, 6]. This leads us and others to the assumption that the absence of H3K9me2 in the paternal genome causes a strong Tet-mediated 5mC oxidation followed by a mostly replication-dependent "passive" demethylation [20, 21]. A correlation between H3K9me2 and 5mC/5hmC has already been shown for maternal chromatin in mouse zygotes [9]. To address the question whether an increase in H3K9me2 on paternal chromatin would influence DNA methylation and hydroxymethylation, we ectopically expressed G9a, an H3K9me2 histone methyltransferase, in mouse zygotes. We first analysed the pattern of endogenous G9a and observed its appearance (and nuclear localization) starting from the four-cell stage, but not at earlier developmental stages (Additional file 6). We first expressed a G9a full-length GFP tagged version (G9aFL-GFP) in the zygote, which lead to only a very minor effect on H3K9me2. We concluded that the N-terminus of G9a interfered with the catalytic function in the zygote, suppressing the G9a methylation function [7]. Indeed, the

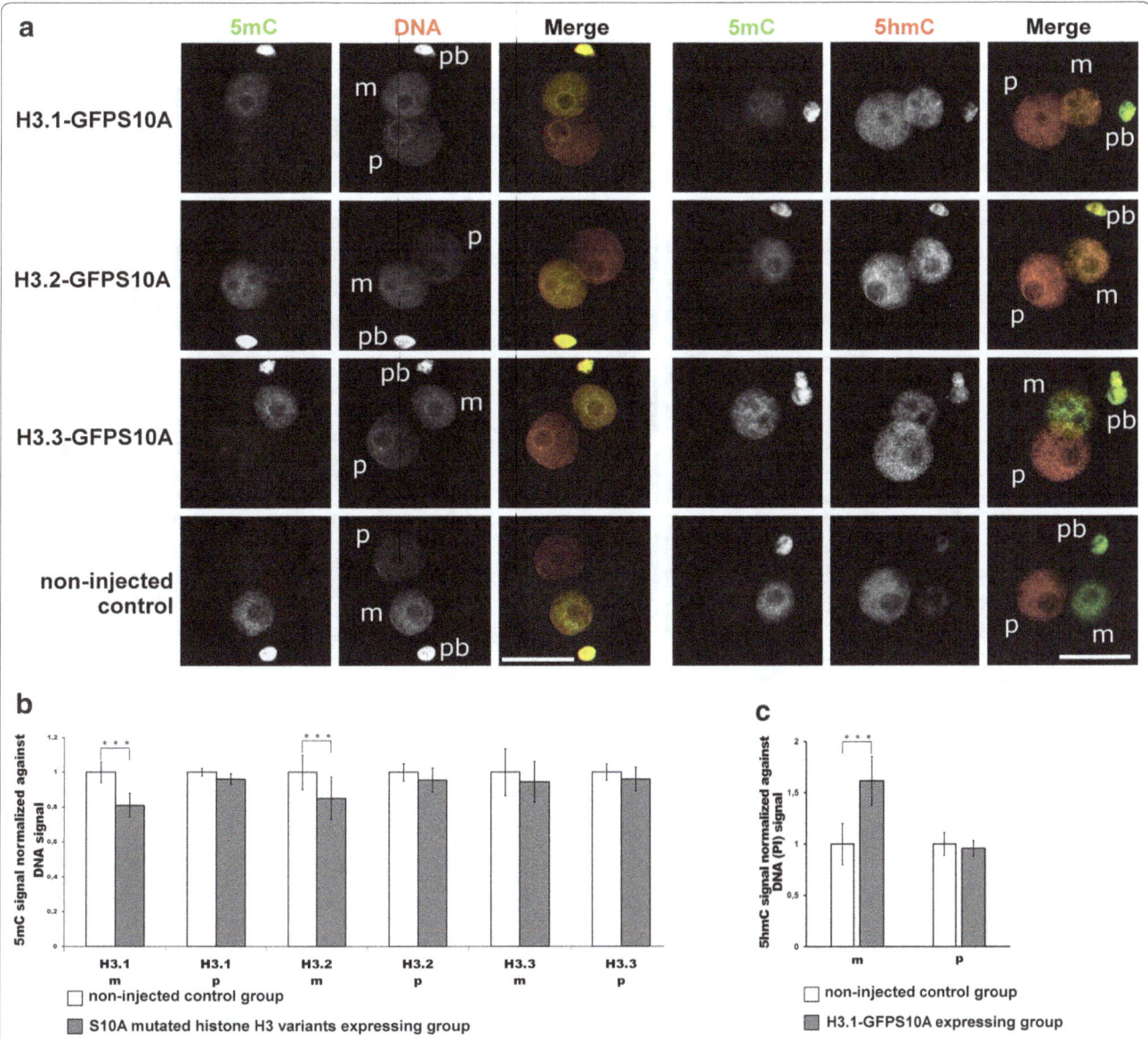

**Fig. 6** Effects of H3.1/2/3-GFPS10A expression in mouse zygotes on 5mC and 5hmC. **a** Shown are the representative images of PN4/5 stage zygotes stained with antibodies against 5mC together with anti-ssDNA antibodies, or together with 5hmC antibodies. *m* Maternal pronucleus, *p* paternal pronucleus, *pb* polar body. *Scale bar* 50 μm. **b** Quantification of 5mC signals, normalized against ssDNA signals in both parental genomes of zygotes at PN4/5. Relative signal intensities in control groups are set to 1. Statistical significance was calculated using *t* test (***P < 0.001). **c** Quantification of 5hmC signals, normalized against DNA (*PI* propidium iodide) signals in both parental genomes of zygotes at PN4/5. Relative signal intensities in control groups are set to 1. Statistical significance was calculated using *t* test (***P ≤ 0.001)

injection of the mRNA encoding a shorter G9aCat-NLS-GFP version overcame this control and efficiently enhanced the H3K9me2 [but not H3K9me3 (see Additional file 7)] signal in both maternal and paternal chromatin (Fig. 9).

However, such strong increase in H3K9me2 signals in both pronuclei did not induce a major increase in (paternal) 5mC (Fig. 10). To examine the overall effect on 5mC/5hmC, we performed hairpin bisulphite sequencing of three repetitive elements: Line1 (L1Tf), intracisternal A-particle element (IAP) and major satellites (mSat) on late stage zygotes (i.e. after replication). Neither of these elements showed a significant increase in total methylation or hemimethylated sites when G9aCat-NLS-GFP injected group was compared with non-injected one (Additional file 8). We conclude that the increase in H3K9me2 alone does not directly control the genome-wide amount and replication-dependent persistence of

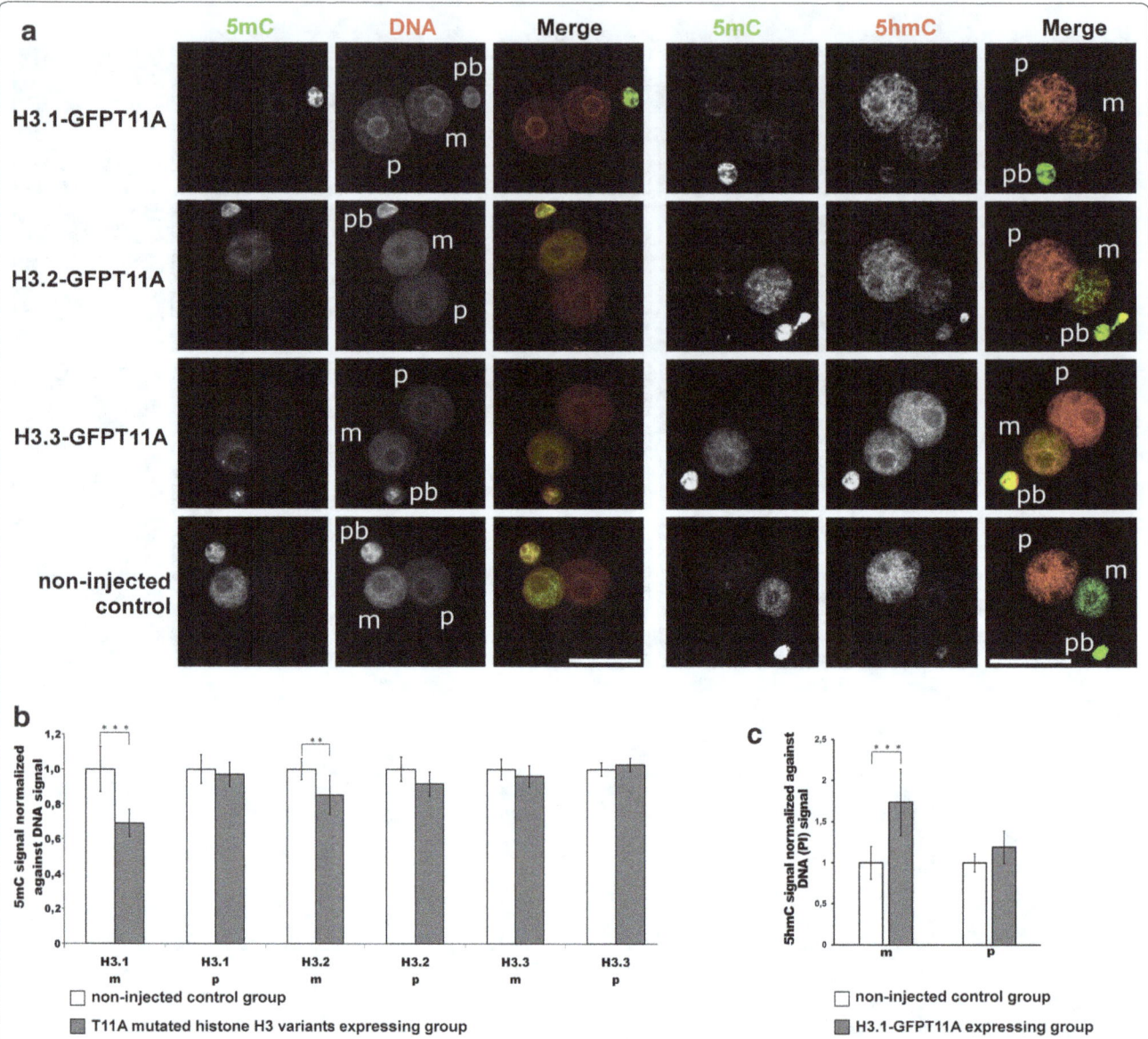

**Fig. 7** Effects of H3.1/2/3-GFPT11A expression in mouse zygotes on 5mC and 5hmC. **a** Shown are the representative images of PN4/5 stage zygotes stained with antibodies against 5mC together with anti-ssDNA antibodies, or together with 5hmC antibodies. *m* Maternal pronucleus, *p* paternal pronucleus, *pb* polar body. *Scale bar* 50 μm. **b** Quantification of 5mC signals, normalized against ssDNA signals in both parental genomes of zygotes at PN4/5. Relative signal intensities in control groups are set to 1. Statistical significance was calculated using *t* test (***$P < 0.001$; **$P < 0.01$). **c** Quantification of 5hmC signals, normalized against DNA (*PI* propidium iodide) signals in both parental genomes of zygotes at PN4/5. Relative signal intensities in control groups are set to 1. Statistical significance was calculated using *t* test (***$P \leq 0.001$)

DNA methylation. Our results are in line with a report by Liu et al., who also observed no visible changes in 5mC level (also visualized by immunostaining), despite the global increase in H3K9me2 on the paternal genome, caused by cycloheximide treatment of mouse zygotes [7].

Having shown that H3.1-GFPT11A mutant can be methylated by G9aCat-NLS-GFP in *E. coli* (see above), we next asked whether the decrease in H3K9me2, caused by the expression of H3T11A mutants, may

be compensated by ectopic G9a overexpression in the mouse zygote. We co-injected mRNA encoding G9aCat-NLS-GFP with either H3.1-GFPT11A or H3.2-GFPT11A or H3.3-GFPT11A mRNAs, respectively. Indeed, the co-injection partially compensates for the H3K9me2 loss from maternal chromatin observed for H3.1-GFPT11A or H3.2-GFPT11A expressing groups alone, while the co-expression with H3.3T11A did not cause this effect and the G9A mediated H3K9me2 methylation was as high,

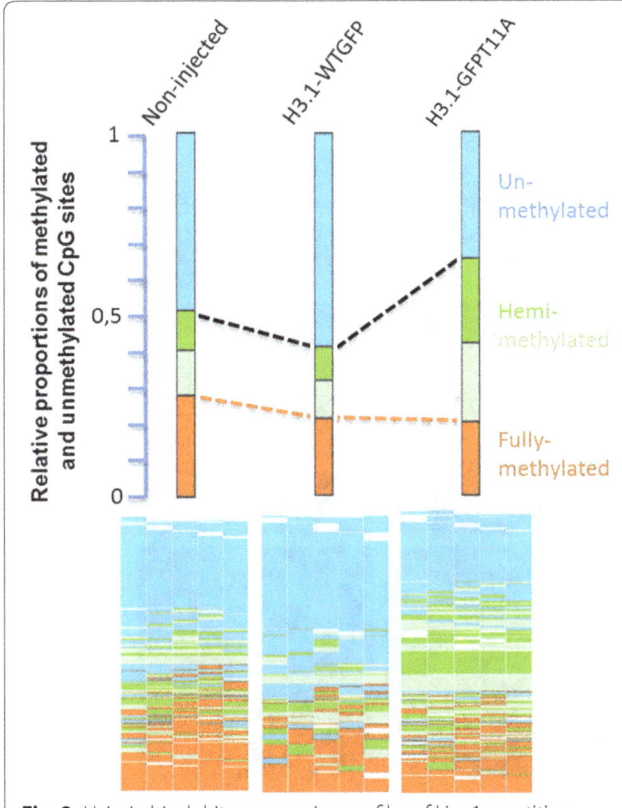

**Fig. 8** Hairpin bisulphite sequencing profiles of Line1 repetitive elements in PN4/5 zygotes expressing either H3.1-GFPWT, H3.1-GFPT11A or in non-injected control group. The *bars* represent the proportions of fully-, hemi- and unmethylated CpG dyads. The *map* below represents the distribution of methylated sites. Each *column* shows individual neighbouring CpG dyads, and each *line* represents one sequence read. The reads in the *map* are sorted according to their methylation status

as in zygotes, injected only with G9aCat-NLS-GFP alone (Additional file 9). The paternal chromatin, which is naturally depleted with H3K9me2, had an increased amount of this modification in all three mutated H3 variants (Additional file 9b). These results support the notion that the maintenance of H3K9me2 on H3.1 and H3.2 (and partially on H3.3) during replication is controlled by the H3T11 phosphorylation status.

## Discussion

In our work, we investigated the possible dependency of the phosphorylation status at H3S10 and H3T11, the establishment and replication-dependent maintenance of H3K9me2 and the conversion of 5mC to 5hmC. Our findings add new facets to the role and crosstalk between histone modifications in epigenetic reprogramming of the zygote. In the zygote, H3K9me2 is a key epigenetic signal mainly inherited from the maternal chromosomes of the oocyte [2, 5, 6]. It remains an open question how

the pronuclear asymmetry of H3K9me2 is maintained throughout the first cell cycle. De novo H3K9 dimethylation activities appear to be largely absent in the zygote [7]. Our findings support this assumption but also indicate some low-level de novo H3K9me2 activity on both paternal and maternal chromatin during late S- or G2-phase [7]. The nature of H3K9me2 modifying enzymes remains unclear. G9a, the main H3K9me2-specific methyltransferase, cannot be detected by immunofluorescence until four-cell stage (Additional file 6), and the inhibition of G9a by the specific inhibitor BIX 01294 does not cause any changes in the amount and distribution pattern of H3K9me2 in the zygote (Fig. 11). We also observe that the maternally inherited H3K9me2 signal is "halved" on nucleosomes after the first round of replication (Fig. 12a, b). We furthermore show that the H3K9me2 signal decreases when we overexpress either H3.1/2-GFPS10A or H3.1/2-GFPT11A mutated histones. Together these observations suggest that the phosphorylation controls the maintenance of H3K9me2. A "phospho-methylation" crosstalk has been proposed for H3K4me3. Here the demethylation by LSD1 is controlled by the absence of H3T6 phosphorylation [22]. A different crosstalk has been proposed for H3T11phos and H3K9me2 where phosphorylation triggers the demethylation in a cancer cell line. The authors suggested that the phosphorylation of H3T11 by PRK1 kinase accelerates the demethylation by JMJD2C histone demethylase [23]. While our findings also suggest a "phospho-methylation" crosstalk, they are obviously distinct from the cancer cell scenario. We observe (similar to the H3K4me3 example) that the absence of phosphorylation and not its presence triggers the reduction in (pre-existing) H3K9me2 in the zygote—predominantly in the maternal pronucleus. In the postreplicative chromatin, the newly assembled nucleosomes contain the "old" modified histones, probably as H3/H4 dimers, assembled with "new" unmodified H3/H4 dimers [24]. The phosphorylation of H3S10 or H3T11, which might occur on free H3 histones or on newly assembled H3/H4 dimers, appears to protect this "old" H3K9me2 probably by blocking a putative H3K9me2 demethylase (Fig. 13). In line with this scenario, we would like to note that the overexpression of non-methylatable H3.1K9R and H3.2K9R mutants strongly decreases T11 phosphorylation in maternal pronuclei (Additional file 10) and in consequence leads to a reduction (approx. 20%) in H3K9me2 as well as H3K9me3 (approx. 15%, see Additional file 11a, b).

We observe that dynamics and the variant-specific distribution of S10 and T11 phosphorylation are distinct and only partially overlapping. H3T11phos is preferentially linked to S-phase and the histone variants H3.1 and H3.2, while H3S10phos is present at G1 and G2 and

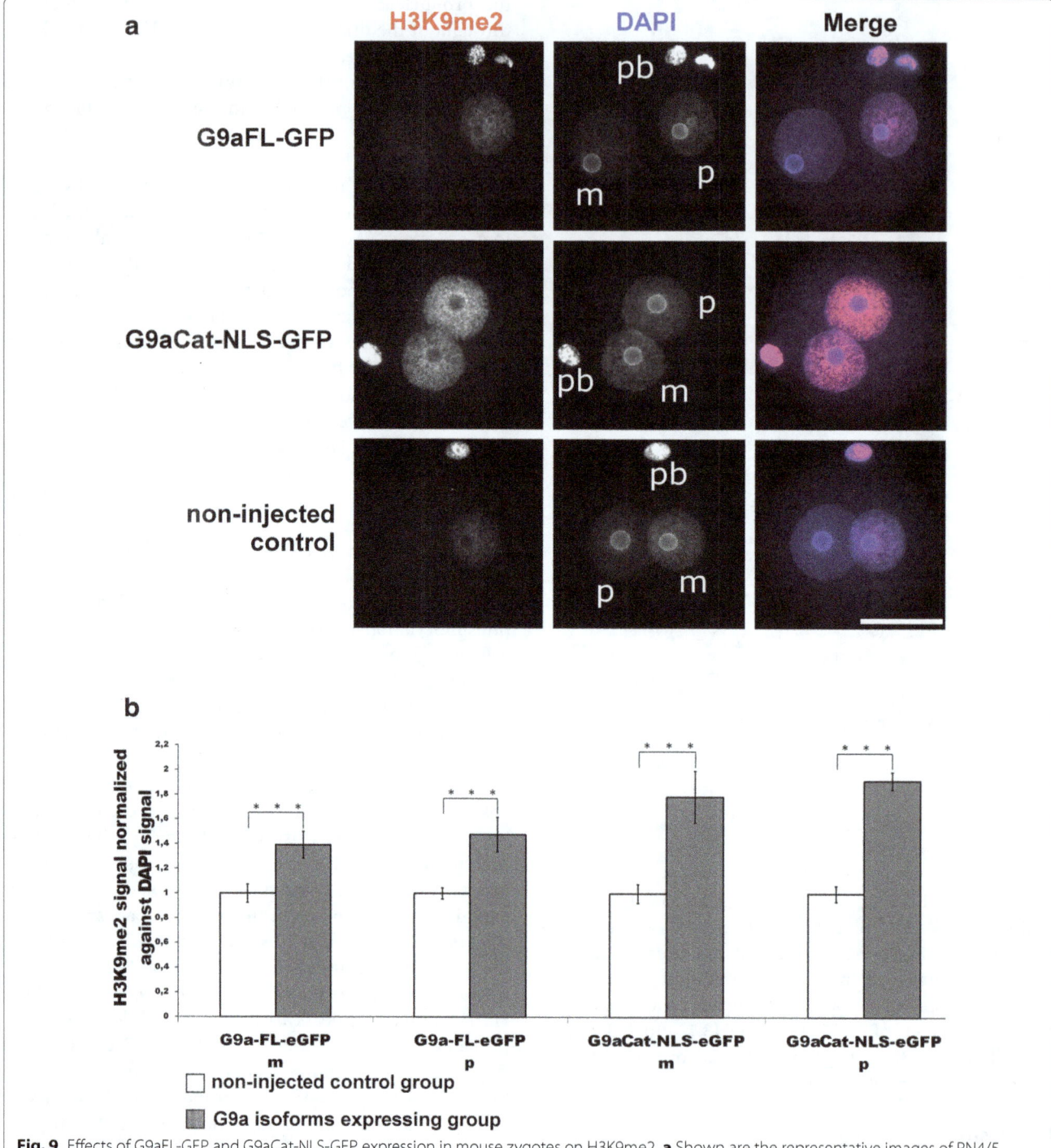

**Fig. 9** Effects of G9aFL-GFP and G9aCat-NLS-GFP expression in mouse zygotes on H3K9me2. **a** Shown are the representative images of PN4/5 stage zygotes stained with antibodies against H3K9me2. DNA is visualized by DAPI. *m* Maternal pronucleus, *p* paternal pronucleus, *pb* polar body. *Scale bar* 50 μm. **b** Quantification of H3Kme2 signals normalized against DNA signals in both parental genomes of zygotes at PN4/5. Relative signal intensities in control groups are set to 1. Statistical significance was calculated using *t* test (***$P < 0.001$)

found on all histone variants (Figs. 2, 3, respectively). Despite this differential dynamics, mutations in S10 and T11 phosphorylation sites cause very similar global effects on both H3K9me2 histone methylation and 5mC/5hmC DNA modifications. It remains to be clarified which regulatory proteins (histone demethylases or

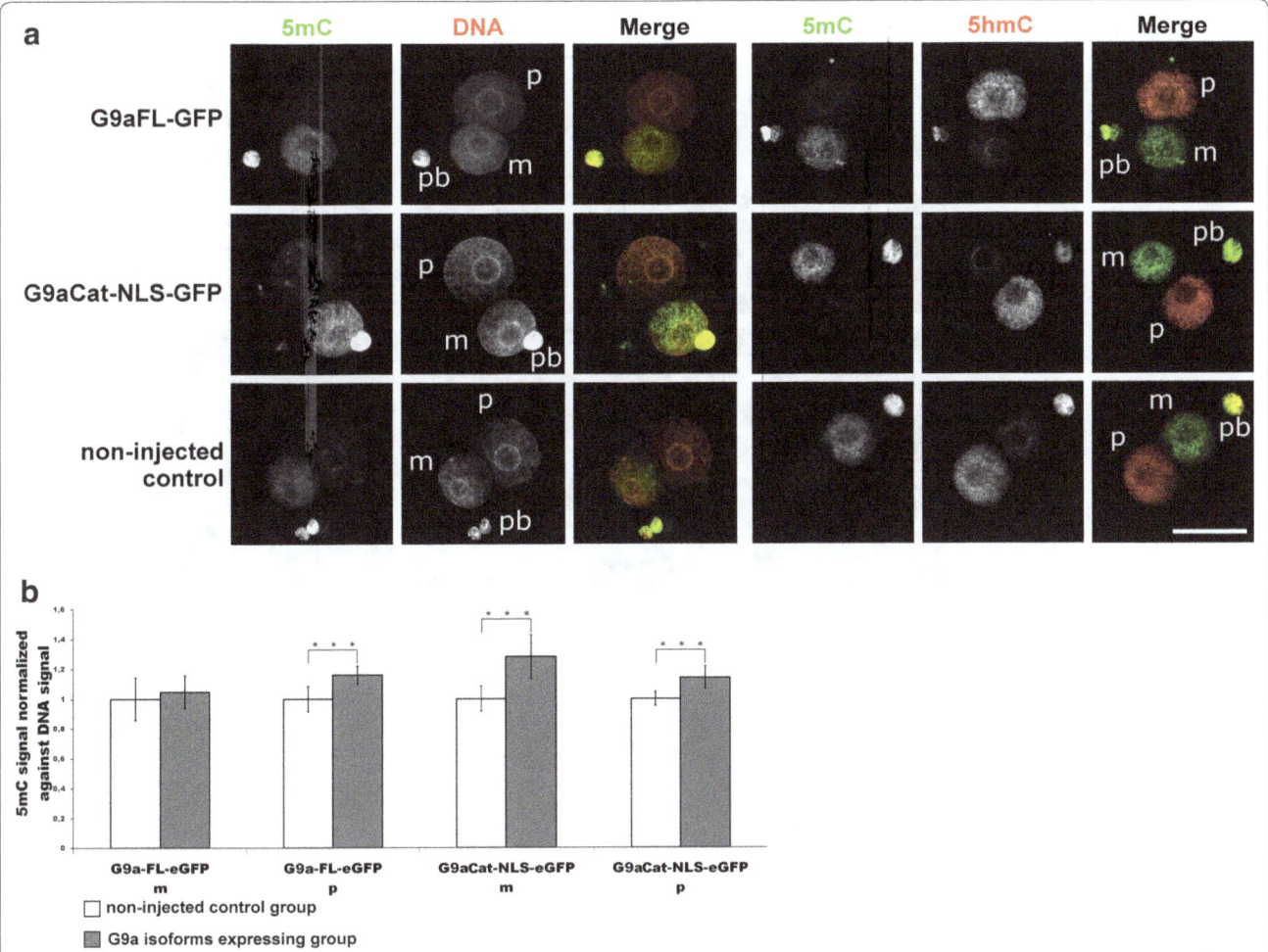

**Fig. 10** Effects of G9aFL-GFP and G9aCat-NLS-GFP expression in mouse zygotes on 5mC and 5hmC. **a** Shown are the representative images of PN4/5 stage zygotes stained with antibodies against 5mC together with anti-ssDNA antibodies, or together with 5hmC antibodies. *m* Maternal pronucleus, *p* paternal pronucleus, *pb* polar body. *Scale bar* 50 μm. **b** Quantification of 5mC signals, normalized against ssDNA signals in both parental genomes of zygotes at PN4/5. Relative signal intensities in control groups are set to 1. Statistical significance was calculated using *t* test (\*\*\**P* < 0.001)

DNA modification controller) are influenced by S10 and T11 phosphorylation and how and where in the genome this occurs for the different histone H3 variants.

Some of our data suggest that the conversion of 5mC into 5hmC is directly linked to the histone phosphomethylation switch model discussed above. The overexpression of both H3.1/2-GFPS10A and H3.1/2-GFPT11A mutated histones indeed induces significant accumulations of 5hmC in the maternal pronucleus (Figs. 6a, 7a and Additional file 5). It is tempting to assume that the altered S10 and T11 phosphorylation influences changes in maternal H3K9me2 levels, which in turn influence the Tet3-mediated conversion of 5mC to 5hmC. However, one experiment argues against such a simple scenario. When overexpressing G9A in the zygote, we could massively increase the levels of H3K9 without comparably

dramatic changes in the presence or the ratio of 5mC and 5hmC in maternal and paternal pronuclei (Fig. 10a, b). Hence, while the decrease in maternal H3K9me2 causes a strong change on maternal 5hmC amounts, the G9aCat-NLS-GFP-induced increase in H3K9me2 mainly in the paternal chromosome had no or little influence on 5hmC (Fig. 10a). This finding indicates that H3K9me2 (in combination with Stella) might not constitute the exclusive chromatin modification protecting chromatin against Tet-mediated oxidation.

In all our IF experiments, we observed a strong decrease in 5mC and an increase in 5hmC in zygotes overexpressing S10 and T11 mutants. When examining the DNA methylation patterns by hairpin bisulphite sequencing after the first DNA replication, we observe a massive increase in hemimethylated sites in zygotes overexpressing H3.1-GFPT11A

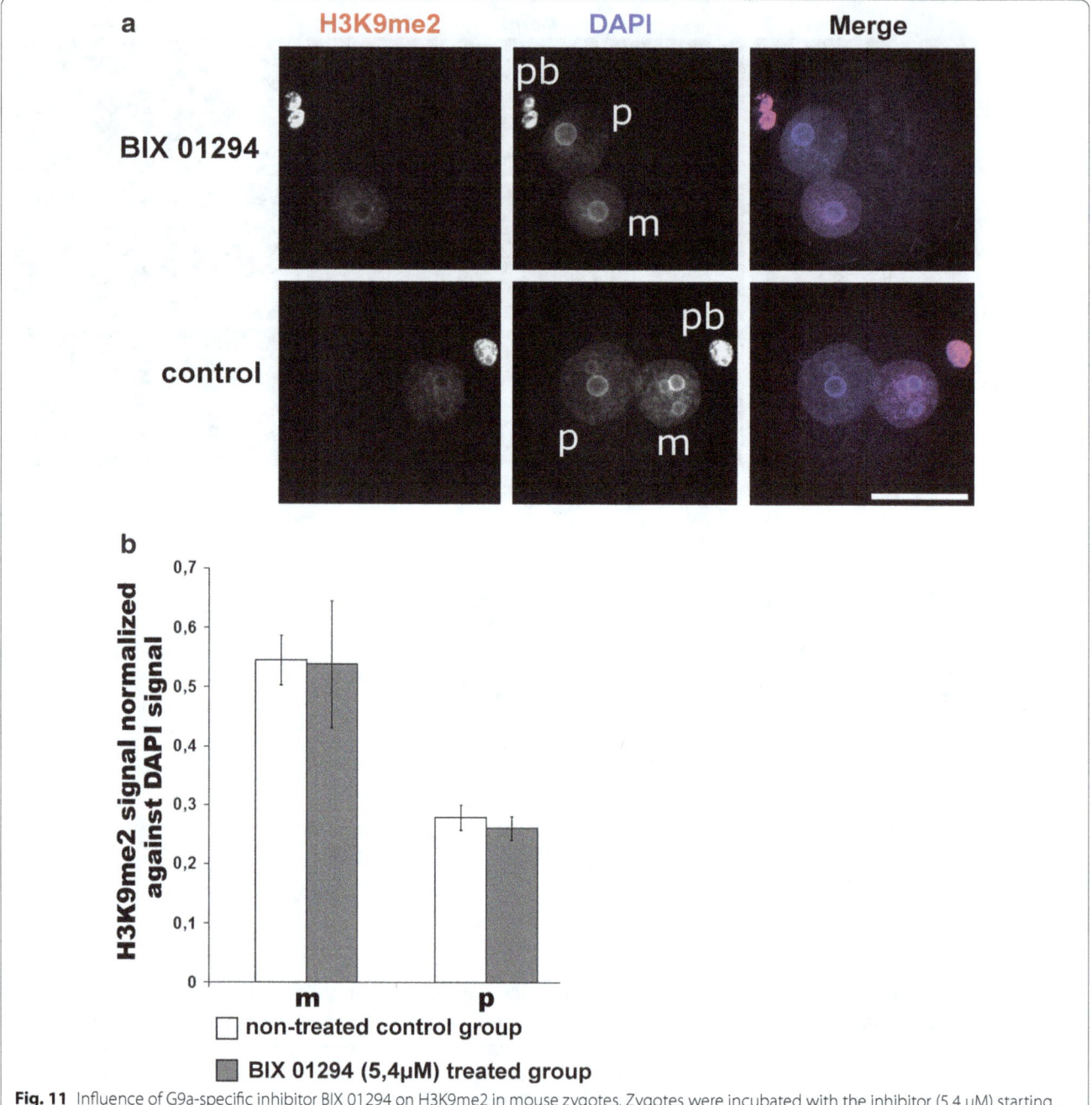

**Fig. 11** Influence of G9a-specific inhibitor BIX 01294 on H3K9me2 in mouse zygotes. Zygotes were incubated with the inhibitor (5,4 μM) starting from 2 h post-fertilization until 12 h and then fixed for immunostaining. **a** Shown are the representative images of PN4/5 stage zygotes stained with antibodies against H3K9me2. DNA is visualized by DAPI. *m* Maternal pronucleus, *p* paternal pronucleus, *pb* polar body. *Scale bar* 50 μm. **b** Quantification of H3K9me2 signals, normalized against DNA signals in both parental genomes of zygotes at PN4/5. Statistical significance was calculated using *t* test (***$P < 0.001$)

mutants (Fig. 8). A lack of T11 (and probably also S10) phosphorylation apparently increases the presence of 5hmC, which in turn affects the maintenance of full methylation after DNA replication. This observation is in line with our previous findings, showing that 5hmC may interfere with DNA methylation maintenance [20, 21].

## Conclusions

Our findings provide novel insights into the crosstalk between DNA and histone modifications in the mammalian zygote. We find that serine 10 and threonine 11 phosphorylation of histone H3 shows a different dynamic and pronuclear appearance during the first

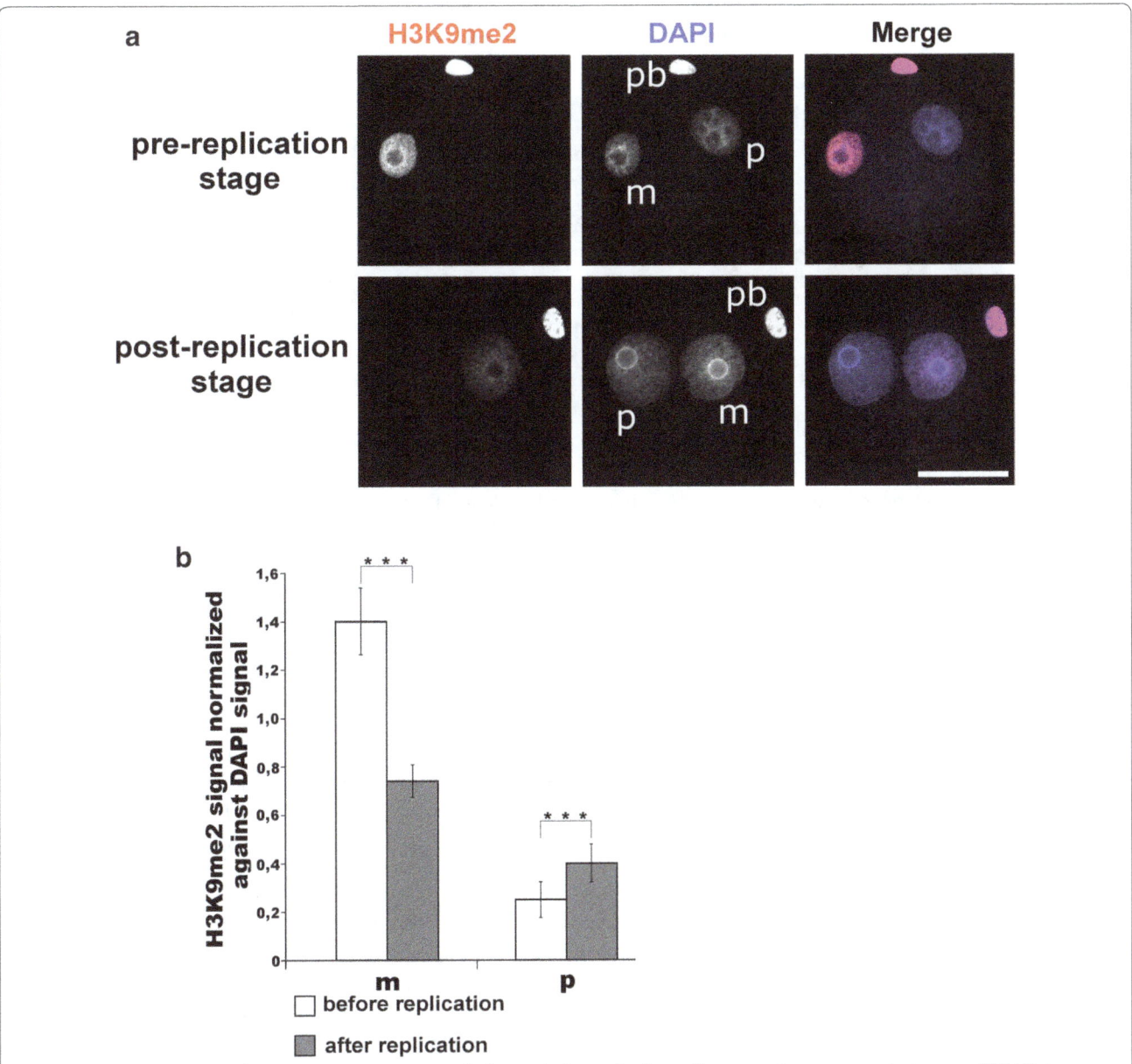

**Fig. 12** Relative abundance of H3K9me2 in mouse zygotes before and after replication. **a** Shown are the representative images of PN4/5 stage zygotes stained with antibodies against H3K9me2. DNA is visualized by DAPI. *m* Maternal pronucleus, *p* paternal pronucleus, *pb* polar body. *Scale bar* 50 μm. **b** Quantification of H3K9me2 signals, normalized against DNA signals in both parental genomes of zygotes at PN4/5. Statistical significance was calculated using *t* test (***$P < 0.001$)

cell cycle. Overexpression of variant-specific serine 10 and threonine 11 mutants reveals that the phosphorylation at both positions is important for the maintenance of the maternally inherited H3K9me2 as well as for the maintenance of symmetric CpG DNA methylation (5mC/5hmC). Mechanistically our data suggest that H3S10phos and H3T11phos help to stabilize H3K9me2 during the first cell cycle most likely by protecting maternal chromatin against a K9-specific demethylation activity.

## Methods

### Mouse superovulation and in vitro fertilization

For superovulation, females were injected with 6 IU pregnant mare serum gonadotropin (PMSG) and 6 IU human chorionic gonadotropin (hCG) with a time interval of around 48 h. Spermatozoa isolation, oocytes collection and IVF procedures were carried out as previously published protocol with small modifications [25]. Briefly, sperm was isolated from the cauda epididymis of male mice and capacitated in pre-gassed modified

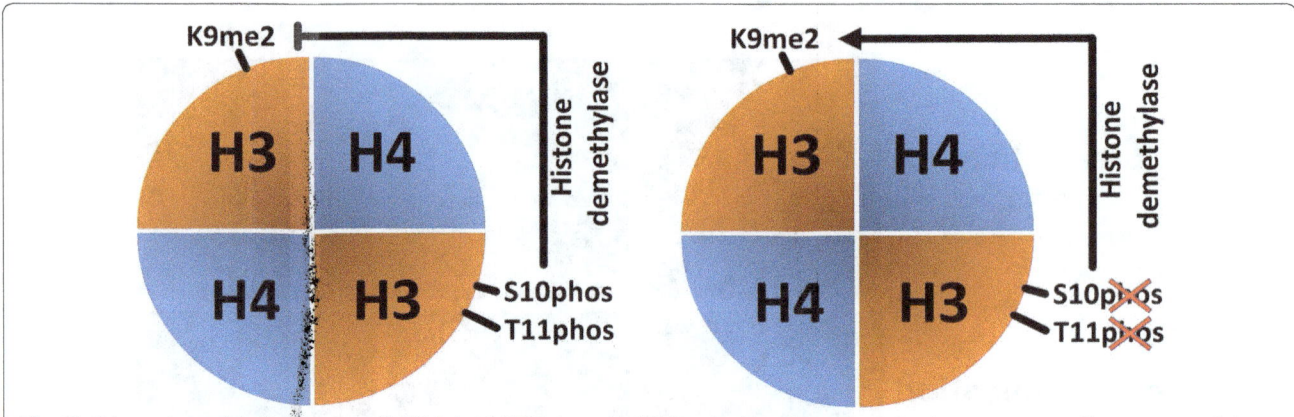

**Fig. 13** Schematic model of interaction of H3S10phos/H3T11phos with H3K9me2 within the nucleosome in mouse zygote. The presence of phosphorylation marks stabilizes H3K9me2, presumably by protecting from H3K9-specific histone demethylase

KSOM medium supplemented with 30 mg/ml BSA for 2 h. Mature oocytes were collected 15 h post-hCG injection of female mice. Cumulus–oocyte complexes were placed into a 400-μl drop of KSOM medium, mixed with capacitated sperm and incubated at 37 °C in a humidified atmosphere of 5% $CO_2$ and 95% air.

### Plasmids preparation

The cDNAs of H3.1, H3.2 (gifts from Prof. Dr. Fugaku Aoki), H3.3 (purchased from ImaGenes GmbH) were cloned to pEGUP1 (a modified pET28b vector containing GFP coding sequence) by using NcoI and XhoI sites. The cDNA of G9a full length (gift from Prof. Dr. Y. Shinkai, Prof. Dr. M. Tachibana, Prof. Dr. M. Brand and Prof. Dr. M. R. Stallcup) was subcloned to pEGUP1. The same strategy was applied to G9aCat-GFP, HDAC1 (obtained from Addgene) and HDAC2 (gift from Prof. Dr. Richard M. Schultz). To create G9aCat-NLS-GFP expressing construct, G9aCat encoding DNA fragment was inserted to pEGUP1-NLS (pEGUP1 version with nuclear localization signal). For co-expression experiments in *E. coli*, G9aCat encoding fragment was cloned into pET15Am (a modified pET28b containing ampicillin resistance marker, p15 replication origin and no His-tag sequence). All positive constructs were verified by Sanger sequencing. All the primers used are listed in Additional file 12, also including the ones for mutagenesis, in vitro transcription as well as hairpin bisulphite sequencing.

### Site-specific mutagenesis

The generation of all histone H3 mutants (K9R, S10A, T11A) was done by overlapping PCR-based mutagenesis. Briefly, two PCRs, namely PCR1 and PCR2, were applied with two sets of primers in which reverse primer for PCR1 and forward primer for PCR2 are complementary

and contain the desired mutation. After gel purification, the cleaned-up fragments from PCR1 and PCR2 were mixed as templates in overlapping PCR. The expected PCR products were then double-digested by NcoI and XhoI, followed by insertion into pEGUP1. The positive clones containing the corresponding mutations were selected and confirmed by Sanger sequencing.

### In vitro transcription of mRNAs

To prepare the DNA templates, highly purified DNA templates were extracted with phenol:chloroform, and capped mRNAs were generated by in vitro transcription using AmpliCap-Max T7 High Yield Message Maker Kit (CellScript, Inc.), followed by purification using RNA Clean & Concentrator™-25 Kit (Zymo research) according to the manufacturers' instructions. Finally, transcribed mRNAs were eluted with nuclease-free water (Life technologies) and stored at −80 °C for later use.

### Expression of G9aCat and histone H3 variants in *E. coli*

For co-expression of G9aCat with histone H3.1-GFPWT, or H3.1-GFPT11A or H3.3-GFPWT, both constructs were transformed into Rosetta™ 2 Competent Cells. The transformed cells were grown in SOC medium with ampicillin (100 mg/mL) and kanamycin (25 mg/mL) until OD value reached 0.5–0.6 at 37 °C, followed by 0.5 mM isopropyl-b-D-thiogalactopyranoside (IPTG) induction for 24 h at 16 °C. Modified proteins of H3.1-GFPWT, H3.1-GFPT11A, H3.3-GFPWT were purified through Ni–NTA resin by taking advantage of the 6xHis tag. The protein expression was confirmed by Coomassie blue staining of SDS-PAGE gel. The presence of H3K9me2 or H3K9me3 on recombinant histones was verified by Western blotting using specific antibodies, which were also used for immunofluorescence.

The detailed information regarding the antibodies used is listed in Additional file 13.

### BIX 01294 treatment of zygotes

At 2 h post-fertilization, zygotes were collected and transferred into the 100 µl drop of KSOM medium containing BIX 01294 (5.4 µM). The zygotes were cultured to post-replication stages (PN4/5), followed by fixation for immunostaining.

### Microinjection of mRNAs into zygotes

At around 1hpf, just before microinjection, zygotes were collected, washed in M2 medium for several times and transferred into 100 µl drops of pre-gassed KSOM medium. Microinjection was performed under a Zeiss Axiovert 200 M inverted microscope (Zeiss) equipped with a FemtoJet microinjector and piezo-driven micromanipulators (Eppendorf). After a quick set-up, zygotes were placed into a 20-µl drop of M2 medium and cytoplasm injection was done with mRNAs encoding for H3.1/2/3-GFPWT/K9R/S10A/T11A, G9aFL-GFP, G9aCat-GFP, G9aCat-NLS-GFP or in combinations. Each microinjection experiment was replicated at least once, and groups of 10–30 zygotes were used for each experiment.

### Live monitoring chromatin incorporation of H3.1/2/3-GFPK9R and H3.1/2/3-GFPWT

After being microinjected with corresponding mRNAs, the zygotes were transferred into a KSOM drop of 10 µl on a glass bottom culture dish and placed into the chamber of Nikon BioStation live cell imaging System under the condition of 37 °C in a humidified atmosphere of 5% $CO_2$ and 95% air. The pictures were captured every 15 min for 97 cycles with the exposure time 1/250 s.

### IF staining

The detailed protocol was described earlier elsewhere [10, 25]. After brief treatment in acidic Tyrode's solution, zona-free zygotes and cleavage stage embryos were fixed by incubating in 3.7% paraformaldehyde solution in PBS for 30 min at RT, followed by permeabilization in 0.2% TritonX100/PBS at RT for 15 min. Then, the embryos were blocked at RT for 4 h or overnight at 4 °C in PBS, containing 0.1% TritonX100 and 3 mg/ml BSA. Later, the embryos were incubated with primary antibody overnight at 4 °C in blocking solution. Note that for 5mC, 5hmC, 5caC and ssDNA staining DNA was denatured: the fixed embryos were incubated in 4 M HCl for 15 min and then neutralized in 100 mM TrisCl pH 8.0 for 10 min at RT. For simultaneous 5hmC detection and PI DNA staining, the denaturation time was reduced to 6 min 30 s. After incubation with primary antibody, the embryos were washed 3–4 times in blocking solution and then incubated at RT for 2 h with fluorescently labelled secondary antibody. Finally, the embryos were washed and mounted on a glass slide in Vectashield mounting medium (Vector Laboratories). All antibodies used in this study are listed in Additional file 13.

### IF microscopy, quantification and statistical analysis

The mounted embryos were analysed on Zeiss Axiovert 200 M inverted microscope equipped with the fluorescence module and B/W digital camera for imaging. The IF images were captured, pseudocoloured and merged using AxioVision software (Zeiss). GIMP and Image J software were used together to complete quantification of the signals of z-stack computed images. For each group, at least 10 zygotes were analysed from at least two repeated experiments. Student's $t$ test was employed for statistical analysis.

### Hairpin bisulphite sequencing

Hairpin bisulphite high-throughput sequencing was performed as previously described with no modifications to the protocol [20].

## Additional files

**Additional file 1.** Localization of the ectopically expressed mutated histone H3 variants and G9a isoforms in zygotes. The direct visualization was enabled by the presence of GFP-tag fused to C-terminal part of the protein of interest. DNA was counterstained with DAPI.

**Additional file 2.** Influence of the overexpression of wild-type (WT) histone H3 variants fused to GFP on H3S10phos and H3T11phos at PN4/5. The figure provides three examples for each injection showing representative examples in which H3S10phos and H3T11phos patterns are either strongly changed (*top* example, few cases only), mildly changed (*middle*, less abundant) and unchanged (*bottom*, the majority of the injected zygotes) as compared to the uninjected control. DNA was visualized by DAPI. *Scale bar 50 µm.*

**Additional file 3.** Effects of H3.1-GFPT11A expression in mouse zygotes on H3K9me2 in direct comparison to H3.1-GFPWT expressing zygotes. Quantification of H3K9me2 signals, normalized against DNA signals in both parental genomes of zygotes at PN4/5. Relative signal intensities in control group are set to 1. Statistical significance was calculated using $t$ test (***$P < 0.001$).

**Additional file 4.** No interference of T11A mutation on H3.1-GFP with the ability of G9a to methylate histone H3 at K9 residue and with H3K9me2-specific antibody binding to its target. The WT or T11A mutated histone H3.1-GFP was expressed in *E. coli* either alone or together with G9aCat. The recombinant histone proteins were partially purified on Ni-NTA sepharose and applied to Western blot. *Lane 1* expression of H3.1-GFPWT alone; *lane 2* co-expression of H3.1-GFPWT and G9aCat; *lane 3* co-expression of H3.1-GFPT11A and G9aCat. *Upper panel* probing with anti-histone H3 antibody; *lower panel* probing with anti-H3K9me2 antibody.

**Additional file 5.** The influence of H3.1/2/3-GFPS10A or H3.1/2/3-GFPT11A expression on 5hmC in PN4/5 zygotes. The mean values (MV) of 5hmC signals in paternal (MVpat) or maternal (MVmat) pronuclei were calculated as integral signal density (ID) to area ratio (MVpat = IDpat/Area; MVmat = IDmat/Area). Plotted are MVpat to MVmat ratios. Statistical significance was calculated using $t$ test (***$P < 0.001$).

**Additional file 6.** Localization of the endogenously expressed G9a histone methyltransferase in pre-implantation embryos from zygote to blastocyst stages. Embryos at different stages were fixed and immunostained with antibodies against G9a. DNA is visualized by DAPI. *Scale bar* 50 µm.

**Additional file 7.** G9aFL-GFP and G9aCat-NLS-GFP expression in mouse zygotes has no effect on H3K9me3. Shown are the representative images of PN4/5 stage zygotes stained with antibodies against H3K9me3. DNA is visualized by DAPI. *m* Maternal pronucleus, *p* paternal pronucleus, *pb* polar body. *Scale bar* 50 µm.

**Additional file 8.** Hairpin bisulphite sequencing profiles of Line1, IAP and mSat repetitive elements in PN4/5 zygotes expressing G9aCat-NLS-GFP or in non-injected control group. *Bars* represent the sum of the DNA methylation status of all CpG dyads. The *map* next to the *bar* represents the distribution of methylated sites. Each *column* shows individual neighbouring CpG dyads, and each *line* represents one sequence read. The reads in the *map* are sorted according to their methylation status.

**Additional file 9.** Effects of H3.1/2/3-GFPT11A co-expression with G9amCat-NLS-GFP in mouse zygotes on H3K9me2. **a** Shown are the representative images of PN4/5 stage zygotes stained with antibodies against H3K9me2. DNA is visualized by DAPI. *m* Maternal pronucleus, *p* paternal pronucleus, *pb* polar body. *Scale bar* 50 µm. **b** Quantification of H3K9me2 signals, normalized against DNA signals in both parental genomes of zygotes at PN4/5. Relative signal intensities in control groups are set to 1. Statistical significance was calculated using *t* test (***$P < 0.001$).

**Additional file 10.** Effects of H3.1/2/3-GFPK9R expression in mouse zygotes on H3T11phos. Shown are the representative images of PN4/5 stage zygotes stained with antibodies against H3T11phos. DNA is visualized by DAPI. *m* Maternal pronucleus, *p* paternal pronucleus, *pb* polar body. *Scale bar* 50 µm.

**Additional file 11.** Effects of H3.1/2/3-GFPK9R or wild-type H3.1/2/3-GFP expression in mouse zygotes on **a** H3K9me2 and **b** H3K9me3. H3K9me2 and H3K9me3 signals were quantified and normalized against DNA signals in either both parental genomes (for H3K9me2) or maternal genomes only (for H3K9me3, due to virtually absent signals in paternal pronuclei) of zygotes at PN4/5. Relative signal intensities in control (non-injected) groups are set to 1. Statistical significance was calculated using *t* test (***$P < 0.001$; **$P < 0.01$; *$P < 0.05$).

**Additional file 12.** The list and sequences of primers used in this study.

**Additional file 13.** The list and detailed information about the antibodies used in this study.

## Abbreviations
H3S10: histone H3 serine 10; H3T11: histone H3 threonine 11; H3S10phos: histone H3 serine 10 phosphorylation; H3T11phos: histone H3 threonine 11 phosphorylation; H3K9me2: histone H3 lysine 9 dimethylation; H3K9me3: histone H3 lysine 9 trimethylation; H3K4me3: histone H3 lysine 4 trimethylation; 5mC: 5-methylcytosine; 5hmC: 5-hydroxymethylcytosine; WT: wild-type; IF: immunofluorescence; G9aCat: catalytic domain of G9a histone methyltransferase; G9aFL: full-length G9a histone methyltransferase; IAP: intracisternal A-particle element; mSat: major satellites; NLS: nuclear localization signal; PMSG: pregnant mare serum gonadotropin; hCG: human chorionic gonadotropin.

## Authors' contributions
JW, KL and JL conceived the project. JW and KL wrote the manuscript with contributions and approval by all authors. JL and KL designed the experiments, performed the experiments and evaluated the data. PG performed the hairpin bisulphite sequencing and evaluated the sequencing data. All authors read and approved the final manuscript.

## Author details
[1] FR 8.3, Biological Sciences, Genetics/Epigenetics, University of Saarland, Campus A2.4, 66123 Saarbrücken, Germany. [2] Present Address: Faculty of Medicine, Free University of Brussels, C.P. 614, Building GE, 5th floor, 808 Route de Lennik, 1070 Brussels, Belgium.

## Acknowledgements
We thank F. Aoki for providing cDNAs of histone H3.1 and H3.2, Y. Shinkai, M. Tachibana, M. Brand and M. R. Stallcup for providing cDNA of G9a.

## Competing interests
The authors declare that they have no competing interests.

## Funding
This work was supported by a Grant from Deutsche Forschungsgemeinschaft (DFG) WA 1029. J.L. was supported by fellowship offered by Chinese Government Scholarship Council. KL was partially supported through the BLUEPRINT project (European Union's Seventh Framework Programme grant agreement 282510).

## References
1. Hake SB, Allis CD. Histone H3 variants and their potential role in indexing mammalian genomes: the "H3 barcode hypothesis". Proc Natl Acad Sci USA. 2006;103:6428–35.
2. van der Heijden GW, Dieker JW, Derijck AA, Muller S, Berden JH, Braat DD, et al. Asymmetry in histone H3 variants and lysine methylation between paternal and maternal chromatin of the early mouse zygote. Mech Dev. 2005;122:1008–22.
3. Torres-Padilla ME, Bannister AJ, Hurd PJ, Kouzarides T, Zernicka-Goetz M. Dynamic distribution of the replacement histone variant H3.3 in the mouse oocyte and preimplantation embryos. Int J Dev Biol. 2006;50:455–61.
4. Akiyama T, Suzuki O, Matsuda J, Aoki F. Dynamic replacement of histone H3 variants reprograms epigenetic marks in early mouse embryos. PLoS Genet. 2011;7:e1002279.
5. Lepikhov K, Walter J. Differential dynamics of histone H3 methylation at positions K4 and K9 in the mouse zygote. BMC Dev Biol. 2004;4:12.
6. Santos F, Peters AH, Otte AP, Reik W, Dean W. Dynamic chromatin modifications characterise the first cell cycle in mouse embryos. Dev Biol. 2005;280:225–36.
7. Liu H, Kim JM, Aoki F. Regulation of histone H3 lysine 9 methylation in oocytes and early pre-implantation embryos. Development. 2004;131:2269–80.
8. Santos F, Hendrich B, Reik W, Dean W. Dynamic reprogramming of DNA methylation in the early mouse embryo. Dev Biol. 2002;241:172–82.
9. Nakamura T, Liu YJ, Nakashima H, Umehara H, Inoue K, Matoba S, et al. PGC7 binds histone H3K9me2 to protect against conversion of 5mC to 5hmC in early embryos. Nature. 2012;486:415–9.
10. Wossidlo M, Nakamura T, Lepikhov K, Marques CJ, Zakhartchenko V, Boiani M, et al. 5-Hydroxymethylcytosine in the mammalian zygote is linked with epigenetic reprogramming. Nat Commun. 2011;2:241.
11. Nakamura T, Arai Y, Umehara H, Masuhara M, Kimura T, Taniguchi H, et al. PGC7/Stella protects against DNA demethylation in early embryogenesis. Nat Cell Biol. 2007;9:64–71.
12. Wang C, Li Y, Cai W, Bao X, Girton J, Johansen J, et al. Histone H3S10 phosphorylation by the JIL-1 kinase in pericentric heterochromatin and on the fourth chromosome creates a composite H3S10phK9me2 epigenetic mark. Chromosoma. 2014;123:273–80.
13. Chin HG, Pradhan M, Esteve PO, Patnaik D, Evans TC Jr, Pradhan S. Sequence specificity and role of proximal amino acids of the histone H3 tail on catalysis of murine G9A lysine 9 histone H3 methyltransferase. Biochemistry. 2005;44:12998–3006.
14. Lohse B, Helgstrand C, Kristensen JB, Leurs U, Cloos PA, Kristensen JL, et al. Posttranslational modifications of the histone 3 tail and their

impact on the activity of histone lysine demethylases in vitro. PLoS ONE. 2013;8:e67653.

15. Sawicka A, Seiser C. Histone H3 phosphorylation—a versatile chromatin modification for different occasions. Biochimie. 2012;94:2193–201.

16. Teperek-Tkacz M, Meglicki M, Pasternak M, Kubiak JZ, Borsuk E. Phosphorylation of histone H3 serine 10 in early mouse embryos: active phosphorylation at late S phase and differential effects of ZM447439 on first two embryonic mitoses. Cell Cycle. 2010;9:4674–87.

17. Lachner M, O'Sullivan RJ, Jenuwein T. An epigenetic road map for histone lysine methylation. J Cell Sci. 2003;116:2117–24.

18. Laird CD, Pleasant ND, Clark AD, Sneeden JL, Hassan KM, Manley NC, et al. Hairpin-bisulfite PCR: assessing epigenetic methylation patterns on complementary strands of individual DNA molecules. Proc Natl Acad Sci USA. 2004;101:204–9.

19. Arand J, Spieler D, Karius T, Branco MR, Meilinger D, Meissner A, et al. In vivo control of CpG and non-CpG DNA methylation by DNA methyltransferases. PLoS Genet. 2012;8:e1002750.

20. Arand J, Wossidlo M, Lepikhov K, Peat JR, Reik W, Walter J. Selective impairment of methylation maintenance is the major cause of DNA methylation reprogramming in the early embryo. Epigenetics Chromatin. 2015;8:1.

21. Giehr P, Kyriakopoulos C, Ficz G, Wolf V, Walter J. The influence of hydroxylation on maintaining CpG methylation patterns: a hidden Markov model approach. PLoS Comput Biol. 2016;12:e1004905.

22. Metzger E, Imhof A, Patel D, Kahl P, Hoffmeyer K, Friedrichs N, et al. Phosphorylation of histone H3T6 by PKCbeta(I) controls demethylation at histone H3K4. Nature. 2010;464:792–6.

23. Metzger E, Yin N, Wissmann M, Kunowska N, Fischer K, Friedrichs N, et al. Phosphorylation of histone H3 at threonine 11 establishes a novel chromatin mark for transcriptional regulation. Nat Cell Biol. 2008;10:53–60.

24. Tagami H, Ray-Gallet D, Almouzni GV, Nakatani Y. Histone H3.1 and H3.3 complexes mediate nucleosome assembly pathways dependent or independent of DNA synthesis. Cell. 2004;116:51–61.

25. Wossidlo M, Arand J, Sebastiano V, Lepikhov K, Boiani M, Reinhardt R, et al. Dynamic link of DNA demethylation, DNA strand breaks and repair in mouse zygotes. EMBO J. 2010;29:1877–88.

# 2

# Fetal testis organ culture reproduces the dynamics of epigenetic reprogramming in rat gonocytes

Arlette Rwigemera, Fabien Joao and Geraldine Delbes*⊙

## Abstract

**Background:** Epigenetic reprogramming is a critical step in male germ cell development that occurs during perinatal life. It is characterized by the remodeling of different epigenetic marks such as DNA methylation (5mC) and methylation of histone H3. It has been suggested that endocrine disruptors can affect the male germline epigenome by altering epigenetic reprogramming, but the mechanisms involved are still unknown. We have previously used an organ culture system that maintains the development of the different fetal testis cell types, to evaluate the effects of various endocrine disruptors on gametogenesis and steroidogenesis in the rat. We hypothesize that this culture model can reproduce the epigenetic reprogramming in gonocytes. Our aim was to establish the kinetics of three epigenetic marks throughout perinatal development in rats in vivo and compare them after different culture times.

**Results:** Using immunofluorescence, we showed that H3K4me2 transiently increased in gonocytes at 18.5 days post-coitum (dpc), while H3K4me3 displayed a stable increase in gonocytes from 18.5 dpc until after birth. 5mC progressively increased from 20.5 dpc until after birth. Using GFP-positive gonocytes purified from GCS-EGFP rats, we established the chronology of re-methylation of *H19* and *Snrpn* in rat gonocytes. Most importantly, using testis explanted at 16.5 or 18.5 dpc and cultured for 2–4 days, we demonstrated that the kinetics of changes in H3K4me2, H3K4me3, global DNA methylation and on parental imprints can generally be reproduced ex vivo with the model of organ culture without the addition of serum.

**Conclusions:** This study reveals the chronology of three epigenetic marks (H3K4me2, H3K4me3 and 5mC) and the patterns of methylation of *H19* and *Snrpn* differentially methylated regions in rat gonocytes during perinatal development. Most importantly, our results suggest that the organ culture can reproduce the process of epigenetic reprogramming and can be used to study the impact of environmental chemicals on the establishment of the male germ cell epigenome.

**Keywords:** Male germ cells, Gonocytes, Rat, Organ culture, Epigenetic reprogramming, 5mC, DNA methylation, Imprinted genes, Histone 3 modifications, H3K4me2, H3K4me3

## Background

Germ cell development in mammals is a unique and complex process encompassing cell migration, proliferation, quiescence, differentiation and meiosis. Germ cell differentiation is initiated in the embryo when a small number of cells from the epiblast, the primordial germ cells (PGC), acquire the germ cell lineage fate and migrate to colonize the genital ridge [1]. These cells subsequently commit to male or female developmental pathway depending on the surrounding somatic environment. In the male gonad, Sertoli cells surround the germ cells, named gonocytes [2], forming seminiferous cords. Gonocytes proliferate before entering a quiescent phase where the cells are in G0 cell cycle arrest [3]. After birth, mitosis resumes and the gonocytes progressively differentiate into spermatogonia, which are responsible

*Correspondence: geraldine.delbes@iaf.inrs.ca
Institut National de la Recherche Scientifique, Centre INRS – Institut Armand-Frappier, 531, boulevard des Prairies, Laval, QC H7V 1B7, Canada

for the continuous sperm production in adulthood [3]. In brief, perinatal development represents a key developmental step for the germ line as male germ cell fate is programmed.

One of the key events that occur in germ cells during that window of development is epigenetic reprogramming: a resetting of the epigenome that is critical for the re-establishment of parental imprints, erasure of epimutations and generation of the transcriptional identity of the germ line [4]. This reprogramming occurs in part through DNA methylation and posttranslational modifications of histones which are interrelated [4]. Briefly, in PGC, there is a unique transient loss of DNA methylation associated with the increased methylation of lysine K27 of histone 3 (H3K27me3) [4–6]. This is followed in gonocytes by a progressive de novo establishment of DNA methylation patterns specific to male germ cells [7, 8]. This event is associated with the methylation of lysines 4 and 9 of histone 3 (H3K4me3 and H3K9me3) [4, 9]. Importantly, it has been shown that the dynamics of epigenetic reprogramming in germ cells is finely tuned by DNA methyltransferases (DNMTs), histone methyltransferases (HMTs) and histone demethylases (HDMs) [8]. Disruption of these processes could have an impact on the subsequent germ cells. In fact, DNA methylation plays an important role in successful spermatogenesis as germ cell-specific (GCS) knockout mice for *Dnmt3a* or *Dnmt3l* or mice mutated for *Dnmt3c* are infertile due to meiotic arrest [8, 10]. In addition, impairment of epigenetic reprogramming during perinatal development has been shown to permanently affect the sperm epigenome, which in turn could lead to adverse progeny outcome [11–13].

It was shown that in utero exposure to man-made chemicals that can mimic steroid hormones such as vinclozolin, bisphenol A, or di-(2-ethylhexyl) phthalate, can affect DNA methylation in mature sperm [14–16]. These compounds are called endocrine disruptors (ED) and in the last 60 years, there has been increasing concern about their impact on male reproductive health [17]. Indeed, correlating with the rising production of ED, an increased frequency of human male reproductive disorders has been observed worldwide, including testicular cancer, disorders of sex development, cryptorchidism, hypospadias and poor semen quality [18, 19]. These disorders may result from impaired testicular development specifically during perinatal life as epidemiological and experimental data have established that exposure to ED early in life can induce permanent defects in male fertility [19–22]. We have shown that there is a specific window of development when gametogenesis is most sensitive to estrogen-like compounds [20]. Nevertheless, there are still gaps in the literature regarding the mechanisms by

which EDs affect fetal germ cells. It was shown that such compounds can affect gene expression in gonocytes [16] potentially affecting germline differentiation, but very little is known about whether and how they affect epigenetic reprogramming.

This lack of knowledge is in part due to a need of a robust study model. Studies in vivo allow assessment of ED effects on male reproductive tract disorders such as hypospadias or cryptorchidism [23] as well as long-term transgenerational effects [16], but they require the use of several animals as opposed to in vitro models. One powerful in vitro system that reproduces the in vivo kinetics of perinatal testicular development is the organ culture of rat fetal testes [24]. This technique allows the preservation of both tissue structure and the interaction between the different cell populations of the testis. As with any model, this technique has limitations. In particular, the maintenance of testicular development can only be achieved for a few days. Nonetheless, we and others have used this approach as a toxicological test to evaluate the direct effects of various ED on gametogenesis and steroidogenesis in rodent and human testes [20, 25, 26]. But, it is still unknown whether the epigenetic reprogramming that occurs in male germ cells at that time is faithfully reproduced in vitro.

We hypothesize that organ culture of rat fetal testes can reproduce the epigenetic reprogramming in gonocytes. To test this hypothesis, we quantified three key epigenetic marks after different culture times: DNA methylation (5mC), dimethylation and trimethylation of histone H3 on lysine 4 (H3K4me2, H3K4me3). These marks were chosen because it was shown that 5mC and H3K4me3 vary during development and reprogramming in vivo [5, 7, 9]; however, little is known about H3K4me2. Importantly, as the chronology of epigenetic reprogramming has mostly been established in mice, this study also quantified the dynamics of these key epigenetic marks in rat gonocytes throughout perinatal development.

## Methods
### Animals

Transgenic Sprague–Dawley rats expressing germ cell-specific GFP (GCS-EGFP) were generated and generously provided by Hammer [27]. The animal studies were conducted in accordance with the guidelines set out by the Canadian Council of Animal Care (CCAC) and as reviewed and approved by the Institutional Animal Care and Use Committee of the INRS (Protocol N°: 1510-06). Females were caged with males for one night, and vaginal smears were done the following day to identify sperm-positive females. That day was counted as 0.5 day post-coitum (dpc). Pregnant rats were euthanized by $CO_2$ asphyxiation and subsequent cervical dislocation. The fetuses were removed from the uterus, decapitated and placed on ice.

## In vivo and in vitro sample collection

To analyze expression levels of different epigenetic marks in vivo, testes were sampled at different stages of development covering the time of DNA re-methylation [7, 9]: 16.5, 18.5, 20.5 dpc and 3 days postpartum (dpp). For organ culture, testes were sampled at 16.5 or 18.5 dpc and maintained in culture for 3–4 days to reproduce the same developing time window as in vivo. Analyses were done on testes retrieved after 2 and 4 days of culture when explanted at 16.5 dpc, and after 3 days of culture when explanted at 18.5 dpc (Fig. 1). When sampled or at the end of the culture period, some testes were fixed while others were pooled for germ cell purification (Fig. 1).

## Organ culture

Testes were sampled from fetuses obtained at 16.5 and 18.5 dpc. Organ culture was done as previously described [24] on Millicell culture inserts (Millipore PICM01250, Etobicoke, Ontario, Canada) floating on culture media (DMEM/F-12, HEPES 15 mM), no phenol red; #11039-021, Life Technologies, Burlington, Ontario, Canada) containing 0.04 mg/mL Gentamicin (Life Technologies, #1570-060). Briefly, each explanted testis was placed in cold HBSS (Life Technologies, #14175095), cut in two or four equal parts when sampled at 16.5 or 18.5 dpc, respectively, and placed on a culture insert floating on 500 μL of media. Culture was maintained at 37 °C and 5% $CO_2$ and media was changed daily.

## Immunofluorescence for H3K4me2 and H3K4me3

Testes sampled in vivo or after organ culture were immersed overnight in freshly prepared 4%

paraformaldehyde fixative (Electron Microscopy Science, Hatfield, PA, USA), dehydrated, embedded in paraffin and cut in 5 μm sections for histological analysis. Because the GFP protein carried in germ cells by our transgenic animals could not be detected after fixation, we used HSP90 as a germ cell-specific marker [28]. Tissue sections were deparaffinized, rehydrated, submerged in the antigen retrieval tris buffer at pH 9 (tris 10 mM, EDTA 1 mM, tween 20 0.05%, pH 9) and microwaved for 3 min at full power or until the solution came to a boil then 7 min at 30% power. Slides were left to cool down in the solution for 15 min at room temperature and then rinsed with double-distilled water ($ddH_2O$). All incubations with antibody/serum were done in a humidified chamber (Simport Scientific, Beloeil, Quebec, Canada) and subsequent washes were in PBS. Tissue sections were blocked with a 5% (w/v) BSA (Sigma-Aldrich, #A4503, Oakville, Ontario, Canada) solution for 30 min before incubation overnight at 4 °C with anti-H3K4me2 [29] (1:500, rabbit, #9726, Cell Signaling, Danvers, MA, USA) or anti-H3K4me3 [30] (1:500, rabbit, Cell Signaling, #9751) antibody in combination with the anti-HSP90 (1:200, mouse, #610419, BD Biosciences, San Jose, CA) diluted with 1% BSA in PBS. Slides were then washed and incubated with goat anti-mouse 488 (1:200, Life Technologies, #A11001) and goat anti-rabbit 594 (1:200, Life Technologies, #A11037) for an hour at room temperature. Following washes, slides were mounted in SlowFade Diamond Antifade Mountant with DAPI (Life Technologies, #S36972). Negative controls were done by omitting the primary antibodies. Moreover, non-cross-reactivity of the antibodies with unmodified histone H3 has been confirmed by Western blot (see Additional file 1: Figure S. 1).

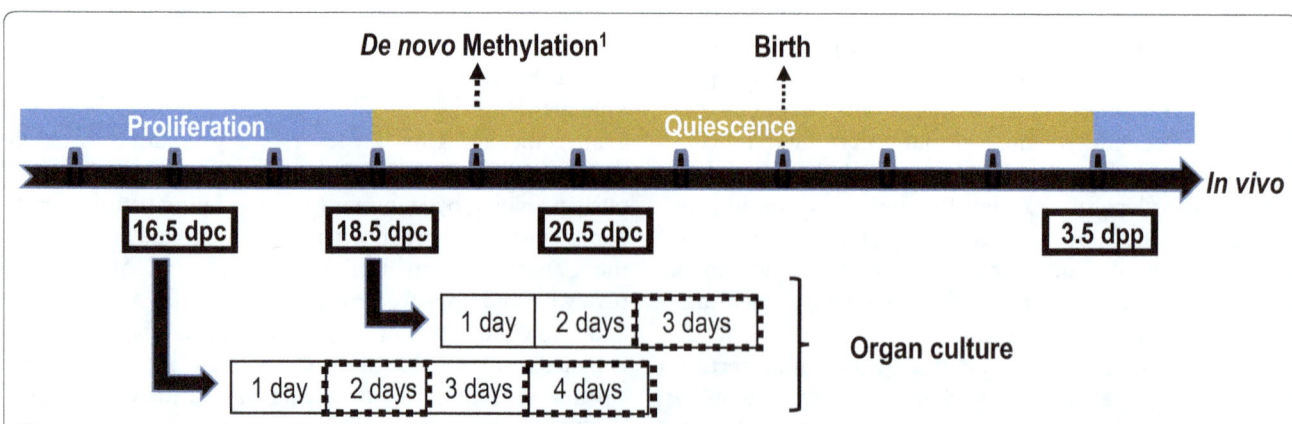

**Fig. 1** Schematic representation of our sampling design throughout gonocyte perinatal development in the rat. Rat testes were collected at four stages of development in vivo (*solid red boxes*) or at three time points after culture (*dashed red boxes*). For organ culture, testes were collected at 16.5 and 18.5 dpc and cultured for 4 and 3 days, respectively. [1]Rose et al. [7]

## Immunofluorescence for 5mC

Following rehydration, tissue sections were incubated 5 min with pepsin (Sigma-Aldrich, #R2283) at 37 °C then washed with PBS before incubation with HCl 0.2 N for 10 min at 37 °C. Slides were washed and incubated 10 min with HCl 2 N at 37 °C. Following further washes, tissue sections were incubated with 5mC antibody [31, 32] (1:150, mouse, #A-1014-100, Epigentek, Farmingdale, NY, USA) diluted with 1% BSA in PBS overnight at 4 °C. Slides were then washed and incubated with goat anti-mouse 594 (1:200, Cell Signaling, #8890) and processed as above. Negative controls were done by omitting the primary antibody.

## Image capture and immunofluorescence quantification

Sections of all stages were placed on the same slide allowing comparison of immunofluorescence intensity. As a reference for normalization, one 16.5 dpc testis section from the same specimen was placed on each slide and used in all staining. Representative images of stained testicular sections were taken using a Nikon eclipse T$i$-S inverted microscope (Mississauga, Ontario, Canada) equipped with a DS-Ri2 camera and the NIS-Elements imaging software. Color pictures were first converted to monochrome using ImageJ software (National Institute of Health). The region of interest was determined as being inside the seminiferous tubules, and the Sertoli cell ring was distinguished from HSP90-immunostained gonocytes. Using the ImageJ 'multi-point' tool, four points were drawn in the nucleus of each gonocyte (HSP90-positive) and each Sertoli cell (HSP90-negative) of that region. This was repeated for 100 cells/section/biological replicate. The mean fluorescence (expressed in pixel intensity) was then calculated for each cell type and normalized using the value obtained for the reference 16.5 dpc testis sample.

## Germ cell purification

We used rats that express GFP (GFP+) specifically in germ cells which allows purification of germ cells at different stages of development or after culture. Explanted testes were pooled per litter and slightly cut to expose the seminiferous tubules, immerged in a solution of 1 mg/mL collagenase (Sigma, #C9891) and 0.02 mg/mL DNase (Sigma, #D4527) for 15 min at 37 °C and then centrifuged 5 min at 500$g$. The pellet was re-suspended in 0.25% trypsin–EDTA (Life Technologies, #25200-056) and incubated 10 min at 37 °C after which 10% fetal bovine serum (FBS) (Life Technologies, #10099-133) was added to neutralize trypsin. Cells were filtered with a 70 μm strainer (#352350, Fisher Scientific, Ottawa, Ontario, CA), centrifuged 5 min at 500 g and suspended in presort buffer (PBS 1X no Ca2+, no Mg2+ supplemented in EDTA 1 mM and HEPES 25 mM and 1% de FBS) prior to cell sorting with FACSJAZZ (BD Biosciences, San Jose, CA) at an event rate of ~2500/s. The sort gates were set on forward and side scatter, trigger pulse width and GFP expression. Cells were interrogated using a 488 nm laser, and emission was measured using 530/40 (FITC/GFP) and collected in FBS. After sorting, both GFP-positive and GFP-negative cell fractions were washed with HBSS and counted. Sorted cells were flash-frozen and stored at −80 °C until DNA extraction. After culture, 4–16 testes were pooled and processed similarly but without trypsin treatment. For each fraction at each stage, we evaluated the purity by calculating the percentage of GFP-positive or GFP-negative cells, respectively, on a hematocytometer (Table 1). GFP-negative fractions were all 100% pure. Importantly, GFP-positive cell fractions after culture showed low purity in terms of GFP signal but we noted that all exhibited typical gonocytes' morphology (see Table 1; Additional file 1: Figure S. 2). This suggested that FACS sorting decreased the GFP signal in these cells.

## DNA extraction

Cell pellets were thawed prior to extracting DNA using the Qiagen kit, QIAamp DNA micro (#56304, Qiagen, Toronto, Ontario Canada). Briefly, 30 μL of buffer ATL and 20 μL of proteinase K were added to cells and

**Table 1  Purity of GFP-positive fractions obtained at sampling or after organ culture**

| Age of sampling | Percentage of GFP-positive cells | Percentage of gonocytes based on morphology |
|---|---|---|
| 16.5 dpc | 92.85 ± 1.13% ($n = 13$) | 96.26 ± 1.09% ($n = 13$) |
| 18.5 dpc | 91.88 ± 1.18% ($n = 4$) | 97.34 ± 1.55% ($n = 4$) |
| 20.5 dpc | 78.33 ± 2.91% ($n = 4$) | 99.27 ± 0.73% ($n = 4$) |
| 3 dpp | 91.59 ± 2.78% ($n = 5$) | 97.06 ± 2.24% ($n = 5$) |
| 16.5 dpc + 2 d | 66.00 ± 4.88% ($n = 5$) | 97.71 ± 2.29% ($n = 5$) |
| 16.5 dpc + 4 d | 61.85 ± 9.04% ($n = 4$) | 83.75 ± 13.89% ($n = 4$) |
| 18.5 dpc + 3 d | 61.27 ± 6.00% ($n = 5$) | 97.01 ± 1.23% ($n = 5$) |

Purity was evaluated on a hemocytometer based on fluorescence and morphology. All data are expressed as mean ± SEM ($n = 4$–13)

incubated for 1 h at 56 °C with pulse-vortexing every 30 min. After incubation, 50 μL of buffer ATL and 100 μL of buffer AL were added and mixed for 15 s. followed by the addition of 100 μL of filtered ethanol at 100%. DNA was then cleaned on columns as per manufacturer's instructions and quantified using a spectrophotometer NanoDrop ND-1000 (Thermo Scientific, Wilmington, DE, USA). DNA was stored at −20 °C until use if not converted on the same day.

## Bisulfite conversion

Because the amount of DNA that was available for bisulfite conversion was limited by the number of sorted cells that could be obtained, especially after culture, we used different conversion kits depending on the amount of available DNA. The kit EpiTect Fast Bisulfite conversion (Qiagen, #59824) was used for samples with more than 500 ng of DNA while the EZ DNA Methylation-Direct (#D5020, Zymo Research, Irvine, CA, USA) was used for samples with less than 500 ng of DNA, with a minimum of 100 ng of DNA per reaction. Conversion was done as per manufacturer's instructions, and converted DNA was stored at −20 °C until use.

## Pyrosequencing

The DNA methylation levels at differentially methylated regions (DMR) of paternally (*H19*) and maternally methylated (*Snrpn*) imprinted genes in both gonocytes and somatic cells were analyzed using the PyroMark Q24 Advanced (Qiagen). We used the CpG-rich repeat regions mapped by Stadnick et al. [33] to locate the studied DMR of *H19*. For *Snrpn*, genomic locations for the DMRs in the mouse established by Shemer et al. [34] were overlapped with the rat sequence (RGSC Rnor_6.0, July 2014) provided at http://genome.ucsc.edu. CpG-rich regions were then located using the CpG prediction program, MethPrimer [35]. The PyroMark Q24 Advanced software was used to design the best primers for the regions of interest: *H19* = forward: 5′-GGTTTTTAGGT GAT-TTGGGA-TATT-3′, biotinylated reverse: 5′-ACATTTA AATTTATAAAA-TAATCCCCTTCT-3′, sequencing: 5′-GT TGAATTTTAG-TTTTTTTTTATGG-3′. *Snrpn* = forward: TGGTGGTTT-GAGGTTGTTGAT, biotinylated reverse: 5′-TCCTAAAACCCAAAAACCATTCAATAAC-3′, sequencing: 5′-GGATGTAGGAG-TTATGT-3′. Assays were designed to analyze the methylation level of 7 and 5 CpGs for *H19* and *Snrpn*, respectively (Fig. 5a). Converted DNA was amplified using HotStarTaq kit (Qiagen, #203443). 50 to 100 ng of converted DNA was used per reaction in a volume of 25–50 μL with the following cycle repeated 50 times: 15 min at 95 °C, 15 s at 95 °C, 30 s at the annealing temperature (*H19* = 51 °C; *Snrpn* = 60 °C), 15 s at 72 °C; this was followed by a 10 min step at 72 °C. PCR products were stored at −20 °C until analysis. Amplified sequences were sequenced

using the PyroMark Q24 Advanced CpG kit (Qiagen, #970922) and the PyroMark Q24 Vacuum Workstation as per manufacturer's protocol.

## Statistical analysis

All values are mean ± SEM of 3 biological replicates obtained from different litters. All statistical analyses were done using GraphPad Prism 6 (GraphPad Software, La Jolla, CA, USA). The level of significance for all statistical tests was set at $p < 0.05$. All parameters obtained in vivo were analyzed using a one-way ANOVA followed by Tukey's test. The significance of the difference between the mean values of testes at the stage of sampling and testes after organ culture was evaluated using a one-way ANOVA followed by Tukey's test at 16.5 dpc and using a Student's unpaired $t$ test at 18.5 dpc.

## Results

### Histone 3 methylation

We assessed the expression level of two histone 3 methylation marks known to vary during male mouse gonocytes' development [4, 5, 9]. To test whether the in vitro kinetics followed the same pattern as in vivo in rat gonocytes, testes were sampled from different key stages of development, equivalent to different times of organ culture (Fig. 1). Using immunofluorescence co-staining, we assessed the expression of H3K4me2 (Fig. 2) and H3K4me3 (Fig. 3) in gonocytes (HSP90-positive cells) and Sertoli cells (HSP90-negative cells within the seminiferous cords).

### *Dimethylation of histone H3 on lysine 4 (H3K4me2)*

H3K4me2 was present in the nucleus of all cell types in the fetal and postnatal testes at all stages studied, with an evident transient increase in gonocytes at 18.5 dpc (Fig. 2a). Further quantification confirmed such variation throughout perinatal development (Fig. 2c). Interestingly, H3K4me2 level in gonocytes was highest at 18.5 dpc and significantly decreased onwards until 3.5 dpp (Fig. 2a, c). In parallel, H3K4me2 levels were stable throughout time in Sertoli cells (Fig. 2d).

In vitro, the semi-quantitative study revealed a slight but significant decrease in H3K4me2 levels in Sertoli cells from testes cultured for 4 days when sampled at 16.5 dpc, a change that was not observed in vivo (Fig. 2b, d). In parallel, the level of H3K4me2 was not changed after 2 days of culture when testes were sampled at 16.5 dpc which suggest that the peak observed at 18.5 dpc in vivo was not reproduced in vitro. Interestingly, however, our data also showed that if the testes were sampled at 16.5 or 18.5 dpc, there was a significant decrease in the level of H3K4me2 in gonocytes in vitro at the end of the culture period (Fig. 2c). Overall, despite the shortcut of the peak

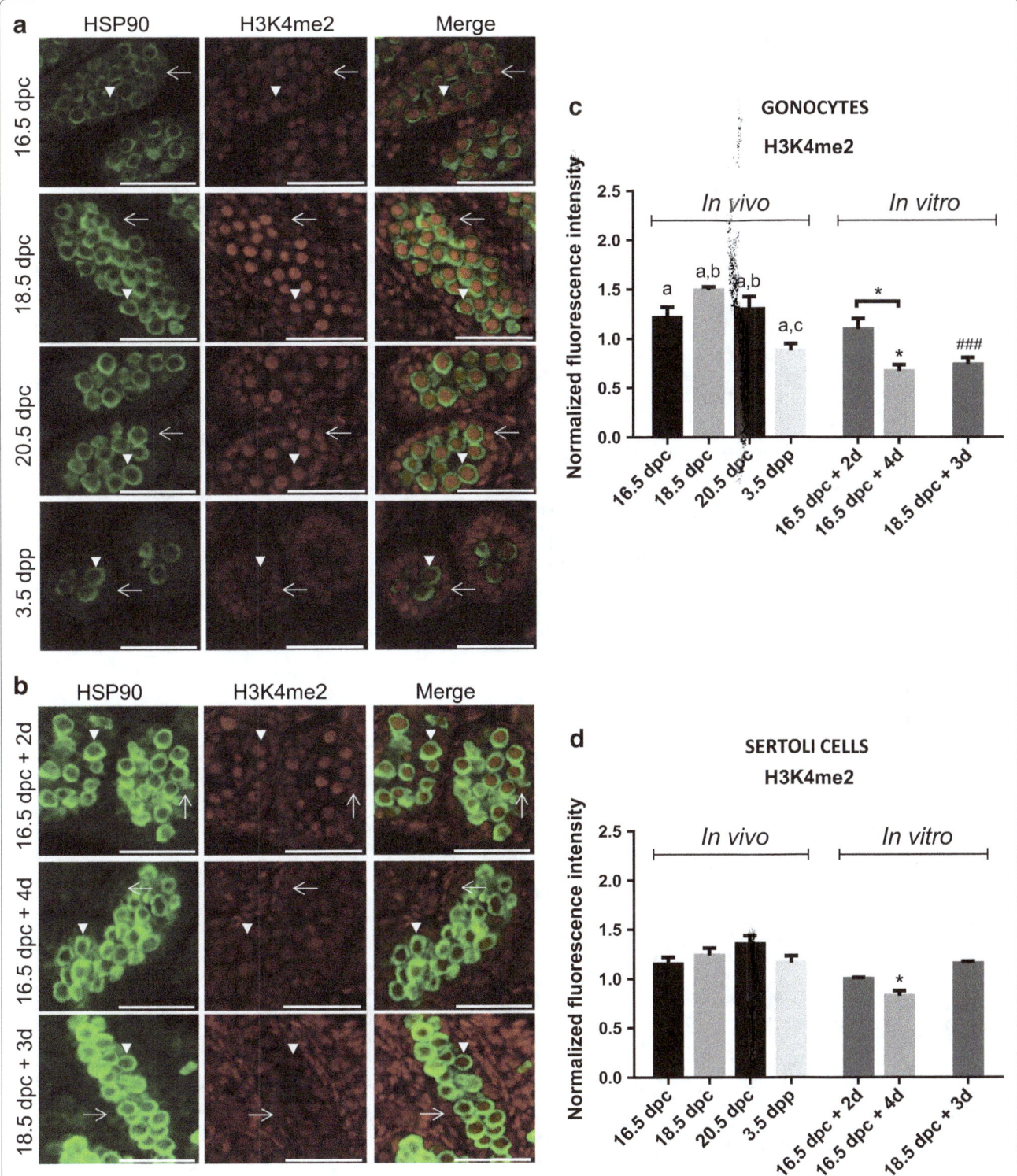

**Fig. 2** Global changes in H3K4me2 in gonocytes and Sertoli cells in vivo and in vitro. Co-staining of H3K4me2 (*red*) with HSP90 (*green*: marker of gonocyte) was done on sections of rat testes sampled in vivo (**a**) or after organ culture (**b**). Gonocytes (*arrow head*) and Sertoli cells (*arrow*) can be observed on sections of testis samples at different stages in vivo and in vitro. Immunofluorescence intensity was quantified in gonocytes (**c**) and in Sertoli cells (**d**). Data represent the average normalized fluorescence intensity ±SEM (*n* = 3/time point). [a,b,c]*p* < 0.05 between different stages in vivo using a one-way ANOVA followed by a Tukey's post hoc test. \**p* < 0.05 between 16.5 dpc and different time of culture using a one-way ANOVA followed by a Tukey's post hoc test. [###]*p* ≤ 0.001 between 18.5 dpc and after culture using a Student unpaired *t* test. *Scale* = 50 μm

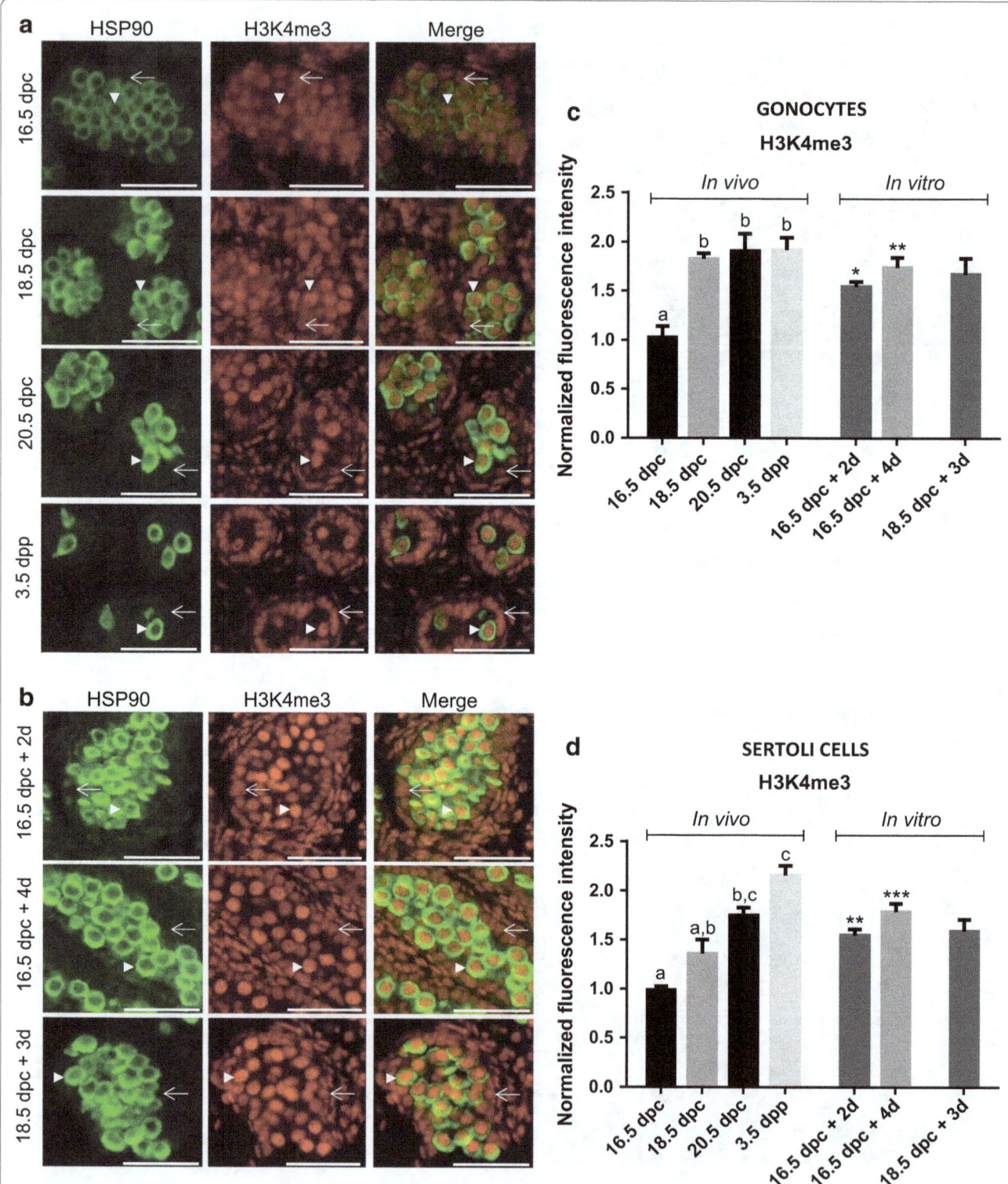

**Fig. 3** Global changes in H3K4me3 in gonocytes and Sertoli cells in vivo and in vitro. Co-staining of H3K4me3 (*red*) with HSP90 (*green*: marker of gonocyte) was done on sections of rat testes sampled in vivo (**a**) or after organ culture (**b**). Gonocytes (*arrow head*) and Sertoli cells (*arrow*) can be observed on sections of testis samples at different stages in vivo and in vitro. Immunofluorescence intensity was quantified in gonocytes (**c**) and in Sertoli cells (**d**). Data represent the average normalized fluorescence intensity ± SEM ($n = 3$/time point). [a,b]$p < 0.05$ between different stages in vivo using a one-way ANOVA followed by a Tukey's post hoc test. *$p < 0.05$, **$p \leq 0.01$, ***$p \leq 0.001$ between 16.5 dpc and different times of culture using a one-way ANOVA followed by a Tukey's post hoc test. *Scale* = 50 μm

observed in vivo, this decrease mirrored the significant decrease observed in vivo between 18.5 and 3.5 dpp, indicating that the kinetics of changes in H3K4me2 levels in gonocytes in vivo is partially reproduced in vitro.

### Trimethylation of histone H3 on lysine 4 (H3K4me3)

H3K4me3 was also present in the nucleus of all cell types in the fetal and postnatal testes at all stages studied (Fig. 3a). Observations on tissue sections suggested an increase in the staining intensity with time especially in gonocytes (Fig. 3a). This was further confirmed by semi-quantitative analysis showing that H3K4me3 levels significantly increased between 16.5 and 18.5 dpc in gonocytes in vivo and remained stable until 3.5 dpp (Fig. 3c). Interestingly, a progressive increase in H3K4me3 level was also quantified in Sertoli cells with a significant difference at 20.5 dpc compared to 16.5 dpc (Fig. 3d).

In vitro, the semi-quantitative study demonstrated a significant increase in H3K4me3 levels in gonocytes and Sertoli cells after culture compared to the day of seeding (Fig. 3b–d). This increase was specifically observed in the testes taken at 16.5 dpc already after 2 days of culture and was then maintained. In addition, levels of H3K4me3 are maintained after 3 days of culture when testes were sampled at 18.5 dpc (Fig. 3b, d). These data demonstrate that the in vivo kinetics of changes in H3K4me3 levels in gonocytes and Sertoli cells are reproduced in vitro.

## DNA methylation (5mC)

### Immunofluorescence

Similar to what was done for histone 3 methylation, we evaluated global methylation of DNA in testicular cells using immunofluorescence for 5mC. Because co-staining with HSP90 was impossible due to antigen retrieval methods to reveal 5mC, we identified cells based on nucleus morphology. Indeed, gonocytes have round nucleus of about 10 μm diameter in the center of the seminiferous cords, whereas Sertoli cells have small lobular nucleus of about 4–5 μm diameter on the periphery of the cords [36]. We detected 5mC in the nucleus of all somatic cell types in the fetal and postnatal testes at all stages studied (Fig. 4a). On the other hand, only very low staining for 5mC could be observed in the nucleus of gonocytes from 16.5 to 20.5 dpc when the signal increased until 3.5 dpp (Fig. 4c). It is interesting to note that from 16.5 to 3.5 dpp, some gonocytes exhibit some staining (Fig. 4a, white arrow) while others did not (Fig. 4a, white arrowhead). Interestingly, when displaying data from individual cells, we observe a continuous variability of staining intensity at each stage of development suggesting that re-methylation does not occur synchronously in rat gonocytes (see Additional file 1: Figure S. 3). The semi-quantitative analysis of DNA methylation

levels in Sertoli cells showed that it remained constant throughout the window of development studied (Fig. 4d).

In vitro, while constant staining could be observed in the nucleus of all somatic cells (Fig. 4b, d), we observed no or very low staining for 5mC in all gonocytes in testes sampled at 16.5 dpc after 2 days of culture (Fig. 4b). After 4 days of culture, some gonocytes became positive for 5mC (Fig. 4b, white arrow) while others remained negative (Fig. 4b, white arrowhead). This observation was confirmed by semi-quantitative analysis showing a decrease in 5mC staining in gonocytes from testes sampled at 16.5 dpc after 2 days of culture, compared to staining level at 16.5 dpc in vivo (Fig. 4c). Furthermore, although nonsignificant, a global increase was observed from 2 to 4 days of culture suggesting a re-gain of methylation that mirrors the in vivo kinetics (Fig. 4c). Interestingly, the level of 5mC significantly increased after 3 days of culture in gonocytes from testes sampled at 18.5 dpc (Fig. 4c), suggesting a re-methylation following the same kinetics as in vivo (Fig. 4b, c). In parallel, 5mC level in Sertoli cells in vitro did not vary at any time of culture studied (Fig. 4d) reproducing what was observed in vivo.

### Pyrosequencing

To test whether normal DNA methylation patterning occurred in our organ culture model following the same pattern as in vivo, we first established the kinetics of re-methylation of the DMRs of *H19* (paternally methylated imprinted gene) and *Snrpn* (maternally methylated imprinted gene) in rat male gonocytes. Using pyrosequencing on genomic DNA extracted from GFP-negative somatic cells, we showed that the DNA methylation levels of the *H19* and *Snrpn* DMRs ranged between 34.0 and 48.7% (Fig. 5b). A similar analysis on genomic DNA extracted from GFP-positive purified gonocytes revealed that *Snrpn* DMR methylation levels remained below 5% at all stages studied (Fig. 5c). Interestingly, the average *H19* DMR methylation level remained low in gonocytes between 16.5 dpc (6.86 ± 0.92%) and 18.5 dpc (2.86 ± 1.13%) but then increased from 20.5 dpc (12.57 ± 5.92%) until 3.5 dpp, where the highest level of methylation was measured (78.64 ± 5.33%; Fig. 5c). We also analyzed the level of DNA methylation in gonocytes at each CpG site for *H19* DMRs and observed that the increase in the average DNA methylation of *H19* DMR at 20.5 dpc was in fact due to two CpG sites (site #2 and #6) that stood out from the others (see Additional file 1: Figure S. 4). This indicated that those two sites are re-methylated earlier than the other five since their level of methylation also increase at 3.5 dpp.

In vitro, analysis of the DNA methylation of *H19* and *Snrpn* DMRs in GFP-negative somatic cells showed

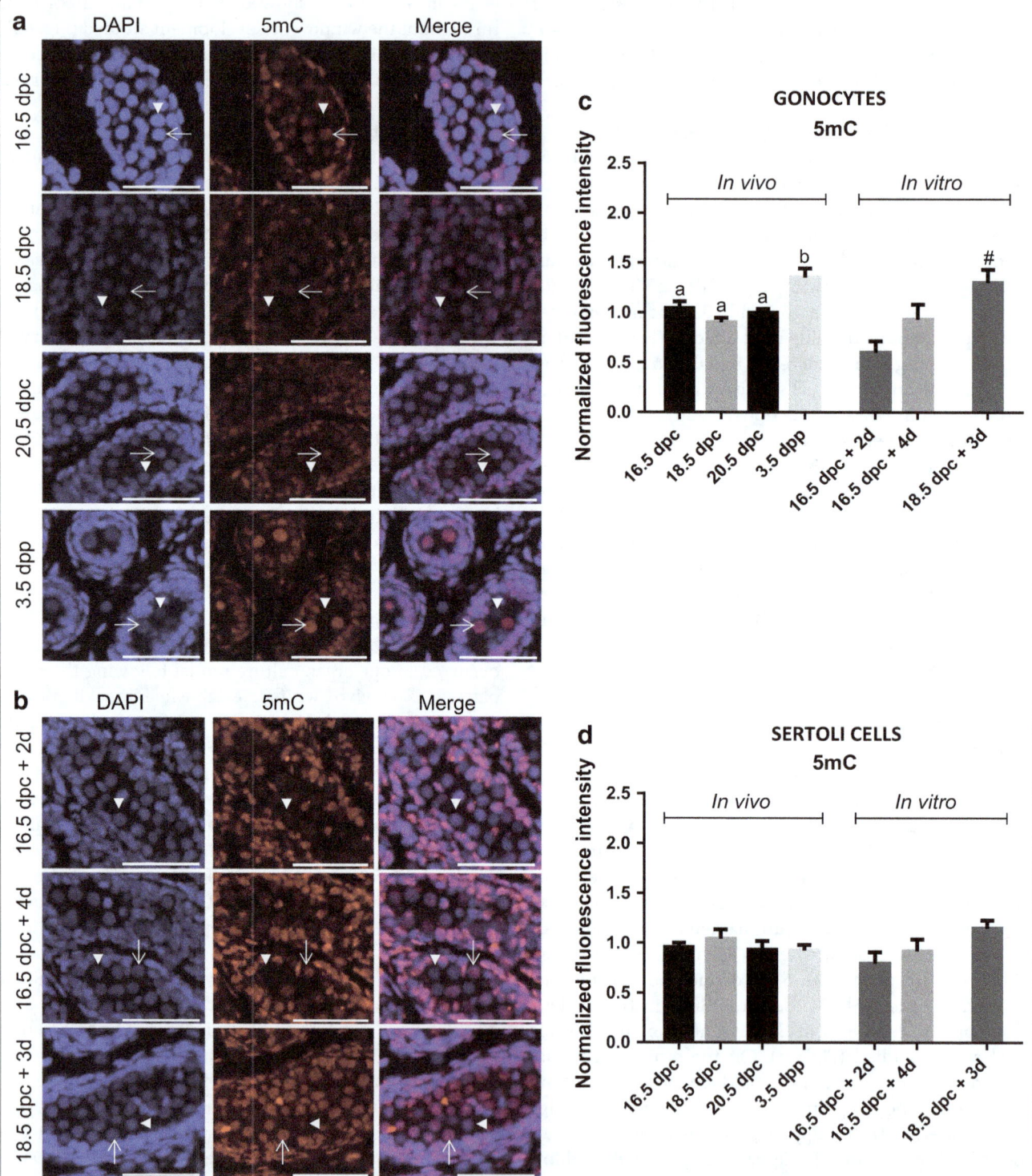

**Fig. 4** Global changes in DNA methylation in gonocytes and Sertoli cells in vivo and in vitro. Co-staining of 5mC (*red*) and DAPI (*blue*) was done on sections of rat testes sampled in vivo (**a**) or after organ culture (**b**). Note that at all age, we observed both unmethylated (*arrow head*) and methylated gonocytes (*arrow*). Immunofluorescence intensity was quantified in gonocytes (**c**) and in Sertoli cells (**d**). Data represent the average normalized fluorescence intensity ± SEM (*n* = 3/time point). [a,b]$p < 0.05$ between different stages in vivo using a one-way ANOVA followed by a Tukey's post hoc test. [#]$p < 0.05$ between 18.5 dpc and after culture using a Student unpaired *t* test. *Scale* = 50 μm

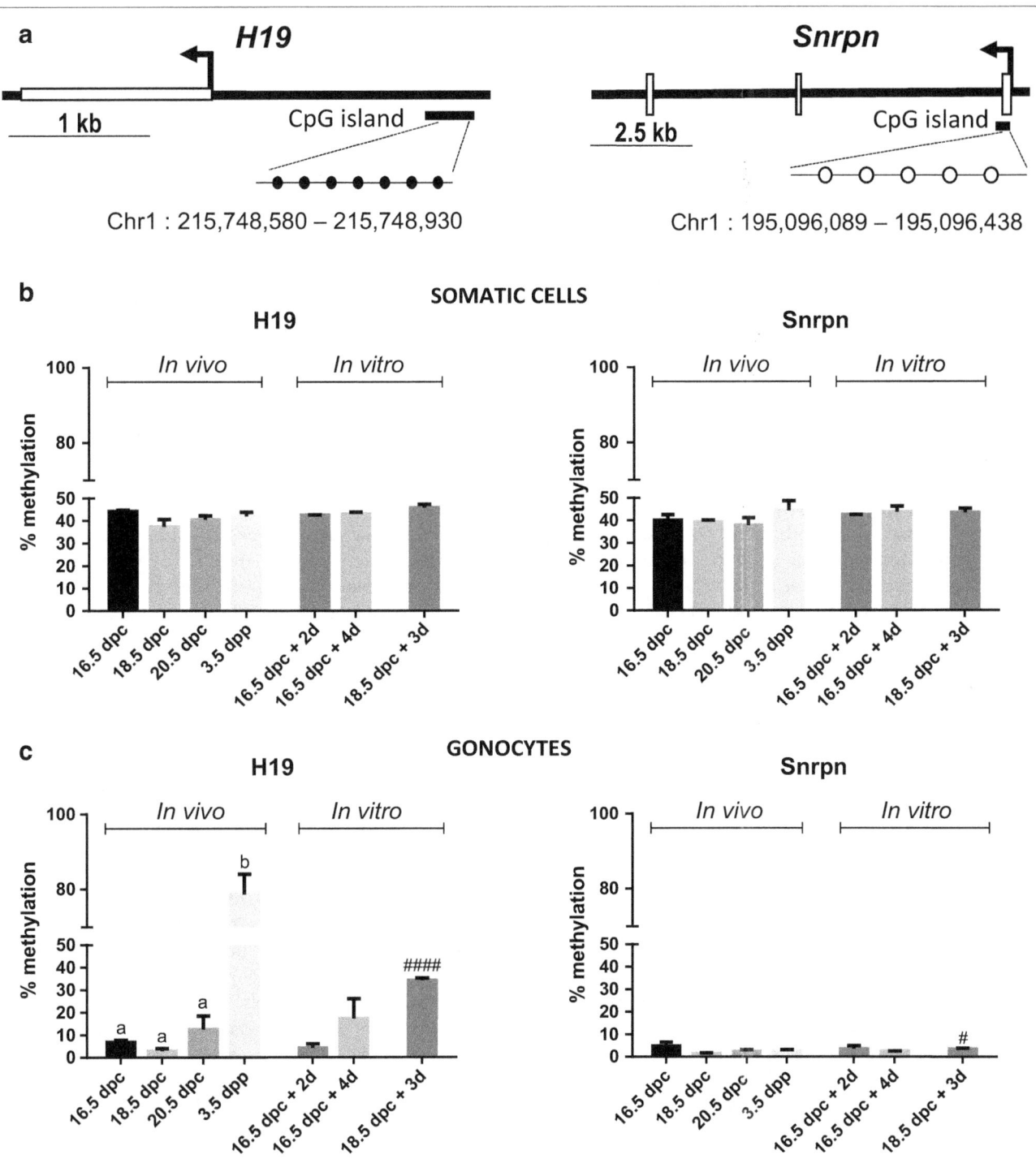

**Fig. 5** *H19* and *Snrpn* DMR methylation profiles in gonocytes and somatic cells in vivo and in vitro. **a** 7 CpG and 5 CpG sites were analyzed in *H19* (*black circles*) and *Snrpn* (*white circles*) DMR, respectively. The average percentage of methylation was obtained in GFP-positive gonocytes (**b**) and GFP-negative somatic cells (**c**) at different stages in vivo and in vitro. Data represent the average % methylation ± SEM ($n = 3$/time point). $^{a,b}p < 0.05$ between different stages in vivo using a one-way ANOVA followed by a Tukey's post hoc test. $^{\#}p < 0.05$, $^{\#\#\#\#}p \leq 0.0001$ between 18.5 dpc and after culture using a Student unpaired *t* test

levels ranging from 41.2 to 47.3% which are similar to those measured in vivo. In addition, *Snrpn* DMR methylation levels in GFP-positive gonocytes did not vary in vitro over time when testes were sampled at 16.5 dpc but slightly increased after 3 days of culture of testes explanted at 18.5 dpc (Fig. 5c). On the other hand, in

GFP-positive gonocytes purified after culture of testes sampled at 16.5 dpc, the level of methylation of *H19* DMR remained as low as the level at 16.5 dpc in vivo after 2 days but increased after 4 days of culture (Fig. 5c). Similarly, we observed that methylation levels of *H19* DMR significantly increased after 3 days of culture of testes explanted at 18.5 dpc (Fig. 5c). Looking at site-specific methylation level in gonocytes, it was interesting to note that sites #2 and #6 that were noted in vivo also stood out as the highest methylated CpG in vitro both at 16.5 dpc after 4 days of culture and at 18.5 dpc after 3 days of culture (see Additional file 1: Figure S. 4). These results show that the in vivo kinetics of *H19* and *Snrpn* DMRs methylation in male gonocytes is reproduced in vitro in rat fetal testis organ culture.

## Discussion

Epigenetic reprogramming is a key event occurring in germ cells during perinatal development and is essential for the establishment of germline fate. The dynamic changes in global DNA methylation and histone modifications associated with epigenetic reprogramming have mostly been described in the mouse [4, 9, 15]. In fact, it is generally considered that these global changes are conserved across mammalian species [37, 38], but the extent to which the exact timing is preserved remains largely unknown. In the present study, we have quantified three epigenetic marks that had been shown to vary in mouse gonocytes during perinatal development [4, 9, 15]. Our study is the first to characterize the in vivo dynamics of the histone modifications H3K4me2 and H3K4me3 in fetal male gonocytes of Sprague–Dawley rats during late gestation. Using immunofluorescence, we were able to quantify the pattern of H3K4me2 and H3K4me3 in gonocytes and Sertoli cells. We showed that H3K4me2 levels display a transient increase in rat gonocytes at 18.5 dpc and decreases thereafter until after birth. On the other hand, we showed that H3K4me3 was detectable in gonocytes at all stages studied and increased from 16.5 to 18.5 dpc to remain constant. While H3K4me2 had not been described in mice, the H3K4me3 pattern described in rats correlates nicely with what was observed in mouse gonocytes where it has been shown to increase between 15.5 and 17.5 dpc. Interestingly, we observed that H3K4me3 levels also increased in Sertoli cells which, to our knowledge, had not been described before. Since H3K4me2 and H3K4me3 are related to transcription activation [5], their dynamic could be associated with the setting-up of the transcriptional identity of the different testicular cell types [39–41].

The global DNA methylation erasure in PGCs followed by a re-methylation in germ cells in a sex-specific manner has been described in mouse, rat, and human [4, 7, 42], but the specific timing might differ between species. For example, while in mouse, the onset of DNA re-methylation occurs at some imprinted loci and at repetitive elements from 15.5 dpc, data from Rose et al. [7] suggest that this occurs later in Wistar rat gonocytes. Indeed, in rats, there is a 2-day difference in gonocyte development so that the equivalent of mouse 15.5 dpc would be 17.5 dpc in rat [43]. However, the onset of re-methylation in rat has been described at 19.5 dpc [7]. Our present data also support a delay in the re-methylation phases in rat gonocytes compared to mouse gonocytes as we characterized DNA re-methylation only from 20.5 dpc onward. This was observed both for global DNA methylation measured by immunofluorescence and, for the first time in rats, on two imprinted genes using pyrosequencing of genomic DNA obtained from purified gonocytes. Interestingly, our data also suggest that DNA re-methylation does not occur synchronously in rat gonocytes as from 16.5 to 3.5 dpp some we observed inter-nucleus variability (Fig. 4a, arrow and arrowhead). One could hypothesize that such changes are due to a change in the chromatin structure associated with cell cycle. But the fact that such variability can be seen at 20.5 dpc, a stage when all gonocytes are in G0 cell cycle arrest [43], suggests that it is not.

It is known that DNA methylation and posttranslational modifications of histones are highly interrelated (reviewed in [44–46]). It was suggested that some histone modifications could be involved in the process of de novo methylation, but the mechanisms are still unclear [47, 48]. As some studies have shown that changes in H3K4me3 and H3K4me2 can precede or accompany de novo DNA methylation, these marks were proposed as good candidates to guide de novo DNA methylation [9, 47, 52]. On the other hand, others suggested that methylation of H3K4 may act as a protective marks from de novo methylation [48–51]. Such interaction in gonocytes has yet to be elucidated. Our findings show that in rat gonocytes, H3K4me2 levels decrease while H3K4me3 increases when 5mC increases (Fig. 6). Such kinetics of change suggests that histone 3 methylation may be involved in guiding de novo DNA methylation. Nevertheless, in the present study, only global changes were quantified which significantly limits the interpretation of such interaction on specific sequence. Further investigation using sequencing-based techniques would be required to elucidate such a dialogue.

Alterations in epigenetic reprogramming by EDs during perinatal life have been suggested to be at the origin of male fertility issues, affecting the DNA methylation pattern in mature sperm [15]. However, the mechanisms involved have not been elucidated yet. We have used the organ culture of rat fetal testes to study the impact of

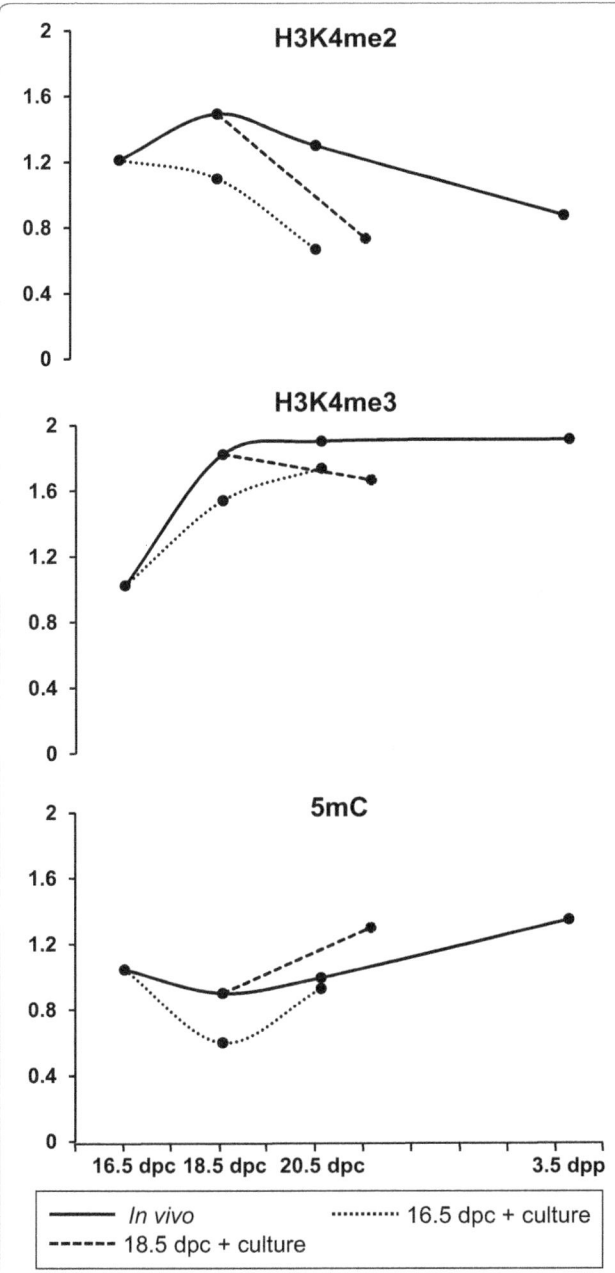

**Fig. 6** Summary of the dynamic changes in the levels of H3K4me2, H3K4me3 and 5mC in rat perinatal gonocytes in vivo (*full line*) and in vitro (*dashed line*)

that the global kinetics of the three epigenetic marks studied were recovered in gonocytes maturing in vitro when compared to in vivo (Fig. 6). Indeed, DNA methylation patterns followed the same global increase as measured in vitro especially in testes explanted at 18.5 dpc after 3 days of culture. Similarly, the early increase and maintenance of H3K4me3 were faithfully reproduced. Only the H3K4me2 patterns were not completely faithful to those observed in vivo. Indeed, signal intensity for H3K4me2 was always lower in vitro than in vivo, in gonocytes and somatic cells, bypassing the peak observed in vivo at 18.5 dpc. This indicates that this specific mark may be more sensitive to culture conditions or that some exogenous factors influencing the pattern of that mark may be missing. Importantly, these results were obtained in media deprived of FBS and very similar data could be obtained when FBS was added to the culture media (see Additional file 1: Figure S. 5). This demonstrates that the epigenetic reprogramming occurs in fetal germ cells without the addition of any exogenous factor. The mechanisms guiding these changes in gonocyte chromatin are therefore driven only through autocrine or paracrine regulations.

Characterizing the methylation level of two imprinted genes in purified rat gonocytes before or after culture by pyrosequencing was made possible thanks to the use of GCS-EGFP rat testis [27]. The accuracy of the methylation levels measured by pyrosequencing largely depends on the purity of the fractions obtained. We have quantified that our GFP-positive fractions contained at least 83.75% of gonocyte-like cells and that the GFP-negative fractions contained 100% of cells negative for GFP signal. We believe our cell sorting method is valid first because in the GFP-negative fractions, we systematically obtained methylation levels that are very similar to the methylation profile expected in somatic cells. Moreover, the methylation level of *Snrpn* that is expected to be low in male germ cells [34] remained below 5% at all age studied in GFP-positive fractions. This confirms that our GFP-negative and GFP-positive fractions indeed represent somatic cells and gonocytes, respectively. Interestingly, our data show that the re-methylation of *H19* DMR that is expected in male gonocytes followed very similar patterns in vitro compared to the one described in vivo. Specifically, we have shown that re-methylation occurs first at specific CpG sites in vivo, a sequenced pattern that is reproduced in vitro. On the other hand, the demethylated status of the *Snrpn* DMR and the maintenance of DNA methylation in somatic cells were also demonstrated. To our knowledge, this is the first time that gonocytes have been purified from rat fetal testes maintained in culture in vitro, which offers great potential to study

various ED on gametogenesis and steroidogenesis during perinatal life [20, 25, 26]. It was proved to be a robust model to reproduce the kinetics of development of the rat fetal testis [24]. In the present study, we have further characterized the in vitro development of gonocytes and Sertoli cells in this culture system, assessing the kinetics of changes in the three epigenetic marks studied in vivo. Our immunofluorescence quantification results show

the molecular mechanisms by which epigenetics are regulated or altered specifically in these cells.

## Conclusions

Together the results of the present study describe the perinatal chronology of three epigenetic marks (H3K4me2, H3K4me3 and 5mC) and the pattern of methylation of *H19* and *Snrpn* DMR in rat gonocytes during perinatal development. Most importantly, we demonstrated that the chronology of those epigenetic marks can be reproduced in vitro with the model of organ culture without the addition of serum. Thus, we are confident that this model can be used to study the process of epigenetic reprogramming in rat fetal germ cells. Furthermore, this model represents a powerful tool to study the impact of environmental chemicals on the establishment of those epigenetic marks in fetal male germ cells.

## Additional file

**Additional file 1: Figure S. 1.** Validation of H3K4me2 and H3K4me3 antibodies cross-reactivity. Western blot was done with 10 μg of recombinant protein H3 (New England Biolabs, #M2503S, Whitby, Ontario, Canada). Membranes were incubated with anti-H3 (1/10,000) (A), anti-H3K4me2 (1/2000) (B) and anti-H3K4me3 (1/2000) (C). The expected band can be visualized with anti-H3 at ~15 kD, but no band could be detected using anti-H3K4me2 or anti-H3K4me3 demonstrating no cross-reactivity against the full unmodified histone H3. **Figure S. 2.** Representative images of dispersed cells from rat fetal testis before and after FACS sorting. Testes were sampled at 18.5 dpc and cultured for 3 days prior to sorting by FACS. In the unsorted cell population (A), GFP-positive gonocytes (arrow) and GFP-negative somatic cells (arrow head) can be discriminated based on fluorescence but also difference in cell morphology. In the GFP-positive fraction (B), all cells exhibited gonocytes' morphology when observed under bright field and some did not express GFP (*). Note that the GFP-negative gonocytes were stained with trypan blue (*), suggesting these cells died after sorting. Scale = 50 μm. **Figure S. 3.** Global changes in DNA methylation in gonocytes in vivo. Immunofluorescence intensity of 5mC was quantified in gonocytes (as in Fig. 4). Data are represented for 100 individual cells per testis at each time point ($n = 3$/time point). **Figure S. 4.** *H19* and *Snrpn* DMR methylation level per CpG site in gonocytes in vivo and in vitro. The percentage of methylation was obtained for each CpG site in GFP-positive gonocytes at different stages in vivo and in vitro. Data represent the average % methylation ± SEM ($n = 3$/time point). *: $p < 0.05$ between 16.5 dpc and different time points of culture using a one-way ANOVA followed by a Tukey's post hoc test. #: $p < 0.05$ between 18.5 dpc and after culture using a Student unpaired $t$ test. **Figure S. 5.** Summary of the dynamic changes in the levels of H3K4me2, H3K4me3 and 5mC, in rat perinatal gonocytes in vivo (full line) and in vitro with (red dashed line) or without FBS (black dashed line). Immunofluorescence intensity was quantified in gonocytes. Data represent the average normalized fluorescence intensity ± SEM ($n = 3$/time point). *: $p < 0.05$, **: $p \leq 0.01$ between 16.5 dpc and different time points of culture using a one-way ANOVA followed by a Tukey's post hoc test. #: $p < 0.05$, ###: $p \leq 0.001$ between 18.5 dpc and after culture using a Student unpaired $t$ test.

## Abbreviations

5mc: 5-methylcytosine; H3K4me2: H3 lysine 4 dimethylation; H3K4me3: H3 lysine 4 trimethylation; GFP: green fluorescent protein; HSP90: heat-shock protein 90; FACS: fluorescence-activated cell sorting; DMR: differentially methylated regions.

## Authors' contributions

AR and GD designed the experiments and wrote the manuscript. AR did all the technical work and analyzed the data. FJ maintained the animal colony and contributed to acquisition of sorted cells in vivo. All authors read and approved the final manuscript.

## Acknowledgements

We gratefully thank Dr. Sarah Kimmins (McGill University) for allowing us access to her pyrosequencer and Dr. Romain Lambrot (McGill University) for his technical assistance and expertise with pyrosequencing. The authors would like to acknowledge Guylaine Lassonde (INRS) for her help with cell sorting and Professor Jacquetta Trasler (McGill University) for her critical review of the manuscript.

## Competing interests

The authors declare that they have no competing interests.

## Funding

These studies were funded by a Discovery Grant from Natural Sciences and Engineering Research Council of Canada (NSERC#04607-2014). Arlette Rwigemera received scholarships from the Fondation Armand-Frappier and the Fond de Recherche en santé du Québec (FRQ-S).

## References

1. Saitou M, Yamaji M. Primordial germ cells in mice. Cold Spring Harb Perspect Biol. 2012;4:a008375.
2. Clermont Y, Perey B. Quantitative study of the cell population of the seminiferous tubules in immature rats. Am J Anat. 1957;100:241–67.
3. Rouiller-Fabre V, Levacher C, Pairault C, Racine C, Moreau E, Olaso R, et al. Development of the foetal and neonatal testis. Andrologia. 2003;35:79–83.
4. Ly L, Chan D, Trasler JM. Developmental windows of susceptibility for epigenetic inheritance through the male germline. Semin Cell Dev Biol. 2015;43:96–105.
5. Cantone I, Fisher AG. Epigenetic programming and reprogramming during development. Nat Struct Mol Biol. 2013;20:282–9.
6. Hajkova P, Ancelin K, Waldmann T, Lacoste N, Lange UC, Cesari F, et al. Chromatin dynamics during epigenetic reprogramming in the mouse germ line. Nature. 2008;452:877–81.
7. Rose CM, van den Driesche S, Sharpe RM, Meehan RR, Drake AJ. Dynamic changes in DNA modification states during late gestation male germ line development in the rat. Epigenetics Chromatin. 2014;7:19.
8. Trasler JM. Epigenetics in spermatogenesis. Mol Cell Endocrinol. 2009;306:33–6.
9. Abe M, Tsai SY, Jin S-G, Pfeifer GP, Szabó PE. Sex-specific dynamics of global chromatin changes in fetal mouse germ cells. PLoS ONE. 2011;6:e23848.
10. Barau J, Teissandier A, Zamudio N, Roy S, Nalesso V, Hérault Y, et al. The DNA methyltransferase DNMT3C protects male germ cells from transposon activity. Science. 2016;354:909–12.
11. Siklenka K, Erkek S, Godmann M, Lambrot R, McGraw S, Lafleur C, et al. Disruption of histone methylation in developing sperm impairs offspring health transgenerationally. Science. 2015;350:aab2006.

12. Radford EJ, Ito M, Shi H, Corish JA, Yamazawa K, Isganaitis E, et al. In utero effects. In utero undernourishment perturbs the adult sperm methylome and intergenerational metabolism. Science. 2014;345:1255903.

13. Frick KM, Zhao Z, Fan L. The epigenetics of estrogen. Epigenetics. 2011;6:675–80.

14. Mattison DR, Karyakina N, Goodman M, LaKind JS. Pharmaco- and toxicokinetics of selected exogenous and endogenous estrogens: a review of the data and identification of knowledge gaps. Crit Rev Toxicol. 2014;44:696–724.

15. Wu H, Hauser R, Krawetz SA, Pilsner JR. Environmental susceptibility of the sperm epigenome during windows of male germ cell development. Curr Environ Health Rep. 2015;2:356–66.

16. Iqbal K, Tran DA, Li AX, Warden C, Bai AY, Singh P, et al. Deleterious effects of endocrine disruptors are corrected in the mammalian germline by epigenome reprogramming. Genome Biol. 2015;16:59.

17. Toppari J. Environmental endocrine disrupters and disorders of sexual differentiation. Semin Reprod Med. 2002;20:305–12.

18. Sharpe RM, Irvine DS. How strong is the evidence of a link between environmental chemicals and adverse effects on human reproductive health? BMJ. 2004;328:447–51.

19. Skakkebaek NE, Rajpert-De Meyts E, Buck Louis GM, Toppari J, Andersson A-M, Eisenberg ML, et al. Male reproductive disorders and fertility trends: influences of environment and genetic susceptibility. Physiol Rev. 2016;96:55–97.

20. Delbès G, Duquenne C, Szenker J, Taccoen J, Habert R, Levacher C. Developmental changes in testicular sensitivity to estrogens throughout fetal and neonatal life. Toxicol Sci Off J Soc Toxicol. 2007;99:234–43.

21. Skakkebaek NE, Rajpert-De Meyts E, Main KM. Testicular dysgenesis syndrome: an increasingly common developmental disorder with environmental aspects. Hum Reprod Oxf Engl. 2001;16:972–8.

22. Storgaard L, Bonde JP, Olsen J. Male reproductive disorders in humans and prenatal indicators of estrogen exposure: a review of published epidemiological studies. Reprod Toxicol. 2006;21:4–15.

23. Yasuda Y, Kihara T, Tanimura T, Nishimura H. Gonadal dysgenesis induced by prenatal exposure to ethinyl estradiol in mice. Teratology. 1985;32:219–27.

24. Livera G, Delbès G, Pairault C, Rouiller-Fabre V, Habert R. Organotypic culture, a powerful model for studying rat and mouse fetal testis development. Cell Tissue Res. 2006;324:507–21.

25. Lassonde G, Nasuhoglu D, Pan JF, Gaye B, Yargeau V, Delbes G. Ozone treatment prevents the toxicity of an environmental mixture of estrogens on rat fetal testicular development. Reprod Toxicol. 2015;58:85–92.

26. Habert R, Muczynski V, Grisin T, Moison D, Messiaen S, Frydman R, et al. Concerns about the widespread use of rodent models for human risk assessments of endocrine disruptors. Reprod Camb Engl. 2014;147:R119–29.

27. Cronkhite JT, Norlander C, Furth JK, Levan G, Garbers DL, Hammer RE. Male and female germline specific expression of an EGFP reporter gene in a unique strain of transgenic rats. Dev Biol. 2005;284:171–83.

28. Ohsako S, Bunick D, Hayashi Y. Immunocytochemical observation of the 90 KD heat shock protein (HSP90): high expression in primordial and pre-meiotic germ cells of male and female rat gonads. J Histochem Cytochem. 1995;43:67–76.

29. Nair VD, Ge Y, Balasubramaniyan N, Kim J, Okawa Y, Chikina M, et al. Involvement of histone demethylase LSD1 in short-time-scale gene expression changes during cell cycle progression in embryonic stem cells. Mol Cell Biol. 2012;32:4861–76.

30. Song N, Liu J, An S, Nishino T, Hishikawa Y, Koji T. Immunohistochemical analysis of histone H3 Modifications in germ cells during mouse spermatogenesis. Acta Histochem Cytochem. 2011;44:183–90.

31. Costa G, Barra V, Lentini L, Cilluffo D, Di Leonardo A. DNA demethylation caused by 5-Aza-2′-deoxycytidine induces mitotic alterations and aneuploidy. Oncotarget. 2016;7:3726–39.

32. Heras S, Forier K, Rombouts K, Braeckmans K, Van Soom A. DNA counterstaining for methylation and hydroxymethylation immunostaining in bovine zygotes. Anal Biochem. 2014;454:14–6.

33. Stadnick MP, Pieracci FM, Cranston MJ, Taksel E, Thorvaldsen JL, Bartolomei MS. Role of a 461-bp G-rich repetitive element in H19 transgene imprinting. Dev Genes Evol. 1999;209:239–48.

34. Shemer R, Birger Y, Riggs AD, Razin A. Structure of the imprinted mouse Snrpn gene and establishment of its parental-specific methylation pattern. Proc Natl Acad Sci. 1997;94:10267–72.

35. Li L-C, Dahiya R. MethPrimer: designing primers for methylation PCRs. Bioinformatics. 2002;18:1427–31.

36. Novi AM, Saba P. An electron microscopic study of the development of rat testis in the first 10 postnatal days. Z Für Zellforsch Mikrosk Anat. 1968;86:313–26.

37. Chen Z, Riggs AD. DNA methylation and demethylation in mammals. J Biol Chem. 2011;286:18347–53.

38. Li E. Chromatin modification and epigenetic reprogramming in mammalian development. Nat Rev Genet. 2002;3:662–73.

39. Orth JM. Proliferation of sertoli cells in fetal and postnatal rats: a quantitative autoradiographic study. Anat Rec. 1982;203:485–92.

40. Nel-Themaat L, Jang C-W, Stewart MD, Akiyama H, Viger RS, Behringer RR. Sertoli cell behaviors in developing testis cords and postnatal seminiferous tubules of the mouse. Biol Reprod. 2011;84:342–50.

41. Boulogne B, Habert R, Levacher C. Regulation of the proliferation of cocultured gonocytes and Sertoli cells by retinoids, triiodothyronine, and intracellular signaling factors: differences between fetal and neonatal cells. Mol Reprod Dev. 2003;65:194–203.

42. von Meyenn F, Reik W. Forget the parents: epigenetic reprogramming in human germ cells. Cell. 2015;161:1248–51.

43. Culty M. Gonocytes, the forgotten cells of the germ cell lineage. Birth Defects Res Part C Embryo Today Rev. 2009;87:1–26.

44. Rose NR, Klose RJ. Understanding the relationship between DNA methylation and histone lysine methylation. Biochim Biophys Acta BBA Gene Regul Mech. 2014;1839:1362–72.

45. Miller JL, Grant PA. The role of DNA methylation and histone modifications in transcriptional regulation in humans. In: Kundu TK, editor. Epigenetics: development and disease. Springer Netherlands; 2013. p. 289–317. doi:10.1007/978-94-007-4525-4_13.

46. Cedar H, Bergman Y. Linking DNA methylation and histone modification: patterns and paradigms. Nat Rev Genet. 2009;10:295–304.

47. Stewart KR, Veselovska L, Kim J, Huang J, Saadeh H, Tomizawa S, et al. Dynamic changes in histone modifications precede de novo DNA methylation in oocytes. Genes Dev. 2015;29:2449–62.

48. Morselli M, Pastor WA, Montanini B, Nee K, Ferrari R, Fu K, et al. In vivo targeting of de novo DNA methylation by histone modifications in yeast and mouse. eLife. 2015;4:e06205.

49. Okitsu CY, Hsieh C-L. DNA methylation dictates histone H3K4 methylation. Mol Cell Biol. 2007;27:2746–57.

50. Singh P, Li AX, Tran DA, Oates N, Kang E-R, Wu X, et al. De novo dna methylation in the male germ line occurs by default but is excluded at sites of H3K4 methylation. Cell Rep. 2013;4:205–19.

51. Weber M, Hellmann I, Stadler MB, Ramos L, Pääbo S, Rebhan M, et al. Distribution, silencing potential and evolutionary impact of promoter DNA methylation in the human genome. Nat Genet. 2007;39:457–66.

52. Smallwood SA, Kelsey G. De novo DNA methylation: a germ cell perspective. Trends Genet TIG. 2012;28:33–42.

# 3

# Stabilization of Foxp3 expression by CRISPR-dCas9-based epigenome editing in mouse primary T cells

Masahiro Okada[*], Mitsuhiro Kanamori, Kazue Someya, Hiroko Nakatsukasa and Akihiko Yoshimura[*]

## Abstract

**Background:** Epigenome editing is expected to manipulate transcription and cell fates and to elucidate the gene expression mechanisms in various cell types. For functional epigenome editing, assessing the chromatin context-dependent activity of artificial epigenetic modifier is required.

**Results:** In this study, we applied clustered regularly interspaced short palindromic repeats (CRISPR)-dCas9-based epigenome editing to mouse primary T cells, focusing on the *Forkhead box P3 (Foxp3)* gene locus, a master transcription factor of regulatory T cells (Tregs). The *Foxp3* gene locus is regulated by combinatorial epigenetic modifications, which determine the Foxp3 expression. Foxp3 expression is unstable in transforming growth factor beta (TGF-β)-induced Tregs (iTregs), while stable in thymus-derived Tregs (tTregs). To stabilize Foxp3 expression in iTregs, we introduced dCas9-TET1CD (dCas9 fused to the catalytic domain (CD) of ten-eleven translocation dioxygenase 1 (TET1), methylcytosine dioxygenase) and dCas9-p300CD (dCas9 fused to the CD of p300, histone acetyltransferase) with guide RNAs (gRNAs) targeted to the *Foxp3* gene locus. Although dCas9-TET1CD induced partial demethylation in enhancer region called conserved non-coding DNA sequences 2 (CNS2), robust Foxp3 stabilization was not observed. In contrast, dCas9-p300CD targeted to the promoter locus partly maintained Foxp3 transcription in cultured and primary T cells even under inflammatory conditions in vitro. Furthermore, dCas9-p300CD promoted expression of Treg signature genes and enhanced suppression activity in vitro.

**Conclusions:** Our results showed that artificial epigenome editing modified the epigenetic status and gene expression of the targeted loci, and engineered cellular functions in conjunction with endogenous epigenetic modification, suggesting effective usage of these technologies, which help elucidate the relationship between chromatin states and gene expression.

**Keywords:** Treg, Foxp3, CRISPR, dCas9, TET1, p300, Epigenome editing

## Background

Epigenetic marks of histone modification and DNA cytosine methylation determine cell identity and function by regulating transcriptional activity at individual loci. Artificial epigenome editing is a novel strategy for manipulating cell fate by altering the specific epigenomic landscape and can help elucidate the mechanisms between chromatin states and gene expression [1]. Epigenome editing tools are fusion proteins consisting of a DNA-binding domain fused with epigenetic-modifying enzymes. Previous reports have shown that DNA-binding proteins, such as zinc finger protein (ZFP) and transcription activator-like effector (TALE) protein, can be used for targeted epigenome editing by fusion with epigenome-modifying enzymes [2, 3]. Although their programmability is verified, there is a disadvantage to designing extensive site-specific constructions. The clustered regularly interspaced short palindromic repeats (CRISPR)-associated protein 9 (Cas9) system (CRISPR-Cas9 system) from *Streptococcus pyogenes* has been used for genome editing by inducing a guide RNA (gRNA) sequence-specific

*Correspondence: okada@z5.keio.jp; yoshimura@a6.keio.jp
Department of Microbiology and Immunology, Keio University School of Medicine, 35 Shinanomachi, Shinjuku-ku, Tokyo 160-8582, Japan

double-strand DNA break. Due to its simple design and high efficiency, the CRISPR-Cas9 system is expected to be utilized extensively in high-throughput and multi-targeted genome editing [4]. Catalytic inactive Cas9 (dCas9) is also recruited to the targeted sequence of the DNA locus, and various fusion proteins with dCas9 can be used for target-specific transcriptional activation and repression [5, 6]. For epigenetic modifications, dCas9 fusion with p300, lysine-specific demethylase 1 (LSD-1), Krüppel-associated box (KRAB), DNA methyltransferase 3a (DNMT3a), and ten-eleven translocation (TET) dioxygenase 1 (TET1) enable gene expression regulation by modifying epigenetic states [7–11]. These biological devices were developed by using cultured cell lines and clearly proposed their versatile performance. However, on the basis that gene transcription is complexly regulated by epigenetic modifications in our body, it is easy to suppose the effectiveness of epigenome editing differs among target loci and cells. Therefore, applying them to primary tissues or cells and evaluation of their activity is expected in the next studies [12]. In primary immune cells, recent research has applied CRISPR-dCas9-based epigenome editing to human primary T lymphocytes, mainly for silencing gene expression [13]. However, only a few studies used epigenome editing mainly for activating gene expression in primary immune cells. Furthermore, little is known about the relationship between artificial epigenome editing and endogenous epigenetic modifications in immune cells.

Regulatory T cells (Tregs) play a pivotal role in regulating immune responses and maintaining immunological tolerance. Treg adoptive transfer therapy is expected to provide a clinical cure for various immunological disorders [14–16]. Tregs are mainly generated via two different routes. The first is through direct development from Treg progenitor cells in the thymus by thymic antigen presentation with high affinity. These Tregs are called naturally occurring Tregs (nTregs) or thymic Tregs (tTregs). The second is through differentiation from naïve CD4 T cells in the periphery by antigen presentation with transforming growth factor (TGF)-β. These Tregs are called induced Tregs in vitro (iTregs) or peripherally induced Tregs (pTregs) [17, 18]. Both Tregs have similar suppression activity and markedly express Forkhead box P3 (Foxp3), a master transcriptional factor for Tregs. Foxp3 expression is required for the differentiation and maintenance of Treg function by expressing Treg signature genes and suppressing effector T cell (Teff) genes [19–23]. The number of available nTregs is limited. It is thought that antigen-specific iTregs could be substituted for nTregs, because iTregs are induced and expanded with antigen specificity in vitro. However, Foxp3 expression is unstable in iTregs owing to the lack of active epigenetic modifications compared with tTregs [24, 25]. Hence, some remaining issues must be resolved prior to the clinical application of ex vivo-expanded iTregs, since iTregs lose Foxp3 expression easily and convert to other pathogenic T cell subsets in vivo [26–28].

The epigenetic modification of the *Foxp3* locus, promoter, and three enhancer regions called conserved non-coding DNA sequences (CNS)1, CNS2, and CNS3, plays pivotal roles in the sustainable expression of Foxp3 [29]. Various transcriptional factors induce active histone modification, such as H3K27 acetylation and H3K4 methylation [30]. Also, the microbial fermentation product butyrate enhances histone acetylation of the *Foxp3* promoter locus and promotes the induction of pTregs in the intestine [31, 32]. In addition to histone modifications, DNA cytosine methylation also effects stable Foxp3 expression. nTregs show a Treg-specific demethylation pattern. Importantly, the *Foxp3* CNS2 locus is also maintained under hypomethylation in nTregs; this hypomethylation contributes to the stable expression of Foxp3 [24, 25, 33]. Recent research has shown that TET family proteins are extensively involved in this demethylation process and maintain Treg stability [34, 35]. In fact, some epigenetic-modifying compounds, such as histone deacetylase (HDAC) inhibitors [36], DNMT inhibitors [37], and TET activators [38], are known for their potential use in effective iTreg induction. However, their target loci are not limited because of low specificity, and there is a risk of undesirable effects like those observed with many of the epigenetic-modifying compounds used to treat cancer [39]. It is essential for functional iTregs to modulate epigenetic modification at necessary locus and not to modulate unnecessary excess locus.

In this study, we established two epigenome-modifying systems based on CRISPR-dCas9 technology and applied them to the *Foxp3* gene locus. We aimed to investigate the cross-talk of epigenome editing and endogenous cellular responses in primary immune cells and to lay a foundation for future clinical development. To stabilize Foxp3 expression in artificially epigenome-edited iTregs: dCas9 fused with TET1CD was targeted to the *Foxp3* CNS2 locus, and dCas9 fused with p300CD to the *Foxp3* promoter locus. We designed 10 gRNA sequences in each locus, screened effective sequences in T cell lines 68-41, and then applied them to mouse primary T cells. We confirmed that both systems with specific gRNAs could induce epigenetic modifications in cultured cell lines. In primary T cells, dCas9-TET1CD partially demethylated the CNS2 locus under iTreg conditions, but Foxp3 expression was not robustly stabilized by inflammatory cytokine stimuli. In contrast, dCas9-p300CD strongly activated and stabilized Foxp3 expression, particularly with TGF-β, even under inflammatory conditions.

dCas9-p300CD epigenome-edited iTregs also showed high expression of Treg signature genes and enhanced suppression activity. Through various T cell culture conditions, we concluded that epigenome editing technology can be used in targeted epigenome research, and effectiveness depends on culture conditions. We expect that our study becomes the premise for broad clinical application in the future.

## Methods

### Mice

Foxp3-hCD2-hCD52-KI mice originated from the laboratory of Dr. S Hori (Laboratory of Microbiology, Graduate School of Pharmaceutical Sciences, the University of Tokyo, Tokyo, Japan).

### Antibodies and reagents

For flow cytometry analysis, fluorescein isothiocyanate (FITC), phycoerythrin (PE), peridinin chlorophyll protein-cyanine 5.5 (PerCP-Cy5.5), allophycocyanin (APC), PE-Cy7, and APC-Cy7-conjugated antibodies were purchased from BioLegend (San Diego, CA, USA) or eBioscience (San Diego, CA, USA). The following clones were used: anti-CD4 (RM4-5), Foxp3 (FJK16s), hCD2 (RPA2.10), CD25 (PC61.5), CTLA-4 (UC10-4F10-11), and CD45.1 (A20). Fixable Viability Dye eFluor 780 (FVD780) was used to remove dead cells. Cytokines (mouse interleukin-2 (IL-2), IL-12, IL-4, and IL-6) were purchased from Peprotech (Rocky Hill, NJ, USA), and human TGF-β1 was purchased from BioLegend. LY2157299 was purchased from Shanghai Biochempartner Co., Ltd (Hubei, China).

### Cell culture

Human embryonic kidney cells 293 (HEK293T cells) were obtained from the American Type Culture Collection (ATCC) and maintained in Dulbecco's modified Eagle medium (DMEM, 4500 mg/l glucose) supplemented with 10% fetal bovine serum (FBS). The 68-41 cells were a gift from Dr. M. Kubo (Division of Molecular Pathology, Research Institute for Biomedical Science, Tokyo University of Science, Tokyo, Japan) [40]. They were maintained in Roswell Park Memorial Institute (RPMI) 1640 medium supplemented with 10% FBS and 55 μM 2-mercaptoethanol (2-ME). Primary T cells were maintained in RPMI 1640 medium supplemented with 10% FBS, 55 μM 2-ME, 1% penicillin/streptomycin, 2 mM L-glutamine, and 100 nM non-essential amino acid solution. All cells were cultured in a humid, 5% $CO_2$, 37 °C incubator.

### Plasmid constructions

LentiCRISPR (Addgene, catalog no. 49535) was mutated at amino acid positions D10A and H840A by mutagenesis polymerase chain reaction (PCR) to construct Flag-dCas9-P2A-puro. Mouse TET1 catalytic domain (TET1CD) or p300CD was amplified from complementary DNA (cDNA) and subcloned into a MIGR vector. TET1CD or p300CD and internal ribosome entry site green fluorescent protein (IRES-GFP) sequences were amplified with a Gly–Gly–Gly–Gly–Ser linker and recombined into Flag-dCas9-P2A-puro instead of P2A-puro. Flag-dCas9-TET1CD or p300CD and IRES-GFP were recombined into pMXs-GW vectors. Amino acid sequences of each construct are detailed in the supplementary material (Additional file 1). For gRNA expression, DsRed was recombined into lentiCRISPR instead of Cas9-P2A-puro. Next, U6-gRNA-EFS-DsRed was recombined into CSII vector. Each gRNA expression vector was generated by the annealing of the oligonucleotides, followed by ligation into BsmBI-digested gRNA expression vectors based on CSII vector for lentiviral infection. In some cases, U6-gRNA-EFS-DsRed was recombined into pMXs-GW vectors for retroviral infection. gRNA off-target predictions were performed by CCTop—CRISPR/Cas9 target online predictor [41]. Max. total mismatches, core length, and max. core mismatches were set to 4, 12, and 2, respectively. Predicted off-target loci were listed in Additional file 2: Table S1.

### Retroviral or lentiviral production

pMXs and pCL-Eco were co-transfected for retroviral production, or CSII, pMDLg/pRRE, and VSV-G/Rsv-Rev were co-transfected for lentiviral production into HEK293T cells using polyethylenimine MAX (PEI-MAX), followed by a medium change to remove the transfection reagents. Virus-containing medium was harvested, filtered (0.45 μm), and then concentrated by centrifugation overnight.

### Establishment of dCas9-TET1CD or p300CD stable 68-41 cells and gRNA transduction

dCas9-TET1CD- or p300CD-expressing retrovirus was transduced into 68-41 cells with 5 μg/ml polybrene with centrifugation at 2500 rpm for 2 h at 35 °C. GFP-positive cells were sorted using a SH800 (SONY, Tokyo, Japan) or ARIA (BD Biosciences, San Jose, CA, USA) cell sorter. GFP-positive cells were further cultured and maintained GFP-positive >85% by sorting again. These stable cell lines were roughly isolated and not from a single clone. Next, gRNA-expressing lentivirus was transduced as above, and further analysis was performed. Transduction efficiency was almost >80%, but in the case of low transduction efficiency, GFP and DsRed double-positive cells were sorted. For intracellular Foxp3 staining, fixation buffer (eBiosciences) and 0.2% Triton-X were used for fixation and permeabilization to retain GFP and DsRed fluorescence.

## Primary T cell culture

Naïve CD4+ T cells (CD4+CD25-hCD2-CD62L+) were isolated using magnetic-activated cell sorting (MACS) from the spleen and lymph nodes of 6- to 8-week-old-male Foxp3-hCD2-hCD52-KI mice. Splenocytes and lymphocytes were red blood lysed and depleted using AutoMACS (Miltenyi Biotec, Tokyo, Japan) with biotin-conjugated anti-B220 (RA3-6B2), CD8a (53-67), CD49b (DX-5), CD11b (M1/70), CD11c (N418), CD25 (PC61), TER119 (TER-119), Ly6G (RB6-8C5), T cell receptor (TCR)-γδ (GL3), and sometimes hCD2 (RPA2.10) (BioLegend or eBioscience), or CD4 isolation kit (Miltenyi Biotec) with anti-CD25 and sometimes hCD2, and then with anti-Biotin microbeads or streptavidin microbeads (Miltenyi Biotec). CD4+ T cells were further incubated with CD62L microbeads (Miltenyi Biotec), and CD4+CD62L+ T cells were positively selected by using AutoMACS. The purity was almost >95%. Isolated naïve CD4+ T cells ($3-4 \times 10^5$ cells) were cultured on 24-well plates under iTreg conditions using anti-CD3e (plate-coated 2C11, 4 µg/ml), anti-CD28 (PV1.17.10, 1.2 µg/ml), anti-IFN-γ (R4-6A2, 5 µg/ml), anti-IL-4 (11B11, 5 µg/ml), recombinant human TGF-β (2 ng/ml), and IL-2 (20 ng/ml). On day 2, dCas9-TETCD or p300CD and gRNA were co-transduced with 6 µg/ml polybrene (Merck Millipore, Billerica, MA, USA) or 10 µg/ml Protransduzin A (Immundiagnostik AG, Bensheim, Germany) (mainly for sorting experiment), with centrifugation at 2500 rpm for 2 h at 35 °C. The next day, co-transduced iTregs were harvested and further cultured for 2 days under Th1, Th2, and Th0 + IL-6 conditions using anti-CD3e (plate-coated, 4 µg/ml), anti-CD28 (1.2 µg/ml), and IL-2 (20 ng/ml). Conditions were as follows: Th1, anti-IL-4 (5 µg/ml), IL-12 (20 ng/ml); Th2, anti-IFN-γ (5 µg/ml), IL-4 (20 ng/ml); Th0 + IL-6, anti-IFN-γ (5 µg/ml), anti-IL-4 (5 µg/ml), IL-6 (20 ng/ml); Th0, anti-IFN-γ (5 µg/ml), anti-IL-4 (5 µg/ml).

For helper T cell subsets skewing, naive CD4+ T cells ($3-4 \times 10^5$ cells) were cultured on 24-well plates under Th1, Th2, and Th17 conditions. Th1 and Th2 were the same as above: Th17, anti-IFN-γ (5 µg/ml), anti-IL-4 (5 µg/ml), human TGF-β (0.5 ng/ml), IL-6 (20 ng/ml).

For TGF-β signal inhibition experiment, indicated concentration of TGF-β was used for iTreg condition in the presence of LY2157299 or anti-TGF-β (1D11).

For intracellular cytokine staining, cells were stimulated with 50 ng/ml phorbol myristate acetate (PMA), 1 µg/ml ionomycin, and Brefeldin A solution for 4 h. Cells were harvested and stained with FVD780, anti-CD4, hCD2, and CD25. For intracellular staining, anti-IFN-γ, anti-IL-2, and anti-CTLA-4 were used after fixation buffer and permeabilization buffer (eBioscience).

## Western blot analysis

The cells were lysed using an immunoprecipitation lysis buffer (50 mM Tris–HCl (pH 7.5), 150 mM NaCl, 10 mM ethylenediaminetetraacetic acid (EDTA, pH 8.0), 1% sodium deoxycholate, 1% Triton X-100, 5 µg/ml leupeptin, and 1 mM PMSF). The cell lysates were centrifuged, and the supernatants were mixed with 5*sodium dodecyl sulfate (SDS) sample buffer (10% SDS, 40% glycerol, 0.2 M Tris–HCl (pH 6.8), 0.025% bromophenol blue, and 50 mM dithiothreitol [DTT]). After boiling, the samples were separated through electrophoresis and transferred to polyvinylidene difluoride (PVDF) membranes. The membranes were probed with anti-Flag (M2; Sigma-Aldrich, St. Louis, MO, USA) and anti-α-tubulin (DM1A; Sigma-Aldrich) and detected using the Chemi-Lumi One system (Nacalai Tesque, Kyoto, Japan).

## Bisulfite sequencing

Cells were lysed using Wizard SV Lysis Buffer (Promega Corporation, Madison, WI, USA). Genomic DNAs were isolated by phenol–chloroform extraction, isopropanol precipitation, and 70% ethanol purification. Genomic DNAs were digested with BamHI, and the same amount of digested DNA (<1.5 µg) was aligned to 19 µl by adding $H_2O$. Next, 1.2 µl of 5 M NaOH was added and incubated at 37 °C for 15 min. Then, 121.2 µl of bisulfite mixture was added and incubated for 1 h at 80 °C. The bisulfite mixture (121.2 µl) consisted of 3.6 M sodium bisulfite (1.92 g in 4.4 ml $H_2O$; 107 µl), 0.57 mM hydroquinone (11 mg in 10 ml $H_2O$; 7 µl), and 0.3 M NaOH (5 M NaOH; 7.2 µl). After purification and extraction using a 50-µl GP3 solution from a column using a FastGene Gel/PCR Extraction Kit (Nippon Genetics, Tokyo, Japan), 3 µl of 5 M NaOH was added and incubated for 5 min at 37 °C to complete the bisulfite reaction. Bisulfite products were precipitated using isopropanol with 10 µg glycogen and rinsed using 70% ethanol. The Foxp3 CNS2 locus was amplified by Quick Taq HS DyeMix (Toyobo Life Science, Tokyo, Japan) polymerase with a bisulfite sequence primer (forward primer: TTTTGGGTTTTTTTGGTATTTAAGA and reverse primer: AACTAACCAACCAACTTCCTA-CACTAT designed by MethPrimer [online]) and then subcloned into pGEM-T EASY Vector (Promega Corporation). Plasmids were purified and sequenced using SP6 primer. Methylation analysis was performed using the quantification tool for methylation analysis (online).

## Real-time PCR analysis

Cells were lysed using RNAiso Plus (Takara Bio Inc., Shiga, Japan). Total RNA was isolated by chloroform extraction, isopropanol precipitation, and 70% ethanol purification. Total RNA from primary T cells was isolated by using ReliaPrep RNA Miniprep Systems (Promega

Corporation) after treatment with RNase inhibitor from human placenta (Nacalai Tesque). cDNA was synthesized by reverse transcription by a High-Capacity cDNA Synthesis Kit (Applied Biosystems, Thermo Fisher Scientific K.K., Kanagawa, Japan) from RNA. Real-time PCR (RT-PCR) analysis was performed using an iCycler iQ multicolor RT-PCR detection system with SsoFast Eva-Green Supermix (Bio-Rad Laboratories, Hercules, CA, USA). Amplification primers (mFoxp3, forward primer: CCCAGGAAAGACAGCAACCTT, reverse primer: TTCTCACAACCAGGCCACTTG, mHPRT1, forward primer: TGAAGAGCTACTGTAATGATCAGTC, reverse primer: AGCAAGCTTGCAACCTTAACCA) were used for quantification.

## Chromatin immunoprecipitation assay

Cells were fixed with 1 ml of 1% formaldehyde for 10 min at room temperature and washed two times using phosphate-buffered saline (PBS). Then, cells were lysed using lysis buffer (1% SDS, 10 mM EDTA (pH 8.0), 50 mM Tris–HCl (pH 8.0), and protease inhibitor cocktail (Nacalai Tesque)). Genomes were sonicated to a mean size of 300 bp using an Acoustic Solubilizer (Covaris, Woburn, MA, USA) with the "300 bp shearing pro" program. Genomic fragment solutions were diluted tenfold with chromatin immunoprecipitation (ChIP) dilution buffer (0.01% SDS, 1.1% Triton-X, 1.2 mM EDTA (pH 8.0), 16.7 mM Tris–HCl (pH 8.0), 167 mM NaCl, and protease inhibitor cocktail). The input samples were collected, and the remainders were immunoprecipitated with 1 μg of Anti-acetyl-Histone H3 Antibody (Merck Millipore) or control rabbit immunoglobulin (Ig) G (Santa Cruz Biotechnology, Inc, Dallas, TX, USA), followed by incubation with dynabeads protein G (Invitrogen, Thermo Fisher Scientific K.K.). Precipitants were washed in order, using low salt wash buffer (0.1% SDS, 1% Triton-X, 2 mM EDTA (pH 8.0), 20 mM Tris–HCl (pH 8.0), and 150 mM NaCl), high salt wash buffer (0.1% SDS, 1% Triton-X, 2 mM EDTA (pH 8.0), 20 mM Tris–HCl (pH 8.0), and 500 mM NaCl), LiCl buffer [0.25 M LiCl, 1% NP-40, 1% sodium deoxycholate, 1 mM EDTA (pH 8.0), and 10 mM Tris–HCl (pH 8.0)], and TE buffer (twice). Chromatin samples were eluted twice by incubation with 100 μl of elution buffer (1% SDS, 0.1 M NaHCO₃, and 10 mM DTT) for 15 min at room temperature. Followed by de-cross-linking using incubation at 65 °C overnight, 8 μl of 5 M NaCl was added to elution products. Next, 20.36 μl proteinase K mixture was added and incubated at 45 °C for 6 h. The proteinase K mixture (20.36 μl) consisted of 0.5 M EDTA (pH 8.0): 4 μl, 0.5 M Tris–HCl (pH 6.8): 16 μl, 20 mg/ml proteinase K: 0.36 μl. Purification and extraction were performed using a 50-μl GP3 solution from a column using a FastGene Gel/PCR

Extraction Kit (Nippon Genetics). ChIP products were analyzed by quantitative PCR. Amplification primers (forward primer: CCCTGCAATTATCAGCACACAC, reverse primer: ATCAGCCTGGCTTGTGGGAAAC) were used for quantification.

## In vitro Treg suppression assay

Responder effector cells (CD4+CD25-) isolated by MACS as described above from Ly5.1 cognate mice were labeled with 2 μM carboxyfluorescein diacetate succinimidyl diester (CFSE) in 37 °C PBS for 10 min and then washed with sufficient RPMI medium, termed as Teff. Splenic CD11c-positive cells isolated by CD11c beads (Miltenyi Biotec) were used as antigen-presenting cells. iTregs co-transduced with dCas9-p300CD and gRNAs, gated on CD4+GFP+DsRed+hCD2+ cells, were sorted using a cell sorter SH800. Next, $4 \times 10^4$ Teff and $2 \times 10^4$ splenic dendritic cells (DCs) with 1 μg/ml soluble anti-CD3e, with or without $1 \times 10^4$ iTreg, were cultured in a 96-well U-bottomed dish for 4 days. CFSE dilution and Foxp3 (hCD2) expression were analyzed.

## Statistical analysis

All values are presented as the means ± standard deviations (SDs). Unpaired Student's $t$ tests were used, and $p < 0.05$ was defined as statistically significant.

## Results

### Constructions and expression

For targeted epigenome editing, we constructed a retroviral expression system for dCas9-TET1CD and dCas9-p300CD fusion proteins. Mouse TET1CD H1620Y, D1622A, and mouse p300CD D1398Y were mutated for catalytic inactive mutants. As these retroviral vectors contain the IRES-GFP sequence, the expression of fusion proteins can be monitored by GFP expression. For gRNA expression, we constructed retroviral or lentiviral expression systems that contain DsRed as the fluorescence marker (Fig. 1a). We confirmed the expression of fusion proteins in HEK293T cells using western blotting methods (Fig. 1b).

### dCas9-TET1CD induced demethylation of the Foxp3 CNS2 locus, but weakly sustained Foxp3 expression

The *Foxp3* CNS2 locus contains 12 CpG sites, and its methylation or demethylation status is extensively involved in the unstable or stable Foxp3 expression phenotype, respectively. To edit the methylation status, we designed 10 gRNA sequences at the *Foxp3* CNS2 locus (Fig. 2a) and transduced them into a 68-41 T cell line that stably expressed dCas9-TET1CD. The *Foxp3* CNS2 locus was heavily methylated in the 68-41 T cell line. After purifying gRNA-positive cells, the methylation status of

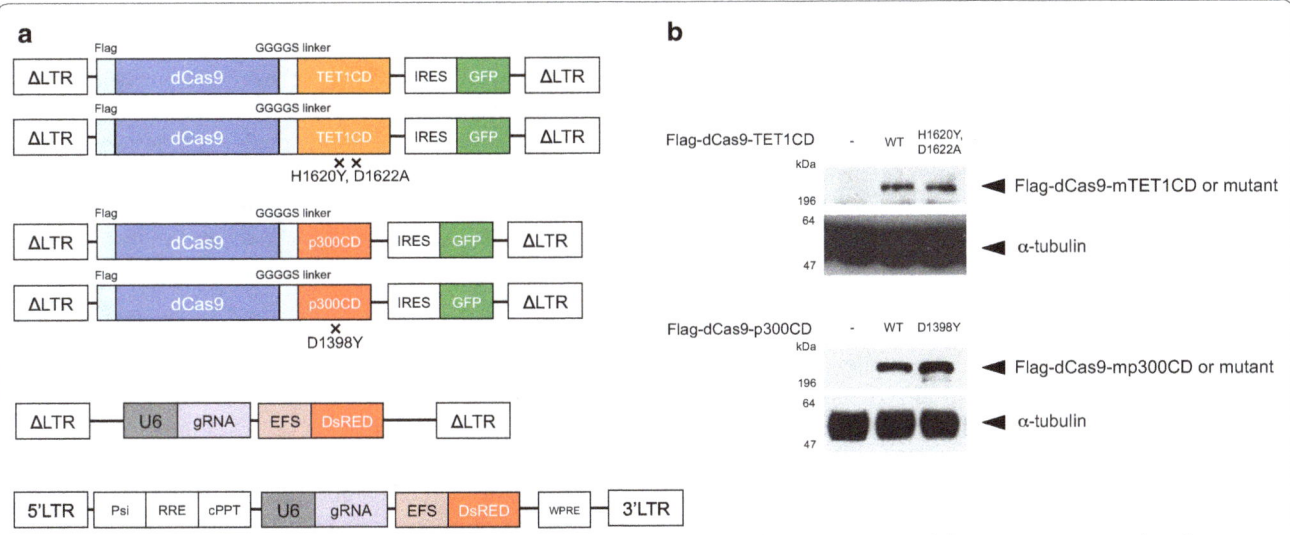

**Fig. 1** CRISPR-dCas9-based epigenome editing for primary T cells. **a** A retroviral vector for the expression of dCas9-epigenome regulator fusion proteins from Moloney murine leukemia virus promoter long terminal repeats (ΔLTRs) and green fluorescent protein (GFP) from an internal ribosomal entry site (IRES). Retroviral and lentiviral vector for bicistronic expression of the gRNA from a U6 promoter (U6) and DsRed from a short EF1a promoter (EFS). **b** Protein expression of dCas9-epigenome regulator fusion proteins in transfected HEK293T cells was detected by western blot against anti-Flag antibody. Anti α-tubulin antibody was used for loading control

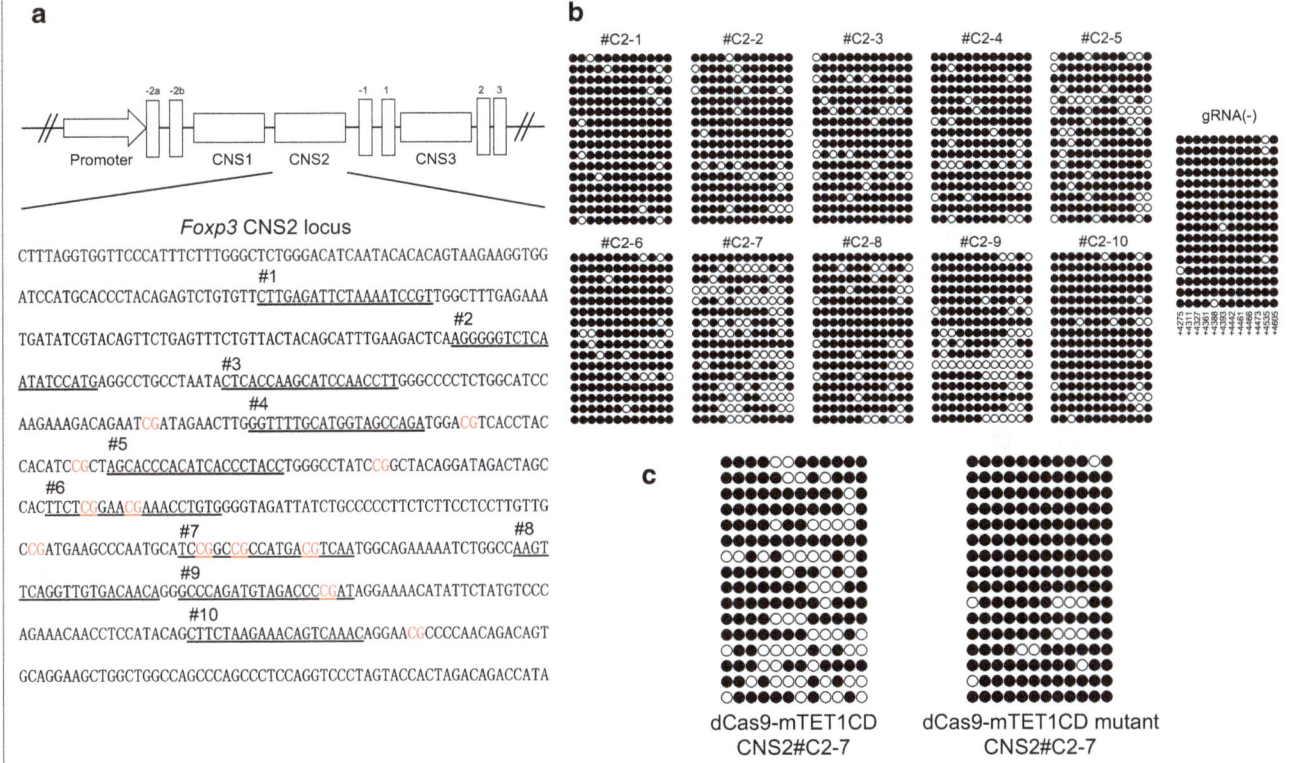

**Fig. 2** dCas9-TET1CD-mediated demethylation of the *Foxp3* CNS2 locus. **a** Sequence at the *Foxp3* CNS2 locus is shown. Each gRNA sequence is *underlined* and *numbered* #C2-1 to #C2-10. Specific CpG sites are *lettered red*. **b** and **c** The methylation status of CpG sites at the *Foxp3* CNS2 locus in dCas9-TET1CD and each gRNA-expressing 68-41 cells (**b**) and dCas9-TET1CD or the TET1CD catalytic mutant and gRNA #C2-7 expressing 68-41 cells (**c**) was determined by bisulfite sequence analysis. The 68-41 cells stably expressing dCas9-TET1CD were transduced with each gRNA expression lentivirus and sorted (**b**). The 68-41 cells were co-transduced with dCas9-TET1CD or TET1CD mutant and gRNA CNS2#C2-7 and sorted (**c**). A *horizontal row* depicts one sequenced clone in which CpGs was methylated (*black*) or demethylated (*white*). Data are pooled from two independent experiments

the *Foxp3* CNS2 locus was analyzed by bisulfite sequencing. The results revealed that several gRNA sequences, such as #C2-7 and #C2-5, could induce demethylation to some extent (approximately 30% by #C2-7), whereas #C2-1 and #C2-10 had little effect (Fig. 2b). The catalytic inactive mutant of TET1CD induced less with #C2-7, indicating that demethylation by dCas9-TET1CD was TET enzyme activity dependent (Fig. 2c). We selected #C2-1 as a negative control gRNA and #C2-7 as a positive control gRNA. Unlike the reported demethylation pattern by TALE-TET1 fusion proteins [42], dCas9-TET1CD fusion proteins could demethylate some CpG sites distant from the designed gRNA sequences. These findings coincide with previous reports [11, 43].

Next, we applied dCas9-TET1CD to primary T cells from male Foxp3-hCD2-hCD52-KI mice under iTreg skewing conditions and confirmed its demethylation activity in iTregs. Unlike 68-41 T cell lines, iTregs showed slight demethylation at the CNS2 locus at the basal level in the absence of gRNAs, and this demethylation was enhanced by co-transduction of dCas9-TET1CD with gRNAs (Fig. 3a). We then examined the promotive effect of dCas9-TET1CD on Foxp3 stability in iTregs. Foxp3 stability under inflammatory conditions was investigated using the following method. Naive CD4+ T cells were cultured under iTreg skewing conditions for 3 days, resulting in >90% Foxp3(+) cells. iTregs were harvested and further cultured under the same iTreg conditions (for positive control) or under inflammatory cytokine (in the presence of IL-12, IL-4, or IL-6) conditions for 2 days. This re-stimulation destabilized Foxp3 expression, which was monitored by surface hCD2 staining correlated with intracellular Foxp3 staining as shown in Additional file 3: Figure S1. The results coincide with previous reports [44–47]. To retain GFP and DsRed fluorescence, we monitored Foxp3 expression by hCD2 without intracellular staining. Using this method, compared with no gRNA-transduced cells (GFP(+)DsRed(−) cells), dCas9-TET1CD and gRNA co-transduction yielded stabilized Foxp3 expression (Additional file 3: Figure S2a). Since demethylation occurred in #C2-1 co-transduced iTregs to some extent, a partial stabilization effect was observed in #C2-7 co-transduced iTregs in comparison with #C2-1 co-transduced iTregs (Fig. 3b), which were confirmed by *Foxp3* mRNA expression (Additional file 3: Figure S2b). Comparable to the dCas9-TET1CD mutant, similar stabilization effects were detected to some extent (Fig. 3c). These data indicated that dCas9-TET1CD for the *Foxp3* CNS2 locus had a certain stabilizing effect for Foxp3 expression, but its effect was weak especially when exposed to inflammatory cytokines (Fig. 7a). A previous report suggested that inflammatory cytokine signals (IL-4/STAT6, IL-6/STAT3) recruit DNMT1 and DNMT3a to the CNS2 locus after stimulation, leading to Foxp3 loss even in nTregs [46]. We speculated that dCas9-TET1CD targeted to the CNS2 locus competes with methyltransferases under inflammatory conditions, resulting in earlier loss of demethylation function than under iTreg conditions. In fact, Foxp3 mean fluorescence intensity (MFI) was greater in dCas9-TET1CD than in TET1CD catalytic inactive mutant under iTreg conditions (Fig. 3c), but was weakened by inflammatory stimuli.

### dCas9-p300CD induced acetylation of the Foxp3 promoter locus and induced stable expression of Foxp3 in a cultured T cell line

Foxp3 expression is induced by histone acetylation of the promoter locus. We designed 10 gRNA sequences at the *Foxp3* promoter locus (Fig. 4a) and transduced them into the 68-41 T cell line that stably expressed dCas9-p300CD. The 68-41 cell line showed little Foxp3 expression. We measured the amount of mRNA expression induced by dCas9-p300CD. We observed that #P-4 and #P-9 strongly activated Foxp3 transcription, and that #P-5, #P-6, #P-1, and #P-10 induced moderately activated Foxp3 transcription (Fig. 4b). Next, we assessed protein expression. A small but significant fraction of endogenous Foxp3 expression could be detected in #P-4 and #P-9 transduced cells (Fig. 4c). This induction was dependent on p300CD autoacetylation activity, since the catalytic inactive mutant could not induce it (Fig. 4d). We selected #P-3 as a negative control gRNA and #P-4 as a positive control gRNA. In #P-4 transduced cells, the histone acetylation of the Foxp3 promoter locus was promoted compared with #P-3, correlating with transcriptional activation (Fig. 4e).

Although similar expression levels of dCas9-p300CD and #P-4 are gated, approximately 10% of the population significantly expressed Foxp3. By limiting dilution, we isolated several clones that stably expressed a high amount of Foxp3, and others that never expressed it (Fig. 4f). Although we could not explain the mechanism of this bipolarization phenomenon, the data suggested that it was not due to the oscillation of the cell population. The data indicated that targeted histone acetylation could strongly maintain epigenetic modification and transcriptional activation in certain specific cells.

### dCas9-p300CD induces stable Foxp3 expression in primary T cells

We applied this system to primary T cells. Under helper T cell culture skewing conditions, we co-transduced dCas9-p300CD and #P-3 and #P-4 into isolated naïve CD4+ T cells and investigated Foxp3 expression. Foxp3 expression was induced under all skewing conditions,

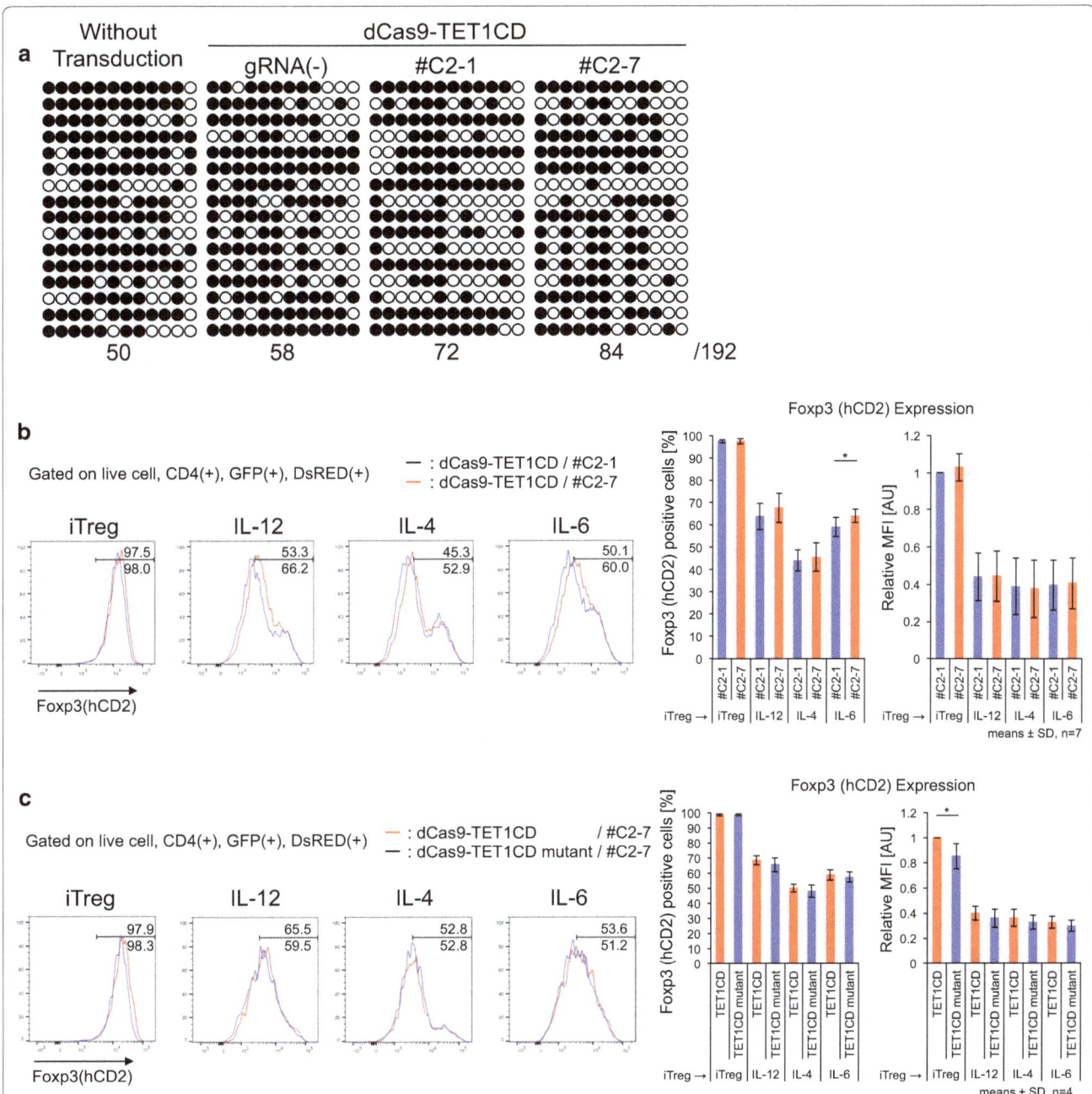

**Fig. 3** Maintenance of Foxp3 expression induced by dCas9-TET1CD-mediated demethylation of the *Foxp3* CNS2 locus. **a** The methylation status of CpG sites at the *Foxp3* CNS2 locus in untransduced, dCas9-TET1CD single, with #C2-1, and with #C2-7 transduced iTregs was determined by bisulfite sequence analysis. GFP/DsRed(−/−), (+/−), and (+/+) cells are sorted, respectively. A *horizontal row* depicts one sequence clone in which CpGs was methylated (*black*) or demethylated (*white*). The *number* below bisulfite sequences indicates demethylated CpG sites. Data are pooled from two independent experiments. (**b** and **c**) Flow cytometry analysis of Foxp3(hCD2) expression in iTregs co-transduced with dCas9-TET1CD and gRNA CNS2 #C2-1 (*blue*) or #C2-7 (*red*) (**b**), dCas9-TET1CD (*red*), or TET1CD mutant (*blue*) and gRNA #C2-7 (**c**) under inflammatory cytokine conditions. Percentages of Foxp3(+) and mean fluorescence intensity (MFI) relative value to iTregs co-transduced with dCas9-TET1CD and #C2-1 (**b**), dCas9-TET1CD and #C2-7 (**c**) were plotted. Data are pooled from seven (**b**) or four (**c**) independent experiments and represent the means ± SDs. *$p < 0.05$

and a notably superior enhancing effect was observed under the Th17 condition (IL-6 and TGF-β), as shown in Fig. 5a. IL-2 was added to all skewing conditions

to improve T cell proliferation and transduction efficiency. Thus, the majority of the population expressed Foxp3, even under Th17 conditions. TGF-β induces

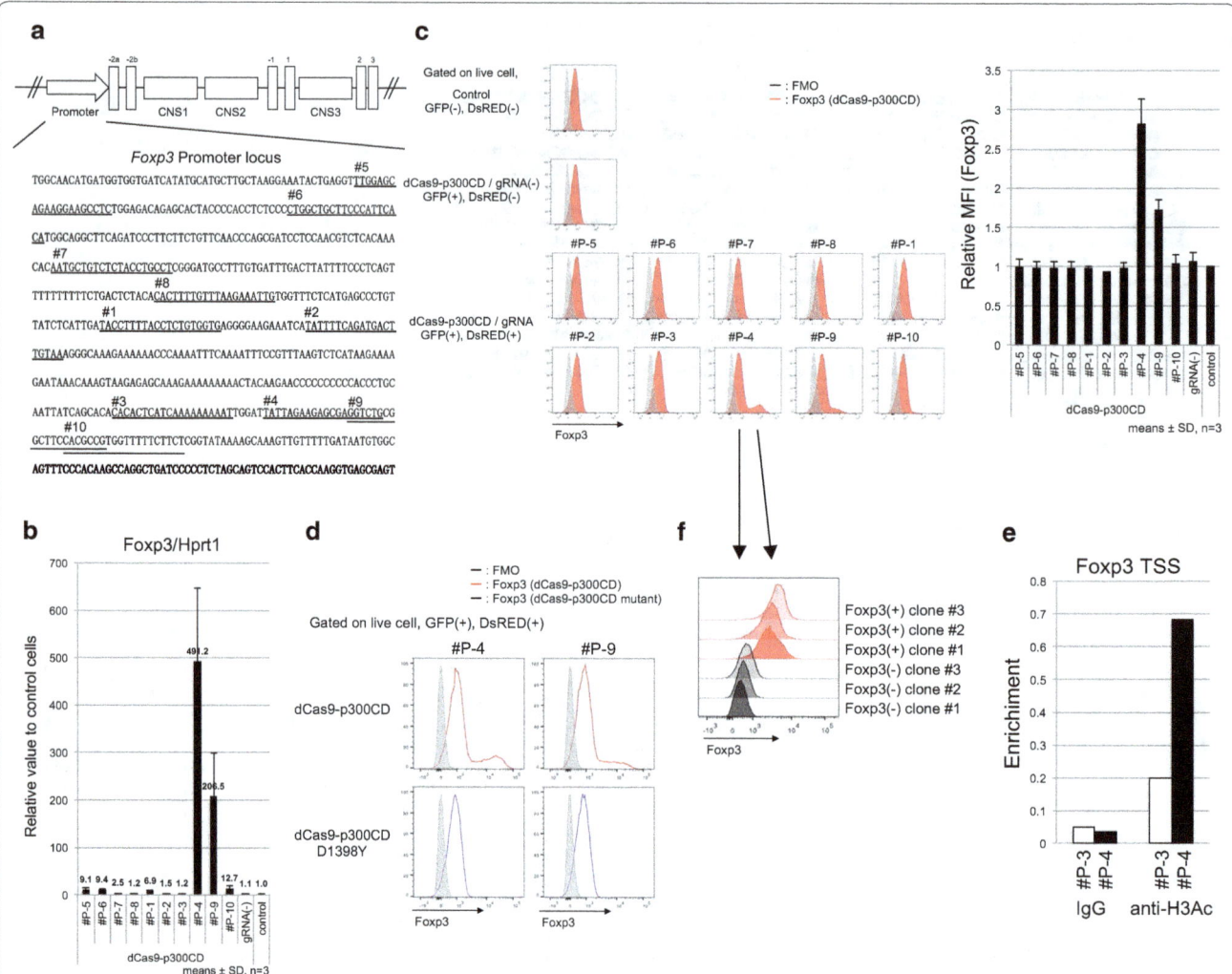

**Fig. 4** dCas9-p300CD-mediated *Foxp3* promoter acetylation and transcriptional activation. **a** Sequence at the *Foxp3* promoter locus is shown. Each gRNA sequence is *underlined* and *numbered* #P-1 to #P-10. Transcription start sites are *lettered bold red*. **b** Foxp3 mRNA expression in each gRNA-transduced 68-41 cell stably expresses dCas9-p300CD relative to control 68-41 cells (control). Data are pooled from three independent experiments and represent the means ± SDs. **c** Flow cytometry analysis of Foxp3 expression in each gRNA-transduced 68-41 cell stably expresses dCas9-p300CD. Foxp3 MFI relative value to control 68-41 cells was plotted. Data are pooled from three independent experiments and represent the means ± SDs. **d** Flow cytometry analysis of Foxp3 expression in each gRNA-transduced 68-41 cell stably expresses dCas9-p300CD or p300 mutant. **e** Enrichment of acetyl histone H3 at *Foxp3* TSS locus in #P-3 or #P-4 transduced 68-41 cell stably expresses dCas9-p300CD. **f** Flow cytometry analysis of Foxp3 expression in each clone isolated by limiting the dilution from dCas9-p300CD and #P-4 co-transduced 68-41 cells

Foxp3 mainly through the *Foxp3* CNS1 enhancer locus by Smad2 and Smad3 signals [29, 48, 49]. This indicates that the artificial histone acetylation of the promoter locus activated transcription from the promoter locus, and it was augmented by the TGF-β signal, which principally activates the enhancer locus. In other words, the TGF-β signal additionally activated Foxp3 transcription, even when the promoter locus was artificially opened. This possible enhanced activity of dCas9-p300CD under TGF-β signal condition was confirmed by blocking its signals. Low-dose TGF-β enhanced Foxp3 expression

within #P-4 co-transduced cells than #P-3, and this effect was cancelled by treating LY2157299 (TGF-β receptor kinase inhibitor) or anti-TGF-β antibody, indicating the involvement of TGF-β signal in enhancing dCas9-p300CD activity (Additional file 3: Figure S3). Then, we speculated whether co-transduction of dCas9-p300CD and #P-4 facilitated TGF-β signal to engage in transactivation by investigating Foxp3 expression in T cell plasticity culture (Th1 to iTregs). Since helper T cell subset plasticity is strictly regulated, Foxp3 cannot be induced by TGF-β in already differentiated Th1 cells [50]. As we

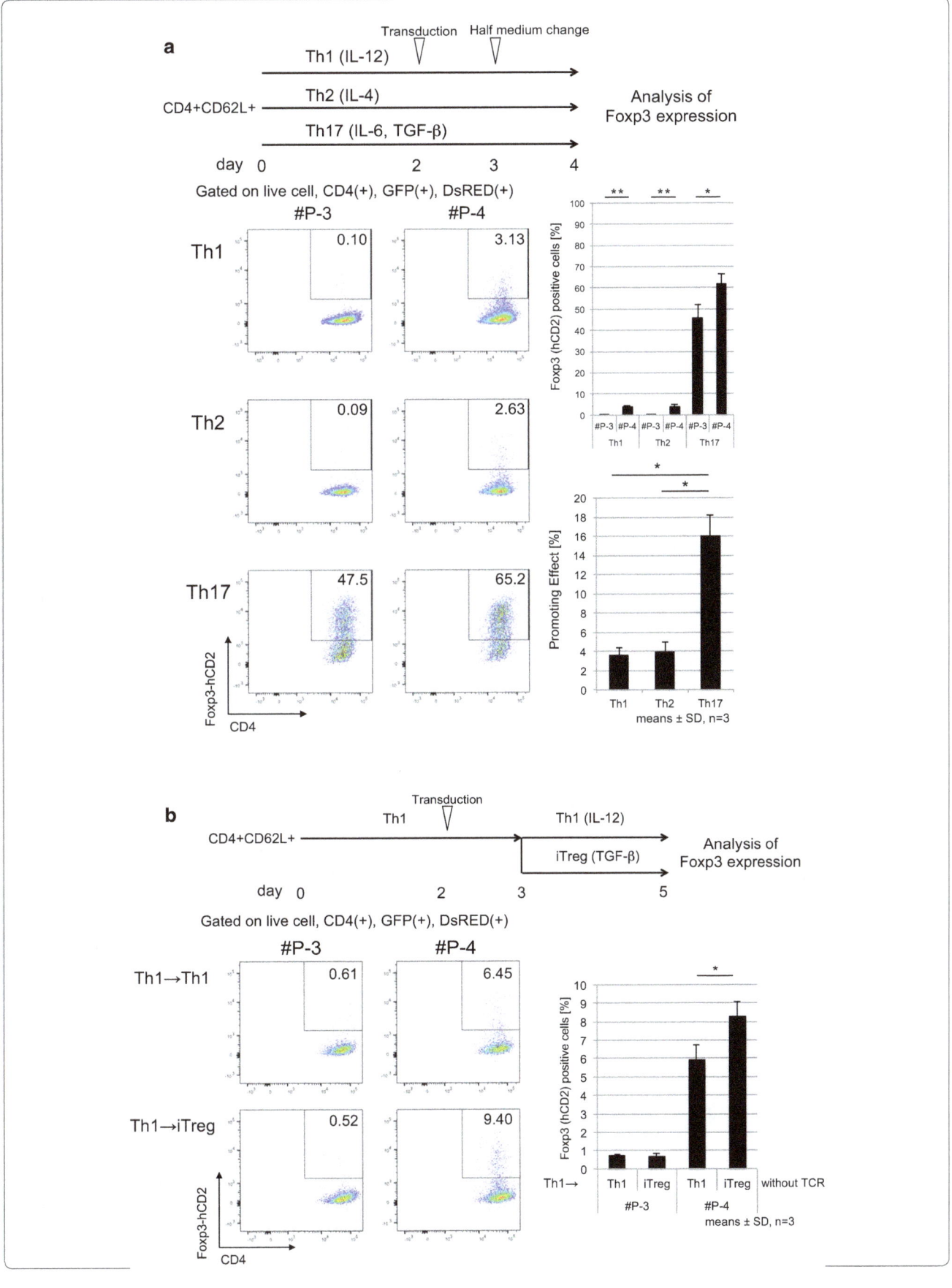

expected, Foxp3 expression was promoted in dCas9-p300CD and #P-4 co-transduced cells, even when differentiated Th1 cells were further treated with TGF-β. This promotion effect never occurred in #P-3 cells (Fig. 5b). Epigenome editing could thus be a novel method for converting the T cell subset.

Then, we examined the maintenance of Foxp3 by dCas9-p300CD in iTregs. The stability of Foxp3 under inflammatory cytokines was investigated (same as Additional file 3: Figure S1). We observed that iTregs co-transduced with dCas9-p300CD and #P-4 retained a high amount of Foxp3 compared with #P-3 under inflammatory conditions (Fig. 6a). We confirmed this maintenance is actually dependent on p300CD autoacetylation activity by co-transduction with p300 catalytic inactive mutant (Fig. 6b). Moreover, the Treg signature genes CD25 and CTLA-4 were slightly but significantly upregulated under IL-12 conditions in #P-4 transduced iTregs (Fig. 6c). Finally, we examined the iTreg suppression activity in vitro. iTregs co-transduced with dCas9-p300CD and gRNA were sorted (Additional file 3: Figure S4a). Splenic DCs were used as antigen-presenting cells, and the proliferation of effector T cells was further suppressed by #P-4 transduced iTregs, which correlated with Foxp3 stabilization (Fig. 6d). Similar tendency was observed in comparison with catalytic activity (Additional file 3: Figure S4b). These data showed that applying dCas9-p300CD to primary T cells, especially iTregs, could modify both transcription and cell function. These data also clarified one aspect of the Foxp3 transcriptional activation mechanisms (Fig. 7b).

### Off-target analysis of selected gRNAs
CRISPR-Cas9 or CRISPR-dCas9-based technologies are constantly at risk of off-target activity [51, 52]. For clinical usage or to validate results, we have to consider off-target effects. We used the CCTop online tool [41], and selected gRNA sequences (#C2-1, #C2-7, #P-3, and #P-4) were investigated for potential off-target sites. We observed that the selected gRNA sequences had at least three mismatches on similar sequences, and most off-target candidate sequences were localized in intergenic

regions (Additional file 2: Table S1). In our study, which mainly focused on mice experiments and revealing the relationships between epigenetics and gene expression, all candidate genes were not strongly involved in direct Foxp3 induction or Treg functions to the best of our knowledge. For future clinical usage, we have to re-select gRNA sequences in the human genome and investigate off-target activity in our next study.

### Discussion
Artificial targeted epigenome editing mediated by CRISPR-dCas9 can be utilized to clarify the relationship between chromatin states and gene expression and to develop novel clinical strategies. Previous research proposed this biological device and demonstrated its universal performance and efficiency at the targeted locus. In this study, we expanded epigenome editing to mouse primary T cells, with a focus on the *Foxp3* locus to elucidate epigenetic regulation mechanisms, and to advance future clinical usage in immunotherapy.

In this study, we applied dCas9-TET1CD and dCas9-p300CD to the *Foxp3* CNS2 and promoter locus, respectively, and attempted to generate Foxp3 stability-enhanced iTregs. We succeeded in epigenome editing at both loci, and the histone acetylation at the promoter locus strongly activated Foxp3 expression, but the DNA demethylation at the CNS2 locus slightly affected for Foxp3 expression.

dCas9-TET1CD demethylated the CNS2 locus, but did not intensely stabilize Foxp3 expression under inflammatory conditions. Although a certain level of stabilization was achieved by co-transduction with gRNA #C2-1 or #C2-7, we could not observe a similar statistically significant stabilization effect in comparison with dCas9-TET1CD catalytic activity. We speculate that demethylation efficiency is not sufficient for stable Foxp3 expression, as seen in nTregs, because even nTregs lose Foxp3 expression under inflammatory conditions, and the CNS2 locus was methylated in a parallel way [46, 53, 54]. Endogenous epigenetic modifiers could have excluded dCas9-TET1CD targeted to the CNS2 locus in iTregs under inflammatory conditions. In addition,

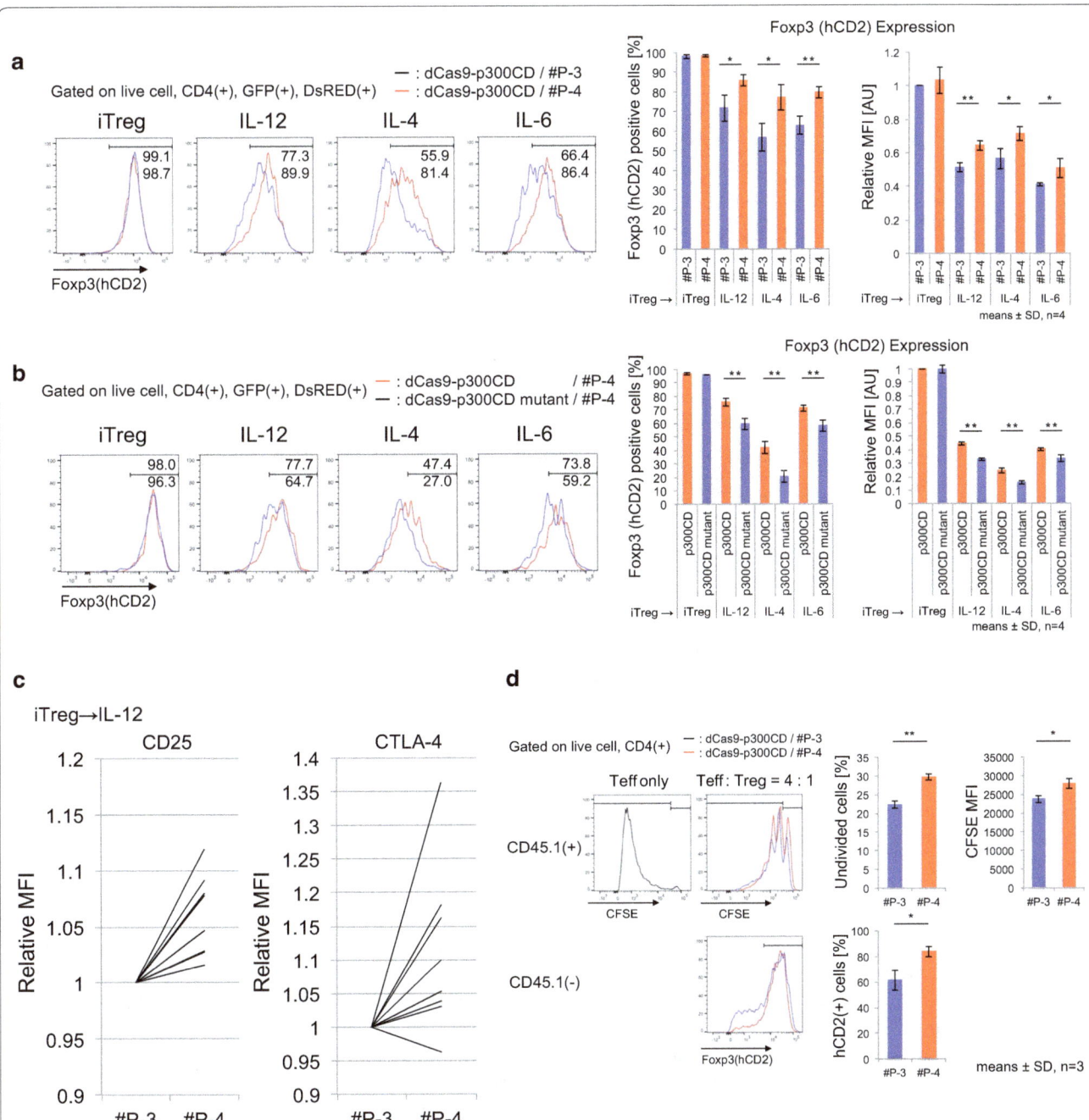

**Fig. 6** dCas9-p300CD-applied iTregs showed higher suppressive activity. **a** and **b** Flow cytometry analysis of Foxp3 (hCD2) expression in iTregs co-transduced with dCas9-p300CD and gRNA #P-3 (*blue*) or #P-4 (*red*) (**a**), dCas9-p300CD (*red*) or p300CD mutant (*blue*) and #P-4 (B). Percentages of Foxp3(+) cells and MFI relative value to iTregs co-transduced with dCas9-p300CD and #P-3 (**a**) or dCas9-p300CD and #P-4 (**b**) were plotted. Data are pooled from four independent experiments and represent the means ± SDs. *p < 0.05; **p < 0.01. **b** Relative expression of CD25 or CTLA-4 of IL-12-treated iTregs co-transduced with dCas9-p300CD and gRNAs. Relative MFI values to #P-3 were plotted. Data represent each experimental value of eight independent experiments. **c** Treg suppression activity was measured by CFSE dilution in labeled Teff. Flow cytometry analysis of CFSE dilution in labeled Teff, co-cultured with or without iTregs co-transduced with dCas9-p300CD and gRNA #P-3 (*blue*) or #P-4 (*red*). Foxp3 (hCD2) expression in iTregs co-transduced with dCas9-p300CD and gRNA #P-3 (*blue*) or #P-4 (*red*) was also analyzed. Data are pooled from three independent experiments and represent the means ± SDs. *p < 0.05; **p < 0.01

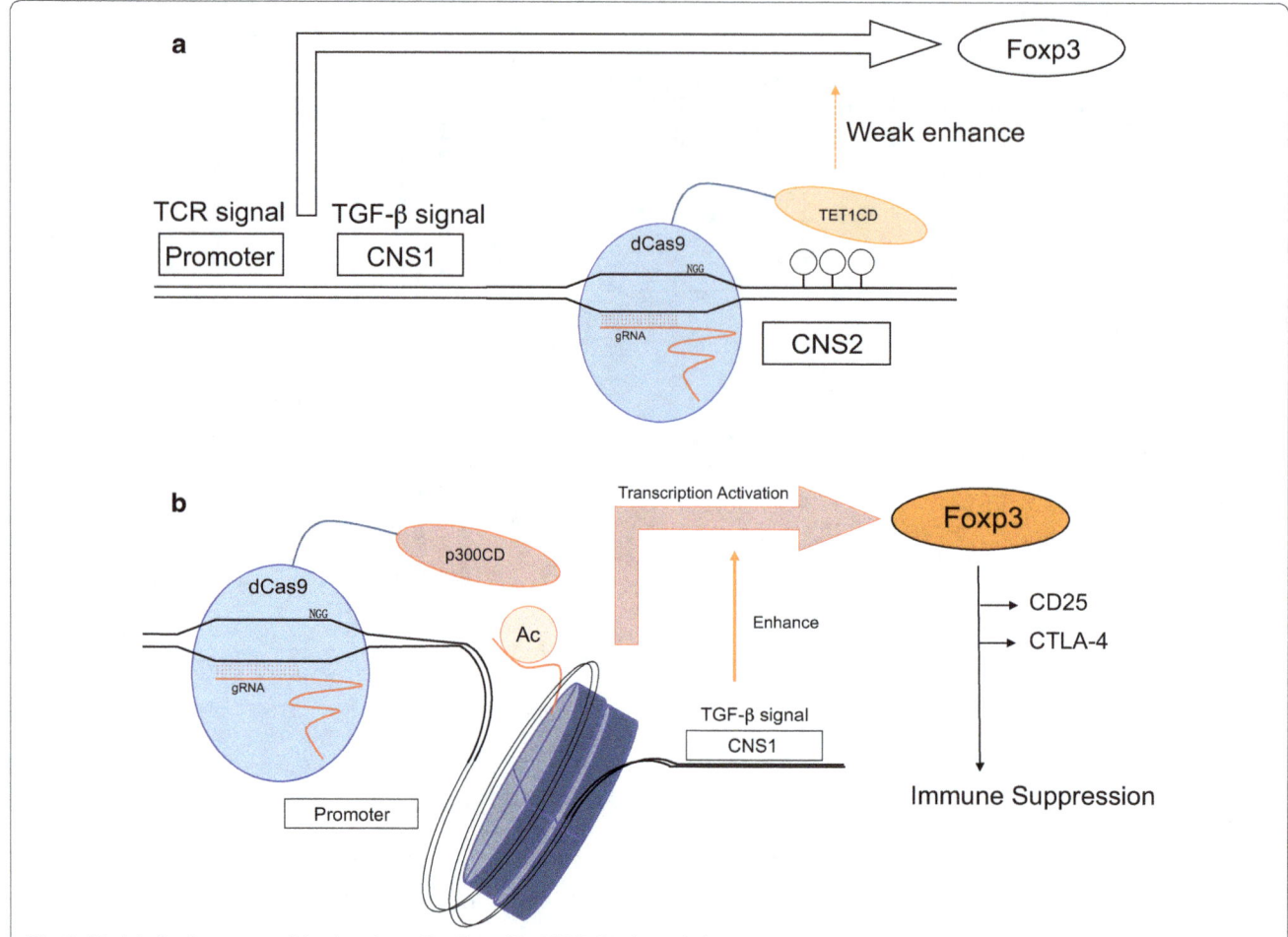

**Fig. 7** Model of epigenome editing in primary T cells. **a** dCas9-TET1CD demethylates the *Foxp3* CNS2 locus and enhances Foxp3 expression weakly. **b** dCas9-p300CD acetylates the *Foxp3* promoter locus, activates transcription in coordination with the TGF-β signal, and promotes immunosuppressive function

dCas9-TET1CD itself impedes interaction of the *Foxp3* CNS2 locus with other endogenous transcriptional factors. As dCas9 itself is reported to inhibit transcription, and it is utilized in CRISPR interference technology [55]. When comparing dCas9-TET1CD with its catalytic inactive mutant, the difference in the Foxp3 stabilization effect was smaller than the difference between #C2-1 and #C2-7. We presume that dCas9-TET1CD (or even dCas9-TET1CD catalytic inactive mutant itself) targeted by #C2-7 was not protected from DNMT1 or DNMT3a recruited by inflammatory signals. In fact, in iTregs, the CNS2 locus was not passively regulated, and TET1CD catalytic inactive mutant decreased Foxp3 expression as measured by MFI, indicating that dCas9-TET1CD inactive mutant had a negative effect on Foxp3 transcription.

Recent research reported that the dCas9-TET1CD system in combination with repeating peptide array SunTag technology or engineered gRNA technology, in which bacteriophage MS2 RNA elements are inserted,

succeeded in upregulating gene expression via considerable targeted demethylation of some promoter loci [43, 56]. Another study proposed that modified dCas9, with its degradation controlled by a chemical compound, could prevent dCas9 fusion proteins from remaining at targeted loci [57]. Additionally, SaCas9, which is smaller than SpCas9, is reported to overcome the size problem [58, 59]. In the future study, we plan to use these modified dCas systems in order to improve demethylation efficiency.

In accordance with a previous report, we confirmed that dCas9-p300CD could induce gene expression through histone acetylation. Like Hilton et al. [7], we observed that a single gRNA sequence is sufficient for transcriptional activation, and that this sequence is located approximately 60-bp upstream from the transcriptional start site. Additionally, we investigated the potential effects of dCas9-p300CD from two points of view. First, we clarified that artificial histone acetylation at the *Foxp3* promoter locus activated Foxp3 transcription, which could be enforced by the TGF-β signal. TGF-β signal is shown to accelerate Foxp3 induction by modifying the CNS1 enhancer locus [60]. It means that dCas9-p300CD targeted to the promoter locus did not mask the other enhancer locus function; rather, it could be activated. To dissect the locus specific regulation clearly, examining the activity of dCas9-p300CD in CNS1 locus-deficient cells is needed in the next study, since TGF-β signal was also reported to effect promoter locus [61, 62]. In addition to enhanced effectiveness of dCas9-p300CD by TGF-β signal, dCas9-p300CD targeted to the promoter locus was not interfered by inflammatory cytokines. The cytokine signals IL-12/STAT4, IL-4/STAT6, and IL-6/STAT3 did not mainly target the *Foxp3* promoter locus for downregulating Foxp3 transcription. Putative STAT6-binding sites even in *Foxp3* promoter locus are located further upstream of gRNA-targeted regions [63]. We presumed that dCas9-p300CD was salvaged from inactivation of the *Foxp3* gene locus by remaining at the *Foxp3* promoter locus under inflammatory conditions. It was expected that applying this system to various gene loci could identify a novel enhancer element and its regulatory stimuli or factors. Second, we showed that artificial histone acetylation not only sustained Foxp3 expression, but also reinforced Treg function. iTregs transduced with dCas9-p300CD and appropriate gRNA #P-4 highly expressed the Treg signature genes CD25 and CTLA-4, resulting in higher suppression activity. This indicates that dCas9-p300CD induced sufficient protein expression for engineering cellular functions.

Considering that dCas9-p300CD-mediated gene activation is observed only in a certain fraction, but not all of the transduced cells, effectiveness of dCas9-p300CD depends on each transduced cell. Examination of original chromatin states or accessibility of epigenetic modifier to the target locus in individual cells will clarify the more effective usage of epigenome editing. For example, H3K27me3, inactive epigenetic modification, is marked at the *Foxp3* promoter locus in conventional T cells [64]. Supposing that dCas9-p300CD has to rewrite this inactive mark with eraser help for transactivation, it is easy to speculate that effectiveness is decreased in such cells than H3K27 unmodified cells. Furthermore, memorization and stabilization of artificially induced epigenetic modification become issue. Our result suggested gene activation is strongly maintained in some cases. Whether this phenomenon was the results of epigenome editing or stable existence of epigenetic modifier is carefully examined in the next study.

Since *Foxp3* locus-targeted epigenome editing worked well in primary T cells for increasing Treg properties to some extent, further epigenetic modifications to other Treg-characteristic gene loci feasibly convert iTregs to nTregs. In fact, Foxp3 alone is not strictly and not sufficient to determine Treg signature, and multiple co-transcription factors have redundant functions for Treg physiology [65]. Primarily, we have to identify these gene loci and modify the epigenetic status in conjunction with the *Foxp3* locus by multiple gRNAs combination. Moreover, for future clinical usage of these epigenome-edited iTregs, optimization for therapeutic effect is required for functional Tregs. Expanding the target genes, suppressive cytokines, or inhibitory molecules seems to be effective for clinical usage. Other transcriptional activation systems can be applied for this purpose [66, 67].

Finally, we could not verify the function of epigenome-edited iTregs in the in vivo mouse model, because the transduction efficiency was not high enough to obtain a sufficient number of iTregs for disease model study. However, it has been reported that, in contrast to mice T cells, human T cells could be expanded using a rapid expansion protocol [68]. Moreover, lentivirus-mediated gene delivery methods have been established for clinical uses [69]. In our future study, we aim to apply our system to the human genome and human T cells, and expect its usage in medicine in the future.

## Conclusions

We proposed that applying epigenome editing to genes of interest would clarify gene expression regulation mechanisms. Our study firstly investigated the crosstalk of CRISPR-dCas9-based epigenome editing and endogenous cellular signaling in mouse primary T cells, focusing on *Foxp3* gene locus. dCas9-TET1CD and dCas9-p300CD edited specific CpG sites and chromatin histone, and we showed that subsequent gene activation was occurred by cooperating with the TGF-β signal (in the case of dCas9-p300CD) and was interfered by inflammatory cytokine signals (in the case of dCas9-TET1CD). It indicated that different epigenetic states at the *Foxp3* locus among cell types and cell culture conditions determined the effectiveness of dCas9-based epigenome editing. We expected that broad application to key genes for cell differentiation or human diseases would clarify epigenetic regulation mechanisms. We concluded that epigenome editing endowed innovation of clinical research and promise to clinical application in the next study.

## Additional files

**Additional file 1.** Amino acid sequences of dCas9-TET1CD and dCas9-p300CD.

**Additional file 2: Table S1.** Off-target analysis of selected gRNAs.

**Additional file 3: Figure S1.** Experimental scheme of Foxp3 stability assay. Upper, naïve CD4+ T cells (CD4+CD62L+hCD2-) were MACS sorted and cultured under iTreg skewing conditions, and on day 2, dCas9-fusion protein and gRNA were transduced with polybrene. The next day, iTregs were harvested and further cultured under iTreg or inflammatory cytokine conditions for 2 days. Foxp3 expression (hCD2) was analyzed by flow cytometry. Lower, representative Foxp3 expression. Flow cytometry plots show expression of Foxp3 (endogenous) and Foxp3(hCD2, surface indicator) in primary T cells from Foxp3-hCD52-hCD2 KI mice. **Figure S2**. dCas9-TET1CD-mediated Foxp3 stabilization. (A) Histogram of Foxp3(hCD2) in dCas9-TET1CD (GFP/DsRed(+/-)) and dCas9-TET1CD with #C2-7 (GFP/DsRed(+/+)) cells under inflammatory conditions. Related to Figure 3b. (B) Foxp3 mRNA expression same as in Figure 3b. Data are pooled from three independent experiments and represent the means ± SDs.
**Figure S3**. TGF-β signal enhanced effectiveness of dCas9-p300CD-mediated Foxp3 induction. Foxp3 expression induced by low-dose TGF-β in the presence of LY2157299 or anti-TGF-β was monitored by Foxp3(hCD2) MFI. **Figure S4**. dCas9-p300CD and gRNA co-transduced iTregs. (A) Sorting strategy and purification. (B) Suppression assay of iTregs comparing dCas9-p300CD and #P-4 with dCas9-p300CD catalytic mutant.

## Abbreviations
Foxp3: Forkhead box P3; TGF-β: transforming growth factor beta; Tregs: regulatory T cells; TET: Ten-eleven translocation dioxygenase; DNMT: DNA methyltransferase; CNS: conserved non-coding DNA sequences; STAT: signal transducer and activator of transcription; CTLA-4: cytotoxic T-lymphocyte associated protein 4.

## Authors' contributions
MO and AY conceived and designed the study. MO, MK, KS, and HN contributed to development of methodology. MK, KS, and HN offered advice on study design, helped with technical or material support. MO performed acquisition and analysis of data. MO and AY wrote the manuscript. MO, MK, KS, HN, and AY helped in manuscript review and revision. AY supervised the study. All authors read and approved the final manuscript.

## Acknowledgements
We thank T. Tamiya, N. Shiino, M. Asakawa, Y. Noguchi, H. Yamane, C. Ohkura, and Y. Hirata for their technical assistance and Y. Ushijima for manuscript preparation.

## Competing interests
The authors declare that they have no competing interests.

## Funding
This work was supported by special Grants-in-Aid from the Ministry of Education, Culture, Sports, Science and Technology of Japan (No. 25221305), Grant-in-Aid for Young Scientists (B) (No. 17K15451), Advanced Research & Development Programs for Medical Innovation (AMED-CREST), the Takeda Science Foundation, Uehara Memorial Foundation, Mochida Memorial Foundation for Medical and Pharmaceutical Research, Kanae Foundation, and Senshin Medical Research Foundation Keio Gijuku Academic Developmental Funds.

## References
1. Thakore PI, Black JB, Hilton IB, Gersbach CA. Editing the epigenome: technologies for programmable transcription and epigenetic modulation. Nat Methods. 2016;13(2):127–37.
2. Liu PQ, Rebar EJ, Zhang L, Liu Q, Jamieson AC, Liang Y, Qi H, Li PX, Chen B, Mendel MC, et al. Regulation of an endogenous locus using a panel of designed zinc finger proteins targeted to accessible chromatin regions. Activation of vascular endothelial growth factor A. J Biol Chem. 2001;276(14):11323–34.
3. Konermann S, Brigham MD, Trevino AE, Hsu PD, Heidenreich M, Cong L, Platt RJ, Scott DA, Church GM, Zhang F. Optical control of mammalian endogenous transcription and epigenetic states. Nature. 2013;500(7463):472–6.
4. Sander JD, Joung JK. CRISPR-Cas systems for editing, regulating and targeting genomes. Nat Biotechnol. 2014;32(4):347–55.
5. Maeder ML, Linder SJ, Cascio VM, Fu Y, Ho QH, Joung JK. CRISPR RNA-guided activation of endogenous human genes. Nat Methods. 2013;10(10):977–9.
6. Perez-Pinera P, Kocak DD, Vockley CM, Adler AF, Kabadi AM, Polstein LR, Thakore PI, Glass KA, Ousterout DG, Leong KW, et al. RNA-guided gene activation by CRISPR-Cas9-based transcription factors. Nat Methods. 2013;10(10):973–6.
7. Hilton IB, D'Ippolito AM, Vockley CM, Thakore PI, Crawford GE, Reddy TE, Gersbach CA. Epigenome editing by a CRISPR-Cas9-based acetyltransferase activates genes from promoters and enhancers. Nat Biotechnol. 2015;33(5):510–7.
8. Kearns NA, Pham H, Tabak B, Genga RM, Silverstein NJ, Garber M, Maehr R. Functional annotation of native enhancers with a Cas9-histone demethylase fusion. Nat Methods. 2015;12(5):401–3.
9. Gilbert LA, Larson MH, Morsut L, Liu Z, Brar GA, Torres SE, Stern-Ginossar N, Brandman O, Whitehead EH, Doudna JA, et al. CRISPR-mediated modular RNA-guided regulation of transcription in eukaryotes. Cell. 2013;154(2):442–51.
10. Vojta A, Dobrinic P, Tadic V, Bockor L, Korac P, Julg B, Klasic M, Zoldos V. Repurposing the CRISPR-Cas9 system for targeted DNA methylation. Nucleic Acids Res. 2016;44(12):5615–28.
11. Choudhury SR, Cui Y, Lubecka K, Stefanska B, Irudayaraj J. CRISPR-dCas9 mediated TET1 targeting for selective DNA demethylation at BRCA1 promoter. Oncotarget 2016;7:46545–46556
12. Park M, Keung AJ, Khalil AS: The epigenome: the next substrate for engineering. Genome Biol. 2016;17(1):183. doi:10.1186/s13059-016-1046-5
13. Amabile A, Migliara A, Capasso P, Biffi M, Cittaro D, Naldini L, Lombardo A. Inheritable silencing of endogenous genes by hit-and-run targeted epigenetic editing. Cell. 2016;167(1):219.
14. Sakaguchi S, Ono M, Setoguchi R, Yagi H, Hori S, Fehervari Z, Shimizu J, Takahashi T, Nomura T. Foxp3+ CD25+ CD4+ natural regulatory T cells in dominant self-tolerance and autoimmune disease. Immunol Rev. 2006;212:8–27.
15. Sakaguchi S, Yamaguchi T, Nomura T, Ono M. Regulatory T cells and immune tolerance. Cell. 2008;133(5):775–87.
16. Liston A, Gray DH. Homeostatic control of regulatory T cell diversity. Nat Rev Immunol. 2014;14(3):154–65.
17. Chen W, Jin W, Hardegen N, Lei KJ, Li L, Marinos N, McGrady G, Wahl SM. Conversion of peripheral CD4+CD25- naive T cells to CD4+CD25+ regulatory T cells by TGF-beta induction of transcription factor Foxp3. J Exp Med. 2003;198(12):1875–86.
18. Josefowicz SZ, Rudensky A. Control of regulatory T cell lineage commitment and maintenance. Immunity. 2009;30(5):616–25.
19. Bennett CL, Christie J, Ramsdell F, Brunkow ME, Ferguson PJ, Whitesell L, Kelly TE, Saulsbury FT, Chance PF, Ochs HD. The immune dysregulation,

polyendocrinopathy, enteropathy, X-linked syndrome (IPEX) is caused by mutations of FOXP3. Nat Genet. 2001;27(1):20–1.

20. Hori S, Nomura T, Sakaguchi S. Control of regulatory T cell development by the transcription factor Foxp3. Science. 2003;299(5609):1057–61.

21. Fontenot JD, Gavin MA, Rudensky AY. Foxp3 programs the development and function of CD4+CD25+ regulatory T cells. Nat Immunol. 2003;4(4):330–6.

22. Bettelli E, Dastrange M, Oukka M. Foxp3 interacts with nuclear factor of activated T cells and NF-kappa B to repress cytokine gene expression and effector functions of T helper cells. Proc Natl Acad Sci USA. 2005;102(14):5138–43.

23. Ichiyama K, Yoshida H, Wakabayashi Y, Chinen T, Saeki K, Nakaya M, Takaesu G, Hori S, Yoshimura A, Kobayashi T. Foxp3 inhibits RORgammat-mediated IL-17A mRNA transcription through direct interaction with RORgammat. J Biol Chem. 2008;283(25):17003–8.

24. Ohkura N, Hamaguchi M, Morikawa H, Sugimura K, Tanaka A, Ito Y, Osaki M, Tanaka Y, Yamashita R, Nakano N, et al. T cell receptor stimulation-induced epigenetic changes and Foxp3 expression are independent and complementary events required for Treg cell development. Immunity. 2012;37(5):785–99.

25. Miyao T, Floess S, Setoguchi R, Luche H, Fehling HJ, Waldmann H, Huehn J, Hori S. Plasticity of Foxp3(+) T cells reflects promiscuous Foxp3 expression in conventional T cells but not reprogramming of regulatory T cells. Immunity. 2012;36(2):262–75.

26. Xu L, Kitani A, Fuss I, Strober W. Cutting edge: regulatory T cells induce CD4+CD25-Foxp3- T cells or are self-induced to become Th17 cells in the absence of exogenous TGF-beta. Journal of immunology. 2007;178(11):6725–9.

27. Zhou X, Bailey-Bucktrout SL, Jeker LT, Penaranda C, Martinez-Llordella M, Ashby M, Nakayama M, Rosenthal W, Bluestone JA. Instability of the transcription factor Foxp3 leads to the generation of pathogenic memory T cells in vivo. Nat Immunol. 2009;10(9):1000–7.

28. Sekiya T, Kondo T, Shichita T, Morita R, Ichinose H, Yoshimura A. Suppression of Th2 and Tfh immune reactions by Nr4a receptors in mature T reg cells. J Exp Med. 2015;212(10):1623–40.

29. Zheng Y, Josefowicz S, Chaudhry A, Peng XP, Forbush K, Rudensky AY. Role of conserved non-coding DNA elements in the Foxp3 gene in regulatory T-cell fate. Nature. 2010;463(7282):808–12.

30. Sekiya T, Kashiwagi I, Inoue N, Morita R, Hori S, Waldmann H, Rudensky AY, Ichinose H, Metzger D, Chambon P, et al. The nuclear orphan receptor Nr4a2 induces Foxp3 and regulates differentiation of CD4+ T cells. Nature communications. 2011;2:269.

31. Furusawa Y, Obata Y, Fukuda S, Endo TA, Nakato G, Takahashi D, Nakanishi Y, Uetake C, Kato K, Kato T, et al. Commensal microbe-derived butyrate induces the differentiation of colonic regulatory T cells. Nature. 2013;504(7480):446–50.

32. Arpaia N, Campbell C, Fan X, Dikiy S, van der Veeken J, deRoos P, Liu H, Cross JR, Pfeffer K, Coffer PJ, et al. Metabolites produced by commensal bacteria promote peripheral regulatory T-cell generation. Nature. 2013;504(7480):451–5.

33. Morikawa H, Ohkura N, Vandenbon A, Itoh M, Nagao-Sato S, Kawaji H, Lassmann T, Carninci P, Hayashizaki Y, Forrest AR, et al. Differential roles of epigenetic changes and Foxp3 expression in regulatory T cell-specific transcriptional regulation. Proc Natl Acad Sci USA. 2014;111(14):5289–94.

34. Yang R, Qu C, Zhou Y, Konkel JE, Shi S, Liu Y, Chen C, Liu S, Liu D, Chen Y, et al. Hydrogen sulfide promotes TET1- and TET2-mediated Foxp3 demethylation to drive regulatory T cell differentiation and maintain immune homeostasis. Immunity. 2015;43(2):251–63.

35. Yue X, Trifari S, Aijo T, Tsagaratou A, Pastor WA, Zepeda-Martinez JA, Lio CJ, Li X, Huang Y, Vijayanand P et al. Control of Foxp3 stability through modulation of TET activity. J Exp Med. 2016;213(3):377–397

36. Tao R, de Zoeten EF, Ozkaynak E, Chen C, Wang L, Porrett PM, Li B, Turka LA, Olson EN, Greene MI, et al. Deacetylase inhibition promotes the generation and function of regulatory T cells. Nat Med. 2007;13(11):1299–307.

37. Lal G, Zhang N, van der Touw W, Ding Y, Ju W, Bottinger EP, Reid SP, Levy DE, Bromberg JS. Epigenetic regulation of Foxp3 expression in regulatory T cells by DNA methylation. J Immunol. 2009;182(1):259–73.

38. Sasidharan Nair V, Song MH, Oh KI. Vitamin C facilitates demethylation of the Foxp3 enhancer in a TET-dependent manner. J Immunol. 2016;196(5):2119–31.

39. Hibino S, Saito Y, Muramatsu T, Otani A, Kasai Y, Kimura M, Saito H. Inhibitors of enhancer of zeste homolog 2 (EZH2) activate tumor-suppressor microRNAs in human cancer cells. Oncogenesis. 2014;3:e104.

40. Kubo M, Kincaid RL, Webb DR, Ransom JT. The Ca2+ calmodulin-activated, phosphoprotein phosphatase calcineurin is sufficient for positive transcriptional regulation of the mouse Il-4 gene. Int Immunol. 1994;6(2):179–88.

41. Stemmer M, Thumberger T, Keyer MD, Wittbrodt J, Mateo JL. CCTop: an intuitive, flexible and reliable CRISPR/Cas9 target prediction tool. PLos One. 2015;10(4): e0124633

42. Maeder ML, Angstman JF, Richardson ME, Linder SJ, Cascio VM, Tsai SQ, Ho QH, Sander JD, Reyon D, Bernstein BE, et al. Targeted DNA demethylation and activation of endogenous genes using programmable TALE-TET1 fusion proteins. Nat Biotechnol. 2013;31(12):1137–42.

43. Xu X, Tao Y, Gao X, Zhang L, Li X, Zou W, Ruan K, Wang F, Xu GL, Hu R. A CRISPR-based approach for targeted DNA demethylation. Cell discovery. 2016;2:16009.

44. Yang XO, Nurieva R, Martinez GJ, Kang HS, Chung Y, Pappu BP, Shah B, Chang SH, Schluns KS, Watowich SS, et al. Molecular antagonism and plasticity of regulatory and inflammatory T cell programs. Immunity. 2008;29(1):44–56.

45. Dominguez-Villar M, Baecher-Allan CM, Hafler DA. Identification of T helper type 1-like, Foxp3+ regulatory T cells in human autoimmune disease. Nat Med. 2011;17(6):673–5.

46. Feng Y, Arvey A, Chinen T, van der Veeken J, Gasteiger G, Rudensky AY. Control of the inheritance of regulatory T cell identity by a cis element in the Foxp3 locus. Cell. 2014;158(4):749–63.

47. Bothur E, Raifer H, Haftmann C, Stittrich AB, Brustle A, Brenner D, Bollig N, Bieringer M, Kang CH, Reinhard K et al. Antigen receptor-mediated depletion of FOXP3 in induced regulatory T-lymphocytes via PTPN2 and FOXO1. Nat Commun. 2015;6:8576

48. Tone Y, Furuuchi K, Kojima Y, Tykocinski ML, Greene MI, Tone M. Smad3 and NFAT cooperate to induce Foxp3 expression through its enhancer. Nat Immunol. 2008;9(2):194–202.

49. Takimoto T, Wakabayashi Y, Sekiya T, Inoue N, Morita R, Ichiyama K, Takahashi R, Asakawa M, Muto G, Mori T, et al. Smad2 and Smad3 are redundantly essential for the TGF-beta-mediated regulation of regulatory T plasticity and Th1 development. J Immunol. 2010;185(2):842–55.

50. Murphy KM, Stockinger B. Effector T cell plasticity: flexibility in the face of changing circumstances. Nat Immunol. 2010;11(8):674–80.

51. Hsu PD, Scott DA, Weinstein JA, Ran FA, Konermann S, Agarwala V, Li YQ, Fine EJ, Wu XB, Shalem O, et al. DNA targeting specificity of RNA-guided Cas9 nucleases. Nat Biotechnol. 2013;31(9):827.

52. Pattanayak V, Lin S, Guilinger JP, Ma EB, Doudna JA, Liu DR. High-throughput profiling of off-target DNA cleavage reveals RNA-programmed Cas9 nuclease specificity. Nat Biotechnol. 2013;31(9):839.

53. Takahashi R, Nishimoto S, Muto G, Sekiya T, Tamiya T, Kimura A, Morita R, Asakawa M, Chinen T, Yoshimura A. SOCS1 is essential for regulatory T cell functions by preventing loss of Foxp3 expression as well as IFN-gamma and IL-17A production. J Exp Med. 2011;208(10):2055–67.

54. Nair VS, Song MH, Ko M, Oh KI. DNA demethylation of the Foxp3 enhancer is maintained through modulation of ten-eleven-translocation and DNA methyltransferases. Mol Cells. 2016;39(12):888–97.

55. Qi LS, Larson MH, Gilbert LA, Doudna JA, Weissman JS, Arkin AP, Lim WA. Repurposing CRISPR as an RNA-guided platform for sequence-specific control of gene expression. Cell. 2013;152(5):1173–83.

56. Morita S, Noguchi H, Horii T, Nakabayashi K, Kimura M, Okamura K, Sakai A, Nakashima H, Hata K, Nakashima K et al. Targeted DNA demethylation in vivo using dCas9-peptide repeat and scFv-TET1 catalytic domain fusions. Nat Biotechnol. 2016;34:1060–1065.

57. Balboa D, Weltner J, Eurola S, Trokovic R, Wartiovaara K, Otonkoski T. Conditionally stabilized dCas9 activator for controlling gene expression in human cell reprogramming and differentiation. Stem cell Rep. 2015;5(3):448–59.

58. Ran FA, Cong L, Yan WX, Scott DA, Gootenberg JS, Kriz AJ, Zetsche B, Shalem O, Wu X, Makarova KS, et al. In vivo genome editing using Staphylococcus aureus Cas9. Nature. 2015;520(7546):186–91.

59. Nishimasu H, Cong L, Yan WX, Ran FA, Zetsche B, Li Y, Kurabayashi A, Ishitani R, Zhang F, Nureki O. Crystal structure of Staphylococcus aureus Cas9. Cell. 2015;162(5):1113–26.

60.  Josefowicz SZ, Niec RE, Kim HY, Treuting P, Chinen T, Zheng Y, Umetsu DT, Rudensky AY. Extrathymically generated regulatory T cells control mucosal T(H)2 inflammation. Nature. 2012;482(7385):395–U1510.

61.  Harada Y, Harada Y, Elly C, Ying G, Paik JH, DePinho RA, Liu YC. Transcription factors Foxo3a and Foxo1 couple the E3 ligase Cbl-b to the induction of Foxp3 expression in induced regulatory T cells. J Exp Med. 2010;207(7):1381–91.

62.  Maruyama T, Li J, Vaque JP, Konkel JE, Wang WF, Zhang BJ, Zhang P, Zamarron BF, Yu DY, Wu YT, et al. Control of the differentiation of regulatory T cells and T(H)17 cells by the DNA-binding inhibitor Id3. Nat Immunol. 2011;12(1):86–U114.

63.  Takaki H, Ichiyama K, Koga K, Chinen T, Takaesu G, Sugiyama Y, Kato S, Yoshimura A, Kobayashi T. STAT6 inhibits TGF-beta 1-mediated Foxp3 induction through direct binding to the Foxp3 promoter, which is reverted by retinoic acid receptor. J Biol Chem. 2008;283(22):14955–62.

64.  Kitagawa Y, Ohkura N, Kidani Y, Vandenbon A, Hirota K, Kawakami R, Yasuda K, Motooka D, Nakamura S, Kondo M, et al. Guidance of regulatory T cell development by Satb1-dependent super-enhancer establishment. Nat Immunol. 2017;18(2):173–83.

65.  Fu WX, Ergun A, Lu T, Hill JA, Haxhinasto S, Fassett MS, Gazit R, Adoro S, Glimcher L, Chan S, et al. A multiply redundant genetic switch 'locks in' the transcriptional signature of regulatory T cells. Nat Immunol. 2012;13(10):972–80.

66.  Gao X, Tsang JC, Gaba F, Wu D, Lu L, Liu P. Comparison of TALE designer transcription factors and the CRISPR/dCas9 in regulation of gene expression by targeting enhancers. Nucleic Acids Res. 2014;42(20):e155.

67.  Chavez A, Tuttle M, Pruitt BW, Ewen-Campen B, Chari R, Ter-Ovanesyan D, Haque SJ, Cecchi RJ, Kowal EJ, Buchthal J, et al. Comparison of Cas9 activators in multiple species. Nat Methods. 2016;13(7):563–7.

68.  Hippen KL, Merkel SC, Schirm DK, Sieben CM, Sumstad D, Kadidlo DM, McKenna DH, Bromberg JS, Levine BL, Riley JL et al. Massive ex vivo expansion of human natural regulatory T cells (T-regs) with minimal loss of in vivo functional activity. Sci Transl Med. 2011;3(83): 83ra41. doi:10.1126/scitranslmed.3001809

69.  Naldini L. Gene therapy returns to centre stage. Nature. 2015;526(7573):351–60.

# 4

# Transcription and chromatin determinants of de novo DNA methylation timing in oocytes

Lenka Gahurova[1,6†], Shin-ichi Tomizawa[2†], Sébastien A. Smallwood[1,7], Kathleen R. Stewart-Morgan[1,8], Heba Saadeh[1,9], Jeesun Kim[3], Simon R. Andrews[4], Taiping Chen[3] and Gavin Kelsey[1,5*] (iD)

## Abstract

**Background:** Gametogenesis in mammals entails profound re-patterning of the epigenome. In the female germline, DNA methylation is acquired late in oogenesis from an essentially unmethylated baseline and is established largely as a consequence of transcription events. Molecular and functional studies have shown that imprinted genes become methylated at different times during oocyte growth; however, little is known about the kinetics of methylation gain genome wide and the reasons for asynchrony in methylation at imprinted loci.

**Results:** Given the predominant role of transcription, we sought to investigate whether transcription timing is rate limiting for de novo methylation and determines the asynchrony of methylation events. Therefore, we generated genome-wide methylation and transcriptome maps of size-selected, growing oocytes to capture the onset and progression of methylation. We find that most sequence elements, including most classes of transposable elements, acquire methylation at similar rates overall. However, methylation of CpG islands (CGIs) is delayed compared with the genome average and there are reproducible differences amongst CGIs in onset of methylation. Although more highly transcribed genes acquire methylation earlier, the major transitions in the oocyte transcriptome occur well before the de novo methylation phase, indicating that transcription is generally not rate limiting in conferring permissiveness to DNA methylation. Instead, CGI methylation timing negatively correlates with enrichment for histone 3 lysine 4 (H3K4) methylation and dependence on the H3K4 demethylases KDM1A and KDM1B, implicating chromatin remodelling as a major determinant of methylation timing. We also identified differential enrichment of transcription factor binding motifs in CGIs acquiring methylation early or late in oocyte growth. By combining these parameters into multiple regression models, we were able to account for about a fifth of the variation in methylation timing of CGIs. Finally, we show that establishment of non-CpG methylation, which is prevalent in fully grown oocytes, and methylation over non-transcribed regions, are later events in oogenesis.

**Conclusions:** These results do not support a major role for transcriptional transitions in the time of onset of DNA methylation in the oocyte, but suggest a model in which sequences least dependent on chromatin remodelling are the earliest to become permissive for methylation.

**Keywords:** Oocytes, DNA methylation, Histone modifications, Transcription, Imprinting

*Correspondence: gavin.kelsey@babraham.ac.uk
†Lenka Gahurova and Shin-ichi Tomizawa contributed equally to this work
[1] Epigenetics Programme, Babraham Institute, Cambridge CB22 3AT, UK
Full list of author information is available at the end of the article

## Background

The establishment of DNA methylation in the female germline in mammals is essential for genomic imprinting and successful development of the embryo following fertilisation [1–3]. Following genome-wide erasure of methylation in primordial germ cells [4], mammalian oocytes acquire a highly structured DNA methylation landscape in which domains of uniform methylation are separated by extensive unmethylated domains [5, 6]; this largely bimodal pattern is unique amongst mammalian cell types. DNA methylation is associated mostly with transcriptionally active gene bodies in oocytes, and these methylated domains contain intragenically located CpG islands (CGIs) that also gain methylation, including the germline differentially methylated regions (gDMRs) of imprinted genes [5–7]. As a result, there is highly programmed methylation of a defined set of ~2000 CGIs in oocytes, mostly on account of their location within active transcription units. We, and others, have shown that transcription is functionally required to define methylation in oocytes: Abrogating specific transcription events prevents methylation of the associated loci, including at imprinted gDMRs [6, 8, 9].

The oocyte represents a pure de novo methylation system, as an entire DNA methylation landscape is established on an essentially unmethylated genome in a non-dividing cell [10]; therefore, it provides a unique opportunity to investigate the extent to which different sequence features acquire methylation as a result of common or distinct mechanisms. Current knowledge is largely limited to the fully established DNA methylome in fully grown oocytes at the germinal vesicle (GV) stage or in ovulated metaphase II (MII) oocytes [5, 11], such that differences in the mechanistic requirements for methylation of various sequence elements or in the kinetics of their methylation are obscured. Thus, investigating methylation at intermediate stages would be informative, but genome-wide studies have not yet been done. Analysis of a limited number of imprinted gDMRs identified that de novo methylation is a function of developmental stage of follicles and oocyte size, with methylation initiated around the time follicles transition into the antral or secondary follicle stage of development. Moreover, locus-specific analysis has shown that the onset and progression of methylation appear to differ between imprinted gDMRs [12–14]. This asynchrony has functional importance, as nuclear transfer experiments have shown that different imprinted domains acquire imprinting competence at different stages of oocyte growth [15].

In view of the rather simple methylation landscape of the oocyte, the differential timing of methylation acquisition at gDMRs is unexpected, and the reasons for this asynchrony are unclear. Understanding its basis is essential for identifying the origin of methylation defects in oocytes that could underlie some errors in imprinting. Such asynchrony also suggests that different factors, or combinations of factors, may be necessary for methylation of different gDMRs, individual CGIs or individual methylated domains, aside from the common requirement for the de novo DNA methyltransferase DNMT3A and its obligate partner DNMT3L [5, 7, 11]. Given the strong association with transcription [6], and major changes in the transcription programme during oocyte growth [16], one possibility is that the timings of transcription events traversing gDMRs and CGIs could account for differences in the onset of methylation at individual elements.

At a mechanistic level, de novo DNA methylation occurs in a chromatin template and, in accordance with the biochemical properties of DNMT3A and DNMT3L [17–19], is predicted to depend upon the acquisition of a permissive histone modification state. Thus, regions destined for DNA methylation are proposed to be marked by histone 3 trimethylated at lysine 36 (H3K36me3) and should lack H3 di- or trimethylated at lysine 4 (H3K4me2/me3) [7, 20]. Evidence in support of this model is the requirement for the H3K4 demethylase KDM1B for DNA methylation of most imprinted gDMRs and CGIs that acquire methylation in oocytes and the increase in H3K36me3 at these elements during oocyte growth [20, 21]. Such chromatin state changes may also be downstream of transcription events: H3K36me3 is deposited by SETD2 in association with elongating RNA polymerase II [22–24], although the role of SETD2 in oocytes has not yet been determined; and removal of H3K4me2 and gain of H3K36me3 at the gDMR of the imprinted locus *Zac1* in oocytes was shown to depend on transcription from an upstream, oocyte-specific promoter [6].

To investigate how transcription influences the kinetics of methylation at gDMRs and throughout the genome, we generated genome-wide DNA methylation and high-resolution transcriptome maps of size-selected populations of growing oocytes spanning the onset of methylation. We find that the major remodelling of the oocyte transcriptome occurs well before the onset of DNA methylation, indicating that initiation of transcription events is not temporally coupled to methylation of specific loci. However, rate of gene body methylation does correlate with transcription level, which could reflect the degree of transcription-coupled chromatin remodelling. CGI methylation timing reflects (1) the H3K4me2 levels found in non-growing and early growing oocytes, (2) dependence on H3K4 demethylases and (3) presence of specific transcription factor motifs, supporting a model in which sequences requiring less chromatin remodelling are the earliest to become permissive for de novo methylation.

## Results

### Capturing the onset of de novo DNA methylation in oocytes

To analyse the onset and progression of de novo methylation at a genome-wide scale, we isolated growing oocytes from pre-pubertal mouse ovaries (post-natal days 7–18) and sorted them into the following, non-overlapping size categories: 40–45, 50–55 and 60–65 μm. Genome-wide methylation maps were generated by bisulphite conversion of oocyte DNA and Illumina sequencing. For unbiased genome coverage to enable interrogation of all sequence features in 60–65 μm oocytes, we applied post-bisulphite adapter tagging (PBAT; [25]); for focussed coverage of CGIs and other GC-rich sequences in all three size classes of oocytes, we applied reduced representation bisulphite sequencing (RRBS; [7]). The 60–65 μm PBAT library yielded 98,951,299 uniquely mapped read pairs, covering 18,651,142 (85.3%) of mappable CpG sites at ≥1 read and 5,731,851 CpGs (26.3%) with ≥5 reads. The RRBS libraries covered between 551,677 and 838,372 CpG sites (≥5 reads) and 13,944–15,799 (60.6–68.7%) of the 23,009 CGIs in the mouse autosomes and X chromosome (CGI coverage threshold ≥5 CpG sites with ≥5 reads; Additional file 1: Table S1). The PBAT and RRBS data were compared with published data sets from non-growing oocytes (NGO) and GV or MII oocytes [5, 7, 11]. In parallel, RNA sequencing (RNA-seq) libraries were made from similar pools of size-selected oocytes (see below).

The overall CpG methylation level of 60–65 μm oocytes determined by PBAT was 22.25%, compared with 2.36% in NGOs and 38.68% in GV oocytes, showing that this stage represents a midpoint in the progression of global de novo methylation (Fig. 1a; Additional file 2: Table S2). We then evaluated whether all genomic features that become methylated in GV oocytes gain methylation at similar rates, including the hypermethylated domains of GV oocytes we previously designated [6]. CpGs in hypermethylated domains have attained on average 48.00 ± 0.02% methylation in 60–65 μm oocytes (Additional file 2: Table S2), although there is a considerable spread in the methylation level of these CpGs at this time (Fig. 1b). We previously showed that 85–90% of hypermethylated domains were associated with transcription units active in oocytes [6]; therefore, we asked whether domains associated with transcription units and those apparently not associated with transcription displayed similar kinetics of methylation. Comparison of CpG methylation rate of transcribed hypermethylated domains and apparently transcriptionally silent hypermethylated domains revealed that CpGs in transcriptionally silent regions are methylated later: average CpG methylation in transcribed domains is 50.1% but 30.0% for transcriptionally silent regions (Fig. 1b). For CGIs that become methylated fully (≥75%) in GV oocytes, mean methylation (37.21 ± 0.69%) in 60–65 μm oocytes was less than most other sequence features (Fig. 1b; Additional file 2: Table S2). An effect of CpG density is also apparent when considering 2-kb genomic windows: regions of highest methylation (≥80%) in 60–65 μm oocytes had on average lower CpG density and GC content (Fig. 1c).

Similar to hypermethylated domains, most classes of transposable element (TEs) that become methylated (≥75%) in GV oocytes are midway in methylation progression (Fig. 1b; Additional file 2: Table S2, Additional file 3: Fig. S1A), although there was interesting variation in the kinetics of specific elements. Some TEs start at a higher level of methylation in NGOs, such as some endogenous retroviral (ERVK) long-terminal repeat (LTR) elements, reflecting incomplete erasure of methylation in primordial germ cells [4]. In addition, there was a significant variation in the rate of methylation of specific TE subfamilies. Notably, of the 20 most abundant LINE-L1 subfamilies, methylation of three of the four L1Md subfamilies was significantly delayed (average methylation of L1Md_A 39.9%, L1Md_F3 44.3% and L1Md_T 42.0%, compared with 48.1–54.4% for the remaining L1 subfamilies). In comparison, there were no differences in the methylation rate of the 20 MaLR subfamilies (Additional file 3: Fig. S1B). L1Md elements are amongst the youngest L1s, with the least degenerated sequence, the most intact transcription factor (TF) binding sites and which have to be actively suppressed [26, 27]. Many of the L1Md subfamilies also retained residual methylation in NGOs (6.5–19.9%, compared with 1.4–3.7% for other L1s). These results indicate that different sequence features acquire methylation with similar overall kinetics, suggesting that the de novo methylation complex is not targeted preferentially to any particular sequence feature. However, the delayed methylation of CGIs and specific L1 subfamilies, as well as at untranscribed regions, points to additional or alternative mechanistic requirements at these elements.

In fully grown oocytes, there is a high level of concordance in methylation of adjacent CpGs across the extensive hypermethylated domains [6]. Having captured oocytes midway in the progression of methylation, we looked at the coherence of ongoing methylation to investigate co-operativity of the de novo methylation complex. For each sequencing read containing multiple CpGs, we asked how often and over what distance CpGs had the same methylation state. For 60–65 μm oocytes, neighbouring CpGs were both methylated 60–70% of the time over 60 bp and at least 50% of the time over 90 bp (Fig. 2a). If CpG sites were being methylated individually without co-operativity, the probability that CpG

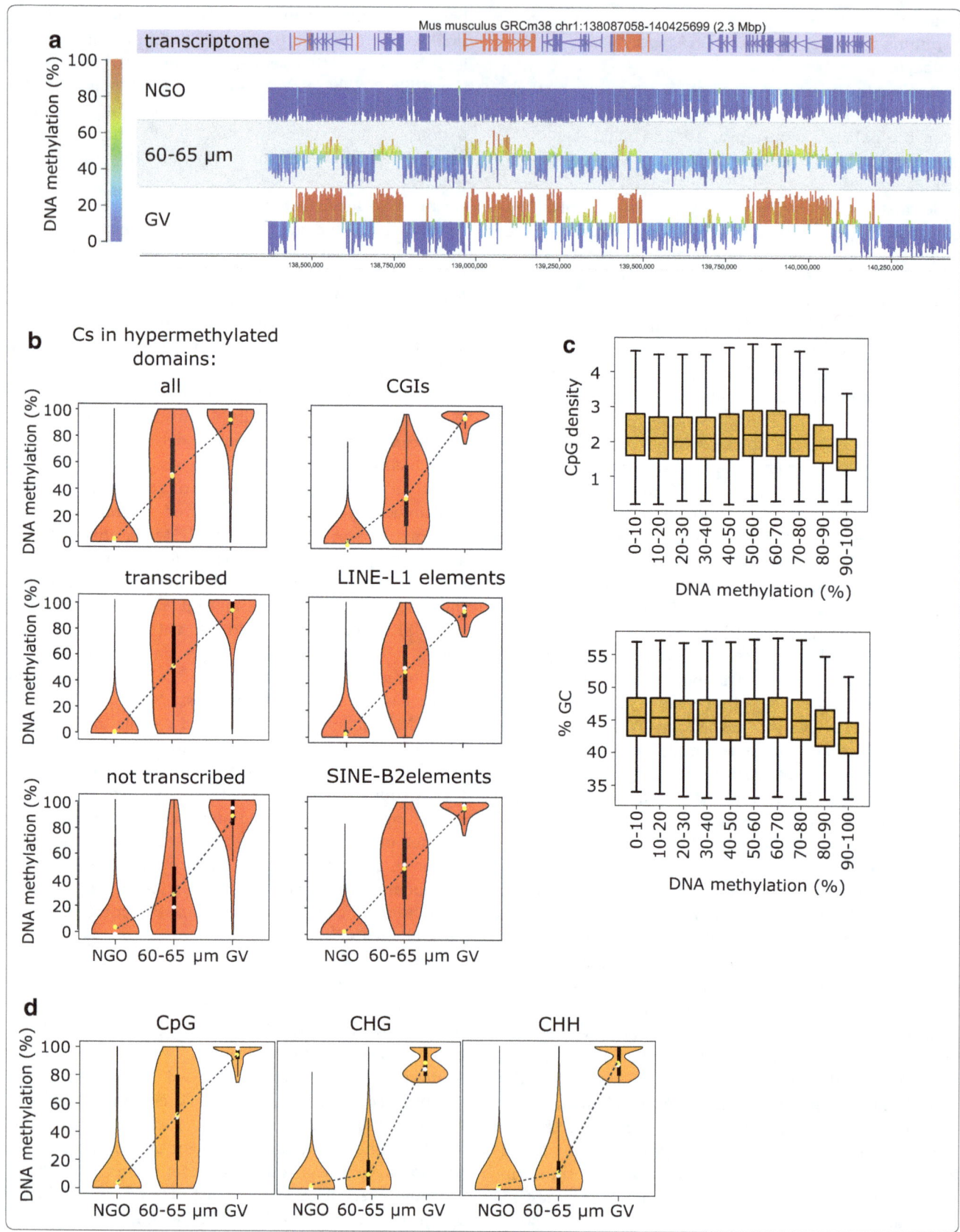

(See figure on previous page.)
**Fig. 1** Rates of de novo DNA methylation of different sequence features in growing oocytes. **a** Screenshot of a 2.3-Mb interval of chromosome 1 depicting methylation in NGOs, 60–65 μm and GV oocytes. *Vertical bars* represent mean methylation of 2-kb windows, with 1-kb steps, *height and colour* denoting % methylation. The *horizontal lines* are set at 50% methylation, with higher levels of methylation *above the line* and lower levels *below the line* and *shaded* according to the *colour scale on the left*. The 60–65 μm data are from PBAT from the current manuscript; NGO and GV data are from [5, 11]. **b** *Violin plots* showing distribution of CpG methylation values in all hypermethylated domains, transcribed hypermethylated domains (≥90% of the length of the domain covered by transcript, domains ≥5 kb), transcriptionally silent hypermethylated domains (≤10% of the length of the domain covered by transcript, domains ≥5 kb), CGIs, LINE L1s and SINE-B2s in NGO, 60–65 μm and GV oocytes. *Shape of the violin plot* represents Kernel density estimation, i.e. probability density of the data at the different values. *White dots* correspond to the median, *yellow dots* to the average, *bold lines* the interquartile range and thin lines adjacent values, i.e. minimum and maximum values within the ×1.5 interquartile range from the first and third quartile, respectively. **c** *Box whisker plots* reporting CpG density and GC content of 2-kb genomic regions that are fully methylated in GV oocytes (≥75% DNA methylation) categorised according to their % DNA methylation in 60–65 μm oocytes (*x* axis). *Boxes*, interquartile range, with bar as median and *whiskers* as ×1.5 interquartile range, outliers not shown. Between 3619 and 30320 2-kb intervals were analysed in each methylation category. **d** *Violin plots* showing methylation levels of Cs in CpG, CHG and CHH contexts in NGOs and 60–65 μm oocytes of Cs that are fully methylated (≥75%) in GV oocytes

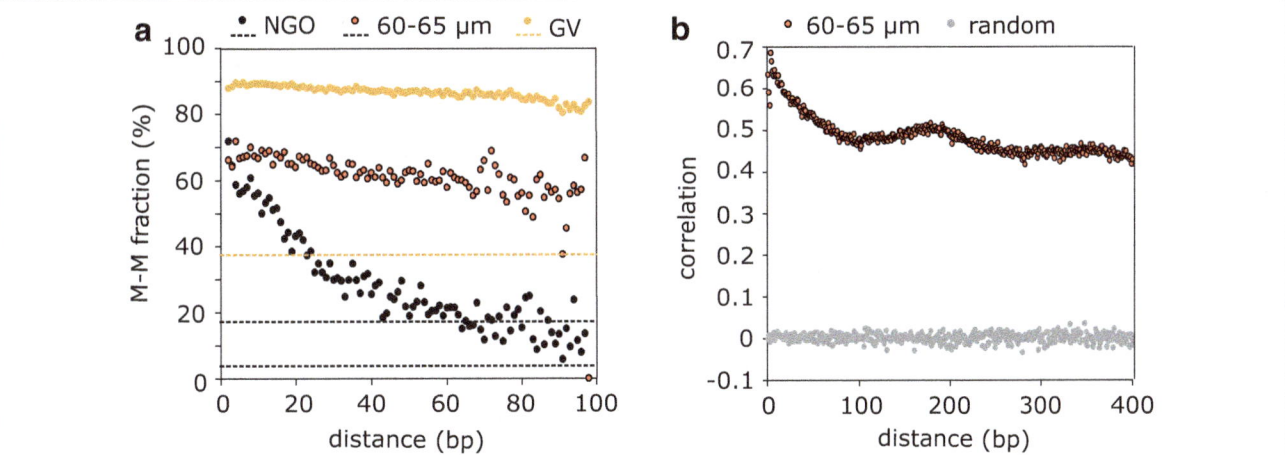

**Fig. 2** Properties of ongoing de novo DNA methylation in growing oocytes. **a** Average proportion of neighbouring CpG pairs where both CpGs are methylated (M–M fraction) by distance of CpG pairs in NGO, 60–65 μm and GV oocytes. The value of the M–M fraction was quantified for each possible distance between two neighbouring CpGs on the same sequencing read using formula M–M pairs/(M–M + M–U pairs), where M–U pairs represent CpG pairs where upstream CpG is methylated and downstream unmethylated. Only reads mapping to chromosome 1 were analysed. The *horizontal lines* represent the genomic average methylation level of each stage. **b** The distant-dependent correlation of methylation between CpG pairs in 60–65 μm oocytes, compared with random-shuffled data

pairs were both methylated would equate to the overall genomic methylation level which, in 60–65 μm oocytes, was 17.56%. Therefore, these data indicate co-operativity in methylation of adjacent CpGs by DNMT3A/DNMT3L during oocyte growth, similar to findings of DNMT3B function in embryonic stem cell (ESC) [28]. We note, also, that although concordance of methylation declines with distance, there is a local maximum in the correlation at ~180 bp (Fig. 2b), which approximates the size of a nucleosome, consistent with a model in which de novo methylation occurs in linker regions, as proposed in ESCs [28]. Finally, oocytes have been shown to have extensive methylation outside of the CpG context, in that methylation of non-CpG sites accounts for more than half of the total amount of methylated cytosine [11, 29]. We looked specifically at all informative cytosines that

become fully methylated (≥75%) in GV oocytes. Strikingly, in 60–65 μm oocytes, CHG and CHH sites (where H = A, T or C) that become methylated in GV oocytes were only 10.77 ± 0.08 and 13.90 ± 0.05% methylated, respectively, compared with 49.31 ± 0.05% for CpG sites, indicating preferential methylation of CpG sites during oocyte development (Fig. 1d; Additional file 4: Table S3).

## CGIs and imprinted gDMRs gain DNA methylation at different rates in oocytes

To look in more detail at the progression of methylation at CGIs, we considered the RRBS datasets. There is very little methylation of CGIs in 40–45 μm oocytes: only three CGIs (of 522 CGIs with sufficient data) that become fully methylated in GV oocytes were methylated ≥25% in 40–45 μm oocytes, and two of these have residual methylation in NGOs [7]. Methylation was first detected in the 50–55 µm size class (22% of CGIs destined for full methylation having ≥25% methylation in this size group) and at least 55% of CGIs showed intermediate (25–75%) to high (≥75%) levels of methylation in 60–65 µm oocytes (Fig. 3a). Overall, there was a very high level of correlation ($R = 0.929$) between the RRBS and PBAT libraries in CGI methylation at the 60–65 µm stage (Additional file 3: Fig. S2A), suggesting that the differences in level of methylation are reproducible and biological in origin. Focusing on imprinted gDMRs, methylation in 60–65 µm oocytes assessed by the two methods ranged from 0 to ~70% (Fig. 3b; Additional file 3: Fig. S2A), again with a high degree of consistency in methylation of individual gDMRs determined by the two methods (noting that RRBS and PBAT will not have identical sequence coverage across each gDMR). For example, the *Igf2r* gDMR had attained 32.5% methylation in 50–55 µm oocytes and 67.9% in 60–65 µm oocytes, while the *Cdh15* igDMR was <5% methylated even in 60–65 µm oocytes (Fig. 3b). This range of methylation is broadly consistent with earlier studies that analysed limited numbers of gDMRs by locus-specific bisulphite sequencing (again, with the caveat that different regions of the gDMRs will have been assayed by the various methods; [12–14]). For a subset of CGIs, we also validated the time of onset by locus-specific bisulphite sequencing (Fig. 3c). The differential onset of CGI methylation is not related to CpG content or GC richness of these CGIs (Additional file 3: Fig. S2B). In conclusion, CGIs destined for methylation in GV oocytes are not co-ordinately methylated but display substantial and reproducible differences in time of onset of methylation in growing oocytes, and this variation is not a simple property of overall sequence composition.

## Mapping changes in the oocyte transcriptome during oocyte growth

We sought to test the relationship between methylation kinetics and changes in transcription during oocyte development and growth. To do so, we generated deep, strand-specific total RNA-seq libraries in duplicate from the same size populations of growing oocytes as used in methylation analysis, as well as an earlier population (10–30 µm) and a GV population (Additional file 5: Table S4). In addition, the data were compared with RNA-seq from NGOs collected at embryonic day E18.5 [20] and an existing GV data set [6]. Although transcriptional changes have been documented during mouse oocyte development before [16], those data were generated using expression microarrays that capture only a fraction of the transcription units actually present in oocytes and cannot be used to infer alternative transcription start site (TSS) use: our previous work has demonstrated the importance of using the correct transcriptome for accurate association with methylation [6]. Although the RNA-seq data sets do not capture nascent transcription events, they do enable us to determine the time during oocyte growth that transcription units are first active, including the use of alternative upstream TSSs that are prevalent in oocytes [6]. Transcript abundance was used as a proxy for transcription rate.

The RNA-seq data sets were compared with the oocyte transcriptome assembly previously generated in our laboratory [6], resulting in the detection of 21,402–32,775 genes (FPKM thresholds 0.017–0.102) in the various oocyte size populations (Additional file 6: Table S5). Principal component (PC) analysis of the global expression patterns showed that data sets from growing and GV oocytes cluster together, with the E18.5 transcriptome being the most distinct; PC2 segregates the growing oocyte populations by size, particularly when the E18.5 data set is excluded (Additional file 3: Fig. S3). It should be noted that E18.5 oocytes were collected using FACS, such that RNA was extracted from fixed samples, whereas all post-natal oocytes were collected manually, and these technical differences could contribute to some differences between the E18.5 transcriptome and the other stages. Nevertheless, most transcripts (68%) were already detected at E18.5, and a further 28% were detected first in 10–30 µm oocytes, with very few appearing for the first time in larger size populations (Fig. 4a). The general stability of gene expression in the growing oocyte populations, even as cytoplasmic volume and mRNA content are increasing substantially, is reflected in the rather small numbers of genes identified as differentially expressed (<4%) between consecutive stages (Additional file 3: Fig. S4). Based on our oocyte transcriptome assembly, we segregated genes into reference genes (i.e. previously annotated genes) and novel genes, either novel multiexonic or monoexonic. For reference genes expressed from their canonical TSSs, 88% were already detected at E18.5; in comparison, most novel genes were detected first in 10–30 µm oocytes (~63% multi- and ~57% monoexonic novel genes), with a small minority first detected in larger oocytes (~8 and ~13% for multi- and monoexonic genes, respectively; Fig. 4a). Similarly, most (~70%) novel upstream TSSs were activated in 10–30 µm oocytes. Therefore, most changes in the oocyte transcriptome occur well in advance of the onset of de novo methylation, which initiates after the 40–45 µm stage. This effect can be seen at individual imprinted loci: all gDMRs are found within transcription units even at the earlier stages, irrespective of whether they are transcribed from alternative promoters or whether methylation is detected early (50–55 µm) or late in oocyte growth (60–65 µm) (Fig. 4b).

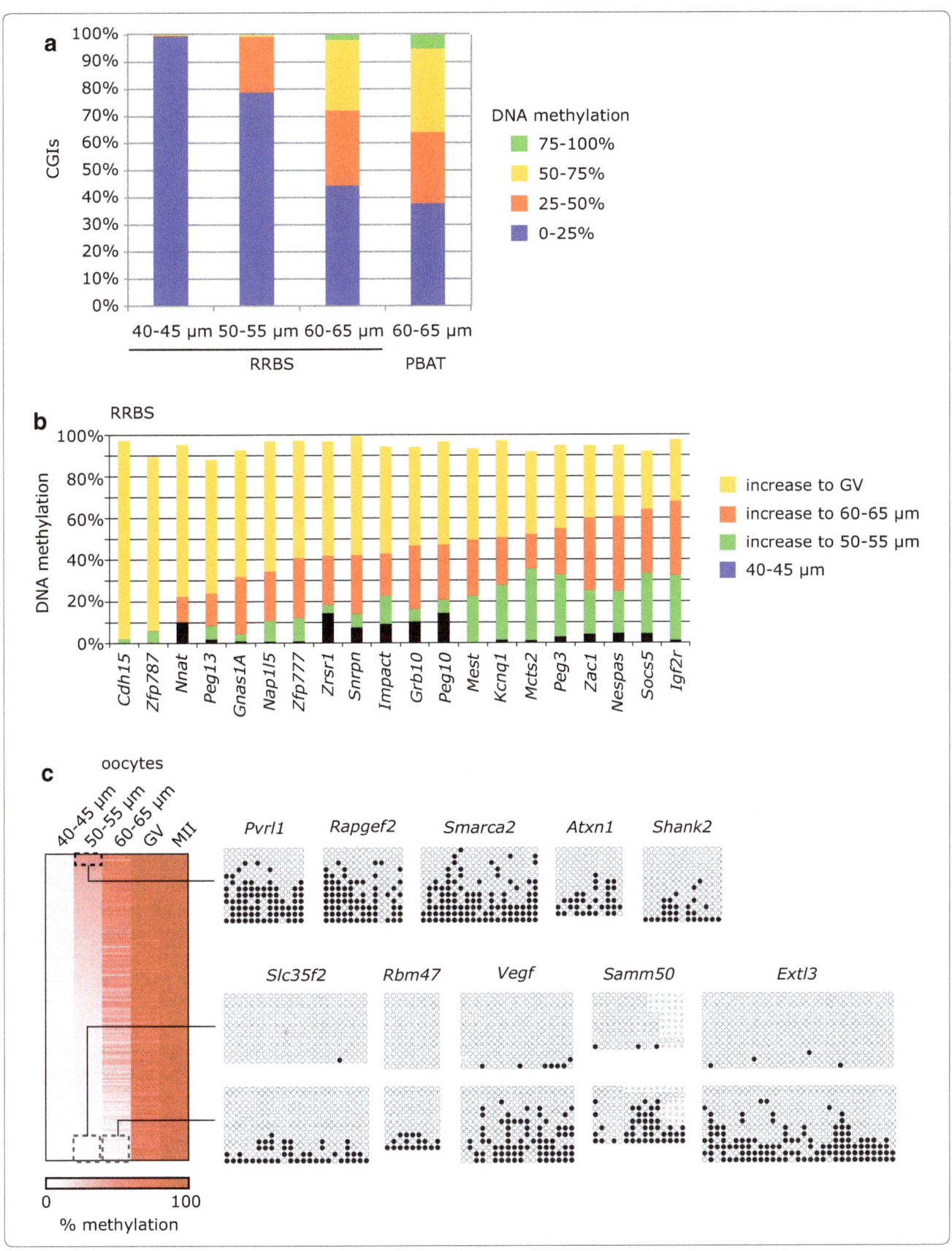

(See figure on previous page.)

**Fig. 3** CpG islands gain DNA methylation at different rates in growing oocytes. **a** Barchart of CGI methylation in the oocyte size populations from the RRBS and PBAT datasets. The number of CGIs covered in each dataset is given in Additional file 1: Table S1. **b** Methylation of gDMRs in RRBS datasets, displaying the basal level in 40–45 μm oocytes, and the increases in methylation to the subsequent size populations. gDMRs are ordered according to their methylation level in 60–65 μm oocytes, which is comparable with PBAT data (see Additional file 3: Fig. S2A). **c** Validation of CGI methylation in different oocyte size populations. Heatmap shows methylation progression at CGIs that become methylated between 40 and 45 μm and MII oocytes (data from published GV and MII RRBS datasets). Five early-methylating CGIs and five late-methylating CGIs were selected, and their methylation in 50–55 μm oocytes (early-methylating CGIs) or both 50–55 and 60–65 μm oocytes was confirmed by locus-specific bisulphite sequencing. *White dots* represent unmethylated CpGs and black dots methylated CpGs

Despite the general stability of gene expression during oocyte growth stages (Additional file 3: Fig. S4), the RNA-seq data sets provide unprecedented detail into the changes in transcript abundance during critical times in oocyte growth and follicular differentiation. We identified 530 genes, mostly protein-coding, up-regulated greater than 50-fold between E18.5 and GV oocytes, and 283 up-regulated >50-fold between E18.5 and 10–30 μm oocytes (Additional file 7: Table S6). Gene ontology (GO) analysis did not reveal particularly strong enrichment terms ("Regulation of reproductive process" containing 10 of the 283 genes had the highest enrichment of 5.53, *p* value $1.27 \times 10^{-5}$, adjusted FDR 0.164), perhaps reflecting the wide diversity of functions required during oogenesis as well as the accumulation of maternal RNA stores for processes in the zygote (Additional file 3: Fig. S5). The set of highly induced transcripts did contain genes for oocyte-specific transcriptional regulators such as OBOX1, 2 and 5, the maternal effect homeobox SEBOX, the zona pellucida proteins 1, 2 and 3 (ZP1, 2, 3), components of the subcortical maternal complex (OOEP, TLE6) and members of the reproduction-related NLRP family (nucleotide-binding oligomerization domain, leucine-rich repeat and pyrin domain-containing proteins), as well as oocyte genes with less well explored functions (*Oas1d, Oosp1, Omt2b*) (Additional file 7: Table S6). We also specifically examined the gene expression dynamics of candidate factors involved in de novo DNA methylation and associated epigenetic modifications, such as DNMT3A and DNMT3L, H3K4 demethylases of the KDM1 and KDM5 families, and the H3K36 methyltransferase SETD2. Although many of the corresponding genes appear to be stably expressed during oocyte growth, there was substantial up-regulation of *Kdm1b, Dnmt1* and particularly *Dnmt3L*, whose transcripts appear first in 10–30 μm oocytes (Additional file 3: Fig. S6). These transcript dynamics are consistent with the reported appearance of KDM1B and DNMT3L proteins during oocyte growth [21, 30].

## DNA methylation kinetics in relation to transcription events

Although the global results above do not support a major role for activation of specific transcription units

in the timing of de novo methylation, we performed several additional analyses to investigate in more detail possible relationships between transcription events and temporal control of methylation. We compared the methylation level of multiexonic reference genes and multiexonic novel genes, reasoning that the reference genes are generally expressed from earlier time points in oocyte growth (Fig. 4a). For this, we selected genes ≥4 kb in length (as shorter genes are unmethylated across much of their length) and set an expression threshold of ≥2 FPKM (to mitigate an effect of expression level). In this comparison, reference genes as a set have accumulated more methylation in 60–65 μm oocytes (Fig. 5a). Level of expression could still contribute to this effect, as novel genes are less highly expressed [6]: for the genes we included above 2 FPKM, median FPKM values were 11.4 and 3.9 for reference and novel genes, respectively. Indeed, there was a positive correlation between gene body methylation and expression level in 60–65 μm oocytes, particularly for reference genes, although the relationship plateaus for more highly methylated gene bodies (Fig. 5b). We also considered whether genes exceeding an expression threshold earlier during oocyte growth acquire methylation sooner, and this appeared to be the case (Fig. 5c). Again, however, it is difficult to separate out an effect of gene expression level, as genes crossing the threshold earlier are also more highly expressed in 60–65 μm oocytes (Fig. 5d). An effect on host gene expression was apparent for intragenic CGIs that gain methylation during oocyte growth, although the differences between groups were not significant (Fig. 5e). We also examined whether the extent of methylation of these CGIs in 60–65 μm oocytes reflected whether they were active TSSs at an earlier stage (E18.5 NGOs). Indeed, CGIs previously acting as TSSs had gained less methylation on average than non-TSS-CGIs (Fig. 5f). This analysis was performed with the PBAT data set, as RRBS data have limited coverage of gene bodies. When we compared DNA methylation of intragenic CGIs in 50–55 and 60–65 μm RRBS data sets with expression levels of overlapping genes in the corresponding RNA-seq datasets, we obtained similar results to the PBAT data (Additional file 3: Fig. S7).

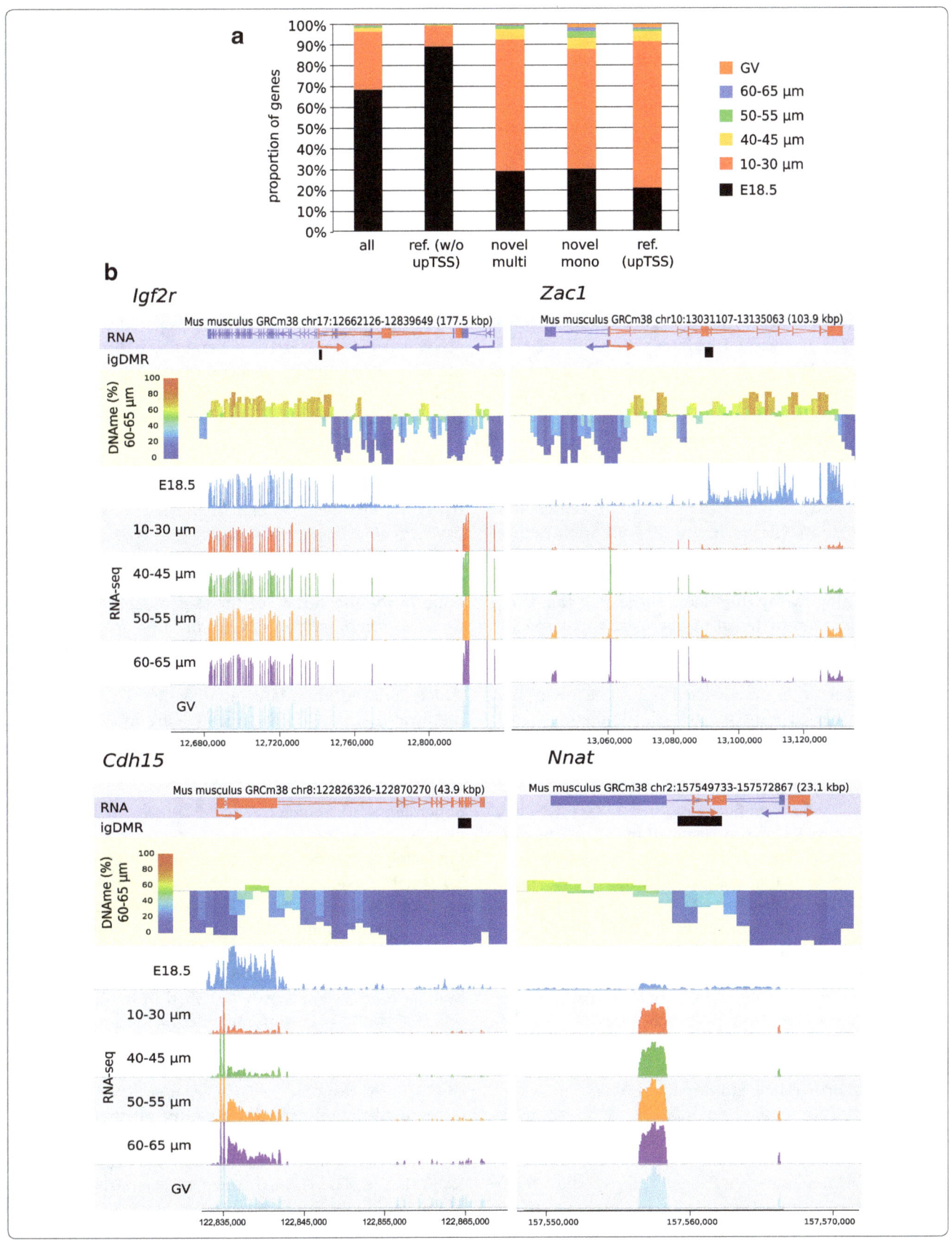

(See figure on previous page.)
**Fig. 4** Transcription dynamics in growing oocytes. **a** *Barchart* showing time of first detection of genes in growing oocytes, according to classification as reference gene, from canonical TSS (w/o upTSS) or novel upstream TSS (upTSS), or novel multi- or monoexonic gene. The total numbers of genes classified as expressed in each RNA-seq datasets are given in Additional file 6: Table S5. **b** Browser screenshots of representative early-methylating (*Igf2r, Zac1*) and late-methylating (*Cdh15, Nnat*) gDMRs in relation to RNA-seq data from different stages of the oocyte growth and DNA methylation acquired in 60–65 μm oocytes. In the RNA annotation track, *red gene* structures are transcribed from *left to right* and *blue gene* structures from *right to left*, with *arrows* showing the most upstream TSSs and direction of transcription. RNA-seq data show that transcriptional pattern is established prior to DNA methylation establishment

Changes in TSS use could reflect changes in binding of sequence-specific TFs at these sites, possibly as a consequence of down- or up-regulation of these factors during oocyte growth. In this context, it has previously been reported that the CGCGC consensus site of E2F1 and E2F2 is enriched in intragenic CGIs that are completely resistant to de novo methylation in oocytes [31]. Accordingly, we used the motif analysis package DREME [32] to identify motifs differentially enriched in CGIs with different levels of methylation in 60–65 μm oocytes. We searched for motifs enriched in late-methylated CGIs ($\leq$25%) compared to CGIs with 25–50, 50–75 and $\geq$75% methylation, as well as for motifs enriched in early-methylated CGIs ($\geq$75 and 50–75% methylation) compared to late-methylated CGIs (Fig. 6a). There were no motifs enriched in early-methylated CGIs compared to CGIs gaining methylation later, suggesting that there is no sequence motif targeting methylation to specific CGIs. On the other hand, we found motifs significantly enriched in late-methylated CGIs. Considering the comparison between $\geq$75% methylated and $\leq$25% methylated CGIs as likely to give the greatest discriminating signal, there were 21 sequence motifs with a significant difference in enrichment, three of which correspond to binding sites of known TFs (Fig. 6b; Additional file 8: Table S7). Of these, the most significant motif C(C/G/T)CCGCC (p value = $7.4 \times 10^{-13}$) was detected in 55% of the late-methylating CGIs but only 9.5% of the early-methylating CGIs. We repeated the analysis with the MEME motif analysis package [33] to search for longer motifs than DREME. Again, the significantly enriched motifs were found only in late-methylated CGIs compared to CGIs with methylation of 50–75 and $\geq$75%. Late-methylated CGIs appear to be enriched in G-rich motifs; however, these motifs are also present in 50% or more of the early-methylating CGIs (Additional file 3: Fig. S8).

## CGI methylation in relation to chromatin state
Since transcription does not appear to be an over-riding factor in the differential timing of CGI methylation, we examined the influence of specific histone post-translational modifications, given the likely importance of chromatin state in recruitment of the DNMT3A:DNMT3L complex. We divided CGIs that become fully methylated ($\geq$75%) in GV oocytes into levels of methylation attained in 60–65 μm oocytes and assessed the enrichment of histone modifications as determined by chromatin immunoprecipitation and sequencing (ChIP-seq) in NGOs (isolated at E18.5) and early growing oocytes (post-natal day p10) [20]. Of the modifications implicated in promoting or antagonising DNA methylation, levels of H3K36me3 showed a positive correlation with DNA methylation level; H3K4me2 and H3K4me3, conversely, were negatively correlated (Fig. 7a, all p values $<1 \times 10^{-10}$). We then looked whether there was a relationship with dependence on the H3K4me2 demethylases KDM1A and KDM1B. We have previously shown that loss of KDM1B, in particular, affects the methylation level acquired in MII oocytes of many CGIs, but there is a considerable variation in the magnitude of the dependency [20]. Therefore, we compared the change in DNA methylation of CGIs in oocytes deficient in KDM1A or KDM1B with level of methylation in wild-type, 60–65 μm oocytes, which showed that later-methylating CGIs (i.e. less methylation in 60–65 μm oocytes) are most dependent on KDM1A or KDM1B to become fully methylated in MII oocytes (Fig. 7b). Examples of early- and late-methylating CGIs in relation to H3K4me2 level and KDM1B dependence are shown in Fig. 7c.

## Modelling factors determining rate of CGI methylation
To test the extent to which the above variables, alone or in combination, account for the differential timing of CGI methylation in growing oocytes, we applied several regression models. We considered up to nine independent variables, including the three transcription factor binding motifs significantly enriched in the late-methylating CGIs (Table 1), with methylation level in 60–65 μm oocytes as response variable. As all the variables except GC content are in statistically significant linear relationship with the response variable, we first tested how much of the methylation variation could be attributed to each of the variables alone in simple linear regression models. H3K4me2 enrichment at p10 and dependence on KDM1B and KDM1A explained the highest proportion of the variability in the methylation data: 11.2, 10.5 and 9.7%, respectively.

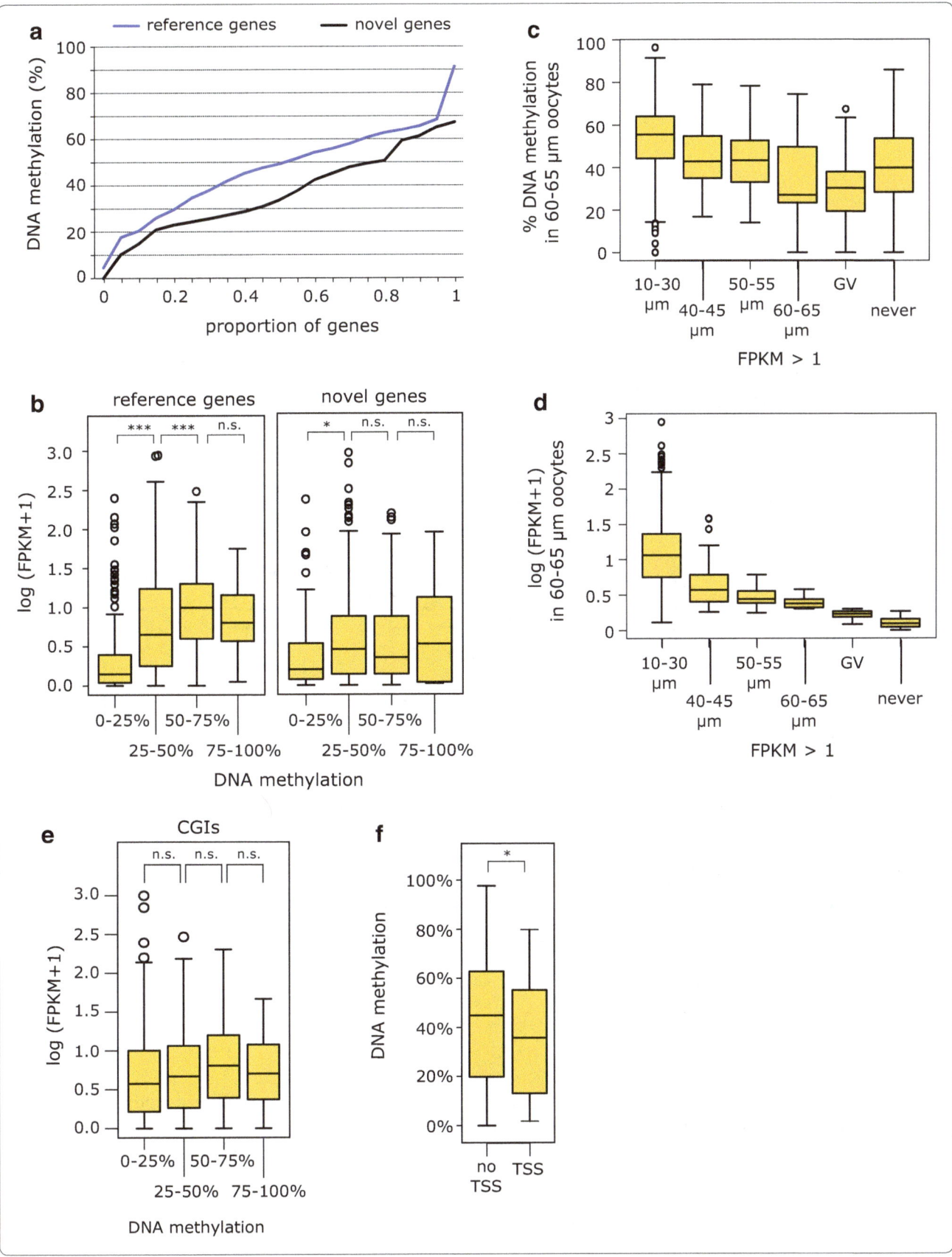

(See figure on previous page.)

**Fig. 5** Gene body and CpG island methylation kinetics in relation to transcription. **a** Cumulative distribution plot of methylation level of reference and novel genes ($\geq$4 kb in length and $\geq$2 FPKM) in 60–65 μm oocytes (PBAT dataset). The numbers of reference and novel genes satisfying the criteria for analysis were 105 and 32, respectively. **b** *Box whisker plots* of methylation of gene bodies of reference (1396) and novel (373) genes in relation to expression level in 60–65 μm oocytes. **c** *Box whisker plot* showing methylation level of CGIs in 60–65 μm oocytes grouped according to the stage in oocyte growth that expression of overlapping gene attained the threshold of >1 FPKM in the RNA-seq datasets. **d** *Box whisker plot* showing the corresponding data from expression level in 60–65 μm oocytes. The numbers of genes in (**c**) and (**d**) are: 1013 for 10–30 μm oocytes, 76 for 40–45 μm, 70 for 50–55 μm, 47 for 60–65 μm, 57 for GV and never 289. **e** Methylation level of intragenic CGIs (CGIs fully methylated in GV oocytes) in relation to expression level of the corresponding gene in 60–65 μm oocytes. The numbers of CGIs analysed in each methylation category (from lowest to highest) are: 269, 210, 281 and 50. **f** Methylation in 60–65 μm oocytes of CGIs (CGIs fully methylated in GV oocytes) according to prior activity as TSS as determined in e18.5 oocytes: 112 TSS-CGIs and 1229 non-TSS-CGIs. *Asterisks* denote *p* values of Student's t test: *0.01–0.001, **0.001–0.0001, ***<0.0001

**a**

| Analysis | No of motifs with DREME E-value $\leq$ 0.05 | Motifs with Tomtom Q-value $\leq$ |
|---|---|---|
| 0-25 vs 25-50 | 0 | 0 |
| 0-25 vs 50-75 | 15 | 4 |
| 0-25 vs 75-100 | 21 | 3 |
| 50-75 vs 0-25 | 0 | 0 |
| 75-100 vs 0-25 | 0 | 0 |

**b**

| Motif | Logo | P-value | E-value | Unerased E-value | Binding sites of known TFs |
|---|---|---|---|---|---|
| CBCCGCC |  | 7.4x10-13 | 3.3x10-8 | 2.1x10-8 | EGR1/2i Smad3i Bcl6bi SP1/2i CH4i Zfp410i Klf4/7/5i ERF1i Ascl2i SUT1i MIG2/3i UGA3i NHP10 |
| CCCMAM |  | 2.7x10-11 | 0.0000012 | 5.8x10-7 | ADR1i YPR022C |
| CBCCGGG |  | 8x10-9 | 0.00033 | 0.000097 | Zic1/2/3i SIP4 |

**Fig. 6** Motifs differentially enriched in early- and late-methylating CpG islands. **a** Summary of results of DREME analysis identifying motifs differentially enriched in CGIs that become fully methylated in GV oocytes grouped according to methylation level in 60–65 μm oocytes. Codes 0–25, 25–50, 50–75 and 75–100 represent CGIs with corresponding percentage methylation in 60–65 μm oocytes. The numbers of CGIs in each category are 470, 329, 384 and 63, respectively. **b** DREME motifs significantly enriched in CGIs methylated $\leq$25% in oocytes compared with $\geq$75% methylated CGIs that correspond to binding site motifs for known TFs. In motif sequence, $B = C/G/T$ and $M = C/A$. *P* value and *E* values are as defined by DREME and binding sites as identified by Tomtom

Because of the multicollinearity amongst independent variables (e.g. high correlation between transcription level and H3K36me3 enrichment, or between H3K4me2 and H3K4me3 enrichments), we could not test the combination of all variables in a classical multiple linear regression model. Instead, we applied linear modelling approaches correcting for multicollinearity—Ridge, Lasso and ElasticNet regressions—and looked for the best fit. Lasso and ElasticNet regression models using all nine variables explain 23.14% of the variability (Fig. 8). However, the cross-validation of models, where individual independent variables are added one by one to the model, in each step adding the variable that explains the highest proportion of the variability, revealed that

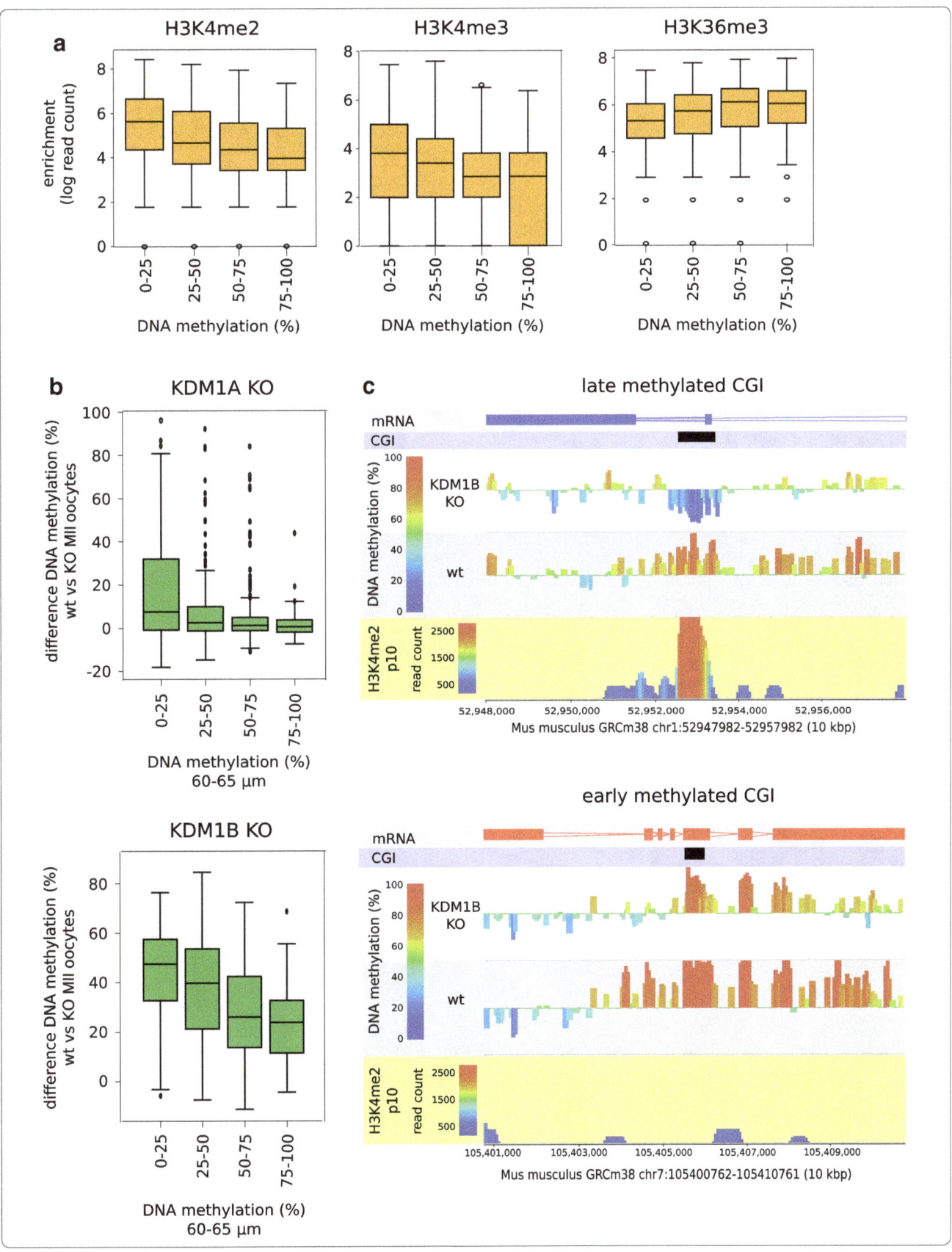

(See figure on previous page.)

**Fig. 7** CpG island methylation kinetics in relation to chromatin parameters. **a** *Box whisker plots* showing enrichment (log-transformed corrected read count) of H3K4me2, H3K4me3 and H3K36me3 at CGIs in relation to DNA methylation in 60–65 μm oocytes (PBAT data). The ChIP-seq data shown are from p10 oocytes; similar trends were observed in ChIP-seq data from e18.5 oocytes. Pearson's correlation coefficients are: −0.293 for H3K4me2, −0.173 for H3K4me3, 0.240 for H3K36me3. The numbers of CGIs analysed in each methylation category (from lowest to highest) were: 464, 327, 382 and 63. **b** *Box whisker plots* showing the degree of DNA methylation change at CGIs in *Kdm1a*- and *Kdm1b*-null MII oocytes in relation to methylation in 60–65 μm oocytes. Pearson's correlation coefficients are: −0.296 for *Kdm1a* and −0.357 for *Kdm1b*. The numbers of CGIs analysed in each methylation category (from lowest to highest) were: 244, 185, 255 and 28 for *Kdm1a*, and 270, 199, 268 and 31 for *Kdm1b*. **c** Browser screenshots of a representative early-methylating and late-methylating CGI (84.2 and 12.2% methylation in 60–65 μm oocytes, respectively) in relation to p10 H3K4me2 enrichment and DNA methylation attained in wild-type (WT) or *Kdm1b*-null MII oocytes

**Table 1 Linear regression models explaining DNA methylation level at CGIs in 60–65 μm oocytes**

| Variable | Simple linear regression | | Lasso regression |
| --- | --- | --- | --- |
| | Significance (*p* value) | % variability explained | Coefficient value in the most regularised model[b, c] |
| H3K4me3 enrichment, p10 ChIP-seq | $2.34 \times 10^{-6}$ | 4.8 | N/A |
| H3K4me2 enrichment, p10 ChIP-seq | $2.88 \times 10^{-13}$ | 11.2 | −0.132 |
| H3K36me3 enrichment, p10 ChIP-seq | $8.21 \times 10^{-8}$ | 6.2 | 0.050 |
| KDM1A dependence | $1.1 \times 10^{-11}$ | 9.7 | −0.132 |
| KDM1B dependence | $1.5 \times 10^{-12}$ | 10.5 | −0.106 |
| CpG density | 0.000767 | 2.5 | N/A |
| %GC content | 0.169199 | 0.9 | N/A |
| Transcription level (log-transformed) | 0.000297 | 2.9 | N/A |
| Enriched motif occurences (CBCCGCC, CCCMAM, CBCCGGG[a]) | $3.97 \times 10^{-8}$ | 6.5 | −0.019 |

Simple linear regressions (variables tested individually) and multiple linear regression (variables tested together) modelling the relationship between explanatory variables and DNA methylation level at CGIs in 60–65 μm oocytes. The outcome of the model is presented as a proportion of the variability in DNA methylation level at CGIs in 60–65 μm oocytes explained by the variables

[a] See Fig. 6a for motifs details. These three motifs were selected as they represent binding sites of known proteins

[b] Coefficients of variables in the model selected after software cross-validation of models as the most regularised model. These coefficients correspond to the values on y axis in Fig. 8. N/A marks variables that are not included in the model

[c] The Lasso regression model including the 5 variables indicated in the column accounts for 18.5% of the variation

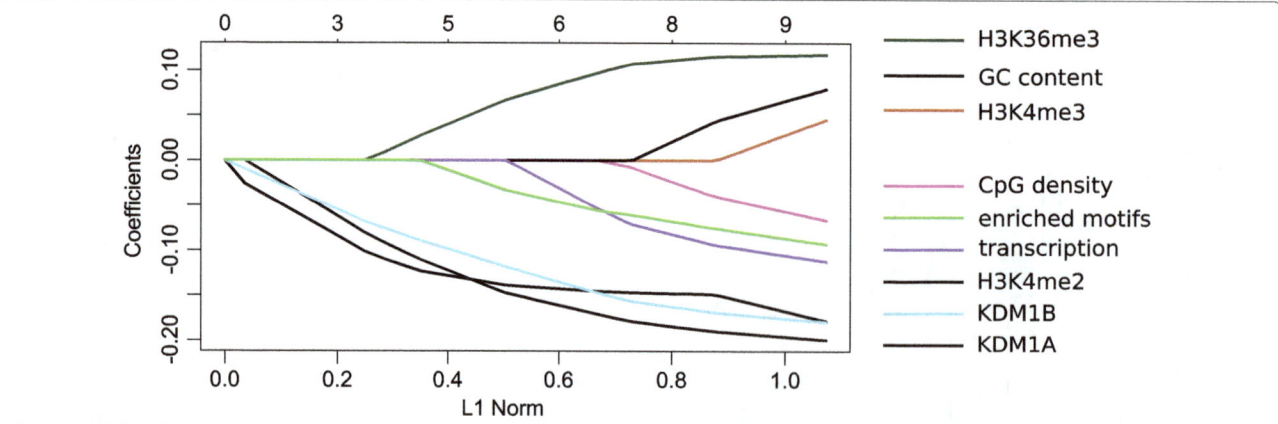

**Fig. 8** Modelling factors determining rate of CpG island methylation. Lasso regression model plot showing the effect of nine independent variables on variability of CGI methylation in 60–65 μm oocytes. *Each line* represents one of the variables. The earlier the line deviates from the *horizontal line* with coefficient 0.0, the more the corresponding variable contributes to the variability of the response variable, and the steeper the slope the greater the effect. If the steepness of the slope of one variable already in the model changes when a new variable comes into the model, it is a sign of correlation between two independent variables

H3K4me2 enrichment, KDM1A and KDM1B dependency, H3K36me3 enrichment and the presence of TF binding motifs are sufficient for the model, explaining 18.5% of the variability (Table 1). Although the remaining variables increase the explained proportion of methylation variability, they also increase the noise level and therefore do not statistically improve the model. We also tested other regression modelling approaches not requiring the linear relationships, such as polynomial regression; however, the fit of the models was not improved.

## Discussion

DNA methylation in the mouse oocyte depends upon DNMT3A and DNMT3L is primarily over gene bodies and largely determined by transcription, but these global dependencies could obscure sequence-specific requirements or the involvement of additional factors at specific elements. By capturing oocytes in the mid-phase of de novo methylation, we find that all sequence features gain CpG methylation at similar rates overall, including most classes of TEs, suggesting a universal rather than a feature-specific targeting mechanism. CGIs as a whole and a subset of L1 elements gain methylation later, however. In relation to CGIs, this relative delay might reflect that they are marked by default with histone modifications antagonistic to DNA methylation, such as H3K4me2/me3, and younger L1 elements may be suppressed by histone modifications inhibitory to DNA methylation. Amongst CGIs, however, there are reproducible differences in time of onset and/or progression in de novo methylation. This finding, at the genome-wide scale, substantially extends earlier studies on limited numbers of imprinted gDMRs [12–14] and suggests that CGIs destined for methylation initially exist in different states of permissiveness. There are a number of factors that could contribute to this asynchrony. Nuclear availability of DNMT3A [34] and DNMT3L is an absolute requirement, and DNMT3L is potently up-regulated during oocyte growth. A study in which DNMT3A2 and DNMT3L were precociously expressed in oocytes was not able to induce methylation of imprinted gDMRs in NGOs however, but did advance methylation of some gDMRs in growing oocytes [30], indicating that some loci are in a state permissive for methylation earlier than others.

Having established a major role for transcription in conferring the DNA methylation landscape of the oocyte, including at CGIs [5–7], we reasoned that timing of transcription events could influence timing of methylation. In fact, we did not find strong evidence to support this proposition. Despite substantial transcriptional changes during initial stages of oocyte growth, most changes occur well in advance of the onset of methylation, indicating that remodelling of the oocyte transcriptome is not a rate-limiting step in determining permissiveness of individual loci. We did find a positive correlation between expression and methylation however, so more highly expressed gene bodies on average gain methylation earlier than less highly expressed genes; this effect could be mediated through transcription-depending chromatin remodelling, including deposition of H3K36me3, whose levels over gene bodies scale with expression in oocytes [20]. A caveat to our analysis is that we used transcript abundance as measured by RNA-seq as a proxy for transcription, rather than directly determining active transcription events. This is because methods have not been developed to allow nascent transcription (such as by NET-seq) to be mapped in small numbers of cells. However, at the very least, the RNA-seq data allow us to determine the time that genes are first transcribed during oocyte growth.

To explain the difference in onset of methylation at CGIs, we considered the contributions of up to nine variables for which data were available. In combination, these variables explain about a fifth of the variation in timing of methylation establishment, with chromatin factors—H3K4me2 enrichment, KDM1A dependence and KDM1B dependence—having the greatest individual effects. There may be several reasons that we are not able to account for more of the variation at this time, apart from unknown factors not included in the modelling. One reason might be the relative imprecision in some of the data types; for example, low-cell ChIP-seq data for histone modifications in growing oocytes are inherently noisy, being at the limits of the capability of this method, and will have missing values at some CGIs. In comparison, PBAT data from *Kdm1a-* and *Kdm1b-*null MII oocytes are likely to be more precise. Therefore, it is reassuring that the magnitude of the individual effects of H3K4me2 enrichment and KDM1B dependence is so similar, since these are likely to be partially dependent variables given that we previously concluded that KDM1B is the major locus-specific H3K4me2 demethylase in oocytes [20]. It was previously suggested that KDM1B may be required to allow methylation of imprinted gDMRs that acquire methylation late in oocyte growth [21]; our genome-wide analysis and modelling partly support this earlier inference.

Gross sequence composition accounts for little of the variation in CGI methylation timing. Although CpG density is a determinant of H3K4me2 enrichment at CGIs, CGIs destined for methylation in oocytes are relatively depleted of H3K4me2 irrespective of CpG density [20]. Several sequence motifs, however, were differentially represented in early- and late-methylating CGIs. Individually, these motifs are not as discriminating as the ZFP57 binding site in imprinted gDMRs [35] that ensures retention of methylation after fertilisation, or the E2F1/E2F2 motif enriched in CGIs that escape DNA methylation in oocytes [31]. When combined, the three motifs for known TFs explain about half as much of the variation in methylation onset as do each of the chromatin factors. These motifs correspond to binding sites for 15 TFs expressed at varying levels in oocytes. Although some of their transcripts are down-regulated during oocyte growth (Additional file 9: Table S8), it is not possible at this stage to conclude whether the dynamics of any of these TFs underlies the differential methylation onset of the CGIs.

By capturing the progression of methylation, we also reveal other important aspects of de novo methylation in an in vivo setting, extending the significance of studies done in models such as ESCs. For example, we identify a co-operativity and nucleosomal pattern of DNMT3A action similar to that observed in ESCs [28]. Non-CpG methylation has been described as a property of oocytes as well as other non-dividing cells [11, 29, 36], but remains an enigmatic modification. Even in oocytes, in which methylation globally at CHG and CHH sites exceeds that at CG sites, few non-CpG sites are methylated (genome-wide average methylation of CHGs is 3.9%, and CHHs are 3.0% compared with 38.7% at CG sites as quantified with our parameters using published data [11]), with sites methylated mostly only to intermediate levels; moreover, CHH/CHG methylation is highly associated with domains of CpG methylation. Combined with its very much later onset, this suggests that CHG/CHH methylation is largely a by-product of sustained DNMT3A activity. Finally, DNA methylated domains not associated with transcribed regions are also late in acquiring methylation, suggesting that they require additional remodelling steps or a distinct mechanism of de novo methylation.

## Conclusions
The mammalian oocyte provides an important model to understand DNA methylation mechanisms, because an entire methylation landscape is established de novo in a non-dividing cell. Epigenetic remodelling events culminate in a distinctive DNA methylation landscape, including the programmed methylation of a defined set of CGIs, mostly associated with transcription units. Despite the simplicity of the methylation landscape, various sequence elements are not co-ordinately methylated, with pronounced asynchrony in methylation of CGIs. In this study, we generated methylation and transcriptome data sets to test whether timing of transcription events explained asynchrony of CGI methylation; however, our results do not support transcriptional transitions as a major factor in time of onset of methylation. By incorporating data on chromatin state, TF binding motifs and the effect of deficiency in H3K4 demethylases, we

could account for a substantial fraction of variation in CGI methylation timing, suggesting that sequences least dependent on chromatin remodelling are the earliest to become permissive for methylation.

## Methods
### Isolation and size selection of growing oocytes
Oocytes were collected from C57BL/6Babr mice. Ovaries were removed and digested for 30 min at 37 °C in 1× PBS containing 2 mg/ml collagenase (Sigma-Aldrich, C2674) and 0.025% trypsin (Sigma-Aldrich, 93615). M2 medium (Sigma-Aldrich, M7167) was added to dilute the digestion mix, and oocytes were picked up with a mouth-controlled drawn-out glass pipette. To eliminate contaminating somatic cells, oocytes were washed extensively in clean drops of M2 medium. A stage micrometre was used in combination with an eyepiece reticle to measure sizes of oocytes. Mice of post-natal days p5–7, p7–12, p9–14 and p13–16 were used to collect oocytes of 10–30, 40–45, 50–55 and 60–65 μm in diameter, respectively; GV oocytes were collected at p20.

### Generation of PBAT and RRBS libraries
RRBS libraries were generated, in duplicate, from ~450 to 550 oocytes per size-selected population, as previously described [7], but without the gel-extraction step. Briefly, DNA was spiked with a small amount of lambda DNA (0.05 pg per 6 ng genomic DNA) for bisulphite conversion control, digested with MspI (Thermo Fisher Scientific, ER0541), end-repaired (Klenow fragment exo-, Thermo Fisher Scientific, EP0421, with 10 nM dATP, 1 nM dCTP and 1 nM dGTP) and ligated with 5mC-adapters (Illumina) with T4 ligase (Thermo Fisher Scientific, EL0014). Bisulphite conversion was done using the EZ DNA Methylation-Direct Kit (Zymo Research, D5020), and DNA was amplified by 18 cycles of PCR using PfuTurbo Cx Hotstart DNA polymerase (Agilent, 600410). Libraries were purified using SPRI beads (Agencourt, A63880) and sequenced 40 bp single end on an Illumina Genome Analyzer IIx. The PBAT library was constructed from 200 60–65 μm oocytes as previously described [20] and sequenced 100 bp paired end on an Illumina HiSeq 1000.

### Generation of strand-specific RNA-seq libraries
Strand-specific RNA-seq libraries were generated as previously described [6] and sequenced 100 bp paired end on an Illumina HiSeq 1000. The numbers of oocytes used per library are listed in Additional file 5: Table S4.

### Conventional bisulphite sequencing
Bisulphite sequencing was performed essentially as previously described [29] using DNA from ~100 to 200 oocytes plus 50 ng lambda DNA spike-in for each

bisulphite conversion using the EZ DNA Methylation Kit (Zymo Research, D5001). Bisulphite-converted DNA from >30 oocytes was used for each PCR amplification; primers are listed in Additional file 10: Table S9. PCR products were cloned using pGEM-T Easy Vector Systems (Promega, A1360) and sequenced with the universal M13 primer. Experiments were done in duplicate for each size group and results combined.

## Mapping sequence reads

RRBS reads were trimmed to remove poor quality calls and adapters using Trim Galore v0.3.5 (parameters – rrbs) and mapped to the mouse genome GRCm38 assembly by Bismark [37] v0.14.0 (options –phred64-quals). For PBAT data, trimmed reads (Trim Galore v0.3.5 using default parameters) were first aligned to GRCm38 in paired-end mode to count overlapping parts of the reads only once while writing out unmapped singleton reads; in a second step remaining singleton reads were aligned in single-end mode. Alignments were carried out with Bismark v0.10.0 with the following parameters: –pbat for paired-end mode, –pbat for single-end mode for read 1 and default parameters for single-end mode read 2. Reads were then deduplicated with Bismark selecting a random alignment for positions covered more than once. CG, CHH and CHG methylation calls were extracted using the Bismark methylation extractor (v0.10.0) with the parameters: –no_overlap –report –ignore 4 –ignore_r2 4 for paired-end mode and –report –ignore 4 for the single-end mode. Published bisulphite-sequencing data were processed as described previously [6]. Raw RNA-seq reads were trimmed to remove poor quality calls and adapters using TrimGalore v.0.2.8 and mapped to GRCm38 using TopHat v.2.0.9 (option –g 1).

## Data analysis and modelling

We used reference and oocyte transcriptomes defined previously [6], including the definition of novel genes and novel TSSs of known genes, coordinates of CGIs, imprinted gDMRs, TEs and methylation domains. CGIs and TEs were used for methylation analyses if the minimum number of reads to count a position/minimum number of positions to count a probe were 5/5 for CGIs, 5/3 for all TEs and 3/3 if only methylated or unmethylated TEs were analysed. Otherwise, informative Cs refers to one covered by a minimum of 5 reads. Concordance of methylation of adjacent CpGs was quantified using custom Perl scripts, using CpGs with ≥5 reads. Expression of transcripts was quantified using Cufflinks v2.1.1 with –G option. Expression of genes was determined as a sum of FPKM values of all transcripts per gene. Expression of upstream TSSs and FPKM cut-off values to discriminate expressed and silent transcripts was defined as previously [6]. A gene was classified as expressed

if at least one of its isoforms was classified as expressed. A gene/TSS was classified as activated at a specific stage if it was classified as expressed in both replicates of that stage, as silent in both replicates in previous stages and as expressed in both replicates in subsequent stages. DNA methylation, RNA-seq and ChIP-seq data were analysed using SeqMonk v0.29.0–0.34.0. PC analysis and statistical analyses were performed in R v.3.0.2. Motif enrichment analysis was performed using DREME [32] and MEME [33] within the MEME suite v4.11.1 with default parameters specifying list of control sequences. Enriched motifs were directly submitted to Tomtom [38] within the MEME suite using default parameters. Regression modelling the relationship between CGI methylation in 60–65 μm oocytes and the variables listed in Table 1 was performed in R v.3.0.2 using CGIs with all information available. Function lm was used for linear regression, package glmnet for Lasso, Ridge and ElasiticNet regression, including cross-validation of models. Values of response and independent variables were normalised to mean 0 and standard deviation 1. GO analysis was performed using GOrilla [39] with specifying all genes expressed in the oocytes as a background.

## Additional files

**Additional file 1: Table S1.** Sequencing statistics for PBAT and RRBS libraries from size-selected oocytes.

**Additional file 2: Table S2.** Methylation levels of all CpGs and various genomic features that become methylated in GV oocytes in NGO, 60–65 μm and GV oocytes.

**Additional file 3: Figure S1.** Progression of DNA methylation at transposable elements (TEs) during oocyte growth, complements Fig. 1. **Figure S2:** Methylation parameters in 60–65 μm oocytes, including comparison of PBAT and RRBS data sets. **Figure S3.** PCA plots for oocyte mRNA-seq libraries. **Figure S4.** Differentially expressed genes between consecutive oocyte size populations. **Figure S5.** GO analysis of genes up-regulated >50-fold between e18.5 and GV oocytes. **Figure S6.** Graph showing expression levels over oocyte growth of transcripts for Dnmts, Kdm1s and Kdm5s, and SetD2. **Figure S7.** Methylation level of intragenic CGIs in RRBS datasets in relation to expression level of the corresponding gene. **Figure S8.** MEME output of search for motifs enriched in late-methylated CGIs.

**Additional file 4: Table S3.** CG, CHG and CHH methylation levels in NGO, 60–65 μm and GV oocytes.

**Additional file 5: Table S4.** Sequencing statistics for ssRNA-seq libraries from NGO, size-selected and GV oocytes.

**Additional file 6: Table S5.** Numbers of genes detected in ssRNA-seq libraries.

**Additional file 7: Table S6.** Genes up-regulated ≥50 from e18.5 to 10–30 μm oocytes and from E18.5 to GV oocytes.

**Additional file 8: Table S7.** Output from DREME analysis of sequence motifs differentially represented in early- and late-methylating CpG islands.

**Additional file 9: Table S8.** Expression levels of transcription factors with binding motifs enriched in CGIs with low methylation level in 60–65 μm oocytes.

**Additional file 10: Table S9.** PCR primers sequences for conventional bisulphite sequencing of selected CGIs.

## Abbreviations

*Cdh15*: cadherin 15 gene; CGIs: CpG islands; ChIP-seq: chromatin immunoprecipitation with high-throughput sequencing; DNMT3A/B/L: DNA methyltransferase 3A, 3B or 3L; DREME: discriminative regular expression motif elicitation; E18.5: embryonic day 18.5; E2F1/2: E2F transcription factor 1 or 2; ERVK: endogenous retrovirus, K family; ESC: embryonic stem cell; FDR: false discovery rate; FPKM: fragments per kilobase per million mapped reads; gDMR: germline differentially methylated region; GO: gene ontology; GV: germinal vesicle; H3K4: histone 3 lysine residue 4; H3K4me2/me3: di- or trimethylated H3K4; H3K36me3: trimethylated H3K36; *Igf2r*: insulin-like growth factor 2 receptor; KDM1A/1B: lysine demethylase 1A or 1B; KDM5: lysine demethylase 5 family; LINE: long interspersed nuclear element; LTR: long-terminal repeat; MII: metaphase II; MEME: multiple expectation maximisation (EM) for motif elicitation; NET-seq: native elongating transcript sequencing; NGO: non-growing oocytes; *Oas1d*: 2',5'-oligoadenylate synthetase 1-like D, gene; OBOX1/2/5: oocyte-specific homeobox 1, 2 or 5; *Omt2b*: oocyte maturation, beta, gene; OOEP: oocyte-expressed protein; *Oosp1*: oocyte-specific protein 1, gene; P10: post-natal day 10; PC: principal component; PBAT: post-bisulphite adapter tagging; RRBS: reduced representation bisulphite sequencing; RNA-seq: RNA sequencing; SEBOX: skin–embryo–brain–oocyte-specific homeobox; SETD2: SET domain-containing 2; SINE: short interspersed nuclear element; TE: transposable element; TF: transcription factor; TLE6: transducing-like enhancer of split; TSS: transcription start site; *Zac1*: zinc finger protein regulating apoptosis and cell cycle arrest, gene; ZFP57: zinc finger protein 57; ZP1, 2, 3: zona pellucida glycoprotein 1, 2 or 3.

## Authors' contributions

LG collected material and performed RNA-seq, transcriptome assembly and much of the data analysis; ST collected material, performed RRBS and contributed to data analysis; SAS performed PBAT and contributed to data analysis; KRS provided material and analysis; JK and TC provided material; HS and SRA performed bioinformatic analyses; GK, ST and SAS initiated and supervised the study; and LG and GK wrote the manuscript. All authors read and approved the final manuscript.

## Author details

[1] Epigenetics Programme, Babraham Institute, Cambridge CB22 3AT, UK. [2] Department of Histology and Cell Biology, School of Medicine, Yokohama City University, Yokohama 236-0004, Japan. [3] Department of Epigenetics and Molecular Carcinogenesis, The University of Texas M.D. Anderson Cancer Center, Smithville, TX 77030, USA. [4] Bioinformatics Group, Babraham Institute, Cambridge CB22 3AT, UK. [5] Centre for Trophoblast Research, University of Cambridge, Cambridge CB2 3EG, UK. [6] Present Address: Laboratory of Developmental Biology and Genetics, Department of Molecular Biology, University of South Bohemia, 37005 Ceske Budejovice, Czech Republic. [7] Present Address: Friedrich Miescher Institute for Biomedical Research, 4058 Basel, Switzerland. [8] Present Address: Biotech Research and Innovation Centre (BRIC), University of Copenhagen, 2200 Copenhagen, Denmark. [9] Present Address: Computer Science Department, KASIT, University of Jordan, Amman, Jordan.

## Acknowledgements

We are grateful to thank Kristina Tabbada for library sequencing, Felix Krueger for additional bioinformatic support and the Babraham Biological Support Unit for mouse husbandry.

## Competing interests

The authors declare that they have no competing interests.

issued by the Home Office (UK) in accordance with the Animals (Scientific Procedures) Act 1986.

## Funding

Work in G.K.'s laboratory was supported by the UK Biotechnology and Biological Sciences Research Council and Medical Research Council (Grants G0800013 and MR/K011332/1) and the Babraham Institute and University of Cambridge graduate student scholarships (to L.G. and K.R. S.-M.). Work in T.C.'s laboratory was supported by a Rising Star Award (R1108) from the Cancer Prevention and Research Institute of Texas (CPRIT).

## References

1. Bourc'his D, Xu G-L, Lin C-S, Bollman B, Bestor TH. Dnmt3L and the establishment of maternal genomic imprints. Science. 2001;294(5551):2536–9.
2. Branco MR, King M, Perez-Garcia V, Bogutz AB, Caley M, Fineberg E, Lefebvre L, Cook SJ, Dean W, Hemberger M, Reik W. Maternal DNA methylation regulates early trophoblast development. Dev Cell. 2016;36(2):152–63.
3. Kaneda M, Okano M, Hata K, Sado T, Tsujimoto N, Li E, Sasaki H. Essential role for de novo DNA methyltransferase Dnmt3a in paternal and maternal imprinting. Nature. 2004;429(6994):900–3.
4. Seisenberger S, Andrews S, Krueger F, Arand J, Walter J, Santos F, Popp C, Thienpont B, Dean W, Reik W. The dynamics of genome-wide DNA methylation reprogramming in mouse primordial germ cells. Mol Cell. 2012;48(6):849–62.
5. Kobayashi H, Sakurai T, Imai M, Takahashi N, Fukuda A, Yayoi O, Sato S, Nakabayashi K, Hata K, Sotomaru Y, Suzuki Y, Kono T. Contribution of intragenic DNA methylation in mouse gametic DNA methylomes to establish oocyte-specific heritable marks. PLoS Genet. 2012;8(1):e1002440.
6. Veselovska L, Smallwood SA, Saadeh H, Stewart KR, Krueger F, Maupetit-Méhouas S, Arnaud P, Tomizawa S, Andrews S, Kelsey G. Deep sequencing and de novo assembly of the mouse oocyte transcriptome define the contribution of transcription to the DNA methylation landscape. Genome Biol. 2015;16:209.
7. Smallwood SA, Tomizawa S, Krueger F, Ruf N, Carli N, Segonds-Pichon A, Sato S, Hata K, Andrews SR, Kelsey G. Dynamic CpG island methylation landscape in oocytes and preimplantation embryos. Nat Genet. 2011;43(8):811–4.
8. Chotalia M, Smallwood SA, Ruf N, Dawson C, Lucifero D, Frontera M, James K, Dean W, Kelsey G. Transcription is required for establishment of germline methylation marks at imprinted genes. Genes Dev. 2009;23(1):105–17.
9. Smith EY, Futtner CR, Chamberlain SJ, Johnstone KA, Resnick JL. Transcription is required to establish maternal imprinting at the Prader–Willi syndrome and Angelman syndrome locus. PLoS Genet. 2011;7(12):e1002422.
10. Stewart KR, Veselovska L, Kelsey G. Establishment and functions of DNA methylation in the germline. Epigenomics. 2016;8(10):1399–413.
11. Shirane K, Toh H, Kobayashi H, Miura F, Chiba H, Ito T, Kono T, Sasaki H. Mouse oocyte methylomes at base resolution reveal genome-wide accumulation of non-CpG methylation and role of DNA methyltransferases. PLoS Genet. 2013;9(4):e1003439.
12. Denomme MM, White CR, Gillio-Meina C, MacDonald WA, Deroo BJ, Kidder GM, Mann MRW. Compromised fertility disrupts Peg1 but not Snrpn and Peg3 imprinted methylation acquisition in mouse oocytes. Front Genet. 2012;3:129.
13. Hiura H, Obata Y, Komiyama J, Shirai M, Kono T. Oocyte growth-dependent progression of maternal imprinting in mice. Genes Cells. 2006;11(4):353–61.
14. Lucifero D, Mann MRW, Bartolomei MS, Trasler JM. Gene-specific timing and epigenetic memory in oocyte imprinting. Hum Mol Genet. 2004;13(8):839–49.
15. Obata Y, Kono T. Maternal primary imprinting is established at a specific time for each gene throughout oocyte growth. J Biol Chem. 2002;277(7):5285–9.

16. Pan H, O'Brien MJ, Wigglesworth K, Eppig JJ, Schultz RM. Transcript profiling during mouse oocyte development and the effect of gonadotropin priming and development in vitro. Dev Biol. 2005;286(2):493–506.

17. Dhayalan A, Rajavelu A, Rathert P, Tamas R, Jurkowska RZ, Ragozin S, Jeltsch A. The Dnmt3a PWWP domain reads histone 3 lysine 36 trimethylation and guides DNA methylation. J Biol Chem. 2010;285(34):26114–20.

18. Ooi SK, Qiu C, Bernstein E, Li K, Jia D, Yang Z, Erdjument-Bromage H, Tempst P, Lin SP, Allis CD, Cheng X, Bestor TH. DNMT3L connects unmethylated lysine 4 of histone H3 to de novo methylation of DNA. Nature. 2007;448(7154):714–7.

19. Zhang Y, Jurkowska R, Soeroes S, Rajavelu A, Dhayalan A, Bock I, Rathert P, Brandt O, Reinhardt R, Fischle W, Jeltsch A. Chromatin methylation activity of Dnmt3a and Dnmt3a/3L is guided by interaction of the ADD domain with the histone H3 tail. Nucleic Acids Res. 2010;38(13):4246–53.

20. Stewart KR, Veselovska L, Kim J, Huang J, Saadeh H, Tomizawa S, Smallwood SA, Chen T, Kelsey G. Dynamic changes in histone modifications precede de novo DNA methylation in oocytes. Genes Dev. 2015;29(23):2449–62.

21. Ciccone DN, Su H, Hevi S, Gay F, Lei H, Bajko J, Xu G-F, Li E, Chen T. KDM1B is a histone H3K4 demethylase required to establish maternal genomic imprints. Nature. 2009;461(7262):415–8.

22. Edmunds JW, Mahadevan LC, Clayton AL. Dynamic histone H3 methylation during gene induction: HYPB/Setd2 mediates all H3K36 trimethylation. EMBO J. 2008;27(2):406–20.

23. Kizer KO, Phatnani HP, Shibata Y, Hall H, Greenleaf AL, Strahl BD. A novel domain in Set2 mediates RNA polymerase II interaction and couples histone H3 K36 methylation with transcript elongation. Mol Cell Biol. 2005;25(8):3305–16.

24. Yoh SM, Lucas JS, Jones KA. The Iws1:Spt6:CTD complex controls cotranscriptional mRNA biosynthesis and HYPB/Setd2-mediated histone H3K36 methylation. Genes Dev. 2008;22(24):3422–34.

25. Miura F, Enomoto Y, Dairiki R, Ito T. Amplification-free whole-genome bisulfite sequencing by post-bisulfite adaptor tagging. Nucleic Acids Res. 2012;40(17):e136.

26. Sookdeo A, Hepp CM, McClure MA, Boissinot S. Revisiting the evolution of mouse LINE-1 in the genomic era. Mob DNA. 2013;4:3.

27. Bulut-Karslioglu A, De La Rosa-Velazquez IA, Ramirez F, Barenboim M, Onishi-Seebacher M, Arand J, Galan C, Winter GE, Engist B, Gerle B, O'Sullivan RJ, Martens JHA, Walter J, Manke T, Lachner M, Jenuwein T. Suv39 h-dependent H3K9me3 marks intact retrotransposons and silences LINE elements in mouse embryonic stem cells. Mol Cell. 2014;55(2):277–90.

28. Baubec T, Colombo DF, Wirbelauer C, Schmidt J, Burger L, Krebs AR, Akalin A, Schubeler D. Genomic profiling of DNA methyltransferases reveals a role for DNMT3B in genic methylation. Nature. 2015;520(7546):243–7.

29. Tomizawa S, Kobayashi H, Watanabe T, Andrews S, Hata K, Kelsey G, Sasaki H. Dynamic stage-specific changes in imprinted differentially methylated regions during early mammalian development and prevalence of non-CpG methylation in oocytes. Development. 2011;138(5):811–20.

30. Hara S, Takano T, Fujikawa T, Yamada M, Wakai T, Kono T, Obata Y. Forced expression of DNA methyltransferases during oocyte growth accelerates the establishment of methylation imprints but not functional genomic imprinting. Hum Mol Genet. 2014;23(14):3853–64.

31. Saadeh H, Schulz R. Protection of CpG islands against de novo DNA methylation during oogenesis is associated with the recognition site of E2f1 and E2f2. Epigenetics Chromatin. 2014;7:26.

32. Bailey TL. DREME: motif discovery in transcription factor ChIP-seq data. Bioinformatics. 2011;27(12):1653–9.

33. Bailey TL, Williams D, Misleh C, Li WW. MEME: discovering and analyzing DNA and protein sequence motifs. Nucleic Acids Res. 2006;34(suppl. 2):W369–73.

34. Ma P, de Waal E, Weaver JR, Bartolomei MS, Schultz RM. A DNMT3A2-HDAC2 complex is essential for genomic imprinting and genome integrity in mouse oocytes. Cell Rep. 2015;13(8):1552–60.

35. Quenneville S, Verde G, Corsinotti A, Kapopoulou A, Jakobsson J, Offner S, Baglivo I, Pedone PV, Grimaldi G, Riccio A, Trono D. In embryonic stem cells, ZFP57/KAP1 recognize a methylated hexanucleotide to affect chromatin and DNA methylation of imprinting control regions. Mol Cell. 2011;44(3):361–72.

36. Lister R, Pelizzola M, Dowen RH, Hawkins RD, Hon G, Tonti-Filippini J, Nery JR, Lee L, Ye Z, Ngo Q-M, Edsall L, Antosiewicz-Bourget J, Stewart R, Ruotti V, Millar AH, Thomson JA, Ren B, Ecker JR. Human DNA methylomes at base resolution show widespread epigenomic differences. Nature. 2009;462(7271):315–22.

37. Krueger F, Andrews SR. Bismark: a flexible aligner and methylation caller for bisulfite-seq applications. Bioinformatics. 2011;27(11):1571–2.

38. Gupta S, Stamatoyannopoulos JA, Bailey TL, Noble WS. Quantifying similarity between motifs. Genome Biol. 2007;8(2):R24.

39. Eden E, Navon R, Steinfeld I, Lipson D, Yakhini Z. GOrilla: a tool for discovery and visualization of enriched GO terms in ranked gene lists. BMC Bioinformatics. 2009;10:48.

# 5

# MeDIP-seq and nCpG analyses illuminate sexually dimorphic methylation of gonadal development genes with high historic methylation in turtle hatchlings with temperature-dependent sex determination

Srihari Radhakrishnan[1,4], Robert Literman[2,4], Beatriz Mizoguchi[3,4] and Nicole Valenzuela[4*] (iD)

**Abstract**

**Background:** DNA methylation alters gene expression but not DNA sequence and mediates some cases of phenotypic plasticity. Temperature-dependent sex determination (TSD) epitomizes phenotypic plasticity where environmental temperature drives embryonic sexual fate, as occurs commonly in turtles. Importantly, the temperature-specific transcription of two genes underlying gonadal differentiation is known to be induced by differential methylation in TSD fish, turtle and alligator. Yet, how extensive is the link between DNA methylation and TSD remains unclear. Here we test for broad differences in genome-wide DNA methylation between male and female hatchling gonads of the TSD painted turtle *Chrysemys picta* using methyl DNA immunoprecipitation sequencing, to identify differentially methylated candidates for future study. We also examine the genome-wide nCpG distribution (which affects DNA methylation) in painted turtles and test for historic methylation in genes regulating vertebrate gonadogenesis.

**Results:** Turtle global methylation was consistent with other vertebrates (57% of the genome, 78% of all CpG dinucleotides). Numerous genes predicted to regulate turtle gonadogenesis exhibited sex-specific methylation and were proximal to methylated repeats. nCpG distribution predicted actual turtle DNA methylation and was bimodal in gene promoters (as other vertebrates) and introns (unlike other vertebrates). Differentially methylated genes, including regulators of sexual development, had lower nCpG content indicative of higher historic methylation.

**Conclusions:** Ours is the first evidence suggesting that sexually dimorphic DNA methylation is pervasive in turtle gonads (perhaps mediated by repeat methylation) and that it targets numerous regulators of gonadal development, consistent with the hypothesis that it may regulate thermosensitive transcription in TSD vertebrates. However, further research during embryogenesis will help test this hypothesis and the alternative that instead, most differential methylation observed in hatchlings is the by-product of sexual differentiation and not its cause.

**Keywords:** Sex-specific thermosensitive DNA methylation, Genome-wide normalized CpG content, MeDIP sequencing, Temperature-dependent and genotypic sex determination, Turtle gonadal embryonic development, Ecological genomics, Epigenetic modification, Phenotypic plasticity, Sexual development, Reptile vertebrate

*Correspondence: nvalenzu@iastate.edu
[4] Department of Ecology, Evolution and Organismal Biology, Iowa State University, 251 Bessey Hall, Ames, IA 50011, USA
Full list of author information is available at the end of the article

## Background

Epigenetic modifications are heritable changes to the DNA that do not change the nucleotide sequence. Among them, DNA methylation is a biochemical process that adds methyl groups to cytosine or adenine nucleotides. Methylated DNA alters gene expression by preventing transcription factor binding [1] or by sometimes favoring the binding of repressors [2, 3]. The regulatory role of DNA methylation is widespread across eukaryotes where it mediates development, environmental responses and disease [4–7]. In animals, the addition of methyl groups occurs on CpG dinucleotides (cytosine linked to a guanine by a phosphate group) within genes in invertebrates [8] and across genic and intergenic regions in vertebrates [9]. Importantly, changes in DNA methylation levels have been linked to the regulation of phenotypic plasticity [10–12]. Temperature-dependent sex determination (TSD) represents a textbook example of phenotypic plasticity (a thermal polyphenism), where individuals with identical genotypes can develop alternative phenotypes (male or female) based on environmental cues [13, 14]. Differential methylation of two genes in the sex-determining pathway has been experimentally identified in a few TSD vertebrates (a fish, a turtle and alligator) [15–17] and in other genes during temperature-induced sex reversal in a fish with a mixed sex-determining system (ZZ/ZW GSD and TSD) [18]. However, the extent to which TSD plasticity is mediated by DNA methylation in turtles remains unclear. As an initial step to address this question, it is necessary to characterize the level of methylation in the genome of TSD turtles and to test whether methylation patterns differ in males and females as would be expected if temperature induces sex-specific methylation profiles [18]. Additionally, if TSD is mediated by DNA methylation of the regulatory network of gonadal development, it would be expected that genes in this network would be the target of differential methylation. And if differential methylation of this network has occurred over evolutionary time, these elements should display a signature left by historic methylation at the DNA sequence level [19, 20].

In silico techniques have been used to estimate historic DNA methylation patterns in animals by measuring the normalized CpG content (or nCpG), i.e., the ratio of the CpG dinucleotide abundance observed at particular genomic regions compared to that expected at random based on the frequency of cytosines and guanines present in the genome [CpG(O/E) = CpG observed/expected] [21]. This value of nCpG is used as a proxy for DNA methylation since (a) DNA methylation is almost entirely targeted to CpG dinucleotides in animals [22], and (b) 5-methylcytosine has the tendency to undergo spontaneous deamination which converts it to thymine leaving a footprint that reflects historic methylation levels [19, 20, 23, 24]. Therefore, nCpG is inversely correlated with the extent of DNA methylation such that in hypermethylated regions (where cytosines within methylated CpGs have been converted to thymine) the nCpG is less than one. On the other hand, an nCpG ratio equal to 1 is indicative of no deviation from random expectation, while a value greater than one indicates hypomethylated regions.

The genome-wide patterns of animal DNA methylation vary within and between invertebrates and vertebrates. For instance, in multiple ant species the genomic distribution of nCpG values is unimodal and centered around 1 [25], whereas the nCpG distribution in honeybee and pea aphid genic regions is bimodal with one peak centered at 0.5 and the other at 1 [26]. Among vertebrates, the promoter regions of eutherian mammals, opossum, chicken, lizard, frog and fish also show a bimodal pattern, such that genes with lower nCpG content (LCG promoters) undergo higher methylation and are linked to tissue-specific expression, and those with higher nCpG content (HCG promoters) are hypomethylated and are linked to broad patterns of gene expression [27, 28]. In contrast, the nCpG content of promoters in platypus (and the urochordate sea squirt) is unimodal [27, 29].

Other studies across various vertebrates reported a genome-wide average CpG ratio of ~25% for mammals and birds and of ~35% for fish and amphibians using alternative approaches such as restriction enzyme assays and HPLC [19, 24, 30]. Because of earlier difficulties to study nCpG distribution in reptilian DNA sequences [31], data at this level are scarce for reptiles. Indeed, pioneering work on reptilian methylation levels focused more heavily on global levels [24, 31], and to our knowledge, only one study examined nCpG content and DNA methylation [27] and another study examined non-methylated islands [32] in anole lizards. Thus, the pattern of genome-wide nCpG distribution in turtles and in TSD vertebrates remains unknown at the DNA sequence level. This is critical given that nCpG not only reflects historic DNA methylation, but is a factor that mediates current DNA methylation levels in ways that influence transcription levels.

Here, we test for broad differences in genome-wide DNA methylation between male and female hatchling gonads of the TSD painted turtle, *Chrysemys picta*, to identify methylated regions of interest that may be important for phenotypically plastic sexual development, which would represent candidates for further analyses in future studies. A similar approach was used to investigate the association of DNA methylation and TSD in a ZZ/ZW + TSD fish by examining DNA methylation in mature individuals [18]. We concentrate our search on genes known to regulate vertebrate primary sexual development [33], as well as genes that by their nature may help mediate TSD sex-specific development, such as genes of the epigenetic machinery [34], hormonal pathways [35] or general sensing responses [36]. Second, we examine the genome-wide nCpG distribution in the painted turtle genome [37, 38] and compare it to that of other vertebrates. We then test whether this index predicts the DNA methylation landscape in turtle gonads and also test for a signature of historic methylation on gene regulators of vertebrate sexual development. Thus, our study provides the first insight into the association between nCpG content and differential methylation in any TSD vertebrate.

## Results

### Genome-wide CpG distribution is bimodal in turtle promoters, introns, exons and intergenic regions

In all genomic regions, the nCpG was much lower than the expected ratio of 1, predicting that a significant fraction of the painted turtle genome is methylated. Notably, for the exons the overall distribution of nCpG was centered at 0.35, whereas the distributions for the rest of the profiled regions were centered at 0.25. Importantly, although at first glance the profiles of genome-wide normalized CpG content (nCpG) appeared unimodal, statistical analyses fitting mixture models to the data [39] revealed higher support for bimodal distributions across gene bodies (exons plus introns), exons and introns individually, and the intergenic regions ($p$ value of likelihood

ratio test <0.00001) (Fig. 1a–d). Interestingly, bimodality was even more obvious in the nCpG profiles of promoter regions measured as 3000, 600, 300 or 150 bases upstream of exon 1 ($p$ value of likelihood ratio test <0.00001) (Fig. 1e–h).

### A substantial portion of the turtle genome is methylated

We used MeDIP-seq to characterize the DNA methylation landscape broadly by profiling differentially methylated regions rather than individual cytosines [40] in two pools of gonads from five male hatchlings each and two pools from five female hatchlings each (Table 1). Over 98% of the MeDIP-seq reads from the male and female hatchling gonads mapped to the *C. picta* genome [37]. The methylome analysis uncovered ~2.95 million methylated 500-bp windows, totaling 1.48 gigabases in size, or ~57% of the assembled genome [37], and overlapping

**Table 1 Illumina library statistics for MeDIP sequencing of *Chrysemys picta* hatchling gonads**

| Sex (incubation temp) | Library size | % Mapped reads |
|---|---|---|
| Male (26 °C) Lib1 | 137,159,464 | 98.85 |
| Male (26 °C) Lib2 | 138,462,674 | 98.80 |
| Female (31 °C) Lib1 | 126,682,102 | 98.83 |
| Female (31 °C) Lib2 | 163,323,313 | 98.87 |

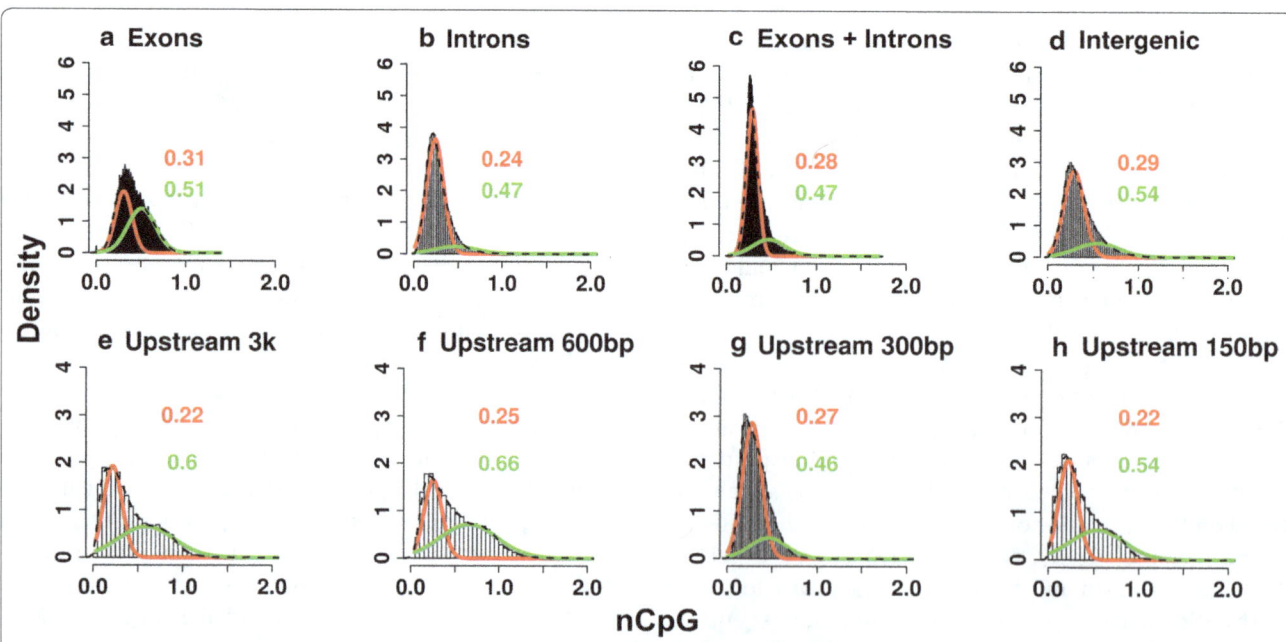

**Fig. 1** Distribution of normalized CpG (nCpG) content in the *Chrysemys picta* genome by region. **a** Exons only (CDS), **b** introns only, **c** exons and introns, **d** intergenic regions and **e–h** at 3000, 600, 300 and 150 bases upstream of exon 1. Fitted Gaussian density curves for the bimodal distribution along with their respective peak values are indicated in *red* and *green*

with 17,646 genes. This corresponds to 78% of the CpG nucleotides in the genome. A total of 40% of the methylated windows fall within gene bodies which is significantly less than the 46% located within 50-kb-upstream sequences that have potential regulatory functions (permutation test $p$ value =0.001). The remaining 14% of methylated windows fall in intergenic regions outside of gene bodies or outside sequences 50 kb upstream of all genes.

### Sexually dimorphic DNA methylation varies by gene region

MeDIP-seq results from the biological replicates revealed strong differences between males and females above and beyond the differences between individual replicates (Fig. 2). Positive (methylated DNA) and negative (unmethylated DNA) controls were used during the MeDIP step and ensure that the MeDIP worked properly,

such that the observed variation between the two replicates of the same sex likely reflects natural population-level variation in methylation among individuals. This was observed also using PCR of DNA of multiple males and females digested or not with a restriction enzyme sensitive to DNA methylation (Fig. 3). Our MeDIP-seq analysis revealed 5647 differentially methylated windows between the sexes. Of these, 3076 windows were hypermethylated in females (in 2414 genes) and 2571 windows in males (in 2086 genes) (Fig. 2a). The log-fold change in methylation between the sexes was highest in introns, followed by promoter sequences and finally exons (Fig. 2c). Additionally, differential methylation of exons was around half that of promoters (there were 536 differentially methylated windows in promoter sequences vs. 281 in exons). Finally, 541 differentially methylated genes contained multiple windows with contrasting

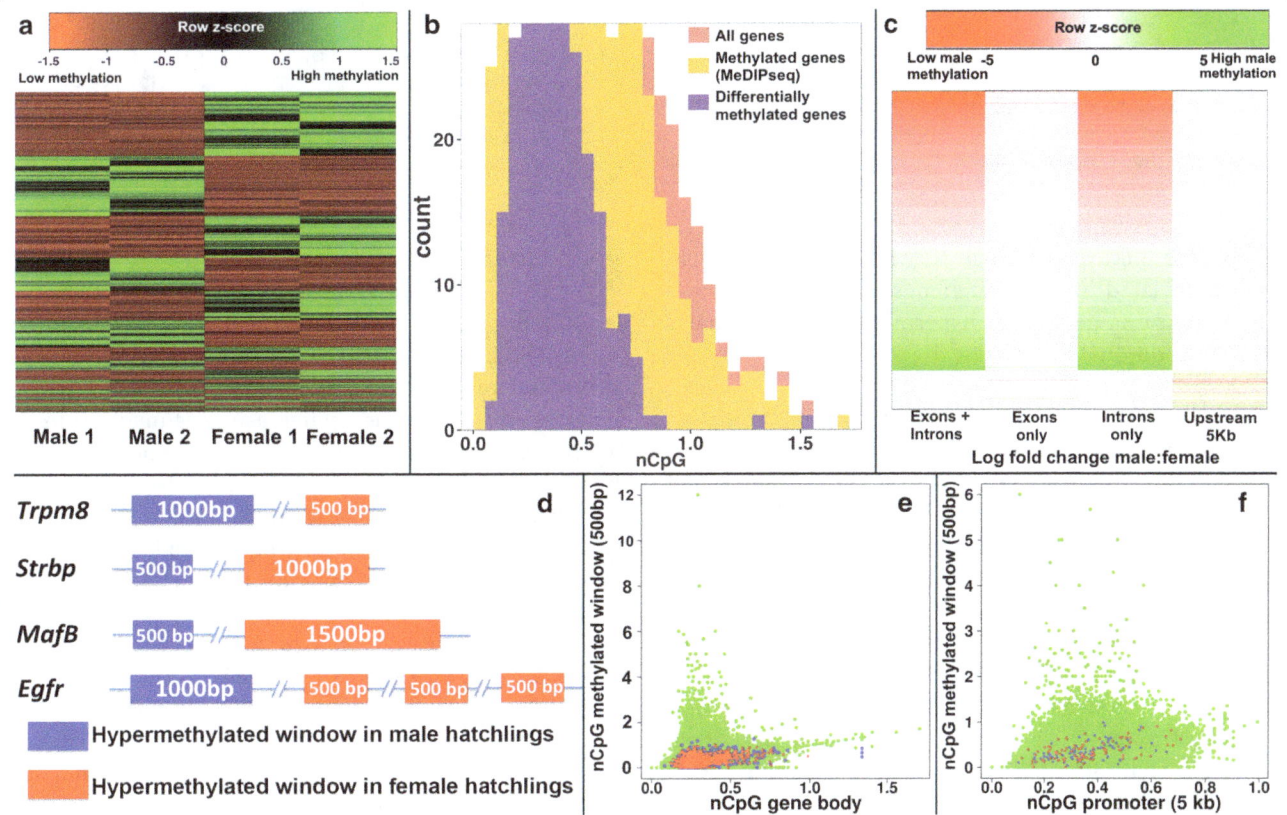

**Fig. 2** **a** RPKM heatmap of differentially methylated genes in *Chrysemys picta* (*rows*) clustered by mean methylation level per gene. Methylation levels were scaled to [−1.5, 1.5] to indicate genes undergoing high (*green*) and low (*red*) relative methylation. **b** Normalized CpG content of all annotated genes (*red*), experimentally verified to be methylated using MeDIP-seq (*yellow*) and differentially methylated (*purple*). **c** Fold change in methylation (*red*: hypermethylated in female; *green* hypermethylated in male) as seen in gene bodies (exons + introns), exons only, introns only and promoters. **d** Examples of genes possessing multiple windows that displayed sex-specific methylation. **e**, **f** Scatterplot of normalized CpG content (nCpG) in methylated windows occurring (**e**) in gene bodies relative to nCpG of gene bodies and **f** in promoters relative to nCpG of the complete promoter sequence (~5 kb upstream). Differentially methylated windows in hatchlings are overlaid, with those hypermethylated in males indicated in *blue*, and those hypermethylated in females indicated in *red*

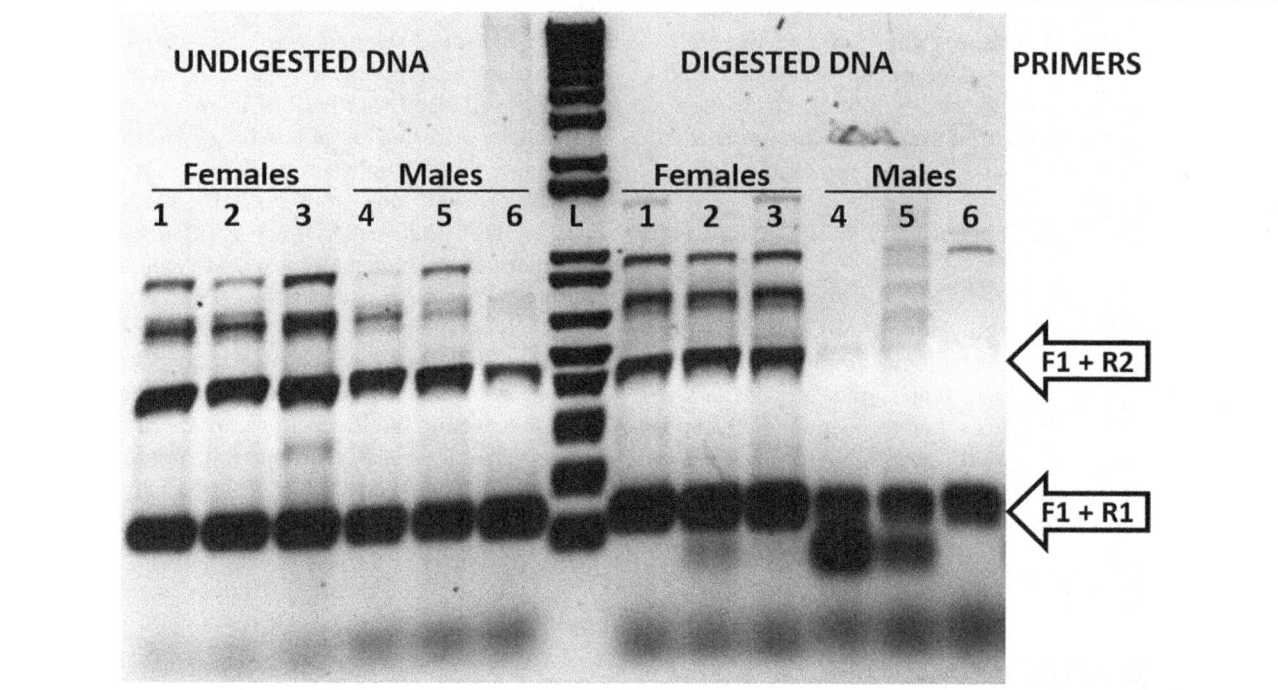

**Fig. 3** Validation of sexually dimorphic DNA methylation of the *fezf2* gene in *Chrysemys picta* hatchling gonads by methylation-sensitive restriction enzyme PCR. DNA from three females (1, 2 and 3) and three males (4, 5 and 6) was digested or not with HpaII (HpaII cuts unmethylated DNA). Expected size of the PCR amplicons from PCR primers F1 plus R1, or F1 and R2 are indicated by the *arrows* (non-specific PCR products were also obtained). L = DNA ladder (1 kb plus). Amplification of the expected fragment in females and not males using the F1 + R2 primers confirms the hypermethylation of *fezf2* in females detected using the MeDIP-seq data

sex-specific methylation, such that some windows were hypermethylated in male hatchlings and other windows within the same gene were hypermethylated in females (Fig. 2d; Additional file 1: Table S1).

### Potential mediators of thermal transduction and regulators of sexual development are differentially methylated

While no particular GO terms were significantly enriched in the MeDIP-seq data after controlling for false discovery (Additional file 1: Tables S2, Additional file 1: Table S3), we detected some interesting pathways and genes with differential methylation. We focused our attention on reptilian homologs of genes that govern mammalian gonadogenesis [33, 41] and on genes involved in painted turtle gonadal development [42] whose biological functions may help convert the temperature signal into the sexual fate of TSD embryos (Additional file 1: Table S4, Additional file 1: Table S5, Additional file 1: Table S6, Additional file 1: Table S7, Additional file 1: Table S8, Additional file 1: Table S9, Additional file 1: Table S10, Additional file 1: Table S11). Full list of differentially methylated genes and GO pathways are presented in Additional file 1: Table S12, Additional file 1: Table S13, Additional file 1: Table S14). Among these were a number of kinases, androgen-/estrogen-related

genes, histone- and ubiquitin-related genes, heat shock and transient potential receptor genes that displayed sexually dimorphic methylation. Interestingly, members of the *Wnt* signaling pathway and genes involved in transcriptional regulation tended to be hypermethylated in males relative to females, whereas genes involved in cell and neuron differentiation tended to be hypermethylated in female hatchlings. Importantly, 13 out of 53 reptilian homologs of genes involved in mammalian gonadogenesis [33, 41] were differentially methylated. For instance, genes that regulate testicular formation, some of which are highly expressed during embryogenesis at male-producing temperature in TSD turtles, including *Amh*, *Ar*, *Gata4*, *Lhx1*, *Lhx9* and *Sf1* [42–46], were significantly hypermethylated in female hatchlings. In contrast, genes important in ovarian formation in mammals, such as *Wnt4* and *Emx2* [47, 48], were hypermethylated in males. Differential methylation also varied by genic region, perhaps reflecting differences in the type of regulation that DNA methylation might exert in several genes in the sexual development network. For instance, while some of these genes exhibited hypermethylation in the promoter regions near the 5′ end, others were hypermethylated in their gene bodies, mostly in their intronic sequences. Among them, *Wt1* exhibited three hypermethylated intronic windows in male hatchlings. Yet other genes, such as *Lhx1* and *Gata4*, show female hypermethylation in both promoter (1 window each) and intronic sequences (two and three windows, respectively).

## CpG content is a good in silico predictor of DNA methylation

There is no significant difference between the number of genes predicted to be methylated using the nCpG index and the number of genes identified as methylated by MeDIP-seq (two-sample Kolmogorov–Smirnov test, $p = 0.1931$), suggesting that in silico predictions from the nCpG index are fairly accurate. Specifically, 94% of all methylated gene bodies detected experimentally via MeDIP-seq had an in silico nCpG value of 0.5 or lower, thereby showing a strong association between CpG depletion and actual methylation. Further, 98.7% (16,884 genes) of the 17,104 genes with nCpG $\leq$ 0.5 are methylated. Only 75 out of 17,646 methylated genes show an nCpG $\geq$ 0.8, including 36 out of 90 methylated tRNA genes.

## Differentially methylated genes suffered greater historic methylation

Interestingly, the nCpG content of the differentially methylated genes was significantly lower (mean = 0.282, range = 0–1.27) than for all methylated genes (mean = 0.306, range = 0–12) identified by MeDIP-seq (resampling test $p$ value =0.001) (Fig. 2b), indicating that genes displaying sexually dimorphic methylation show a signature of higher historic DNA methylation. This pattern was observed in gene bodies (comprising of exons and introns) as well as promoters (Fig. 2e, f), suggesting that differential methylation of hatchling gonads was restricted to the genomic regions of greater CpG depletion.

## DNA methylation appears associated with expression of some genes but not all

In the absence of hatchling transcriptomes, we leveraged gonadal transcriptomes of painted turtle embryos incubated at male-producing temperature (MPT) and female-producing temperature (FPT) that we obtained in another study [42], and uncovered potential candidates for future study when we compared the methylome signatures in hatchlings and differential transcription patterns in late-stage embryos (stage 22). Namely, some genes overexpressed in male embryos were hypermethylated in female hatchlings (58 out of 394), and some genes overexpressed in female embryos were hypermethylated in male hatchlings (40 out of 754) (Additional file 1: Table S15). Thus, indirect evidence was detected suggesting that DNA methylation of numerous genes might mediate

their sexually dimorphic transcription and that this influence may not be global but a gene-by-gene effect where some genes may be affected while others are not. A direct test also revealed the lack of such global effect, as the Fisher exact test on the contingency table of genes methylated or not that were differentially expressed or not was not significant ($p > 0.9999$). However, we note that while intriguing, the finding of gene-by-gene effects should be taken with caution given the difference in life stage between the transcriptomic and methylomic data.

## Repeat elements abound nearby methylated genes and share identical pattern of differential methylation

Because repetitive DNA sequences such as transposable elements can be subject to silencing by DNA methylation [49] which could affect nearby genes, we analyzed the repeat content of the methylome. RepeatMasker analyses revealed that around 40% of the methylome consists of repeats (which represent 9.25% of the CPI genome—Additional file 1: Table S16), with significant representation from the CR1 and HAT repeat categories (~45% of the methylome repeats). CR1 repeats were also the most abundant in the *C. picta* gonadal transcriptome [42] (Fig. 4a; Additional file 1: Table S16). Furthermore, the relative abundance among repeat categories as a fraction of the genome (Additional file 1: Table S16) did not differ significantly between the genome, methylome and transcriptome (pairwise Kruskal–Wallis tests; all $p$ values >0.453, Fig. 4a), even though the absolute abundance of each category and of all repeats combined varied (e.g., 25, 9 and 1% of the genome, methylome and transcriptome consisted of repeats, respectively—Additional file 1: Table S16). Additionally, methylated repeats were significantly more concentrated around 95% of all methylated genes (resampling test $p$ value =0.001) and relatively scarce around non-methylated genes (Table 2). Of these, HAT repeats were the most abundant. Methylated repeats were also common (although significantly less so) in the vicinity of differentially methylated genes (~80% instead of 95%; permutation test $p$ value =0.001). Of these, DIRS repeats were the most common. Interestingly, the direction of the sex-specific methylation was identical for 70% of the differentially methylated genes and their neighboring repeats (Table 2).

Importantly, significantly more genes that were differentially methylated in hatchlings and differentially transcribed in stage 22 embryos were nearby methylated repeats than genes equally expressed in male and females (permutation test $p$ value =0.001). The distribution of categories of methylated repeats did not differ between the differentially and non-differentially expressed genes. Using regression, we evaluated the effect of DNA methylation on repeat silencing, by assessing the repeat

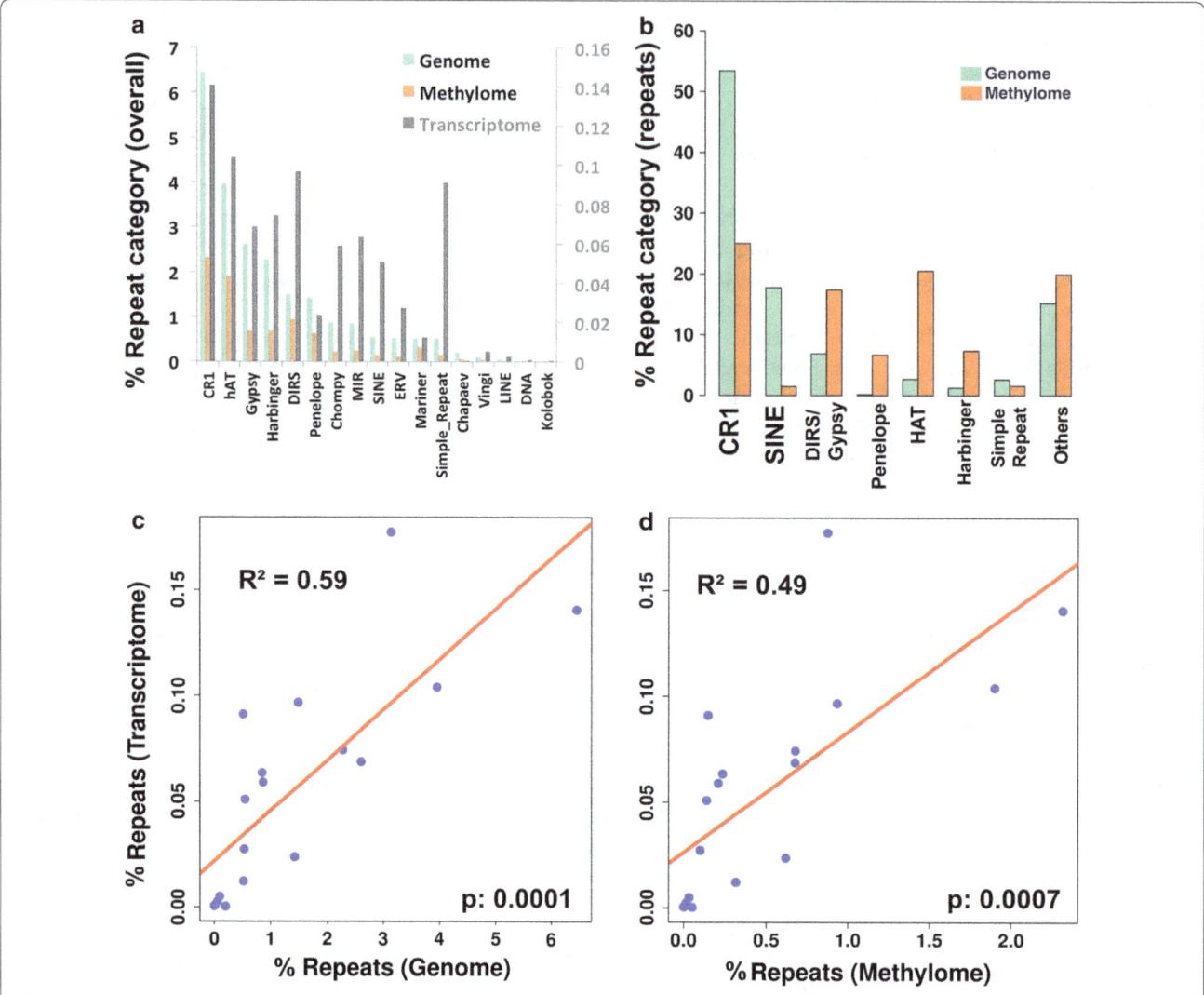

**Fig. 4** *Chrysemys picta* repeat abundance. **a** Relative abundance of repeat categories in the *Chrysemys picta* genome [37], and their relative abundance in the hatchling gonadal methylome (this study) and embryonic gonadal transcriptome [42] as a fraction of the genome. Note that repeat abundance in the transcriptome is plotted in *gray* and scaled by the right-hand axis for visualization purposes. **b** Relative abundance of various repeat categories within the fraction of repeats present in the *C. picta* genome [37] versus the methylome (this study). **c** Regression of transcriptomic repeat abundance as a function of repeat abundance in the genome ($p = 0.0001$) and **d** in the methylome ($p = 0.0007$). Abundance of repeats in the transcriptome is slightly better explained by their genomic abundance ($R^2 = 0.59$) than by their abundance in the methylome ($R^2 = 0.49$)

transcription level as a function of repeat abundance in the genome and as a function of repeat abundance in the methylome. In both cases, the relationship was highly significant and explained a significant proportion of variation in repeat transcription level (repeat abundance in genome: slope = 0.0239, $p = 0.0001$, $r^2 = 0.59$; repeat abundance in methylome: slope = 0.0569, $p = 0.0007$, $r^2 = 0.49$), although the variation in repeat expression itself was small (Fig. 4c, d). Further, multiple regression analyses with all variables combined did not improve the explanatory power significantly, implying that repeat

transcription, repeat methylation status and repeat genomic abundance are tightly linked.

## Discussion

Genomic approaches are advancing our understanding of phenotypic plasticity at unprecedented rates, including the role that DNA methylation plays in mediating plastic responses to environmental inputs [10, 11, 21, 50]. Here we tested whether regulators of vertebrate sexual development (and of other sensing and epigenetic responses) are subjected to differential methylation in male and

**Table 2** Overabundance of methylated repeats upstream of the transcription starting site of differentially and non-differentially methylated genes compared to non-methylated genes in *Chrysemys picta* hatchlings

| Number (#) and percentage (%) of methylated repeats | Number of gene bodies with methylated repeats at three distances of start codon | | |
|---|---|---|---|
| | 1 kb | 5 kb | 10 kb |
| Among all 17,646 methylated genes | 16,791 (95.1%) | 17,030 (96.5%) | 17,202 (97.5%) |
| Among all 433 non-methylated genes | 29 (6.7%) | 90 (20.8%) | 161 (37%) |
| Among the 2086 male-hypermethylated genes | 1650 (79%) | 1656 (79.3%) | 1662 (79.7%) |
| Among the 2414 female-hypermethylated genes | 1949 (80.7%) | 1959 (81.1%) | 1961 (81.2%) |
| Among the 840 methylated genes of interest (Additional file 1: Table S1, Additional file 1: Table S2, Additional file 1: Table S3, Additional file 1: Table S4, Additional file 1: Table S5, Additional file 1: Table S6, Additional file 1: Table S7, Additional file 1: Table S8) | 822 (97.8%) | 828 (98.6%) | 831 (98.9%) |
| Male-hypermethylated repeat windows located near male-hypermethylated genes | 681/946 (72%) | 706/973 (72.6%) | 718/989 (72.6%) |
| Female-hypermethylated repeat windows located near female-hypermethylated genes | 801/1032 (77.6%) | 816/1050 (77.7%) | 825/1064 (77.5%) |

Repeats were overrepresented upstream of all methylated regions examined (Chi-square test $P < 0.00001$ in all cases) but not upstream of non-methylated genes

female hatchlings of a turtle with phenotypically plastic sex determination (*C. picta*) lacking sex chromosomes [51]. Based on these data, we identified candidate genes that may mediate TSD epigenetically. We also characterized the genome-wide nCpG distribution in the painted turtle genome, the first such analysis in reptiles and TSD vertebrates, and found that this proxy predicts reasonably well the methylation levels estimated using MeDIP-seq. Further, nCpG profiles helped assess historic methylation levels of genes regulating vertebrate sexual development. Below we highlight our most important observations and propose working hypotheses to guide further research. Our sex-specific methylomes represent an important genomic resource to aid investigations of the epigenetic regulation of environmental responses.

Although bisulfite sequencing is the gold standard for methylation studies requiring single-nucleotide resolution, MeDIP-seq is appropriate for studies seeking to profile broader patterns of DNA methylation rather than individual cytosines [40], as was our goal. MeDIP-seq provides methylation profiles at a resolution of 150–200 bp, signals tend to concentrate on CpG-rich genomic regions with high methylation levels, and the methylome information can be comprehensive [52], less sequence-biased and fairly concordant with bisulfite-sequencing data for genomes similar in size to those of turtles [37, 52, 53]. The reduction in genomic complexity afforded by MeDIP-seq is also advantageous for taxa with large genomes as is the case of the painted turtle [37, 38]. It should be noted that methylation levels are correlated with 1 kb of neighboring CpG sites [54]. Nonetheless, future bisulfite sequencing will permit evaluating methylation patterns at higher resolution, particularly in non-CpG genomic regions, which was precluded in our study. But importantly, given that methylation status may be more stable at the level of DNA domains rather than at

single nucleotides [55], MeDIP-seq snapshots do contain highly valuable information as discussed below.

## Bimodal distribution of nCpG turtle genomes matches most vertebrate promoters but not introns

Our data revealed that nCpG values follow a bimodal distribution at the promoter regions of genes in painted turtles (Fig. 1), consistent with most major vertebrates lineages studied ([56], Table 3), but not with platypus [29] or tunicates [27]. However, some interesting differences exist among bimodal patterns between turtles and other vertebrates. For instance, the CpG bimodality is less pronounced in turtle promoters than in human and chicken (Fig. 1 and [27]). Namely, the low- and high-CpG modes of the turtle promoter distribution overlap more extensively and are centered around lower CpG values than those of human and chicken promoters (Fig. 1 and [27]), suggesting a potentially larger role for DNA methylation as transcriptional regulator in reptiles than previously anticipated (i.e., greater historic methylation in turtle than in human and chicken). In human, high-CpG promoters are more abundant than low-CpG promoters, whereas the opposite is true in turtles (Fig. 1 and [27]). Human high-CpG promoters are hypermethylated less frequently whereas low-CpG promoters tend to be hypermethylated more often [28, 29], whereas differentially methylated genes in turtles had lower CpG content (Fig. 2b). The full implications of the observed methylation patterns in turtles are unknown. For instance, human promoters are methylated in somatic as well as germline cells and thus could be heritable [55]. Further, the CpG observed/expected ratio [CpG(O/E)] affects gene expression levels and breath in humans [57]. But further research is needed to test if the same is true in turtles. Unlike promoters, the bimodal nCpG content of

**Table 3** **Summary of exemplar studies exploring the diversity of nCpG distributions in vertebrate genomes.** *Sources:* 1 = Weber et al. [28]; 2 = Elango et al. [27]; 3 = Yang et al. [29]

| Group | Species | Region profiled | nCpG distribution | Source |
|---|---|---|---|---|
| Mammals | Homo sapiens | Promoters | Bimodal | 1, 2, 3 |
| | | Introns | Unimodal | 2 |
| | Pan troglodytes | Promoters | Bimodal | 3 |
| | Gorilla gorilla | Promoters | Bimodal | 3 |
| | Pongo abelii | Promoters | Bimodal | 3 |
| | Macaca mulatta | Promoters | Bimodal | 3 |
| | Monodelphis domestica | Promoters | Bimodal | 3 |
| | Mus musculus | Promoters | Bimodal | 3 |
| | Ornithorhynchus anatinus | Promoters | Unimodal | 3 |
| Birds | Gallus gallus | Promoters | Bimodal | 2, 3 |
| | | Introns | Unimodal | 2 |
| Reptiles | Chrysemys picta | Promoters | Bimodal | This study |
| | | Introns | Bimodal | This study |
| | | Exons | Bimodal | This study |
| | | Intergenic | Bimodal | This study |
| Amphibians | Xenopus tropicalis | Promoters | Bimodal | 2 |
| | | Introns | Unimodal | 2 |
| Fish | Danio rerio | Promoters | Bimodal | 2 |
| | | Introns | Unimodal | 2 |
| Tunicates | Ciona intestinalis | Promoters | Unimodal | 2 |
| | | Introns | Bimodal | 2 |

turtle introns (Fig. 1) differs from other vertebrates, as it is unimodal in fish, amphibian, lizard, bird and human [27], and more strikingly bimodal in tunicates [27]. Subtle bimodality can pass undetected during qualitative evaluations (e.g., platypus [29]), unless it is explicitly tested statistically (this study and [27]). Notably, recent evidence suggests bimodal gene body methylation exists in mammals, birds and fish [58].

The overall turtle CpG depletion agrees generally with vertebrates (where it is correlated with high DNA methylation) compared to invertebrates [19, 59, 60]. The overall distribution of nCpG values in the *C. picta* genome, disregarding bimodality, is higher for exons than for other regions (introns, promoters and intergenic sequences), suggesting lower exon methylation. In contrast, human exons are more highly methylated than introns [61]. CpG depletion in exons could affect transcription as in humans where it downregulates genes [62, 63]. Thus, we hypothesize that the higher nCpG content of *C. picta*'s exons relative to other regions (i.e., lower exon CpG depletion relative to overall depletion) might be the result of natural selection to preserve gene expression given that nCpG content is much lower in turtles than in other vertebrates, and perhaps also to prevent the accumulation of CpG to TpG mutations [19] that could produce non-functional proteins [64].

### nCpG content is a reasonable predictor of DNA methylation, except for tRNAs, and matches other vertebrates

Our results link DNA methylation at the resolution measured by MeDIP-seq and CpG depletion, consistent with humans [28] and insects [26]. Indeed, 94% of all the methylated genes revealed by MeDIP-seq had an nCpG $\leq$ 0.5, and 98.7% of all genes with nCpG $\leq$ 0.5 were actually methylated. Thus, CpG content is a reasonable indicator of methylation status, except for tRNAs. Namely, 90 of 182 annotated tRNAs were methylated, 40% of which (36/90) have an nCpG $\geq$ 0.8, indicating that CpG depletion has been suppressed historically in many tRNAs, perhaps to prevent CpG to TpG mutations [19]. This agrees with the high conservation of tRNA sequences in all domains of life [65]. Comparable methylation levels were found here (57% of the genome, including 78% of all CpG dinucleotides) as in fish, salamander, snake, birds and mammals using various approaches [24, 27, 31, 59, 66].

### Turtle methylation patterns suggest *cis*-regulation via DNA methylation

Sequences up to 50 kb upstream of genes in painted turtle, which should have potential regulatory roles, were methylated at significantly higher levels (46%) than gene bodies (40%) in hatchling gonads. Of these upstream windows, 16% fall within the boundaries of another gene upstream of the focal gene (within an intron in 94% of those cases), suggesting that perhaps some methylated genes may serve as alternative upstream promoters for downstream genes, as in humans [67, 68]. Importantly, DNA methylation in promoters is linked to gene silencing, whereas methylation in gene bodies may modulate transcription [8, 69], but further research is needed to test these hypotheses directly in turtles.

## DNA methylation varies by sex, gene and gene region

Results from our biological duplicates, which combined five males or five females each, were fairly consistent (Fig. 2), strengthening the inferences of sex-specific methylation. However, differences were also detected within sex, likely revealing genetic variation among individuals associated with DNA methylation (Fig. 3), but without obscuring some evident sex-specific methylation patterns.

Interestingly, 3% (541) of all methylated genes contained multiple differentially methylated windows, some hypermethylated in females while others in the same gene were hypermethylated in males (Fig. 2d). Introns showed the highest log-fold change in methylation between the sexes, followed by promoters and finally exons (Fig. 2c), but it is unclear whether turtle intronic methylation modulates transcription rather than silencing genes [8, 69], or whether it is linked to the transcription of antisense RNA, as observed in genes important for urogenital development in other vertebrates [70].

Because the males and females studied here were obtained at contrasting temperatures that produce a single sex (only males at 26 °C and only females at 31 °C), perhaps the sex-specific methylation in hatchlings is also temperature specific. Thermosensitive methylation of distinct windows within the same gene could potentially lead to alternative splicing of sex-specific transcripts, as in humans [71]. Alternatively, the observed differential methylation could be a consequence of sexual differentiation. Further studies are needed to test these alternative hypotheses.

## Known regulators of sexual development are differentially methylated in TSD turtles

We investigated 53 genes in the mammalian urogenital regulatory network, whose reptilian homologs are differentially expressed in TSD turtles, including *Wt1* [72, 73], *Sf1* [74, 75], *Dax1* [41, 76], *Sox9* [76–79], *Aromatase* [17, 80], *Dmrt1* [76, 78, 81], *Esr1* [82, 83] and *Rspo1* [79]. Of these, methylation has been studied only in *aromatase* and *Sox9* and demonstrated to influence transcription

in developing TSD reptiles (slider turtle [17], American alligator [84]) and a TSD fish [15]. Our results provide evidence (1) that many more genes in this regulatory network are differentially methylated between male and female gonads, (2) that sexually dimorphic methylation persists post-hatching and (3) that it affects some genes that are differentially expressed during gonadal development.

Genes such as *Amh*, *Ar*, *Gata4*, *Lhx9* and *Sf1* that govern testicular formation are upregulated at male-producing temperature (MPT) late in the thermosensitive period of *C. picta* [42, 43] and displayed hypermethylated promoters in female hatchlings (Table 4). The exception was *Sf1*, whose introns were differentially methylated (Table 4). Additionally, three hypermethylated intronic windows in male hatchlings were observed in *Wt1*, a gene important for the formation of the bipotential gonad and later testicular development [85] that is upregulated at MPT early in the thermosensitive period but not by stage 22, and is a candidate TSD master gene [43, 72, 85]. The consequences of turtle intronic methylation remain unknown for alternative splicing, the use of alternative promoters, transcription of antisense RNA or transcription modulation [8, 64, 69, 70]. Data for *Amh*, *Ar*, *Gata4* and *Lhx9* suggest that upregulation in one sex may result from silencing by promoter methylation in the opposite sex, which was also true for *Wnt4* and *Emx2* that govern ovarian formation [47, 48] and were hypermethylated in male hatchlings. These types of regulation do not appear to be ubiquitous as some other genes in the gonadal development network were differentially expressed but showed no differential methylation. Interestingly, *Lhx1* and *Gata4* displayed female hypermethylation in promoters (1 window each) and introns (2 and 3 windows, respectively) suggesting a potentially more complex regulation by methylation of these elements (Table 4).

Our findings suggest that methylation marks may be stable and persist post-hatching, perhaps longer-term as in mammals [86, 87]. Yet notably, *aromatase* showed no differential methylation in hatchlings, counter to observations in slider turtle embryos [17]. Assuming that embryonic methylation in slider and painted turtles is similar as is *aromatase* transcription [43, 75], our finding in *C. picta* hatchlings suggests that embryonic *aromatase* methylation may be transient, or alternatively, that it passed undetected by unknown technical limitations. Further research will help elucidate between these hypotheses directly.

A contrast of our turtle methylomes with those from the ZZ/ZW + TSD tongue sole fish [18] revealed the overlap of 22 genes (Additional file 1: Table S17) out of 56 genes differentially transcribed or methylated in tongue sole gonads (Additional file 1: Table S8 in [18]). Fish

**Table 4 Summary of differentially methylated (FDR cut-off: 0.05) genes in hatchlings that are putatively involved in reptilian gonadogenesis**

| Gene | Sex of hatchlings showing hyper-methylation | Hypermethylated region | Sex of stage 22 embryos showing upregulation |
|------|------|------|------|
| Amh | Female | P, I | Male |
| Ar | Female | P, D | Male |
| Gata4 | Female | P, I (3) | Male |
| Lhx9 | Female | P | Male |
| Sf1 | Female | I | Male |
| Lhx1 | Female | P, I (2) | – |
| Emx2 | Male | E | – |
| Insr | Male | I | – |
| Wnt4 | Male | I | – |
| Apc | Male | I | – |
| Igf1r | Male | P | – |
| Tcf21 | Male | D | – |
| Six1 | Female | D | – |

Amh, Ar, Gata4, Lhx9, Sf1 are genes upregulated at the male-producing temperature (26 °C) during stage 22 of embryonic development [42] and hypermethylated at 31 °C. All other cells denote differentially methylated genes that are not upregulated in the opposite sex

P promoter, I intron, E exon, D downstream of last exon; () indicates the number of methylated windows if >1

and painted turtle differed in overall patterns of methylation and expression, non-surprisingly given that fish transcription and that in other vertebrates has diverged profoundly [43]. The only exception was *Lhx9* which was upregulated at MPT in turtle embryos and was hypermethyled in both the tongue sole and painted turtle. Lastly, two other differentially methylated genes in turtle are worth mentioning, the epidermal growth factor receptor (*Egfr*) which has been previously implicated in sexual dimorphism in *Drosophila* [88], and *Mafb*, a gene with sexually dimorphic expression responsible for masculinization of male genitalia in mice [89].

## Potential thermal transmitters are differentially methylated in TSD turtles

Sexually dimorphic methylation was present in hatchling gonads in several gene candidates that may help convert the environmental temperature into sexual development signals to establish the sexual fate in TSD taxa [13, 36, 43, 90] (Additional file 1: Table S4, Additional file 1: Table S5, Additional file 1: Table S6, Additional file 1: Table S7, Additional file 1: Table S8, Additional file 1: Table S9, Additional file 1: Table S10, Additional file 1: Table S11), rendering them targets for future study. These include several heat shock genes, transient receptor potential genes, a number of kinases, androgen-/estrogen-related genes and histone-related genes.

Namely, several *heat shock protein* genes (e.g., *Hspa4*, *Hspa12a* and *Hspa12b*) were hypermethylated in males, and they are differentially expressed by temperature in alligators and could play a role in TSD [36]. Further, the *cold-inducible RNA-binding protein* (*Cirbp*), a putative key TSD gene that is upregulated at FPT in snapping turtles *Chelydra serpentina* (TSD) [91] and *C. picta* [42], showed hypermethylated exons in females (Additional file 1: Table S4), perhaps causing differential splicing as in humans [71]. Additionally, *transient receptor potential genes* (e.g., *Trpm1*, *Trpm2*, *Trpm3*, *Trpm7* and *Trpm8*) can respond to temperature stimuli [92] and were differentially methylated. Moreover, many *kinases* such as members of the *Mapk* signaling family which helps activate *Sry* in mice [93, 94] were differentially methylated, along with many *androgen* and *estrogen* signaling genes. Also, genes involved in *histone modification* directly regulate transcription and can act in a sex-specific manner. For instance, *aromatase* transcription in embryonic gonads of *T. scripta* turtle embryos increases by demethylation [17] which depends on local histone acetylation in mammals [95]. We found hypermethylation of *histone acetyltransferases* (*Kat2a* and *Kat6a*) in males and of *deacetylases* (*Hdac4*, *Hdac7* and *Hdac8*) in females, but it is unclear whether differential methylation of histone modifiers is linked to sexually dimorphic transcription in turtles.

## Differentially methylated genes have undergone greater historic methylation

Intriguingly, differential methylation in *C. picta* hatchlings coincided with regions that showed significantly lower normalized CpG content relative to the genome-wide methylation levels, a pattern observed in both gene bodies and promoters (Fig. 2e, f) that indicates stronger historic methylation in these genomic regions. This pattern includes various genes in the sex determination/differentiation network and supports the notion that differential methylation of elements in this network has occurred over long periods during turtle evolution, leaving a footprint of historic methylation at the DNA sequence level [19, 20].

## Differential repeat methylation is linked to methylation of adjacent genes

Importantly, over 95% of methylated genes (but not unmethylated genes) were nearby methylated repeats. Further, repeat methylation was associated with sexually dimorphic methylation of neighboring genes, as >70% of repeat windows nearby differentially methylated genes were methylated in the same direction as the genes (Table 2). This suggests that repeat methylation could promote DNA methylation in adjacent genes perhaps mediating their transcription in *C. picta* as occurs in *Drosophila* and human [96, 97]. Repeat methylation may vary by repeat category and developmental stage as observed in fish [98], but further research is needed to test for DNA methylation in developing turtles.

However, we found that DNA methylation targets repeat elements in general more often than expected by their overall genome abundance, perhaps as a result of repeat silencing. Namely, 10% of the *C. picta* genome assembly is composed of transposable elements [37] [and 25% of the genome consists overall repeats (Additional file 1: Table S16)], whereas repeats composed 40% of all methylated regions. This result is consistent with findings in rat using our same approach where the high proportion of repeats in the methylome was attributed to repeat silencing (41% repeats in the rat genome and 53% in the rat methylome) [99]. Notably, insights about the relative abundance of repeat categories vary depending on whether the entire turtle genome was considered or only the genomic fraction that was composed of repeats (Fig. 4a, b). On the one hand, relative abundance among repeat categories in the methylome followed their relative abundance in the entire genome (Fig. 4a). For instance, CR1 repeats were the most abundant in the methylome, followed by HATs, DIRS, Gypsy and Harbinger repeats, just as in the painted turtle genome [37]), and in the transcriptome [42] (Fig. 4a). On the other hand, some repeat categories were overrepresented and others underrepresented in the methylome compared to their abundance in just the fraction of the genome composed of repeats (Fig. 4b). Thus, not fully conclusive evidence was found for overall repeat silencing by DNA methylation [64] in turtles as occurs in humans [100]. Interestingly, CR1 repeats concentrate at centromeres of a few *C. picta* chromosomes [38] such that their high methylome abundance may reflect chromosome stabilization by DNA methylation of centromeric repeats [64].

## Conclusions

Ours is the first genome-wide assessment of DNA methylation in reptiles and the first study of sexually dimorphic methylation levels in a purely TSD vertebrate [51]. As such, this study sheds light on the epigenetic modifications that may play a role in the development of phenotypically plastic vertebrates, complementing recent work on a fish with mixed sex determination [18]. Our MeDIP-seq data provide empirical validation of in silico predictions obtained from nCpG content never done before in reptiles and show that nCpG content is a reasonable predictor of actual methylation status at the resolution of MeDIP-seq. We found that painted turtles possess a unique pattern with nCpG values below those of other vertebrates, indicative of global and extensive historic methylation. In contrast,

actual methylation levels given the turtle genome CpG content agree with those described for most vertebrates. Our data helped us identify several genes whose methylation status renders them candidates for a putative function in regulating transcription in a thermosensitive manner, a pattern that appears associated with methylation status of neighboring repeat elements. Based on this correlational pattern, we speculate that methylation of some of these elements may play a key role in mediating the sexual fate of TSD reptiles. Further research is warranted to test this hypothesis and the prevalence of DNA methylation in governing the sexual outcome in TSD turtles as it does in sex reversal of fish with mixed sex determination [18], or whether instead, most differential methylation observed in turtle hatchlings is the consequence of sexual differentiation and not its cause.

## Methods

### Sample collection

Eggs from several freshly laid clutches were obtained from a turtle farm, and equal numbers from each clutch were assigned randomly to incubators set at 26 and 31 °C that produce 100% males and 100% females, respectively, under standard incubation conditions [101]. Turtle embryos are at gastrula stage at oviposition [102]. All hatchlings were raised for 3 months in tanks with water heated to ~26 °C and fed ad libitum, to allow full gonadal differentiation prior to sexing. Individual sex was diagnosed by gonadal inspection [51, 103]. All procedures were approved by the IACUC of Iowa State University.

### DNA isolation and sequencing

We extracted DNA from the gonads of 3-month-old *C. picta* hatchlings (ten males and ten females, and pooled in groups of five hatchlings for replication) using the Gentra Puregene DNA extraction kit (Gentra) following the manufacturer's instructions. DNA was processed by BGI Americas following the MeDIP protocol [104] where DNA was fragmented and denatured, and the methylated DNA was immunoprecipitated using the MagMEDI kit (Diagenode). A positive control (methylated DNA) and a negative control (unmethylated DNA) were used at the MeDIP step to ensure that the MeDIP worked properly. The methylated DNA was size selected (100–300 bp) and sequenced using the Illumina HiSeq paired-end protocol. We obtained between 126 million and 163 million 50-bp reads per library amounting to a total of ~564 million reads (Table 1).

### Methylome construction and differential methylation analysis

We mapped the sequencing reads to the *C. picta* genome version 3.0.1 [37] using Bowtie2 version 2.2.5 [105]. We filtered out the unmapped reads using Samtools [106]. We used the MEDIPS package [107] in four steps: (a) to build an index for the *C. picta* genome and to ensure fast querying of the alignment files. (b) To model read counts under a negative binomial distribution. (c) To quantify mapped read counts per 500-bp windows in RPKM (reads per kilobase of genes per million mapped reads). Mean methylation levels were calculated for each 500-bp window by averaging the corresponding RPKM levels in the replicates. A gene was considered as methylated if it contained one or more methylated windows within its start and end coordinates. (d) To merge methylated windows and compute differential methylation by sex, while controlling for false discoveries [108] at a *q*-value cutoff =0.05. Any gene with a *q*-value <0.05 was considered as differentially methylated. Besides the positive and negative controls used during the MeDIP procedure described above, primers were also designed to validate by PCR the differential methylation identified by MeDIP-seq for the gene *Fezf2* that was hypermethylated in females compared to males [109]. The primer cocktail consisted of one forward primer (F1: 5′-GGG GTG AAA AAC CAC AG-3′) and two reverse primers, one closer to the F1 (R1: 5′-CAC ACA CAA GGA GG-3′) and one further away (R2: 5′-CAG CAA CAA CTT GAT TTG G-3′) (Fig. 3). Primers F1 + R1 flank a shorter non-methylated area that should amplify in any individual, which is nested within the larger region encompassed by F1 + R2 that contains the differentially methylated area, and which amplifies only when methylation protects the DNA from digestion by a methylation-sensitive restriction enzyme (HpaII in our case). Gonadal DNA from three male and three female hatchlings was digested with HpaII for 30 min at 37 °C. PCR was carried out in 15ul reactions, using 1ul of 100 ng/ul undigested DNA (control) or digested DNA (experimental) as template, and 5ul PCR products along with 1ul 1 kb plus DNA ladder (Invitrogen) were visualized in 1% agarose gels stained with ethidium bromide. The PCR cocktail contained 1.5 mM MgCl2, 0.2 mM dNTPs, 0.4uM of each primer (F1, R1, R2), 0.4U Taq polymerase, ~100 ng DNA, 1.5ul 10X buffer and 10.5 μl water. PCR conditions included initial denaturing at 94 °C for 3 min, followed by 35 cycles of denaturing at 94 °C for 30 s, annealing at 58 °C for 30 s, and extension at 72 °C for 1 min.

We used Kolmogorov–Smirnov tests to determine whether the nCpG content distribution of all annotated genes was comparable to those identified by the MEDIPS package. To test whether the nCpG of the differentially methylated genes varied significantly from that of all the methylated genes, we performed a resampling test [110] by iteratively drawing a random subset of genes (equal to the number of differentially methylated genes) from the

entire set of methylated genes. We used RepeatMasker v3.3.0 [111] to identify repeats in the *C. picta* genome. We used pairwise Kruskal–Wallis tests [112] to test whether repeat abundances from the genome, methylome and transcriptome came from the same population. We used Bedtools v2.17.0 [113] to compute repeats overlapping with methylated regions identified by MEDIPS. We used permutation tests to evaluate differences in the abundance of methylated DNA repeats occurring in vicinity of methylated genes versus non-methylated genes. In the absence of transcriptional data from hatchlings, we used our transcriptomic dataset from late-developing embryos (stage 22) from another study [42] to test for an association between methylation patterns in hatchlings and transcription patterns using Fisher's exact test. For this purpose, a contingency table was generated of genes that were differentially expressed (or not) and differentially methylated (or not), and a hypergeometric distribution was used to then determine the p value. A p value of ~1 would indicate no significant association between differential methylation and differential expression. We used R [114] to perform regression tests to model transcriptome abundance using genomic abundance of repeats and methylated repeats. We used the DAVID Bioinformatics knowledgebase [115] to assess the enrichment of functional categories to which the differentially methylated genes belong.

**Analysis of normalized CpG content**

The normalized CpG content (nCpG) is calculated as:

$$nCpG = \frac{\left(\frac{cg}{l}\right)}{\left(\frac{c}{l}\right) \times \left(\frac{g}{l}\right)}$$

for a sequence of length *l*, where *c* is the number of occurrences of cytosine, *g* is the number of occurrences of guanine and *cg* is the number of times cytosine is bordered by guanine linked by a phosphate group (CpG) [21]. In theory, nCpG values can range from 0 to +infinity for an infinitely long sequence, with a value of 1 when the number of CpG dinucleotides observed is equal to the expected based on the sequence length and abundance of C and G. Values <1 denote CpG depletion from what is expected by chance, and values >1 represent overabundance of CpGs from random expectation. The methylation of CpG dinucleotides initiates the deamination of the cytosine, transforming it to thymine, thus lowering the nCpG to <1. Studies conducted in vertebrate and invertebrate animals reveal an upper limit between 2 and 2.5 for various genomic regions [21, 25–27, 29, 116].

We used Bedtools [113] to parse the exon, intron, promoter and intergenic coordinates from the *C. picta* genome [37], given the annotations in gff3 format. We used in-house perl scripts to compute the CpG contents by genomic region. For the nCpG distributions, the R package Mclust [39] was used to assess likelihood of a mixture model with one $(G = 1)$ and two $(G = 2)$ components, with the better fit model decided by a likelihood ratio test.

## Additional file

**Additional file 1. Table S1:** Genes with distinct hypermethylated sex-specific windows within the same gene. **Table S2:** Categories enriched $(p = 0.05)$ in hypermethylated genes at male-producing temperature (26 °C). **Table S3:** Categories enriched $(p = 0.05)$ in hypermethylated genes at female-producing temperature (31 °C). Tables S4 through S11—differentially methylated genes in *C. picta* hatchling methylomes along with RPKM, fold change and edgeR *p* value: **Table S4:** Heat shock genes. **Table S5:** Androgen-/estrogen-related genes. **Table S6:** Kinases. **Table S7:** Histone-related genes. **Table S8:** Ubiquitin-related genes. **Table S9:** Transient receptor potential genes. **Table S10:** Genes involved in cell proliferation. **Table S11:** Germline-related genes. **Table S12:** Number of differentially methylated genes by GO term present in testis and ovaries of *C. picta* hatchlings. **Table S13:** Comprehensive list of differentially methylated genes and associated GO pathways between testis and ovaries of *C. picta* hatchlings. **Table S14:** Statistics of all differentially methylated genes between testis and ovaries of *C. picta* hatchlings. **Table S15:** Genes differentially expressed in the *C. picta* transcriptome (stage 22) [42] and differentially methylated in the hatchling methylome (this study). **Table S16:** Relative abundance distribution of repeat categories as a fraction of the genome. **Table S17:** Overlapping genes between *C. picta* hatchling methylomes and the methylomes of tongue sole [18].

## Abbreviations

*Amh*: anti-mullerian hormone; *Ar*: androgen receptor; *Cirbp*: cold-inducible RNA-binding protein; CR1: chicken repeat 1; *Dax1*: dosage-sensitive sex reversal, adrenal hypoplasia critical region, on chromosome X, gene 1; *Dmrt1*: doublesex and mab-3-related transcription factor 1; *Egfr*: epidermal growth factor receptor; *Emx2*: empty spiracles homeobox 2; *Esr1*: estrogen receptor 1; *Gata4*: gata binding protein 4; GSD: genotypic sex determination; HAT: transposon superfamily named after Hobo from *Drosophila melanogaster*, Ac from maize and Tam3 from the snapdragon; *Hdacx*: histone deacetylase number x; HiSeq: high-throughput sequencing technology by Illumina; HPLC: high-performance liquid chromatography; *Hspx*: heat-shock protein number x; *Lhx1*: lim homoeobox 1; *Lhx9*: lim homoeobox 9; *Mafb*: MAF BZIP transcription factor B; *Mapk*: mitogen-activated protein kinase; MeDIP-seq: methyl DNA immunoprecipitation sequencing; nCpG: normalized CpG content; RPKM: reads per kilobase of genes per million mapped reads; *Rspo1*: R-spondin1; *Sf1*: steroidogenic factor 1; *Sox9*: SRY (sex-determining region Y)-box 9; *Sry*: sex-determining region Y; tRNA: transfer RNA; *Trpmx*: transient receptor potential cation channel subfamily M member number x; TSD: temperature-dependent sex determination; *Wnt4*: wingless-type MMTV integration site family, member 4; *Wt1*: Wilms tumor protein 1.

## Authors' contributions

NV conceived of the project, guided data analyses and wrote the manuscript. SR performed all bioinformatics analyses, contributed to the interpretation of results and wrote the manuscript. RL collected samples and contributed to the interpretation of results. BM contributed to data collection. All authors read and approved the final manuscript.

## Author details

[1] Bioinformatics and Computational Biology Program, Iowa State University, Ames, IA 50011, USA. [2] Ecology and Evolutionary Biology Program, Iowa State University, Ames, IA 50011, USA. [3] Interdepartmental Genetics and Genomics Program, Iowa State University, Ames, IA 50011, USA. [4] Department of Ecology, Evolution and Organismal Biology, Iowa State University, 251 Bessey Hall, Ames, IA 50011, USA.

## Acknowledgements

We thank members of the Valenzuela, Adams and Serb Labs at Iowa State University for comments.

## Competing interests

All authors declare that they have no competing interests.

## Funding

His work was partially funded by NSF Grants MCB 1244355 and IOS 1555999 to NV.

## References

1. Watt F, Molloy PL. Cytosine methylation prevents binding to DNA of a hela-cell transcription factor required for optimal expression of the adenovirus major late promoter. Genes Dev. 1988;2(9):1136–43.
2. Boyes J, Bird A. DNA methylation inhibits transcription indirectly via a methyl-CpG binding-protein. Cell. 1991;64(6):1123–34.
3. Hendrich B, Bird A. Identification and characterization of a family of mammalian methyl-CpG binding proteins. Mol Cell Biol. 1998;18(11):6538–47.
4. Varriale A. DNA methylation, epigenetics, and evolution in vertebrates: facts and challenges. Int J Evolut Biol. 2014;2014:475981.
5. Bergman Y, Cedar H. DNA methylation dynamics in health and disease. Nat Struct Mol Biol. 2013;20(3):274–81.
6. Cantone I, Fisher AG. Epigenetic programming and reprogramming during development. Nat Struct Mol Biol. 2013;20(3):282–9.
7. Head JA. Patterns of DNA methylation in animals: an ecotoxicological perspective. Integr Comp Biol. 2014;54(1):77–86.
8. Suzuki MM, Bird A. DNA methylation landscapes: provocative insights from epigenomics. Nat Rev Genet. 2008;9(6):465–76.
9. Zemach A, et al. Genome-wide evolutionary analysis of eukaryotic DNA methylation. Science. 2010;328(5980):916–9.
10. Szyf M, et al. Maternal care, the epigenome and phenotypic differences in behavior. Reprod Toxicol. 2007;24(1):9–19.
11. Kucharski R, et al. Nutritional control of reproductive status in honeybees via DNA methylation. Science. 2008;319(5871):1827–30.
12. Weiner SA, et al. A survey of DNA methylation across social insect species, life stages, and castes reveals abundant and caste-associated methylation in a primitively social wasp. Naturwissenschaften. 2013;100(8):795–9.
13. Valenzuela N, Lance VA, editors. Temperature dependent sex determination in vertebrates. Washington, DC: Smithsonian Books; 2004.
14. Tree of Sex Consortium, et al. Tree of sex: a database of sexual systems. Sci Data. 2014;1:140015.
15. Navarro-Martin L, et al. DNA methylation of the gonadal *aromatase* (*cyp19a*) promoter is involved in temperature-dependent sex ratio shifts in the European sea bass. PLoS Genet. 2011;7(12):e1002447.
16. Parrott BB, et al. Gonadal DNA methylation patterning is affected by incubation temperature in the American alligator, a species undergoing temperature-dependent sex determination. Integr Comp Biol. 2014;54:E327.
17. Matsumoto Y, et al. Epigenetic control of gonadal aromatase *cyp19a1* in temperature-dependent sex determination of red-eared slider turtles. PLoS ONE. 2013;8(6):e63599.

18. Shao C, et al. Epigenetic modification and inheritance in sexual reversal of fish. Genome Res. 2014;24(4):604–15.

19. Bird AP. DNA methylation and the frequency of CpG in animal DNA. Nucleic Acids Res. 1980;8(7):1499–504.

20. Coulondre C, et al. Molecular-basis of base substitution hotspots in *Escherichia coli*. Nature. 1978;274(5673):775–80.

21. Elango N, et al. DNA methylation is widespread and associated with differential gene expression in castes of the honeybee, *Apis mellifera*. Proc Natl Acad Sci USA. 2009;106(27):11206–11.

22. Jabbari K, Bernardi G. Cytosine methylation and CpG, TpG (CpA) and TpA frequencies. Gene. 2004;333:143–9.

23. Shen JC, et al. The rate of hydrolytic deamination of 5-methylcytosine in double-stranded DNA. Nucleic Acids Res. 1994;22(6):972–6.

24. Jabbari K, et al. Evolutionary changes in CpG and methylation levels in the genome of vertebrates. Gene. 1997;205(1–2):109–18.

25. Simola DF, et al. Social insect genomes exhibit dramatic evolution in gene composition and regulation while preserving regulatory features linked to sociality. Genome Res. 2013;23(8):1235–47.

26. Glastad KM, et al. DNA methylation in insects: on the brink of the epigenomic era. Insect Mol Biol. 2011;20(5):553–65.

27. Elango N, Yi SV. DNA methylation and structural and functional bimodality of vertebrate promoters. Mol Biol Evol. 2008;25(8):1602–8.

28. Weber M, et al. Distribution, silencing potential and evolutionary impact of promoter DNA methylation in the human genome. Nat Genet. 2007;39(4):457–66.

29. Yang H, et al. Relating gene expression evolution with CpG content changes. BMC Genomics. 2014;15:693.

30. Aissani B, Bernardi G. CpG islands—features and distribution in the genomes of vertebrates. Gene. 1991;106(2):173–83.

31. Varriale A, Bernardi G. DNA methylation in reptiles. Gene. 2006;385:122–7.

32. Long HK, et al. Epigenetic conservation at gene regulatory elements revealed by non-methylated DNA profiling in seven vertebrates. Elife. 2013;2:e00348.

33. Eggers S, et al. Genetic regulation of mammalian gonad development. Nat Rev Endocrinol. 2014;10(11):673–83.

34. Kuroki S, et al. Epigenetic regulation of mouse sex determination by the histone demethylase *Jmjd1a*. Science. 2013;341(6150):1106–9.

35. Carmi I, et al. The nuclear hormone receptor *Sex-1* is an X-chromosome signal that determines nematode sex. Nature. 1998;396:168–73.

36. Kohno S, et al. Potential contributions of heat shock proteins to temperature-dependent sex determination in the American alligator. Sex Dev. 2010;4(1–2):73–87.

37. Shaffer HB, et al. The western painted turtle genome, a model for the evolution of extreme physiological adaptations in a slowly evolving lineage. Genome Biol. 2013;14(3):21–2.

38. Badenhorst D, et al. Physical mapping and refinement of the painted turtle genome (*Chrysemys picta*) inform amniote genome evolution and challenge turtle-bird chromosomal conservation. Genome Biol Evol. 2015;7(7):2038–50.

39. Fraley C, Raftery A. MCLUST version 3 for R: Normal mixture modeling and model-based clustering. Technical Report No. 504 2006, Department of Statistics, University of Washington.

40. Beck S, Rakyan VK. The methylome: approaches for global DNA methylation profiling. Trends Genet. 2008;24(5):231–7.

41. Valenzuela N. Evolution of the gene network underlying gonadogenesis in turtles with temperature-dependent and genotypic sex determination. Integr Comp Biol. 2008;48(4):476–85.

42. Radhakrishnan S, et al. Transcriptomic responses to environmental temperature by turtles with temperature-dependent and genotypic sex determination assessed by RNAseq inform the genetic architecture of embryonic gonadal development. PLoS ONE. 2017;12:e0172044.

43. Valenzuela N, et al. Transcriptional evolution underlying vertebrate sexual development. Dev Dyn. 2013;242(4):307–19.

44. Yeh SY, et al. Generation and characterization of androgen receptor knockout (ARKO) mice: an in vivo model for the study of androgen functions in selective tissues. Proc Natl Acad Sci USA. 2002;99(21):13498–503.

45. Oreal E, et al. Different patterns of *anti-Mullerian hormone* expression, as related to *Dmrt1, Sf-1, Wt1, Gata-4, Wnt-4,* and *Lhx9* expression, in the chick differentiating gonads. Dev Dyn. 2002;225(3):221–32.

46. Manuylov NL, et al. Conditional ablation of Gata4 and Fog2 genes in mice reveals their distinct roles in mammalian sexual differentiation. Dev Biol. 2011;353(2):229–41.

47. Tripathi V, Raman R. Identification of Wnt4 as the ovary pathway gene and temporal disparity of its expression vis-a-vis testis genes in the garden lizard, *Calotes versicolor*. Gene (Amst). 2010;449(1–2):77–84.

48. Pellegrini M, et al. Emx2 developmental expression in the primordia of the reproductive and excretory systems. Anat Embryol. 1997;196(6):427–33.

49. Weisenberger DJ, et al. Analysis of repetitive element DNA methylation by MethyLight. Nucleic Acids Res. 2005;33(21):6823–36.

50. Bonasio R. The role of chromatin and epigenetics in the polyphenisms of ant castes. Brief Funct Genomics. 2014;13(3):235–45.

51. Valenzuela N, et al. Molecular cytogenetic search for cryptic sex chromosomes in painted turtles *Chrysemys picta*. Cytogenet Genome Res. 2014;144:39–46.

52. Clark C, et al. A comparison of the whole genome approach of MeDIP-seq to the targeted approach of the infinium humanmethylation450 Beadchip for methylome profiling. PLoS ONE. 2012;7(11):e50233.

53. Su Y, et al. Genome-wide DNA methylation profile of developing deciduous tooth germ in miniature pigs. BMC Genom. 2016;17(1):1–9.

54. Eckhardt F, et al. DNA methylation profiling of human chromosomes 6, 20 and 22. Nat Genet. 2006;38(12):1378–85.

55. Bird A. DNA methylation patterns and epigenetic memory. Genes Dev. 2002;16(1):6–21.

56. Jiang N, et al. Conserved and divergent patterns of DNA methylation in higher vertebrates. Genome Biol Evol. 2014;6(11):2998–3014.

57. Park J, et al. What are the determinants of gene expression levels and breadths in the human genome? Hum Mol Genet. 2012;21(1):46–56.

58. Keller TE, et al. Evolutionary transition of promoter and gene body DNA methylation across invertebrate–vertebrate boundary. Mol Biol Evol. 2016;33(4):1019–28.

59. Bird AP, Taggart MH. Variable patterns of total DNA and rDNA methylation in animals. Nucleic Acids Res. 1980;8(7):1485–97.

60. Yi SV, Goodisman MAD. Computational approaches for understanding the evolution of DNA methylation in animals. Epigenetics. 2009;4(8):551–6.

61. Laurent L, et al. Dynamic changes in the human methylome during differentiation. Genome Res. 2010;20(3):320–31.

62. Bauer AP, et al. The impact of intragenic CpG content on gene expression. Nucleic Acids Res. 2010;38(12):3891–908.

63. Krinner S, et al. CpG domains downstream of TSSs promote high levels of gene expression. Nucleic Acids Res. 2014;42(6):3551–64.

64. Jones PA. Functions of DNA methylation: islands, start sites, gene bodies and beyond. Nat Rev Genet. 2012;13(7):484–92.

65. Widmann J, et al. Stable tRNA-based phylogenies using only 76 nucleotides. RNA. 2010;16(8):1469–77.

66. Tweedie S, et al. Methylation of genomes and genes at the invertebrate-vertebrate boundary. Mol Cell Biol. 1997;17(3):1469–75.

67. Kimura K, et al. Diversification of transcriptional modulation: large-scale identification and characterization of putative alternative promoters of human genes. Genome Res. 2006;16(1):55–65.

68. Maunakea AK, et al. Conserved role of intragenic DNA methylation in regulating alternative promoters. Nature. 2010;466(7303):U131–253.

69. Huh I, et al. DNA methylation and transcriptional noise. Epigenetics Chromatin. 2013;6:9.

70. Dallosso AR, et al. Alternately spliced WT1 antisense transcripts interact with WT1 sense RNA and show epigenetic and splicing defects in cancer. RNA. 2007;13(12):2287–99.

71. Maunakea AK, et al. Intragenic DNA methylation modulates alternative splicing by recruiting MeCP2 to promote exon recognition. Cell Res. 2013;23(11):1256–69.

72. Valenzuela N. Relic thermosensitive gene expression in a turtle with genotypic sex determination. Evolution. 2008;62(1):234–40.

73. Spotila LD, et al. Sequence and expression analysis of WT1 and Sox9 in the red-eared slider turtle, *Trachemys scripta*. J Exp Zool. 1998;284:417–27.

74. Valenzuela N, et al. Comparative gene expression of steroidogenic factor 1 in *Chrysemys picta* and *Apalone mutica* turtles with temperature-dependent and genotypic sex determination. Evol Dev. 2006;8(5):424–32.

75. Ramsey M, et al. Gonadal expression of *Sf1* and *Aromatase* during sex determination in the red-eared slider turtle (*Trachemys scripta*), a reptile with temperature-dependent sex determination. Differentiation. 2007;75:978–91.

76. Torres Maldonado LC, et al. Expression profiles of *Dax1*, *Dmrt1*, and *Sox9* during temperature sex determination in gonads of the sea turtle *Lepidochelys olivacea*. Gen Comp Endocrinol. 2003;129:20–6.

77. Barske LA, Capel B. Estrogen represses SOX9 during sex determination in the red-eared slider turtle *Trachemys scripta*. Dev Biol. 2010;341(1):305–14.

78. Valenzuela N. Multivariate expression analysis of the gene network underlying sexual development in turtle embryos with temperature-dependent and genotypic sex determination. Sex Dev. 2010;4(1–2):39–49.

79. Matsumoto Y, et al. Changes in gonadal gene network by exogenous ligands in temperature-dependent sex determination. J Mol Endocrinol. 2013;50(3):389–400.

80. Valenzuela N, Shikano T. Embryological ontogeny of Aromatase gene expression in *Chrysemys picta* and *Apalone mutica* turtles: comparative patterns within and across temperature-dependent and genotypic sex-determining mechanisms. Dev Genes Evol. 2007;217:55–62.

81. Kettlewell JR, et al. Temperature-dependent expression of turtle Dmrt1 prior to sexual differentiation. Genesis. 2000;26(3):174–8.

82. Bergeron JM, et al. Cloning and in situ hybridization analysis of *estrogen receptor* in the developing gonad of the red-eared slider turtle, a species with temperature-dependent sex determination. Dev Growth Differ. 1998;40(2):243–54.

83. Chávez B, et al. Cloning and expression of the estrogen receptor-alpha (*Esr1*) from the Harderian gland of the sea turtle (*Lepidochelys olivacea*). Gen Comp Endocrinol. 2009;162:203–9.

84. Parrott BB, et al. Influence of tissue, age, and environmental quality on DNA methylation in *Alligator mississippiensis*. Reproduction. 2014;147(4):503–13.

85. Wilhelm D, Englert C. The Wilms tumor suppressor WT1 regulates early gonad development by activation of Sf1. Genes Dev. 2002;16(14):1839–51.

86. Cedar H, Bergman Y. Linking DNA methylation and histone modification: patterns and paradigms. Nat Rev Genet. 2009;10(5):295–304.

87. Lande-Diner L, Cedar H. Silence of the genes—mechanisms of long-term repression. Nat Rev Genet. 2005;6(8):648–54.

88. Foronda D, et al. *Drosophila* Hox and sex-determination genes control segment elimination through EGFR and extramacrochetae activity. PLoS Genet. 2012;8(8):e1002874.

89. Suzuki K, et al. Sexually dimorphic expression of Mafb regulates masculinization of the embryonic urethral formation. Proc Natl Acad Sci USA. 2014;111(46):16407–12.

90. Morrish BC, Sinclair AH. Vertebrate sex determination: many means to an end. Reproduction. 2002;124:447–57.

91. Rhen T, Schroeder A. Molecular mechanisms of sex determination in reptiles. Sex Dev. 2010;4(1–2):16–28.

92. Dhaka A, et al. TRP ion channels and temperature sensation. Annu Rev Neurosci. 2006;29:135–61.

93. Bogani D, et al. Loss of mitogen-activated protein kinase kinase kinase 4 (*Map3k4*) reveals a requirement for MAPK signalling in mouse sex determination. PLoS Biol. 2009;7:e1000196.

94. Warr N, et al. *Gadd45y* and *Map3k4* interactions regulate mouse testis determination via p38 mapk-mediated control of *Sry* expression. Dev Cell. 2012;23:1020–31.

95. Cervoni N, Szyf M. Demethylase activity is directed by histone acetylation. J Biol Chem. 2001;276(44):40778–87.

96. Garrison BS, et al. Postintegrative gene silencing within the Sleeping Beauty transposition system. Mol Cell Biol. 2007;27(24):8824–33.

97. Cridland JM, et al. Gene expression variation in *Drosophila melanogaster* due to rare transposable element insertion alleles of large effect. Genetics. 2015;199(1):U85–490.

98. McGaughey DM, et al. Genomics of CpG methylation in developing and developed zebrafish. G3. 2014;4(5):861–9.

99. Sati S, et al. High resolution methylome map of rat indicates role of intragenic DNA methylation in identification of coding region. PLoS ONE. 2012;7(2):e31621.

100. Choi SH, et al. Changes in DNA methylation of tandem DNA repeats are different from interspersed repeats in cancer. Int J Cancer. 2009;125(3):723–9.

101. Valenzuela N. Egg incubation and collection of painted turtle embryos. Cold Spring Harbor Protoc. 2009;4(7):1–3.

102. Ewert M. Embryology of turtles. In: Gans C, editor. Biology of the reptilia, development A, vol. 14. New York: Wiley; 1985.

103. Ewert MA, Nelson CE. Sex determination in turtles: diverse patterns and some possible adaptive values. Copeia. 1991;1991(1):50–69.

104. Weber M, et al. Chromosome-wide and promoter-specific analyses identify sites of differential DNA methylation in normal and transformed human cells. Nat Genet. 2005;37(8):853–62.

105. Langmead B, Salzberg SL. Fast gapped-read alignment with Bowtie 2. Nat Methods. 2012;9(4):U354–7.

106. Li H, et al. The sequence alignment/map format and SAMtools. Bioinformatics. 2009;25(16):2078–9.

107. Lienhard M, et al. MEDIPS: genome-wide differential coverage analysis of sequencing data derived from DNA enrichment experiments. Bioinformatics. 2014;30(2):284–6.

108. Benjamini Y, Hochberg Y. Controlling the false discovery rate—a practical and powerful approach to multiple testing. J R Stat Soc Ser B Methodol. 1995;57(1):289–300.

109. Caetano LC, Ramos ES. MHM assay: molecular sexing based on the sex-specific methylation pattern of the MHM region in chickens. Conserv Genet. 2008;9(4):985–7.

110. Crowley PH. Resampling methods for computation-intensive data-analysis in ecology and evolution. Annu Rev Ecol Syst. 1992;23:405–47.

111. Smit AFA et al. RepeatMasker Open-3.0. 1996–2010.

112. Kruskal WH, Wallis WA. Use of ranks in one-criterion variance analysis. J Am Stat Assoc. 1952;47(260):583–621.

113. Quinlan AR, Hall IM. BEDTools: a flexible suite of utilities for comparing genomic features. Bioinformatics. 2010;26(6):841–2.

114. R Core Development Team: R: a language and environment for statistical computing. In. Version 3.2.2 edn. Vienna: R Foundation for Statistical Computing. Available via http://cran.R-project.org; 2012.

115. Huang DW, et al. DAVID Bioinformatics Resources: expanded annotation database and novel algorithms to better extract biology from large gene lists. Nucleic Acids Res. 2007;35:W169–75.

116. Park J, et al. Comparative analyses of DNA methylation and sequence evolution using *Nasonia* genomes. Mol Biol Evol. 2011;28(12):3345–54.

# 6

# Dosage compensation and sex-specific epigenetic landscape of the X chromosome in the pea aphid

Gautier Richard[1*] [iD], Fabrice Legeai[2,3], Nathalie Prunier-Leterme[1], Anthony Bretaudeau[2,4], Denis Tagu[1], Julie Jaquiéry[5*†] and Gaël Le Trionnaire[1*†]

## Abstract

**Background:** Heterogametic species display a differential number of sex chromosomes resulting in imbalanced transcription levels for these chromosomes between males and females. To correct this disequilibrium, dosage compensation mechanisms involving gene expression and chromatin accessibility regulations have emerged throughout evolution. In insects, these mechanisms have been extensively characterized only in Drosophila but not in insects of agronomical importance. Aphids are indeed major pests of a wide range of crops. Their remarkable ability to switch from asexual to sexual reproduction during their life cycle largely explains the economic losses they can cause. As heterogametic insects, male aphids are X0, while females (asexual and sexual) are XX.

**Results:** Here, we analyzed transcriptomic and open chromatin data obtained from whole male and female individuals to evaluate the putative existence of a dosage compensation mechanism involving differential chromatin accessibility of the pea aphid's X chromosome. Transcriptomic analyses first showed X/AA and XX/AA expression ratios for expressed genes close to 1 in males and females, respectively, suggesting dosage compensation in the pea aphid. Analyses of open chromatin data obtained by Formaldehyde-Assisted Isolation of Regulatory Elements (FAIRE-seq) revealed a X chromosome chromatin accessibility globally and significantly higher in males than in females, while autosomes' chromatin accessibility is similar between sexes. Moreover, chromatin environment of X-linked genes displaying similar expression levels in males and females—and thus likely to be compensated—is significantly more accessible in males.

**Conclusions:** Our results suggest the existence of an underlying epigenetic mechanism enhancing the X chromosome chromatin accessibility in males to allow X-linked gene dose correction between sexes in the pea aphid, similar to Drosophila. Our study gives new evidence into the comprehension of dosage compensation in link with chromatin biology in insects and newly in a major crop pest, taking benefits from both transcriptomic and open chromatin data.

**Keywords:** X chromosome, Dosage compensation, Transcriptomics, Open chromatin, Non-model organism, Pea aphid, *Acyrthosiphon pisum*, Formaldehyde-Assisted Isolation of Regulatory Elements (FAIRE)

*Correspondence: gautier.richard@inra.fr; julie.jaquiery@univ-rennes1.fr; gael.le-trionnaire@inra.fr
†Julie Jaquiéry and Gaël Le Trionnaire participated equally
[1] EGI, UMR 1349, INRA, Institut de Génétique, Environnement et Protection des Plantes (IGEPP), Domaine de la Motte, BP 35327, Le Rheu, France
[5] CNRS, UMR 6553, EcoBio, University of Rennes 1, 35042 Rennes, France
Full list of author information is available at the end of the article

## Background

The sex of individuals relies in many organisms upon morphologically differentiated sex chromosomes. Those chromosomes are referred to as X, Y, Z and W chromosomes depending on which sex is heterogametic. In organisms with male heterogamety, females encompass two X chromosomes, while males possess only one X, accompanied or not by a Y chromosome (XX/XY or XX/X0 systems) [1]. In these organisms showing a degenerated or an absence of Y chromosome, there are an imbalanced number of X-linked alleles between males and females that might induce differential transcription levels for those genes between sexes [2]. Such a disequilibrium needs to be corrected, especially for X-linked genes that interact with autosomal genes, since reduced gene dose in the heterogametic sex might have deleterious phenotypic consequences [2–5]. Dosage compensation mechanisms are thought to have evolved to correct such disequilibrium. These mechanisms tend to generate equilibrated X-linked and autosomal transcript levels, often resulting in XX/AA and X/AA expression ratios equal to 1 in both the homogametic (XX) and the heterogametic (XY or X0) sexes, and consequently of XX/X ratios also equal to 1, as described in several model organisms such as Eutherian mammals [2, 6–9], *Caenorhabditis elegans* [9–11] and the insect model *Drosophila melanogaster* [12–14]. Recently, complete dosage compensation in other male heterogametic insect species (Fig. 1)—namely *Anopheles stephensi* [15], *Anopheles gambiae* [16] and *Manduca sexta* [17]—has been demonstrated using transcriptomic data. A partial dosage compensation was found in *Strepsiptera* [18], and no compensation was detectable in *Teleopsis dalmanni* [19].

Various mechanisms have evolved to counteract the deleterious effects of different doses of X-linked genes in males and females. These mechanisms are all based on the modulation of chromatin accessibility of the X

chromosome(s) in one sex in order to ensure that both sex's somatic—and sometimes germ cells [20]—show similar transcription levels for X-linked genes [21]. In mammals, dosage compensation mechanisms take place in the female where one of the two X chromosomes—either the maternal or paternal one—is entirely inactivated, thus balancing the X/A transcription level in males and females. This transcriptional regulation is achieved by the progressive depletion of active marks such as H3K4me1 and H3K9ac [22] and the enrichment of histone macroH2A1 [23] on the female's inactive X chromosome. Moreover, the histone repressive mark H3K27me3 is enriched on this inactive chromosome through the action of the X-inactive specific transcript (Xist), a long noncoding RNA (lncRNA) [24]. In *C. elegans* (a XX/X0 system), dosage compensation also takes place in the homogametic sex, where the two X chromosomes display halved transcription levels [25]. Reduced transcription is allowed by an enrichment of the repressive H4K20me1 histone mark and a depletion of the active H4K16ac histone mark [26] on the two hermaphrodites X chromosomes through the action of the dosage compensation complex (DCC) resulting in a global reduction in RNA Pol II recruitment [27]. On the other hand, in *D. melanogaster* (XX/XY system) the transcription of the single males' X chromosome is doubled by an overall increase in the chromatin accessibility by the DCC [28, 29]. This complex is composed of five proteins: male-specific lethal 1, 2 and 3 (MSL1, MSL2 and MSL3), males absent on the first (MOF), maleless (MLE) and two lncRNAs (*roX1* and *roX2*) [29]. The DCC binds to the males X chromosome at several genomic locations where it modifies histones chemistry with the active histone mark H4K16ac [30, 31] that enhances the chromatin accessibility of the X in males compared to females, hence allowing an increased transcription of X-linked genes in males.

Aphids are major crop pests that can cause severe damages on a wide range of crops. Their success as pests is largely explained by their remarkable adaptive potential to their environment, and especially the phenotypic plasticity they display during their annual life cycle where they can switch from clonal to sexual reproduction [32, 33] in response to environmental cues. Viviparous parthenogenetic females reproduce clonally from spring to summer and, in response to photoperiod shortening at fall arrival, can generate sexual females and males that will mate and produce overwintering eggs [34–36]. The pea aphid *Acyrthosiphon pisum*—for which the genome is sequenced and partially assembled [37]—is a male heterogametic species with X0 males and XX females. A recent transcriptomic analysis of the different pea aphid sexual morphs by Jaquiéry et al. [34] revealed a general trend where X-linked genes are on average more transcribed

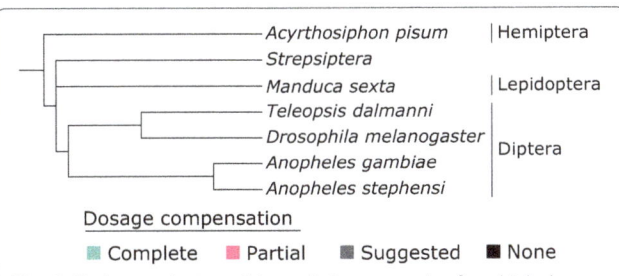

**Dosage compensation**

🟩 Complete 🟥 Partial ⬛ Suggested ⬛ None

**Fig. 1** Phylogenetic tree of the main insect species for which dosage compensation has been studied. This tree has been generated using PhyloT (http://phylot.biobyte.de/) and the NCBI taxonomy. The *color* defines the dosage compensation type (complete dosage compensation in *green*, partial dosage compensation in *pink*, suggested dosage compensation in *gray*, and no dosage compensation in *black*)

in males than in females. Nevertheless, the global expression of X-linked and autosomal genes and consequently XX/AA, X/AA and XX/X ratios were not assessed at that time. Here, our study first aimed at addressing this point taking benefits from a much larger set of genes—being generated overtime [38]—assigned as X-linked or autosomal. We report that expression profiles of expressed X-linked and autosomal genes yield X/AA and XX/AA ratios close to 1, suggesting a chromosome-wide regulation of the expression of X-linked genes, and consequently a potential differential X chromosome chromatin accessibility between sexes. To test this hypothesis, we applied the FAIRE (Formaldehyde-Assisted Isolation of Regulatory Elements) methodology to extract and then sequence the open chromatin from whole male and parthenogenetic female (here after termed as "females") individuals [39, 40]. This open chromatin mainly corresponds to nucleosome-depleted regions (NDR) likely to be more accessible to transcriptional (RNA polymerase II) or regulatory elements (mainly enhancers or insulators). Statistical analyses revealed that the transcription start sites (TSSs) from expressed genes were significantly more accessible on the X chromosome in males compared to females. More specifically, X-linked genes expressed at the same level in males and females and, thus likely to be compensated, have significantly higher chromatin accessibility in males than in females, notably in the TSS. These results suggest that a potential global regulation of chromatin accessibility might occur on the X chromosome of aphids to compensate for the gene dose in males.

## Results
### Expression of X-linked and autosomal genes in males and females

A new assignation of the genomic scaffolds of the *A. pisum* genome to autosomes and X chromosome has recently been performed [38]. We thus reanalyzed previously published *A. pisum* RNA-seq data from whole-individual males and parthenogenetic females (and to a lesser extent, sexual females, since FAIRE has been performed on asexual females and males; see the next result part concerning the FAIRE for a more in-depth explanation) [34] in regard to this new gene assignation to characterize the expression level of the 19,232 and 13,711 genes located on the X and autosomes, respectively. X/A expression ratios were calculated at different expression levels thresholds [15–17] (Table 1). When all genes are taken into account, XX/AA and X/AA ratios close to 0 are observed since over 80% of the X-linked genes are weakly or not expressed in both males and asexual females. When increasing the minimum expression levels thresholds, males X/AA ratio increases and approaches 1

when genes with a mean male and female RPKM (considered as mean RPKM hereafter) superior to 2 are considered. Above this expression level threshold, X-linked and autosomal genes are evenly expressed in males (Wilcoxon rank sum tests with $p > 0.05$). On the other hand, asexual females XX/AA ratios are in all cases lower than 1 (ranging from 0.45 to 0.69, $p < 10^{-16}$ in all cases), suggesting that females autosomal genes are in average more expressed than X-linked genes.

Similar results were obtained using an increasing minimum transcripts per million (TPM) filter separately in sexual females, asexual females and males, ranging from a minimum TPM of 1 to 100 (Fig. 2a–d). The X/AA ratio of males quickly goes up to 0.8, when very lowly expressed genes are filtered on both autosomes and the X chromosome (Fig. 2a). Dosage compensation for males, i.e., when their X/AA ratio is between 0.95 and 1.05 for Wilcoxon rank sum tests with $p > 0.05$), is attained with a minimum TPM threshold of 34–74 (overshadowed in gray, Fig. 2a, b). Both asexual and sexual females share a very similar expression pattern with a maximum XX/AA expression ratio of 0.8 (Fig. 2a) and with Wilcoxon rank sum tests that are always significant ($p < 0.05$), meaning that the distribution of expression of X-linked and autosomal genes is significantly different, despite the two X chromosomes of females (Fig. 2b).

To further investigate the uneven expression of X-linked and autosomal genes in the females, we classified genes (with mean RPKM >2 in males or asexual females) into four non-exclusive classes (*all, female-biased, male-biased and unbiased* genes) based on differential expression analysis (see "Methods," Fig. 2) between males and asexual females (referred to as females hereafter), since FAIRE has been performed on these two sexes. When considering *all* genes (Fig. 2i), violin plots for females display a large group of lowly expressed genes on the X chromosome that pulls down the females' median expression of X-linked genes. This lowly expressed gene group on the females X corresponds to *male-biased* genes (Fig. 2k), and when those are not taken into account, such as in the *unbiased* gene class (Fig. 2l), the XX/AA ratio is approaching 1 (ranging from 0.79 to 0.92 for genes with a mean RPKM >1–4). This suggests an almost balanced transcription in the females of the X-linked and autosomal *unbiased* genes, although the difference in expression between chromosomes is still significant (Wilcoxon rank sum test, $p$ ranging from $7.10^{-7}$ to 0.03 for genes with a mean RPKM >1–4). Such group of lowly expressed genes is absent in males (Fig. 2e–h), thus leading to an equal expression of X-linked and autosomal genes for *all* (Fig. 2e) and *unbiased* (Fig. 2h) gene classes. Moreover, the number of each gene class on autosomes and on the X (Fig. 2m, n,

**Table 1 Ratio of expression between X chromosome(s) and autosomes for males and females using minimum RPKM threshold**

| Genes taken into account | Number of genes retained | | Female | | Male | |
|---|---|---|---|---|---|---|
| | X | A | XX/AA ratio | p value* | X/AA ratio | p value* |
| All genes | 13,708 | 19,230 | 0.00 | <2.20E−16 | 0.00 | <2.20E−16 |
| Genes with RPKM >1 | 1886 | 9087 | 0.45 | <2.20E−16 | 0.83 | 1.16E−04 |
| Genes with RPKM >2 | 1484 | 7871 | 0.58 | <2.20E−16 | 0.94 | 0.372 |
| Genes with RPKM >3 | 1260 | 7010 | 0.62 | <2.20E−16 | 0.99 | 0.891 |
| Genes with RPKM >4 | 1069 | 6275 | 0.69 | <2.20E−16 | 0.91 | 0.119 |

All genes or different mean RPKM cutoffs have been considered to filter expressed genes

* p values were calculated using Wilcoxon rank sum tests to compare X-linked and autosomal genes expression for a given morph

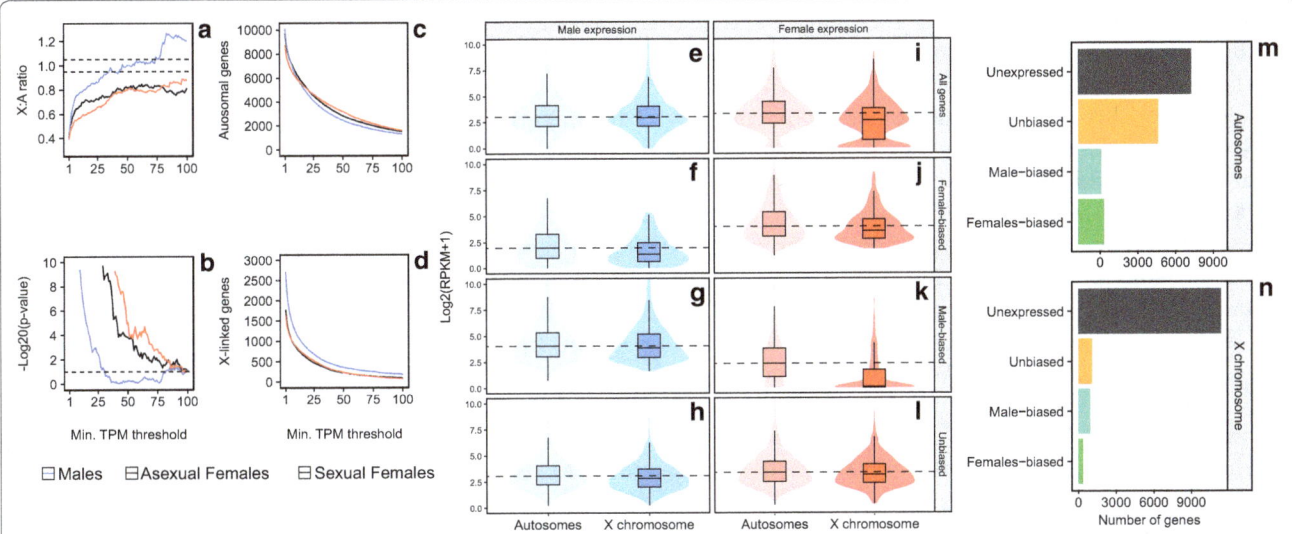

**Fig. 2** Expression data analysis to investigate dosage compensation in *Acyrthosiphon pisum*. **a–d** Comparison of the X-linked and autosomal genes transcription in males and both asexual and sexual females using increasing minimum TPM threshold to filter expressed genes. **a** The X/AA and XX/AA expression ratio of males (*blue*) and both asexual (*red*) and sexual (*black*) females, respectively. The *dashed lines* represent the dosage compensation criteria (X/A ratio between 0.95 and 1.05). **b** −Log20 (*p* value) of the Wilcoxon rank sum test comparing the expression of X-linked and autosomal genes. The *dashed line* represents *p* value = 0.05; all values below that line correspond to not significant Wilcoxon rank sum test. **c**, **d** Number of analyzed genes of A and B considering the expression filtration. Filters corresponding to dosage compensation for males are highlighted in *gray*, i.e., Wilcoxon rank sum test *p* value >0.05 and X/A ratio between 0.95 and 1.05. **e–l** Distribution of expression of autosomal and X-linked genes. Expression is represented for *all* (**e**, **i**), *female-biased* (**f**, **j**), *male-biased* (**g**, **k**) and *unbiased* (**h**, **l**) genes for males (*blue*) and females (*red*). These classes of expression have been determined using EdgeR (see "Methods"). The violin plots represent the gene density at each log2(RPKM + 1) level. *Dashed lines* correspond to the median of expression of autosomal gene in each category. **m**, **n** Number of genes belonging to each gene class on the X chromosome and on the autosomes

respectively) reveals that *male-biased* genes are almost three times more represented than *female-biased* genes on the X while even numbers are observed on the autosomes for these two classes. This further explains the fact that the expression of the X chromosome in females is driven down compared to autosomes, and thus the 0.8 XX/AA ratio. When taking the results of both males and females into account, our analyses show that the XX/AA and X/AA ratios are approaching 1, especially for

*unbiased* genes (Fig. 2h, l). Since males carry only one X per cell, and females two, these results suggest a global readjustment of X-linked genes transcription in males or females.

### Raw FAIRE-seq data analysis

To compare chromatin accessibility in males and females, we used the Formaldehyde-Assisted Isolation of Regulatory Elements (FAIRE) procedure to extract the DNA associated with NDR from whole individuals and for three pools of males and parthenogenetic females [39, 40], as well as a pool of control DNA for each sex. FAIRE experiments on sexual females were unsuccessful, which might be related to the important quantity of yolk contained in the eggs, interfering with the FAIRE procedure. We thus focused on asexual females and males for the FAIRE experiment. An average FAIRE ratio (measured as in 36) of 1.43% ($\pm$0.54) and 2.96% ($\pm$0.63) was obtained for males and asexual females, respectively (referred to as females hereafter). After sequencing the six FAIRE and the two Control DNA samples, FAIRE and Control Reads were mapped onto pea aphid genome and filtered to only conserve uniquely mapped reads.

The coverage of autosomes and X chromosomes was assessed in the control DNA libraries for both males and females (Additional file 1) yielding 14.3 million and 28.0 million 100-bp paired-end uniquely mapped reads for females and males, respectively. The female control library showed an equal coverage for autosomes and X chromosome which is expected since the female is diploid at both X chromosome and autosomes. In males—that are diploid for autosomes and haploid for the X chromosome—autosomes display an expected twofold higher coverage than the haploid X chromosome in males.

The reproducible FAIRE biological replicates for males and females were pooled according to their reproducibility using the MACS2 peakcaller [41] followed by irreproducible discovery rate (IDR) [42, 43] peaks ranking analysis. For males, these algorithms resulted in the discrimination of one of the three male replicates as lowly correlated with the other two. After pooling the relevant FAIRE replicates, three female and two male libraries were then conserved which ended up with 20.5 million and 23.6 million 100-bp paired-end reads mapping to unique positions in the genome for female and male pools, respectively. MACS2 [41] and IDR [42] also allowed us to retain a set of reproducible FAIRE peaks across the biological replicates for each sex: 8143 FAIRE peaks for males and 6369 FAIRE peaks for females were then identified; 39% of these are overlapping between males and females and are thus non-sex-specific (Fig. 3a). In order to assess the level of correlation between the control and retained FAIRE replicates, we concatenated these peaks coordinates into a set of 10,433 FAIRE peaks using BEDTools [44] and performed a Pearson correlation and a hierarchical clustering using deepTools2 [45] (Fig. 3b). Control libraries are alike (Pearson's $R^2$ of 0.83) and segregate together. Female and male FAIRE libraries are separated in the hierarchical clustering, and Pearson's $R^2$ values within each sex are high: 0.88 for males and more than 0.95 for female libraries comparisons.

To describe the overall distribution of the identified FAIRE peaks across the genome, we calculated the abundance of FAIRE peaks overlapping with different genomic features (promoters, TSSs, UTRs, exons, introns and intergenic regions) (Fig. 4). 5'UTRs and TSSs are characterized by high FAIRE peaks densities (Fig. 4a): Although these genomic features represent, respectively, only 1.5 and 0.8% of the genome (Fig. 4b), they contain a large proportion of FAIRE peaks compared with the other genomic features that represent larger parts of the genome. The FAIRE peaks density is thus much lower in all other genomics features (Fig. 4a). Examples of genomic regions displaying some of these open chromatin peaks specific of each sex or in common between sexes near sex-specific or housekeeping genes are observable in Additional file 2.

## Open chromatin signal in males and females

To assess the pea aphid's chromatin opening profile, we calculated the FAIRE signal at the genome scale by dividing the normalized FAIRE coverage by the normalized Control coverage on 10-bp windows (hereafter named bins, as described by deepTools [45]). In Fig. 5a, the FAIRE signal is represented for each gene of the pea aphid genome and RNA-seq data have been used to rank genes according to their level of expression in males or in females. The strongest FAIRE signals are observed in regions upstream the TSS. The more a gene is expressed, the more its FAIRE signal upstream the TSS is high, indicating a positive correlation between gene expression and chromatin accessibility. The heatmaps representing the FAIRE signal along X-linked genes suggest a stronger FAIRE enrichment in males than in females, in comparison with the autosomes that share similar profiles between sexes (Fig. 5a).

We then calculated the mean genic FAIRE signal and 99% confidence interval (CI) for both males and females on autosomal and X-linked genes (Fig. 5b). According to the previous observation, autosomal genes share similar accessibility levels in males and females, with a slightly higher accessibility near the TSS in males than in females based on 99% CI (2.94 $\pm$ 0.13 in males and 2.59 $\pm$ 0.09 in females on the most accessible bin). Contrastingly, the mean FAIRE signal along and around X-linked genes is almost two times higher in males than in females, especially near the TSS (2.40 $\pm$ 0.18 for males and 1.45 $\pm$ 0.07 for females on the most accessible bin). In order to verify whether the FAIRE signal normalization on Control DNA could introduce a bias for the X chromosome, we calculated the mean FAIRE coverage normalized by sequencing depth around all autosomal and X-linked genes in males and females (Additional file 3). It appears that genes on the autosomes and on the X chromosome are equally accessible between males and females, despite the single X in males compared to the two X

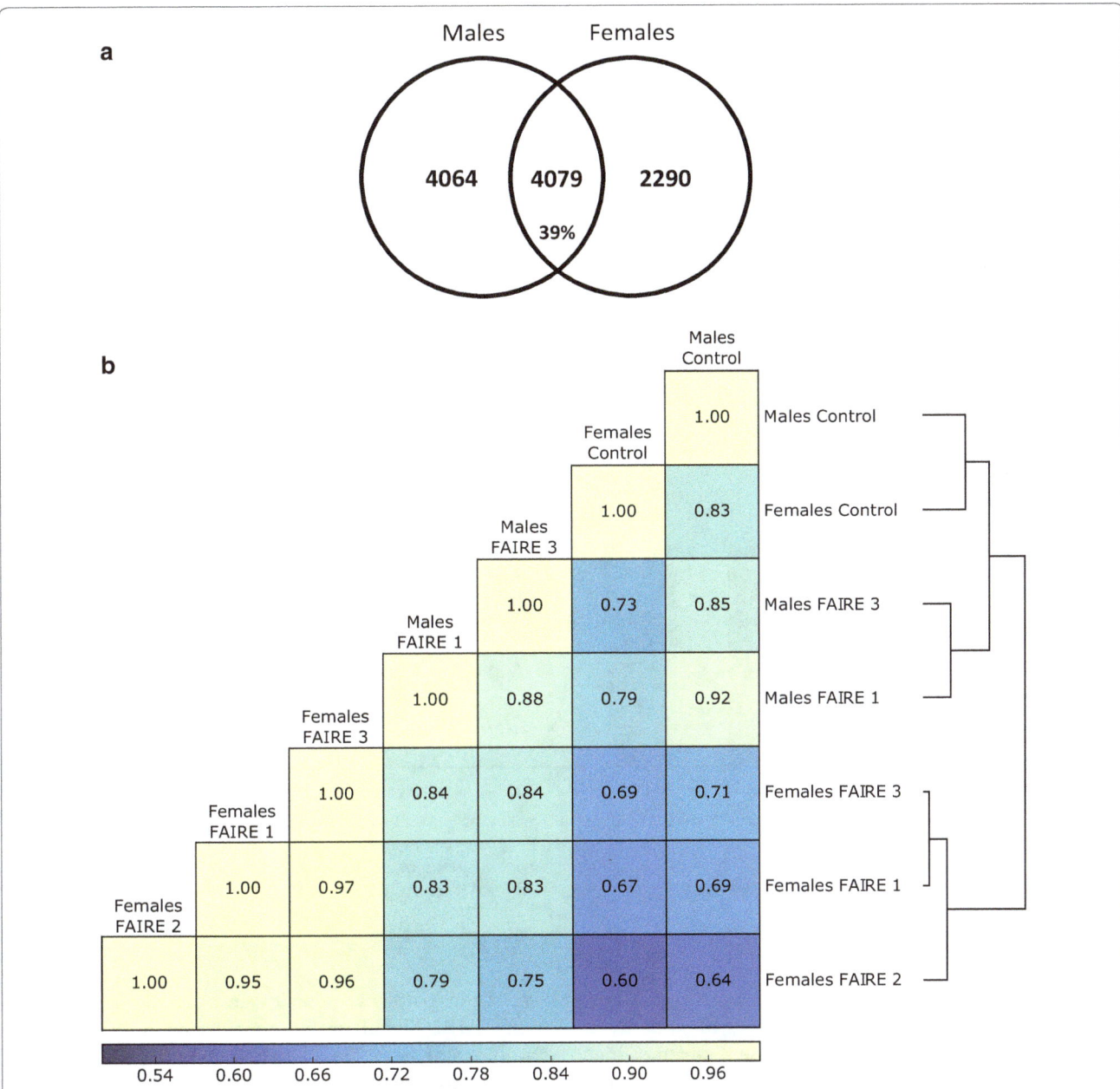

**Fig. 3** FAIRE peaks obtained after MACS2 and IDR analyses and clustering of FAIRE and Control libraries. **a** Number of FAIRE peaks in both males and females (total of 8143 and 6369, respectively). Their specificity between the two sexes is represented in the Venn diagram. **b** Clustering of the FAIRE and Control samples based on Pearson's correlations. Correlation coefficients are shown and were calculated from the reads coverage of 10-bp bins along all open chromatin regions retained by MACS2 followed by IDR. The Males FAIRE 2 library is not included since it has been discarded by IDR because of its low reproducibility compared to Males FAIRE 1 and 3

chromosomes of females. This profile does not invalidate the results observed using input normalized FAIRE signal, since the single X chromosome of males is still more accessible than each female X chromosome, and is thus differentially accessible.

We then investigated the FAIRE signal within intergenic regions in males and females. The assessment of the chromatin accessibility in these intergenic regions was performed around the summit of the intergenic FAIRE peaks retained by IDR. FAIRE peaks that do not overlap any annotation feature were thus extracted and defined as intergenic FAIRE peaks. The mean FAIRE signal and 99% CI were calculated 1000 bp around the summits of those intergenic FAIRE peaks for males and females (Fig. 6). On

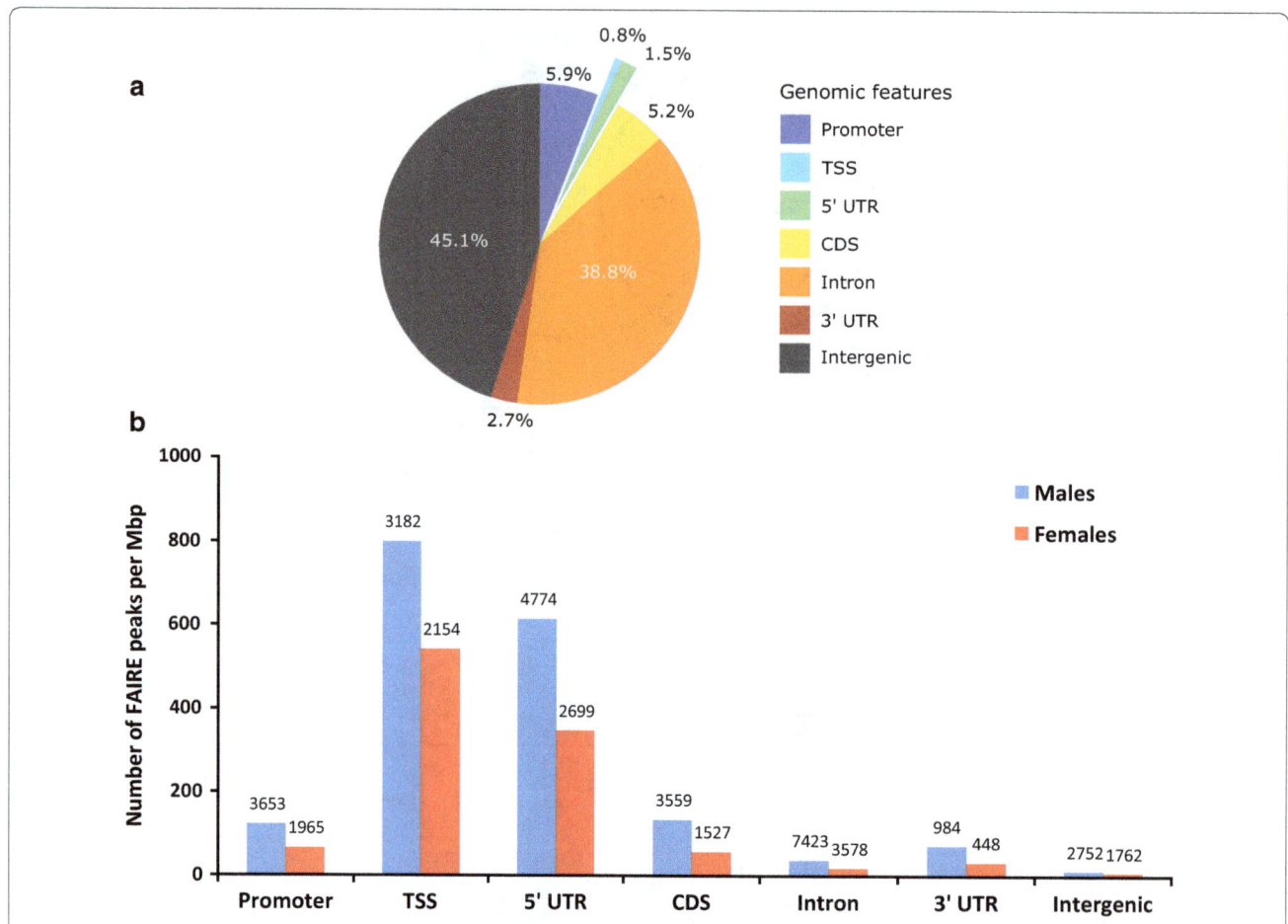

**Fig. 4** FAIRE peaks density for the different genomic features of *A. pisum*. **a** Overlapping of the identified FAIRE peaks using MACS2 and IDR with the various genomic features of *A. pisum* genome annotation v2.1 expressed as the number of peaks per million of base pairs (peaks per Mb) for male (*blue*) and female (*red*) individuals. The raw number of peaks is *indicated above each bar*. The sum of the raw FAIRE peaks is not equal to the total peaks presented in 2. A since one peak can overlap multiple genomic features. **b** Percentage of the different genomic features in the pea aphid genome annotation (aphidbase v2.1)

autosomal intergenic FAIRE peaks, no significant differences of chromatin accessibility were identified between males and females. Contrastingly, for the X chromosome, intergenic FAIRE peaks in males are significantly more accessible than in females, especially near the peaks summits (mean FAIRE signal of $14.90 \pm 2.46$ for males and $9.35 \pm 0.82$ for females on the most accessible bin). The results for intergenic regions are thus similar to those observed for genic regions and demonstrate a comparable accessibility of the autosomes in the two sexes and an increased accessibility of the single X in males compared to the two X chromosomes in females.

**Link between genes expression and chromatin accessibility**
To explore the link between gene expression and chromatin accessibility, and especially around the TSSs, we grouped genes into four classes depending on their

expression patterns in males and females and studied their FAIRE signal profile considering both sexes and chromosome type (Fig. 7): *unexpressed* (Fig. 7a, e) (46.8% on the autosomes, 82.7% on the X), *male-biased* (Fig. 7b, f) (9.6% on the autosomes, 6.8% on the X), *female-biased* (Fig. 7c, g) (10.6% on the autosomes, 2.6% on the X) and *unbiased* genes (Fig. 7d, h) (33.0% on the autosomes, 7.9% on the X). These classes correspond to the ones defined during the RNA-seq experiment (Fig. 2). On the autosomes, *unexpressed*, *male-biased* and *female-biased* gene classes share a similar chromatin accessibility profile in males and females (Fig. 7a–c), and only *unbiased genes* (Fig. 7d) are slightly more accessible in the males TSS compared to females ($4.41 \pm 0.32$ in males and $3.68 \pm 0.22$ in females on the most accessible bin). Contrastingly, on the X chromosome (Fig. 7e–h), two main profiles can be distinguished. The first group is comprised

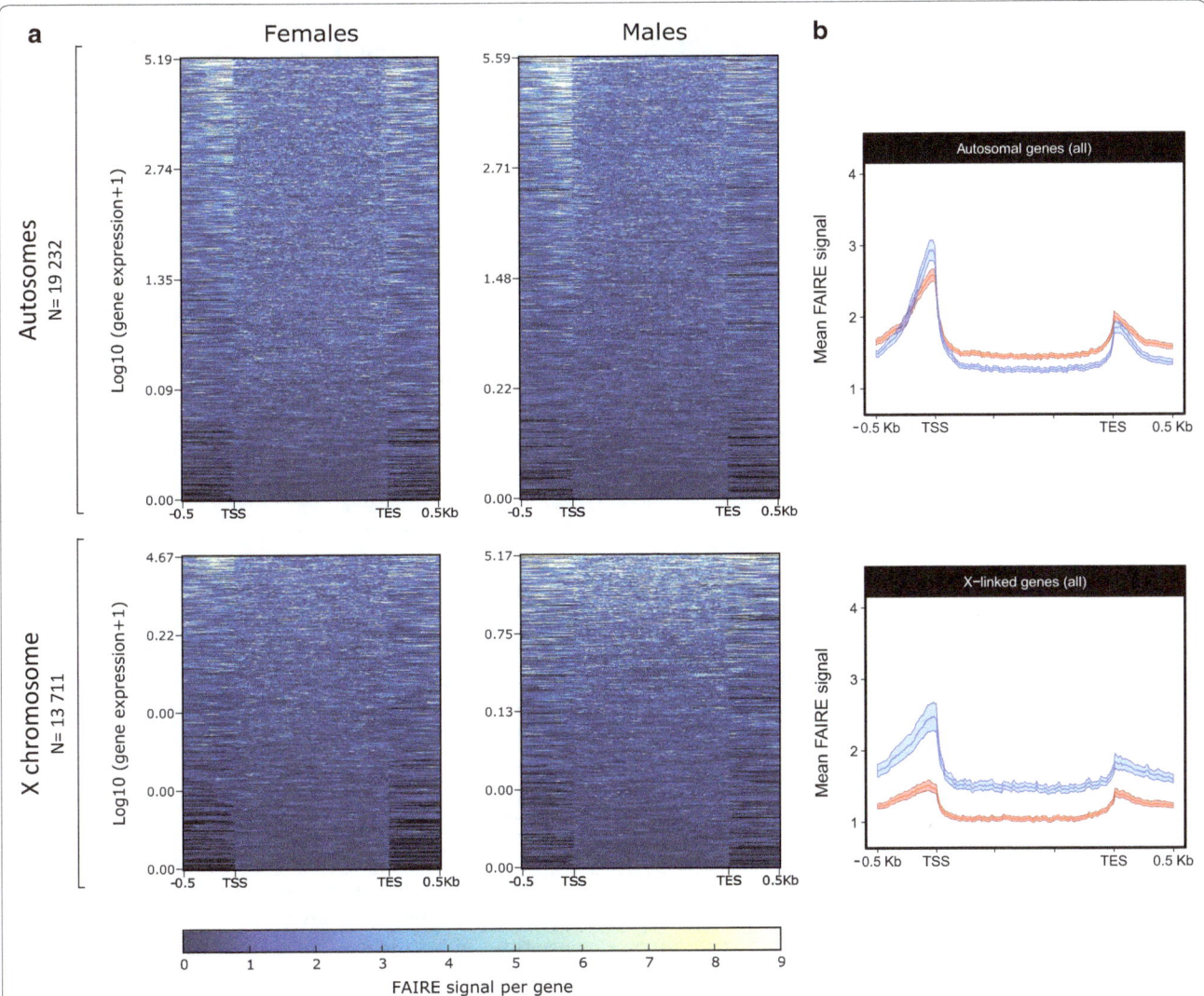

**Fig. 5** FAIRE signal profiling of *A. pisum* females and males for all autosomal and X-linked genes. The input normalized FAIRE signal has been calculated by dividing FAIRE reads by Control reads in bins of 10 bp around (500 pb upstream the TSS and 500 bp downstream the TES) and in the genes body that all have been equalized to 1500 bp. **a** FAIRE signal is shown for every gene carried by the autosomes (*top, black*) or by the X chromosome (*bottom, gray*). Genes are ranked from the most expressed to the least expressed depending on the sex and on the chromosome type. **b** Mean FAIRE signal calculated for all autosomal genes (*top, black*) and X-linked genes (*bottom, gray*) for males (*blue*) and females (*red*). 99% CI based on 1000 bootstrap is shown

of *unexpressed* and *male-biased* genes (Fig. 7e, f) that display comparable—yet significantly different—FAIRE signal in males and in females: 1.56 ± 0.09 for males and 1.17 ± 0.05 for females regarding *unexpressed* genes and 3.27 ± 0.72 for males and 2.05 ± 0.33 for females concerning the *male-biased* genes. The second group is composed of *female-biased* and *unbiased* genes (Fig. 7g, h) that display a stronger chromatin accessibility difference between sexes. Mean FAIRE signals of 10.72 ± 3.06 in males and 4.41 ± 0.84 in females are observed for *female-biased genes* and those of 8.80 ± 1.55 in males and 3.95 ± 0.52 in females are observed for *unbiased*

*genes*. These results suggest that *unbiased* and *female-biased* genes participate in majority to the observed global enhanced accessibility of the X chromosome in males compared to females (Fig. 5), even if they represent only 10% of X-linked genes. Interestingly, we can also observe that in females for *unbiased* (Additional file 4H) and *female-biased* (Additional file 4G) gene classes, chromatin accessibility is similar between the X chromosome and autosomes. On the contrary in males and for the same gene classes, chromatin accessibility is higher on the X chromosome than on autosomes (Additional file 4C, D). This suggests that autosomal genes in males

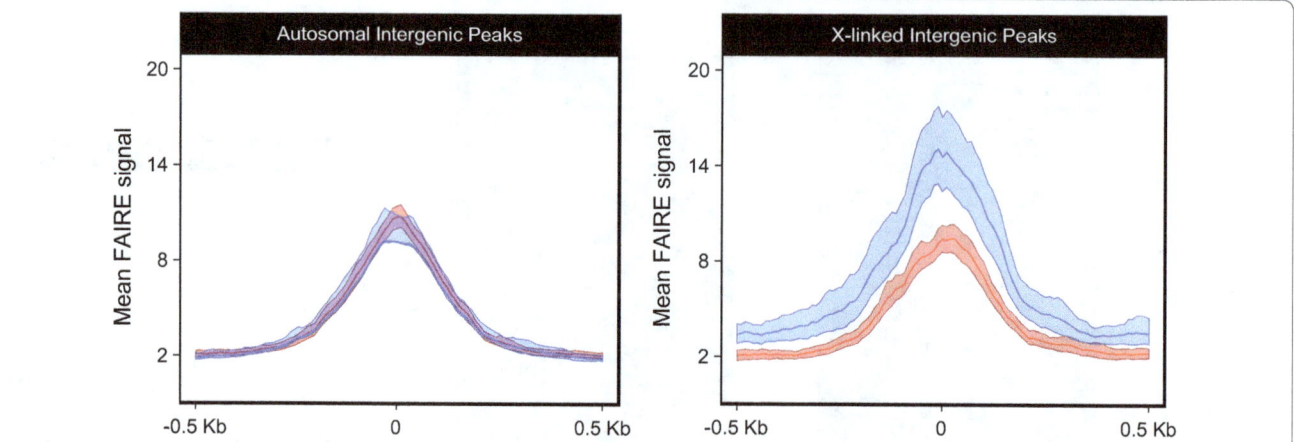

**Fig. 6** Mean FAIRE signal of X-linked and autosomal intergenic peaks. The input normalized FAIRE signal has been calculated around intergenic FAIRE-enriched peaks identified using MACS2 and IDR for males (*blue*) and females (*red*) on the autosomes (*black*) and X chromosome (*gray*). Overlapping peaks are enriched FAIRE regions in common between males and females, while sex-specific FAIRE peaks do not show any overlap between the two morphs

and females and X chromosome in females share similar chromatin patterns and that the differential X chromosome chromatin pattern between sexes is explained by an enhancement of chromatin accessibility in males rather than a reduced accessibility in females.

## Discussion

In this study, we first reanalyzed whole-body RNA-seq data of pea aphid males and females and evidenced a potential dosage compensation mechanism. We then generated an overview of the open chromatin structure (nucleosome-depleted regions) of whole-body pea aphid males and females using the Formaldehyde-Assisted Isolation of Regulatory Elements. This first genome-wide epigenetic study in aphids demonstrated an enhanced chromatin accessibility of the males' single X chromosome compared to the two X of the females.

### Global transcriptomic profiles suggest potential dosage compensation in pea aphid males

Using a total of 3712 genes assigned to autosomes or the X chromosome, Jaquiéry et al. [34] outlined that in the pea aphid, autosomal genes display similar transcription levels between males and females while X-linked genes are slightly more expressed in males than in females. Based on the same data, Pal and Vicoso [46] outlined different results as they found that males and females share similar expression levels on both X and autosomes. These contradictory results can, however, be explained by a misassignment of genes to chromosomes due to a widespread scaffold misassembly in the pea aphid genome [38]. Here, we used a new assignation of scaffolds to chromosomes [38] to analyze X-linked and autosomal gene expression patterns in females and males.

In males, we showed X/AA expression ratios almost equal to 1 when taking into account consistently expressed genes (mean RPKM >2). Similar expression patterns in organisms showing complete dosage compensation have been demonstrated such as in *Manduca sexta* heads [17] where the Z/AA ratio is approaching 1 when lowly expressed genes are not considered. These results suggest a potential dosage compensation mechanism in the male pea aphids.

Contrastingly, the females XX/AA expression ratios are inferior to 1 which is not a common feature within male heterogametic species with complete dosage compensation. Nevertheless, such an unusual ratio can also be observed in the flour beetle (*Tribolium castaneum*). In this species, females display *a* XX/AA ratio superior to 1 which is explained by the fact that females show a global X-linked genes overexpression (supported by an enrichment of female-biased genes on the X), while males show a X/AA ratio equal to 1, ending up with a XX/X ratio different from 1 [47]. In the pea aphid, the X chromosome is predicted to be masculinized [34], and we indeed observed that X-linked *male-biased* genes are almost three times more represented than X-linked *female-biased* genes and that most of the X-linked *male-biased* genes are male specific (i.e., expressed in males but not at all in females, as shown in Fig. 2k). As a result, XX/AA female ratio and female-to-male XX/X ratio are inferior to 1. It has to be noted that mRNA extractions were performed on whole individuals and thus contain mRNAs from both somatic and germ cells. The fact that most X-linked *male-biased* genes are almost male specific suggests that these genes could be testis specific or more generally involved in male phenotypic traits. GO terms enrichment analyses on *female-biased*, *male-biased* and

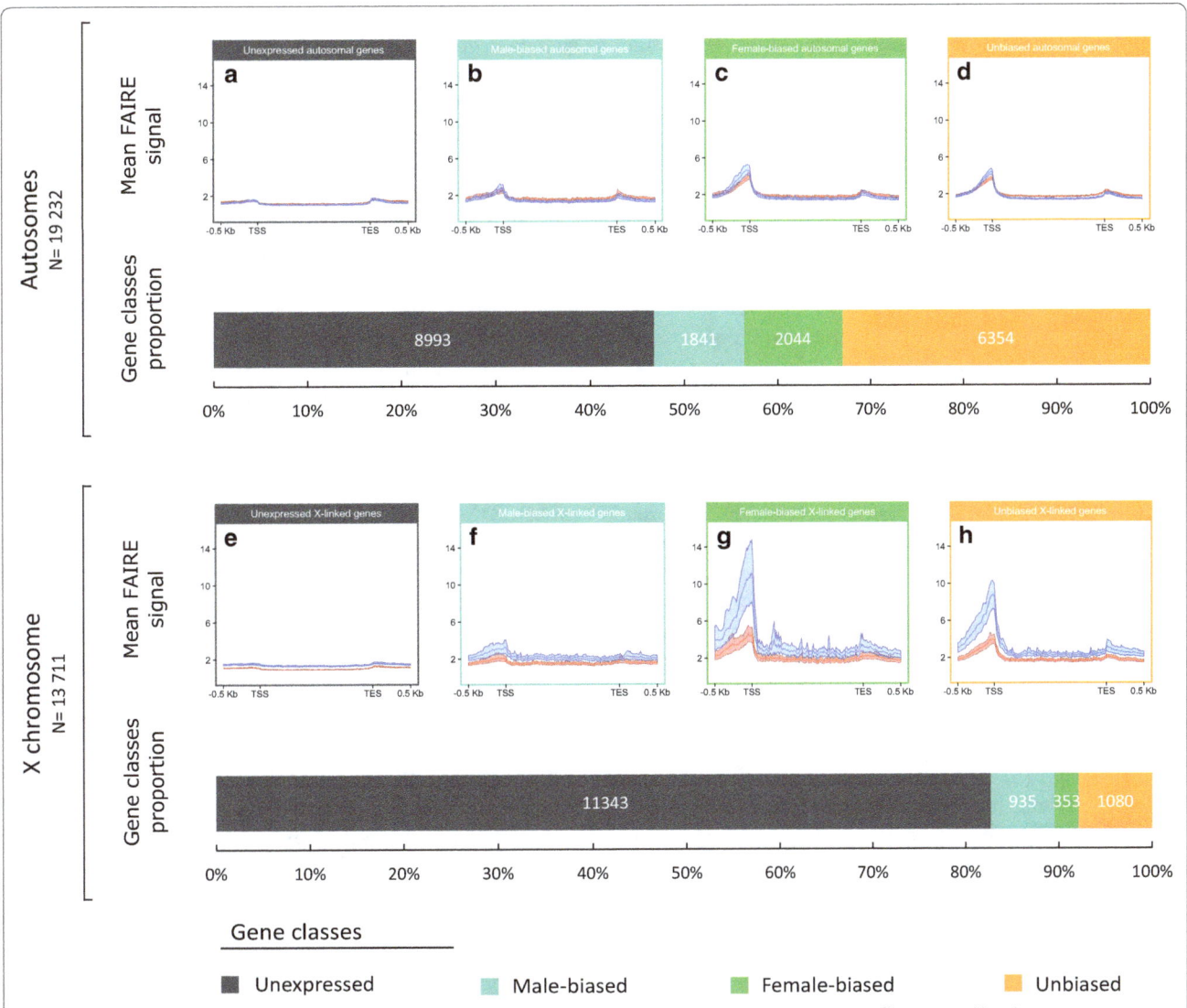

**Fig. 7** Mean FAIRE signal and proportion of X-linked and autosomal genes depending on their expression profile, grouped by chromosome type. Genes have been categorized into four different classes: *unexpressed* (**a**, **e**), *male-biased* (**b**, **f**), *female-biased* (**c**, **g**) and *unbiased* (**d**, **h**) genes between the two sexes based on gene expression data. The mean input normalized FAIRE signal has been calculated around (500 bp) each gene class and in the genes body (scaled) depending on the sex (males in *blue* and females in *red*) and on the chromosome type (autosomes, *top* and X chromosome, *bottom*). 99% CI based on bootstraps is also shown around the mean. The proportion and the number of the four gene classes are represented for autosomes and for the X chromosome. The same results grouped by sex instead of chromosome type are available as Additional file 4

*unbiased* gene classes on the X and autosomes revealed that only *male-biased* and *female-biased* genes on the X display reproductive-related functions. *Male-biased* X-linked genes display an enrichment of biological processes related to male-specific reproduction such as cilium movement, microtubule-based movement, sperm competition and multi-organism reproductive process (Additional file 5). Because the pea aphid's spermatozoids are flagellated [48], such functions are likely to correspond to testis-specific genes, thus reinforcing the hypothesis of their absence of expression in females

because of their male germ cells specificity. The unusual females XX/AA ratio (<1) for a male heterogametic species can thus be explained and is compatible with complete dosage compensation in the male pea aphids.

In mammals and in Drosophila [20, 49], dosage compensation mechanism activity on the X chromosome can be different between germ cells and somatic cells. Mammals do not show any dosage compensation in the germ cells where the X chromosome inactivation in females is withdrawn in developing primordial germ cells, allowing X-linked genes expression in female gametes [20].

Contrastingly, *D. melanogaster* germ cells compensate for gene dose in germlines with, however, distinct mechanisms than the well-known somatic cells dosage compensation [49]. These studies suggest that dosage compensation does not always occur in germ cells. Since the RNA-seq data analyzed in our study comprise mRNAs from germ cells and somatic cells, we cannot conclude whether the potential dosage compensation in the male pea aphids occurs only in somatic cells or in both somatic and germ cells.

## Dosage compensation in the pea aphid might be achieved by an enhanced chromatin accessibility of the X chromosome in males

Dosage compensation is often controlled by epigenetic mechanisms mediating the X or Z chromosome chromatin accessibility. Since our RNA-seq analyses suggest potential complete dosage compensation in the male pea aphids, we aimed at assessing the global chromatin accessibility of the autosomes and the X chromosome(s) in both males and females using FAIRE-seq. The FAIRE methodology allows the identification of open chromatin (nucleosome-depleted) regions. It has been performed for the first time by Giresi et al. [39] and since then has been used in various organisms and tissues [39, 40, 50]. Methods such as assay for transposase-accessible chromatin (ATAC) [51–53] are becoming alternatives to the FAIRE overtime; however, the latter remains a reliable way to assess open chromatin regions at the genome scale when coupled to high-throughput sequencing.

To validate the FAIRE-seq approach used here, we checked the presence of a high FAIRE peaks frequency in TSS and 5′UTR regions, as expected since TSSs are known as the most accessible regions in the genome in order to allow the binding of the RNA polymerase II and thus the transcription of the gene [40, 54–56]. We also observed a positive correlation between genes expression and chromatin opening upstream their TSS, which corresponds to a typical profile of a successful FAIRE experiment [40].

With our FAIRE-seq approach, we then showed a similar mean chromatin accessibility of autosomes in males and females over all genes (Fig. 5b; Additional file 3) and also when genes are classified according to their expression patterns in males and females (Fig. 7; Additional file 4). Similar chromatin accessibility in males and females is also observed within open intergenic regions. These results suggest that the autosomes of the pea aphid are equally accessible in males and females.

Contrastingly, the FAIRE signal analysis over *all* genes (Fig. 5b) revealed that X-linked genes are more accessible in males than in females, or evenly accessible despite the single X of males using sequencing depth normalization

alone (Additional file 3). *Female-biased* and *unbiased* X-linked genes (Fig. 7g, h) display the most important difference of chromatin accessibility between males and females and thus contribute at most to the global male-enhanced chromatin accessibility of the X chromosome, suggesting that the dosage compensation in males might be partial. Since *unbiased* genes have (by construction) similar expression pattern between males and females, X-linked *unbiased* genes correspond to potentially compensated genes. A higher chromatin accessibility of these genes in males than in females thus supports the hypothesis of potential epigenetic mechanism underlying dosage compensation in males. Surprisingly, X-linked *female-biased* genes chromatin accessibility is comparatively higher in males than in females (Fig. 7g). This cannot be used to reject this hypothesis since in *D. melanogaster* males, some female-biased genes also show an enrichment of the active histone mark H4K16ac mediated by the DCC, thus resulting in an enhanced chromatin accessibility for these genes [57]. Additionally, X-linked intergenic FAIRE peaks, which represent only a very small proportion of intergenic regions, are significantly more accessible in males than in females. Altogether, these data suggest a global regulation of the X chromosome chromatin accessibility in order to compensate for gene dose in males.

Interestingly, the chromatin accessibility of X-linked *male-biased* genes is only slightly enhanced in males, despite their high level of expression in that morph. As suggested by our GO analyses, these genes could correspond in majority to testis-specific genes (hence including germ cells-specific genes). The relatively low chromatin accessibility of these genes could be explained on the one hand by the fact that FAIRE is a population assay and that male-specific cells make up only a small percentage of the cell population in the analyzed whole-body male individuals. The FAIRE signal could thus be low for the particular class of *male-biased* genes, hence resulting in a low difference in FAIRE signal between males and females for that gene class. On the other hand, these results could suggest that the epigenetic mechanism underlying the male-enhanced X chromosome chromatin accessibility does not take place in male germ cells, like in the female mammals for example [20]. *Female-biased* and *unbiased* genes might then be part of a given X chromosome territory compensated for the lack of a second X chromosome in the males' genome, while the *male-biased* and *unexpressed* genes might be part of another X chromosome territory where the chromatin accessibility is not submitted to a given mechanism underlying dosage compensation. Such hypothesis would resemble what has been identified in the Drosophila where only 75% of the X chromosome territory is regulated by the DCC [29].

The high chromatin accessibility of the male single X chromosome might be the consequence of an underlying epigenetic mechanism taking place either in male (enhanced chromatin accessibility) or in female individuals (reduced chromatin accessibility). When taking *unbiased* and *female-biased* gene classes into account (Fig. 7c, d, g, h), we can observe that autosomal genes in males and females as well as X-linked genes in females are equally accessible, whereas X-linked genes are more accessible only in males. The differential X chromosome chromatin accessibility observed between sexes can thus be explained by an augmentation of such accessibility in males rather than a reduction in females. This suggests the existence in pea aphids males of an epigenetic mechanism promoting the enhancement of the X chromosome chromatin accessibility associated with dosage compensation. These epigenetic patterns resemble the Drosophila model where an overall increase in chromatin accessibility is observed on the males X chromosome through a male-specific H4K16ac histone posttranslational modification enrichment mediated by the DCC in the gene bodies (H4K16ac being enriched in the genes' TSS and promoter by the NSL complex) [28, 57]. Interestingly, the pea aphid genome contains protein coding for genes homologous to the five proteins composing the *D. melanogaster* DCC (Additional files 6, 7). The DCC is also constituted by two additional *roX* lncRNAs that were not found in the pea aphid genome, but it is noteworthy that lncRNA sequences are rarely conserved between organisms [58]. Interestingly, homologs identities and similarities between *D. melanogaster* and *A. pisum* DCC proteins are found particularly in functional domains. The most conserved proteic domains between the Drosophila and the pea aphid are MOF's MOZ-SAS domain and all MLE's proteic domains, namely dsrm, DEAD, helicase C, HA2 and OB NTP BIND (Additional file 7). MOZ-SAS domain from Drosophila's MOF is involved in the acetylation of lysine 16 of histone 4 (H4K16ac) and is thus responsible for the enhanced chromatin accessibility along the Drosophila males X chromosome. The high proteic domain conservation observed between the pea aphid and the Drosophila could lead to conserved functions of the MOF and MLE proteins in these two organisms. However, functional analyses are required to validate this hypothesis. Moreover, some of the proteins composing the Drosophila's DCC, including MOF and MLE, has been identified as conserved in mammals without, however, playing a role in dosage compensation [29, 59–63]. The pea aphid could then, independently of a DCC complex, display a different epigenetic mechanism supporting dosage compensation by globally enhancing chromatin accessibility of the single X chromosome in males, an alternative that cannot be ruled out taking into account the rather weak conservation of MSL2 protein sequence between Drosophila and the pea

aphid (Additional file 7), particularly in the RING domain which plays a role in the DCC assembly [64].

## Conclusions

This study gives a first insight into *A. pisum* chromatin accessibility patterns in relation to a possible dosage compensation mechanism. The males single X chromosome is globally more accessible than the two female X chromosomes. More importantly, X-linked genes showing similar expression levels between females and males and that could potentially be compensated in the latter are more accessible in males. Further experiments—especially chromatin immunoprecipitation followed by high-throughput sequencing targeting specific histone marks—must be conducted in order to characterize the underlying epigenetic mechanism involved in this overall enhanced chromatin accessibility of the males single X chromosome.

## Methods
### Aphids rearing

*Acyrthosiphon pisum* individuals from the clone LSR1 (the reference clone that was used for genome sequencing [37]) were reared on broad bean *Vicia faba* at low density (less than five individuals per plant) to prevent the production of winged morphs for at least two generations. Parthenogenesis was maintained under a 16-h photoperiod and a temperature of 18 °C. At the third generation, 20 asexual females were then directly frozen into liquid nitrogen for FAIRE extraction. The production of male individuals was initiated by transferring larvae from a 16-h to a 12-h photoperiod at the same temperature of 18 °C [34]. Two generations later, males were produced. A total of 100 adult males were then directly frozen into liquid nitrogen for FAIRE extraction. No RNA extraction was performed since already published RNA-seq data were used [34].

### RNA high-throughput sequencing

We reanalyzed six RNA-seq libraries used by Jaquiéry et al. [34]. Briefly, these six libraries correspond to three male libraries and three parthenogenetic female libraries of clone LSR1. Details regarding aphid rearing, RNA extraction, libraries preparation and sequencing are provided in Jaquiéry et al. [34]. Libraries were mapped on the version 2 of the pea aphid genome assembly using TopHat2 (RRID:SCR_013035) default parameters [65]. The number of reads covering each CDS of the gene prediction v2.1 was then counted using HTSeq-count (RRID:SCR_011867) [66] with the following parameters: −m intersection-strict −s no −t exon. The numbers of mapped reads per library ranged from 7.3 to 20 million, with an average over libraries of 14.3 million reads. Raw read counts were normalized in R (RRID:SCR_001905) [67] with edgeR (RRID:SCR_012802) [68] package by sequencing depth using the TMM method and by genes length (RPKM and TPM calculation). Genes were filtered using increasing minimum TPM threshold ranging from 1 to 100 using a step of 1 in order to remove the less expressed genes. Within each TPM filtration step, X/A ratio in males and females, Wilcoxon rank sum test $p$ value and the number of retained X-linked and autosomal genes have been calculated. Differential expression between males and parthenogenetic females for each gene was tested with edgeR, considering the different libraries for each morph as replicates using genewise exact tests for differences in the means between two groups of negative-binomially distributed counts, based on normalized read counts. Genes have then been grouped into four classes: *unexpressed*, *male-biased*, *female-biased* and *unbiased* genes. *Unexpressed* genes have been determined as such if they displayed less than 1 count per million in at least three libraries as in [69]. *Male-biased* and *female-biased* genes have been determined as such using a FDR <0.05. *Unbiased* genes comprise differentially expressed genes with a FDR >0.05 and genes non-differentially expressed.

### Formaldehyde-Assisted Isolation of Regulation Elements, sequencing and bioinformatics analyses

The FAIRE extraction of three frozen pools of male and parthenogenetic female individuals was performed following the protocol of frozen tissues proposed by Simon et al. [40] with the following parameters. Tissues were first ground using a Biospec Bio-pulverizer and then fixed by the addition of 3% of Thermo Scientific Pierce formaldehyde during 8 min. The fixation was then stopped by the addition of glycine at 125 mM. Subsequently to the pellet rinsing and resuspension, the tissues were ground with a Tissue Lyser, using Qiagen metallic beads, during five cycles of 5 min, each being interrupted during 2 min. The sonication steps were performed using a Bioruptor Plus during 12 cycles of 30 s, each being interrupted during 30 s. Phenol chloroform extraction steps were performed using Sigma phenol chloroform. Subsequently to its extraction, the FAIRE DNA was purified the using ZYMO ChIP DNA Clean & Concentrator. Finally, FAIRE and Control DNA were quantified using a Quantus with the Quantifluor dsDNA kit. The reproducibility of the replicates was assessed before sequencing by calculating the FAIRE/Control ratio described in Simon et al. [40].

FAIRE and Control DNA were sequenced using the Illumina Hiseq 2000 instrument and ChIP TrueSeq kit. Three FAIRE samples for each morph were sequenced, while only a pool of the three samples from the same morph individuals was sequenced for control DNA. The eight different samples (six FAIRE and two Control samples) were 100-pb paired-end sequenced on a single lane. The raw sequenced data for these eight samples are available online at the NCBI under the BioProject accession number PRJNA348188. The reads were mapped using bowtie2 with default parameters [70, 71]. Only uniquely mapped reads with a mapping quality over or equal to 30 were kept using SAMtools (RRID:SCR_002105) [72], following the IDR recommendations [42, 43]. For peak calling, we analyzed separately the reads mapping on the X and on autosomes to avoid bias related to the difference of coverage of the X chromosome (haploid in males) and autosomes (diploid in males and females). In addition, Jaquiéry et al. [38] showed that over 50% of scaffolds greater than 150 kb are chimeras of X and autosomes, which could induce artificial peaks at breakpoints. Following IDR recommendations, MACS2 (RRID:SCR_013291) [41] was used to perform the peak calling with the following parameters: –f BAM –nomodel –extsize 200 –slocal 1000 –llocal 10,000 –p 0.05. The –gsize parameter (which corresponds to the total size of the genome analyzed) has been adapted depending on the size of chromosome types (166,018,072 bp for the X and 340,086,956 bp for autosomes). IDR analyses were then performed as suggested in [43] using a threshold of 0.04 for original replicates, of 0.04 for self-consistency replicates and of 0.01 for pooled pseudoreplicates. Once the final peak set for males and females was generated, HOMER suite (RRID:SCR_010881) [73] was used to calculate Venn Diagrams of males and females FAIRE peaks specificity. Segtools (RRID:SCR_004394) [74] with the aphidbase (RRID:SCR_001765) version 2.1 annotation of the pea aphid genome were used to assign the FAIRE peaks to various genomic features such as promoters (1300 bp upstream TSS regions), TSS (defined as 200 bp upstream the first base of 5′UTRs), 5′UTR, coding regions, introns, 3′UTR and intergenic regions. DeepTools2 [45] was used to calculate and represent the FAIRE signal as the ratio between pooled FAIRE reads over Control reads on bins of 10 pb on whole genome. FAIRE signal data were first retrieved along every gene and represented in heatmaps with deepTools2 [45]. The mean FAIRE signal was calculated using R and represented using the ggplot2 package [67, 75]. In order to estimate a confidence interval for each mean FAIRE signal calculated, bootstrap was done using a custom R script that performed 1000 random resampling of the genes by taking into account the number of genes in each resampled set. 99% confidence interval, which depends on gene number and FAIRE signal values, was then calculated and represented using the ggplot2 [75] package in the R software [67].

### Drosophila BLAST and GO terms enrichment of molecular processes in the pea aphid

Since *A. pisum* gene functions are poorly characterized, the most efficient way to perform GO terms enrichment in this organism was to compare unique *D. melanogaster* homologs of a given list of pea aphid genes of interest against all the unique *D. melanogaster* homologs found in the pea aphid annotation (aphidbase v2.1). In order to find *D. melanogaster* homologs in the pea aphid, BLASTp (RRID:SCR_001010) [76] analyses were performed using BLAST+ [77] with default parameters. GO terms enrichment of biological processes has then been done using GOrilla (RRID:SCR_006848) with default parameters [78, 79] using unique *D. melanogaster*'s homologs coming from a gene list of interest as "target set" and all the unique *D. melanogaster* homologs found in the pea aphid annotation v2.1 as "background set".

## Additional files

**Additional file 1.** FAIRE and Control libraries coverages. Number of reads and coverage of X chromosome (X) and autosomes (A) for male and female control libraries.

**Additional file 2.** Genome browser view of remarkable autosomal and X-linked regions displaying sex-specific and non-specific FAIRE-seq and RNA-seq signal. **A**, **D**: female-specific regions around the genes ACYPI003071 (uncharacterized protein) and ACYPI001644 (cuticular protein 44). **B**, **E**: male-specific regions around the genes ACYPI080359 (uncharacterized protein) and ACYPI081672 (uncharacterized protein). **C**, **F**: regions in common between males and females for the genes ACYPI000061 (ATP synthase subunit beta) and ACYPI006656 (molybdate-anion transporter). The RNA-seq and FAIRE-seq signals have been made equal between males and females for each region.

**Additional file 3.** Mean FAIRE coverage calculated for all autosomal genes (left, black) and X-linked genes (right, gray) for males (blue) and females (red). 99% CI based on 1000 bootstrap is shown around the mean. The FAIRE coverage has been normalized by read-depth in order to allow the comparison between males and females data.

**Additional file 4.** Mean FAIRE signal and proportion of X-linked and autosomal genes depending on their expression profile, grouped by sex. Genes have been categorized in four different classes: *unexpressed* (**A**, **E**), *male-biased* (**B**, **F**), *female-biased* (**C**, **G**) and *unbiased* (**D**, **H**) genes between the two sexes based on gene expression data. The mean FAIRE signal has been calculated around each gene class (500 bp) and in their gene body (scaled), depending on the chromosome type (autosomes in dark color, X chromosome in light color) and depending on the sex (males, top in blue and females, bottom in red. 99% CI based on bootstraps is also shown around the mean.

**Additional file 5.** GO enrichment of biological processes for autosomal and X-linked, male- and female-biased genes.

**Additional file 6.** Protein sequence comparison of the Drosophila's DCC in the pea aphid. Clustal multiple sequence alignment by MUSCLE (3.8) of *A. pisum*'s homologs of *D. melanogaster* DCC. Proteic domains of the Drosophila are represented in full color boxes above the alignments. *A. pisum* gene names found by BLASTp are indicated in parenthesis.

**Additional file 7.** Protein conservation of the Drosophila's DCC in the pea aphid. Proteic domains percentage of identity between *A. pisum* homologs of the five proteins composing the DCC of *D. melanogaster*.

## Abbreviations
H3K4me1: monomethylated lysine 4 of histone 3; H3K9ac: acetylated lysine 9 of histone 3; macroH2A1: histone variant macroH2A1; H3K27me3: trimethylated lysine 27 of histone 3; lncRNA: long noncoding RNA; H4K20me1: monomethylated lysine 20 of histone 4; H4K16ac: acetylated lysine 16 of histone 4; DCC: dosage compensation complex; RNA Pol II: RNA polymerase II; MSL1, MSL2 and MSL3: male-specific lethal 1, 2 and 3; MOF: males absent on the first; MLE: maleless; *roX1*, *roX2*: RNA on X chromosome 1 or 2; FAIRE: Formaldehyde-Assisted Isolation of Regulatory Elements; NDR: nucleosome-depleted regions; TSS: transcription start site; RPKM: reads per kilobase of transcript per million mapped reads; TPM: transcripts per million; IDR: irreproducible discovery rate; MACS2: model-based analysis for ChIP-seq; CI: confidence interval; 5'/3'UTR: five or three prime untranslated region; GO: gene ontology; ATAC: assay for transposase-accessible chromatin; MOZ-SAS: MOZ is a monocytic leukemia Zn_finger protein and the SAS protein from *Saccharomyces cerevisiae*; FDR: false discovery rate; ChIP: chromatin immunoprecipitation; HOMER: Hypergeometric Optimization of Motif EnRichment; BLAST: basic local alignment search tool; NCBI: National Center for Biotechnology Information; MUSCLE: MUltiple Sequence Comparison by Log-Expectation.

## Authors' contributions
GR managed the aphids rearing for the FAIRE-seq part of the current study, performed all the FAIRE-seq and RNA-seq bioinformatics analyses with the help of FL, all the statistical analyses with the help of JJ, supervised and performed all the graphical representations and data visualization and fully drafted the manuscript of the current study. FL and AB helped in providing computing resources, analysis tools and software implementation and FL designed some custom scripts. NPL developed and performed the FAIRE methodology to generate the samples used in the current study. GLT and DT acquired the financial support for the project leading to this study. GLT conceptualized the project leading to this study with the help of JJ. GLT mentored GR and supervised the research activity leading to the current study with the help of JJ. GLT and JJ made the major manuscript corrections, verifications and validation with the participation of DT. All authors read and approved the final manuscript.

## Author details
[1] EGI, UMR 1349, INRA, Institut de Génétique, Environnement et Protection des Plantes (IGEPP), Domaine de la Motte, BP 35327, Le Rheu, France. [2] BIPAA, UMR 1349, INRA, Institut de Génétique, Environnement et Protection des Plantes (IGEPP), Campus Beaulieu, Rennes, France. [3] Genscale, INRIA, IRISA, Campus Beaulieu, Rennes, France. [4] Genouest, INRIA, IRISA, Campus Beaulieu, Rennes, France. [5] CNRS, UMR 6553, EcoBio, University of Rennes 1, 35042 Rennes, France.

## Acknowledgements
We acknowledge Bernard Chaubet and Evelyne Turpeau for the photography of mating *A. pisum* males and females used as the personal cover of the current study.

## Competing interests
The authors declare that they have no competing interests.

## Funding
Gautier Richard's PhD is funded by a Young Scientist Contract from INRA, The Division for Plant Health and Environment (SPE) Department. This project was funded by The French National Research Agency Project MiRNAdapt (ANR-11-BSV6-01701) and by a French National Institute for Agricultural Research (INRA), The Division for Plant Health and Environment (SPE) Department Project Grant.

## References

1. Ellegren H. Sex-chromosome evolution: recent progress and the influence of male and female heterogamety. Nat Rev Genet. 2011;12:157–66.

2. Mank JE, Hosken DJ, Wedell N. Some inconvenient truths about sex chromosome dosage compensation and the potential role of sexual conflict. Evolution. 2011;65:2133–44.

3. Ohno S. Sex chromosomes and sex-linked genes. 1st ed. Berlin: Springer; 1967.

4. Charlesworth B. Model for evolution of Y chromosomes and dosage compensation. Proc Natl Acad Sci USA. 1978;75:5618–22.

5. Marín I, Siegal ML, Baker BS. The evolution of dosage-compensation mechanisms. BioEssays. 2000;22:1106–14.

6. Julien P, Brawand D, Soumillon M, Necsulea A, Liechti A, Schütz F, et al. Mechanisms and evolutionary patterns of mammalian and avian dosage compensation. PLoS Biol. 2012. doi:10.1371/journal.pbio.1001328.

7. Pessia E, Makino T, Bailly-Bechet M, McLysaght A, Marais GAB. Mammalian X chromosome inactivation evolved as a dosage-compensation mechanism for dosage-sensitive genes on the X chromosome. Proc Natl Acad Sci. 2012;109:5346–51.

8. Lin F, Xing K, Zhang J, He X. Expression reduction in mammalian X chromosome evolution refutes Ohno's hypothesis of dosage compensation. Proc Natl Acad Sci. 2012;109:11752–7.

9. Deng X, Hiatt JB, Nguyen DK, Ercan S, Sturgill D, Hillier LW, et al. Evidence for compensatory upregulation of expressed X-linked genes in mammals, Caenorhabditis elegans and Drosophila melanogaster. Nat Genet. 2011;43:1179–85.

10. Kramer M, Rao P, Ercan S. Untangling the contributions of sex-specific gene regulation and X-chromosome dosage to sex-biased gene expression in Caenorhabditis elegans. Genetics. 2016;204:355–69.

11. Albritton SE, Kranz A-L, Rao P, Kramer M, Dieterich C, Ercan S. Sex-biased gene expression and evolution of the x chromosome in nematodes. Genetics. 2014;197:865–83.

12. Straub T, Becker PB. Dosage compensation: the beginning and end of generalization. Nat Rev Genet. 2007;8:47–57.

13. Gelbart ME, Kuroda MI. Drosophila dosage compensation: a complex voyage to the X chromosome. Development. 2009;136:1399–410.

14. Lucchesi JC. Dosage compensation in Drosophila. Annu Rev Genet. 1973;7:225–37.

15. Jiang X, Biedler JK, Qi Y, Hall AB, Tu Z. Complete dosage compensation in Anopheles stephensi and the evolution of sex-biased genes in mosquitoes. Genome Biol Evol. 2015;7:1914–24.

16. Rose G, Krzywinska E, Kim J, Revuelta L, Ferretti L, Krzywinski J. Dosage compensation in the African malaria mosquito Anopheles gambiae. Genome Biol Evol. 2016;8:411–25.

17. Smith G, Chen Y-R, Blissard GW, Briscoe AD. Complete dosage compensation and sex-biased gene expression in the moth Manduca sexta. Genome Biol Evol. 2014;6:526–37.

18. Mahajan S, Bachtrog D. Partial dosage compensation in Strepsiptera, a sister group of beetles. Genome Biol Evol. 2015;7:591–600.

19. Wilkinson GS, Johns PM, Metheny JD, Baker RH. Sex-biased gene expression during head development in a sexually dimorphic stalk-eyed fly. PLOS ONE. 2013. doi:10.1371/journal.pone.0059826.

20. Brockdorff N, Turner BM. Dosage compensation in mammals. Cold Spring Harb Perspect Biol. 2015. doi:10.1101/cshperspect.a019406.

21. Lucchesi JC, Kelly WG, Panning B. Chromatin remodeling in dosage compensation. Annu Rev Genet. 2005;39:615–51.

22. Payer B, Lee JT. X chromosome dosage compensation: how mammals keep the balance. Annu Rev Genet. 2008;42:733–72.

23. Costanzi C, Stein P, Worrad DM, Schultz RM, Pehrson JR. Histone macroH2A1 is concentrated in the inactive X chromosome of female preimplantation mouse embryos. Development. 2000;127:2283–9.

24. Engreitz JM, Pandya-Jones A, McDonel P, Shishkin A, Sirokman K, Surka C, et al. The Xist lncRNA exploits three-dimensional genome architecture to spread across the X chromosome. Science. 2013. doi:10.1126/science.1237973.

25. Ercan S. Mechanisms of X chromosome dosage compensation. J Genomics. 2015;3:1–19.

26. Wells MB, Snyder MJ, Custer LM, Csankovszki G. Caenorhabditis elegans dosage compensation regulates histone H4 chromatin state on X chromosomes. Mol Cell Biol. 2012;32:1710–9.

27. Kruesi WS, Core LJ, Waters CT, Lis JT, Meyer BJ. Condensin controls recruitment of RNA polymerase II to achieve nematode X-chromosome dosage compensation. eLife. 2013. doi:10.7554/eLife.00808.

28. Prestel M, Feller C, Straub T, Mitlöhner H, Becker PB. The activation potential of MOF is constrained for dosage compensation. Mol Cell. 2010;38:815–26.

29. Conrad T, Akhtar A. Dosage compensation in Drosophila melanogaster: epigenetic fine-tuning of chromosome-wide transcription. Nat Rev Genet. 2012;13:123–34.

30. Bell O, Schwaiger M, Oakeley EJ, Lienert F, Beisel C, Stadler MB, et al. Accessibility of the Drosophila genome discriminates PcG repression, H4K16 acetylation and replication timing. Nat Struct Mol Biol. 2010;17:894–900.

31. Shogren-Knaak M, Ishii H, Sun J-M, Pazin MJ, Davie JR, Peterson CL. Histone H4-K16 acetylation controls chromatin structure and protein interactions. Science. 2006;311:844–7.

32. Le Trionnaire G, Jaubert-Possamai S, Bonhomme J, Gauthier J-P, Guernec G, Le Cam A, et al. Transcriptomic profiling of the reproductive mode switch in the pea aphid in response to natural autumnal photoperiod. J Insect Physiol. 2012;58:1517–24.

33. Le Trionnaire G, Wucher V, Tagu D. Genome expression control during the photoperiodic response of aphids. Physiol Entomol. 2013;38:117–25.

34. Jaquiéry J, Rispe C, Roze D, Legeai F, Le Trionnaire G, Stoeckel S, et al. Masculinization of the X chromosome in the pea aphid. PLoS Genet. 2013. doi:10.1371/journal.pgen.1003690.

35. Jaquiéry J, Stoeckel S, Rispe C, Mieuzet L, Legeai F, Simon J-C. Accelerated evolution of sex chromosomes in aphids, an X0 system. Mol Biol Evol. 2012;29:837–47.

36. Wilson ACC, Sunnucks P, Hales DF. SHORT PAPER random loss of X chromosome at male determination in an aphid, Sitobion near fragariae, detected using an X-linked polymorphic microsatellite marker. Genet Res. 1997;69:233–6.

37. International Aphid Genomics Consortium. Genome sequence of the pea aphid Acyrthosiphon pisum. PLoS Biol. 2010;8:e1000313. doi:10.1371/journal.pbio.1000313pmid:20186266.

38. Jaquiery J, Peccoud J, Ouisse T, Legeai F, Prunier-Leterme N, Gouin A, et al. Disentangling the causes for faster-X evolution in aphids. bioRxiv. 2017;125310.

39. Giresi PG, Kim J, McDaniell RM, Iyer VR, Lieb JD. FAIRE (Formaldehyde-Assisted Isolation of Regulatory Elements) isolates active regulatory elements from human chromatin. Genome Res. 2007;17:877–85.

40. Simon JM, Giresi PG, Davis IJ, Lieb JD. Using Formaldehyde-Assisted Isolation of Regulatory Elements (FAIRE) to isolate active regulatory DNA. Nat Protoc. 2012;7:256–67.

41. Zhang Y, Liu T, Meyer CA, Eeckhoute J, Johnson DS, Bernstein BE, et al. Model-based analysis of ChIP-Seq (MACS). Genome Biol. 2008;9:R137. doi:10.1186/gb-2008-9-9-r137.

42. Li Q, Brown JB, Huang H, Bickel PJ. Measuring reproducibility of high-throughput experiments. Ann Appl Stat. 2011;5:1752–79.

43. Kundaje A. ENCODE: TF ChIP-seq peak calling using the Irreproducibility Discovery Rate (IDR) framework (2012). https://sites.google.com/site/anshulkundaje/projects/idr. Accessed 5 Oct 2016.

44. Quinlan AR, Hall IM. BEDTools: a flexible suite of utilities for comparing genomic features. Bioinformatics. 2010;26:841–2.

45. Ramírez F, Ryan DP, Grüning B, Bhardwaj V, Kilpert F, Richter AS, et al. deepTools2: a next generation web server for deep-sequencing data analysis. Nucleic Acids Res. 2016;44:W160–5. doi:10.1093/nar/gkw257.

46. Pal A, Vicoso B. The X chromosome of hemipteran insects: conservation, dosage compensation and sex-biased expression. Genome Biol Evol. 2015;7:3259–68.

47. Prince EG, Kirkland D, Demuth JP. Hyperexpression of the X chromosome in both sexes results in extensive female bias of X-linked genes in the flour beetle. Genome Biol Evol. 2010;2:336–46.

48. Dagg J. Strategies of sexual reproduction in aphids (2002). https://ediss.uni-goettingen.de/handle/11858/00-1735-0000-0006-ABF2-B. Accessed 22 Dec 2016.

49. Gupta V, Parisi M, Sturgill D, Nuttall R, Doctolero M, Dudko OK, et al. Global analysis of X-chromosome dosage compensation. J Biol. 2006;5:3.

50. Simon JM, Giresi P, Davis IJ, Lieb JD. Addendum: using Formaldehyde-Assisted Isolation of Regulatory Elements (FAIRE) to isolate active regulatory DNA. Nat Protoc. 2014;9:501–3.

51. Buenrostro J, Wu B, Chang H, Greenleaf W. ATAC-seq: a method for assaying chromatin accessibility genome-wide. In: Ausubel FM, editor. Current protocols in molecular biology, vol. 109. London: Wiley; 2015. p. 21.29.1–21.29.9.

52. Buenrostro JD, Wu B, Litzenburger UM, Ruff D, Gonzales ML, Snyder MP, et al. Single-cell chromatin accessibility reveals principles of regulatory variation. Nature. 2015;523:486–90.

53. Buenrostro JD, Giresi PG, Zaba LC, Chang HY, Greenleaf WJ. Transposition of native chromatin for fast and sensitive epigenomic profiling of open chromatin, DNA-binding proteins and nucleosome position. Nat Methods. 2013;10:1213–8.

54. Li B, Carey M, Workman JL. The role of chromatin during transcription. Cell. 2007;128:707–19.

55. Thomas S, Li X-Y, Sabo PJ, Sandstrom R, Thurman RE, Canfield TK, et al. Dynamic reprogramming of chromatin accessibility during Drosophila embryo development. Genome Biol. 2011;12:R43.

56. Song L, Zhang Z, Grasfeder LL, Boyle AP, Giresi PG, Lee B-K, et al. Open chromatin defined by DNaseI and FAIRE identifies regulatory elements that shape cell-type identity. Genome Res. 2011;21:1757–67.

57. Kharchenko PV, Alekseyenko AA, Schwartz YB, Minoda A, Riddle NC, Ernst J, et al. Comprehensive analysis of the chromatin landscape in *Drosophila melanogaster*. Nature. 2011;471:480–5.

58. Johnsson P, Lipovich L, Grandér D, Morris KV. Evolutionary conservation of long non-coding RNAs; sequence, structure, function. Biochim Biophys Acta. 2014;1840:1063–71.

59. Hartman TR, Qian S, Bolinger C, Fernandez S, Schoenberg DR, Boris-Lawrie K. RNA helicase A is necessary for translation of selected messenger RNAs. Nat Struct Mol Biol. 2006;13:509–16.

60. Robb GB, Rana TM. RNA helicase A interacts with RISC in human cells and functions in RISC loading. Mol Cell. 2007;26:523–37.

61. Mendjan S, Taipale M, Kind J, Holz H, Gebhardt P, Schelder M, et al. Nuclear pore components are involved in the transcriptional regulation of dosage compensation in Drosophila. Mol Cell. 2006;21:811–23.

62. Smith ER, Cayrou C, Huang R, Lane WS, Côté J, Lucchesi JC. A human protein complex homologous to the Drosophila MSL complex is responsible for the majority of histone H4 acetylation at lysine 16. Mol Cell Biol. 2005;25:9175–88.

63. Taipale M, Rea S, Richter K, Vilar A, Lichter P, Imhof A, et al. hMOF histone acetyltransferase is required for histone H4 lysine 16 acetylation in mammalian cells. Mol Cell Biol. 2005;25:6798–810.

64. Copps K, Richman R, Lyman LM, Chang KA, Rampersad-Ammons J, Kuroda MI. Complex formation by the Drosophila MSL proteins: role of the MSL2 RING finger in protein complex assembly. EMBO J. 1998;17:5409–17.

65. Kim D, Pertea G, Trapnell C, Pimentel H, Kelley R, Salzberg SL. TopHat2: accurate alignment of transcriptomes in the presence of insertions, deletions and gene fusions. Genome Biol. 2013;14:R36.

66. Anders S, Pyl PT, Huber W. HTSeq—a Python framework to work with high-throughput sequencing data. Bioinformatics. 2015;31:166–9.

67. R Core Team. R: a language and environment for statistical computing. Vienna: R Foundation for Statistical Computing; 2014. http://www.R-project.org/.

68. Robinson MD, McCarthy DJ, Smyth GK. edgeR: a bioconductor package for differential expression analysis of digital gene expression data. Bioinform Oxf Engl. 2010;26:139–40.

69. Law CW, Alhamdoosh M, Su S, Smyth GK, Ritchie ME. RNA-seq analysis is easy as 1-2-3 with limma, Glimma and edgeR. F1000Research. 2016;5:1408.

70. Langmead B, Trapnell C, Pop M, Salzberg SL. Ultrafast and memory-efficient alignment of short DNA sequences to the human genome. Genome Biol. 2009;10:R25.

71. Langmead B, Salzberg SL. Fast gapped-read alignment with Bowtie 2. Nat Methods. 2012;9:357–9.

72. Li H, Handsaker B, Wysoker A, Fennell T, Ruan J, Homer N, et al. The sequence alignment/map format and SAMtools. Bioinform Oxf Engl. 2009;25:2078–9.

73. Heinz S, Benner C, Spann N, Bertolino E, Lin YC, Laslo P, et al. Simple combinations of lineage-determining transcription factors prime cis-regulatory elements required for macrophage and B cell identities. Mol Cell. 2010;38:576–89.

74. Buske OJ, Hoffman MM, Ponts N, Le Roch KG, Noble WS. Exploratory analysis of genomic segmentations with Segtools. BMC Bioinform. 2011;12:415.

75. Wickham H. ggplot2: elegant graphics for data analysis. New York: Springer; 2009. http://ggplot2.org.

76. Altschul SF, Gish W, Miller W, Myers EW, Lipman DJ. Basic local alignment search tool. J Mol Biol. 1990;215:403–10.

77. Camacho C, Coulouris G, Avagyan V, Ma N, Papadopoulos J, Bealer K, et al. BLAST + : architecture and applications. BMC Bioinform. 2009;10:421.

78. Eden E, Lipson D, Yogev S, Yakhini Z. Discovering motifs in ranked lists of DNA sequences. PLoS Comput Biol. 2007;3:e39.

79. Eden E, Navon R, Steinfeld I, Lipson D, Yakhini Z. GOrilla: a tool for discovery and visualization of enriched GO terms in ranked gene lists. BMC Bioinform. 2009;10:48.

# 7

# Pho dynamically interacts with Spt5 to facilitate transcriptional switches at the *hsp70* locus

Allwyn Pereira[1] and Renato Paro[1,2]* (iD)

## Abstract

**Background:** Numerous target genes of the Polycomb group (PcG) are transiently activated by a stimulus and subsequently repressed. However, mechanisms by which PcG proteins regulate such target genes remain elusive.

**Results:** We employed the heat shock-responsive *hsp70* locus in *Drosophila* to study the chromatin dynamics of PRC1 and its interplay with known regulators of the locus before, during and after heat shock. We detected mutually exclusive binding patterns for HSF and PRC1 at the *hsp70* locus. We found that Pleiohomeotic (Pho), a DNA-binding PcG member, dynamically interacts with Spt5, an elongation factor. The dynamic interaction switch between Pho and Spt5 is triggered by the recruitment of HSF to chromatin. Mutation in the protein–protein interaction domain (REPO domain) of Pho interferes with the dynamics of its interaction with Spt5. The transcriptional kinetics of the heat shock response is negatively affected by a mutation in the REPO domain of Pho.

**Conclusions:** We propose that a dynamic interaction switch between PcG proteins and an elongation factor enables stress-inducible genes to efficiently switch between ON/OFF states in the presence/absence of the activating stimulus.

**Keywords:** Heat shock, Pol II pausing, Polycomb proteins, Protein–protein interactions, Transcriptional activators

## Background

Polycomb group (PcG) proteins are a highly conserved class of epigenetic regulators that play an essential role during development and adulthood in many organisms [1]. In any given cell type, while PcG proteins maintain stable silencing at a certain subset of genes, other PcG target genes are capable of being activated in the presence of the appropriate stimulus. Thus, by conferring both stability and plasticity upon the transcriptome, PcG proteins bestow competence upon the cells to respond to varied stimuli over their lifetime, in order to maintain cellular homoeostasis [2]. However, the underlying mechanisms that enable PcG-regulated loci to dynamically switch between ON and OFF states remain elusive.

In *Drosophila*, the targeting of PcG proteins to chromatin occurs mainly through the recognition of cis-regulatory elements, known as Polycomb group response elements (PREs), by Pleiohomeotic (Pho) and other DNA-binding proteins [3, 4]. Pho interacts with dSfmbt to form the Pho repressive complex (PhoRC), which acts as a recruitment platform for the Polycomb repressive complex 1 and 2 (PRC1 and PRC2) [5, 6]. Subsequently, PRC1 and PRC2 generate a chromatin structure that is incompatible with gene expression, through their enzymatic and chromatin compaction activities [7, 8]. The transcriptionally silent state is then faithfully inherited through cell division, which involves PRE-dependent mechanisms [9, 10]. Although the mechanisms by which PcG proteins establish and maintain stable silencing at their targets have been subject to intense investigation, little is known about PcG-mediated regulation of target genes with dynamic expression patterns.

In this study, we employed the heat shock-responsive *hsp70* gene in *Drosophila* as a model to study dynamic

*Correspondence: renato.paro@bsse.ethz.ch
[1] Department of Biosystems Science and Engineering, ETH Zurich, 4058 Basel, Switzerland
Full list of author information is available at the end of the article

gene regulation by PcG proteins (Additional file 1: Figure S1a). The *hsp70* locus is an attractive inducible gene model as it is targeted by PRC1 [11]. It is regulated at the level of transcriptional elongation by promoter-proximal RNA polymerase II pausing [12]. At normal growth temperature (25 °C), RNA polymerase II is maintained in the paused state by the DSIF complex (composed of Spt4–Spt5) and the NELF complex [13, 14]. The heat shock stimulus leads to the recruitment of the transcription factor, HSF, to chromatin [15]. This is followed by the recruitment of P-TEFb, which phosphorylates Spt5 and converts it to function as a positive elongation factor [16, 17]. This leads to rapid up-regulation of the *hsp70* locus in response to the heat stress [18].

While the transcriptional response and the chromatin dynamics at the *hsp70* locus have been well characterised in the acute phase of the heat shock response, the effect of the withdrawal of the activating stimulus on the transcription and chromatin dynamics remains poorly understood [18]. By following the chromatin dynamics of PRC1 and Pho during the recovery phase of the heat shock, we sought to dissect the role of these two factors in the re-establishment of silencing at the *hsp70* locus. Pho was previously found to co-localise with RNA polymerase II upon heat shock [19]. Furthermore, Pho interacts with Spt5 [20], and Spt5 is involved in both, promoter-proximal pausing and transcriptional elongation [21]. We thus asked whether the conversion of Spt5 from a pausing factor into an elongation factor upon gene activation and its interaction with Pho could mechanistically explain the spreading of Pho in the body of active genes. Genetic analyses revealed the necessity of Pho gene function in the re-establishment of silencing at the *hsp70* locus in the absence of the activating stimulus [19]. We tested the hypothesis whether Pho could dynamically interact with repressors (PRC1 via dSfmbt) and activators (Spt5) during the heat shock response and the significance of its dynamic protein–protein interactions on transcriptional regulation of the *hsp70* locus.

By combining multiple read-outs, we identified a role for the dynamic protein–protein interaction between Pho and Spt5 in the transcriptional regulation of the *hsp70* locus. Our results highlight how a dynamic interaction switch between PcG proteins and an elongation factor could facilitate PcG-regulated genes, which are responsive to inductive signals, to efficiently switch between ON/OFF states.

## Results

### Antagonistic binding patterns of HSF and PRC1 at the *hsp70* locus

To characterise the transcriptional and chromatin dynamics of the heat shock response, we focused on the binding kinetics of RNA polymerase II (Pol II), HSF and Polyhomeotic (Ph, a component of PRC1) at the *hsp70* locus using *Drosophila* S2 tissue culture cells. Our analysis revealed a rapid increase in *hsp70* transcripts immediately upon heat shock, followed by a steady decrease in mRNA levels as the locus is re-repressed, during the recovery phase in the absence of the heat shock stimulus (Fig. 1b). We investigated the binding dynamics of two forms of Pol II, the unphosphorylated C-terminal domain (CTD) form and the hyperphosphorylated (S2P) form of Pol II, at the promoter-proximal region and within the gene body of the *hsp70* locus, respectively. We detected a rapid increase in both forms of Pol II at the promoter-proximal region and within the gene body of the *hsp70* locus upon heat shock (Fig. 1c, d). This was accompanied by a decrease in Pol II binding levels at the *Act42A* gene (Additional file 2: Figure S2a, b), which is in accordance with the previously described phenomenon of the global decrease in Pol II levels genome-wide in *Drosophila* upon heat shock [22]. However, levels of bound Pol II at the *hsp70* locus steadily decreased as the cells recovered from the heat shock, with the locus returning to the paused state 90 min into the recovery phase (Fig. 1c, d). In contrast, Pol II binding at the *Act42A* gene increased over the recovery phase, indicating that the cells resumed a normal transcriptional programme within 90 min of the heat shock stimulus (Additional file 2: Figure S2a, b).

HSF binding is critical to release paused Pol II into productive elongation at the *hsp70* locus [23]. In agreement, we detected a rapid increase in HSF binding upon heat shock followed by steady decrease as the cells recovered from the heat shock (Fig. 1e). Thus, the chromatin binding dynamic of Pol II and HSF during the heat shock response was similar, emphasising a role for HSF binding at chromatin for the transcriptional activation of the *hsp70* locus [23].

Recent ChIP-seq studies in *Drosophila* and mammalian cell lines have revealed PRC1 binding at genes regulated by paused Pol II, leading to the hypothesis that PRC1 could regulate paused genes [11, 24, 25]. We, therefore, assessed the binding dynamics of Ph during the heat shock response. In keeping with PRC1's role as a repressor, Ph occupancy at the *hsp70* locus decreased significantly upon activation. However, as the locus recovered from the heat shock and occupancy of Pol II and HSF decreased, Ph binding to chromatin was steadily restored to pre-activation levels (Fig. 1f).

Taken together, these results reveal an antagonism between HSF and Ph (PRC1) for occupancy at the *hsp70* locus.

### Co-localisation of Pleiohomeotic (Pho) and Pol II at the heat shock locus is independent of transcriptional elongation

We confirmed the previously identified co-localisation of Pho with Pol II at active genes by co-staining these factors

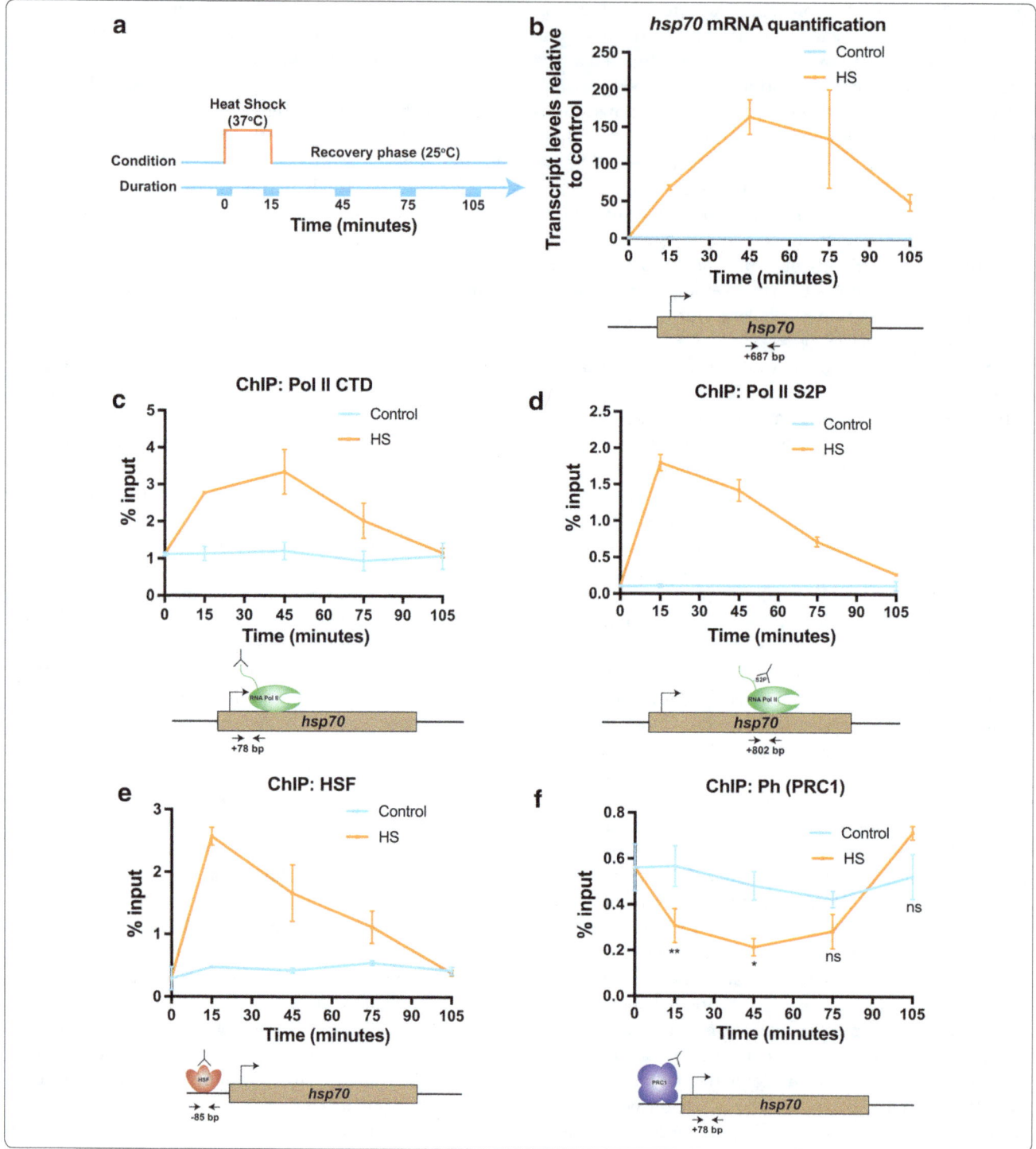

(See figure on previous page.)
**Fig. 1** Antagonistic binding patterns of HSF and PRC1 at the *hsp70* locus. **a** Experimental scheme used throughout the study to measure the dynamics of mRNA/nascent RNA production, chromatin binding patterns of proteins and protein–protein interactions during the heat shock response. S2 DRSC cells were heat shocked at 37 °C for 15 min and thereafter allowed to recover from the heat shock at 25 °C for up to 90 min. At time points indicated in the scheme, cells were harvested and used for the following experimental procedures: mRNA quantification, ChIP-qPCR, co-IP and nascent RNA quantification. **b** Quantitative RT-PCR (qRT-PCR) detection of *hsp70* transcript levels during the heat shock response detailed in **a**. Distance of the location of the primers used from the *hsp70* transcription start site (TSS) is + 687 bp and has been depicted in the form of a cartoon below the data figure. The control line represents cells that were maintained at 25 °C for the entire duration of the time course detailed in **a**. **c–f** ChIP-qPCR measurements of occupancy levels of RNA polymerase II CTD, S2P form of RNA polymerase II, heat shock binding factor (HSF) and Polyhomeotic (Ph, a component of PRC1) at the *hsp70* locus during the heat shock response detailed in **a**. Distances of the location of the primers used from the *hsp70* TSS are as follows: for **c** and **f** (+ 78 bp), for **d** (+ 802 bp) and for **e** (− 85 bp) has been depicted in the form of a cartoon below the data figure. The control line represents cells that were maintained at 25 °C for the entire duration of the time course detailed in **a**. Data information: In **b–f**, data are presented as mean ± SEM for $n = 2$ (except for **f** where $n = 3$). Statistical significance was determined by performing a paired two-tailed $t$ test, $**P \leq 0.01$, $*P \leq 0.05$ and ns = non-significant

on polytene chromosomes from larval salivary glands. Pho and Pol II co-localised at the majority of bands under basal conditions (Fig. 2a). Upon heat shock, Pol II binding was most prominent at the heat shock puffs (Fig. 2b). In addition, Pol II signal intensity decreased outside the heat shock loci, in accordance with a genome-wide decrease in Pol II occupancy upon heat shock (Fig. 2b). Interestingly, we observed a striking co-localisation of Pho and Pol II at all activated heat shock genes (Fig. 2b).

To determine whether this co-localisation is dependent upon transcription, we blocked transcriptional elongation by treatment with flavopiridol, a P-TEFb inhibitor, prior to heat shock [26]. The striking overlap in binding patterns of Pho and Pol II was not perturbed by flavopiridol treatment, suggesting that the interaction between these two factors could occur independently of transcriptional elongation (Fig. 2c). Furthermore, inhibition of P-TEFb followed by heat shock resulted in a marked decrease in the size of the heat shock puffs (Fig. 2d) and strongly reduced *hsp70* transcript levels (Additional file 3: Figure S3d). However, co-localisation of Pho and Pol II was not perturbed by flavopiridol (Fig. 2d). This result indicated that the co-localisation between Pho and Pol II was intact during the paused state as well as during the elongation phase of the transcriptional cycle. It is conceivable that Pho might co-localise with the paused form of Pol II. However, we found it intriguing that a DNA-binding protein like Pho was also capable of co-localising with actively elongating Pol II at chromatin. In order to understand whether the interaction between Pho and elongating Pol II was dependent upon the DNA-binding activity of Pho, we employed native ChIP (N-ChIP) qPCR to measure the occupancy levels of Pho over the time course detailed in Fig. 1a. We employed N-ChIP over formaldehyde cross-linked ChIP to measure the occupancy levels of Pho since N-ChIP only detects proteins that are associated with chromatin due to their direct interaction with DNA [27]. Indeed, while Pho was

bound at the *hsp70* locus under control conditions, its occupancy decreased upon heat shock. However, as the cells recovered from the heat shock, the binding of Pho to the *hsp70* locus via DNA was restored (Additional file 3: Figure S3a). This suggested that the association of Pho to chromatin upon heat shock cannot be explained by its DNA-binding ability and could possibly be due to a protein–protein interaction with RNA Pol II or an elongation factor.

Previous observations established that flavopiridol treatment leads to considerable decrease in mature *hsp70* transcript production, even though the density of Pol II in the gene body remains normal [28]. We have now shown that chromatin binding levels of Ph, a component of PRC1, anti-correlate with mature *hsp70* transcript levels (Fig. 1f). Thus, we hypothesised that HSF recruitment in the presence of flavopiridol could be used to uncouple its effect on the occupancy of Ph at chromatin from the production of mature *hsp70* transcripts.

We, therefore, examined the consequence of flavopiridol treatment on the binding of PRC1 at the *hsp70* locus. We treated cells with flavopiridol prior to heat shock (hereafter referred to as FP + HS) followed by ChIP-qPCR. In agreement with previous observations, flavopiridol treatment led to a significant decrease in S2P Pol II levels in the *hsp70* gene body upon heat shock (Additional file 3: Figure S3c), whereas unphosphorylated Pol II and HSF binding levels were not perturbed (Additional file 3: Figure S3b and Fig. 2e). Ph binding at the *hsp70* locus decreased under FP + HS conditions, once again demonstrating that HSF and PRC1 binding patterns are mutually exclusive at the *hsp70* locus (Fig. 2e, f).

These data revealed that the co-localisation between Pho and Pol II was intact during the paused as well as during the elongation phase of the transcriptional cycle. Interestingly, while Pho continued to localise at the *hsp70* locus upon heat shock, Ph (PRC1) was evicted from the locus. We, thus, concluded that HSF binding could be

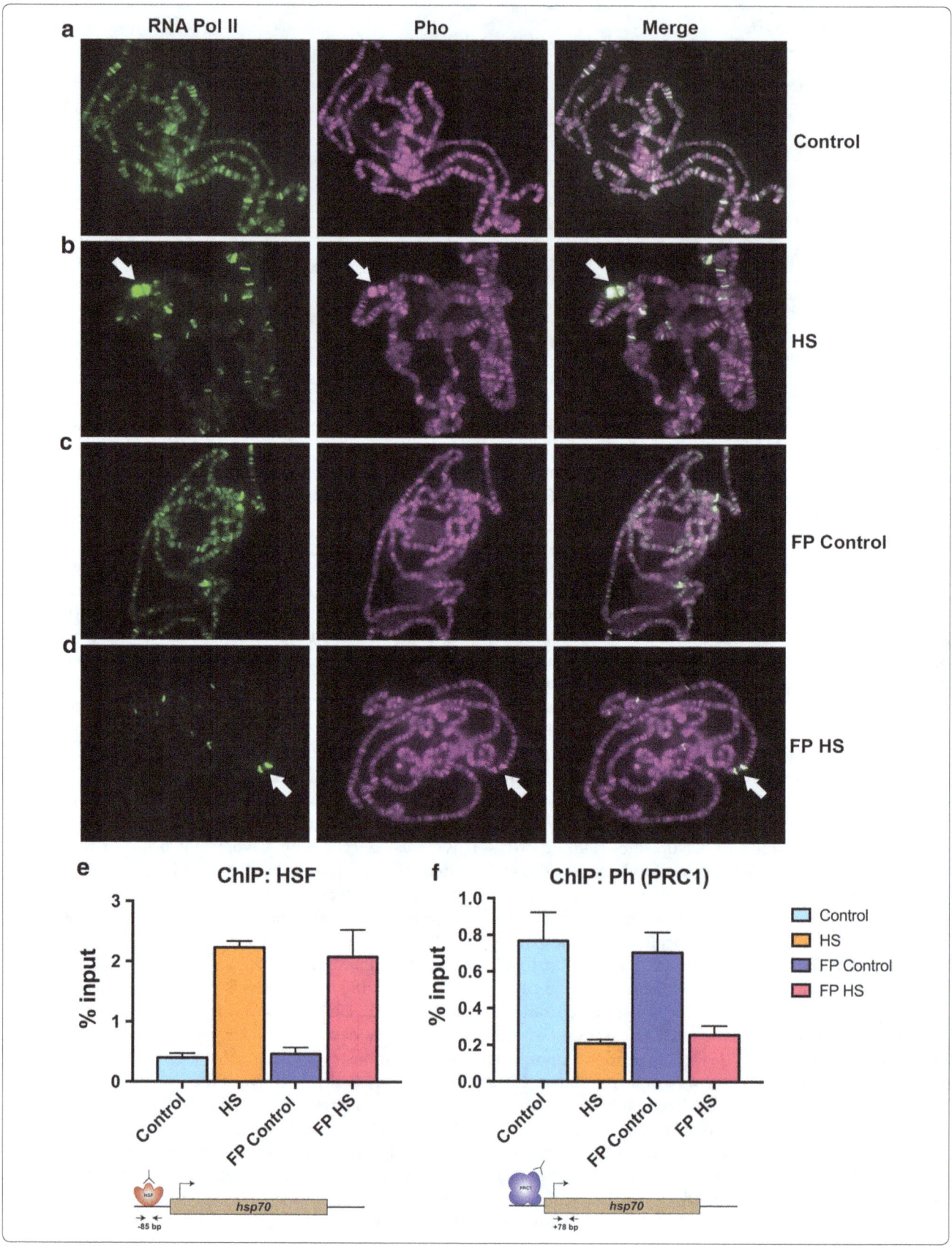

(See figure on previous page.)
**Fig. 2** Co-localisation of Pleiohomeotic (Pho) and RNA polymerase II at the heat shock loci upon activation is independent of transcriptional elongation. **a–d** Double polytene immunostaining for RNA polymerase II (green) and Pho (magenta) under control, HS (heat shock), FP (flavopiridol) control and FP HS (flavopiridol + heat shock), respectively. The white block arrows point at the double heat shock puffs (87A/C) that were observed under HS and FP HS conditions. Magnification used 63×. **e, f** ChIP-qPCR measurements of occupancy levels for HSF and Ph, respectively, at the *hsp70* locus in S2 DRSC cells. Distances of the location of the primers used from the *hsp70* TSS are as follows: for **e** (− 85 bp) and for **f** (+ 78 bp) has been depicted in the form of a cartoon below the data figure. The control bar represents cells that were maintained at 25 °C for 15 min, the HS bar represents cells that were maintained at 37 °C for 15 min, the FP control bar represents cells that were treated with 500 nM flavopiridol for 40 min and then maintained at 25 °C for 15 min and the FP HS bar represents cells that were treated with 500 nM flavopiridol for 40 min and then maintained at 37 °C for 15 min. Data information: In **e, f**, data are presented as mean ± SEM for n = 2

capable of disrupting the ability of Pho to recruit PRC1 at the *hsp70* locus.

### Pho dynamically interacts with Spt5 and dSfmbt

Since we determined that the co-localisation of Pho and Pol II at the *hsp70* locus upon activation was independent of the DNA-binding ability of Pho, we hypothesised that this observation could possibly be explained by the reported genetic and biochemical interaction between Pho and Spt5 [20]. We also determined by ChIP-qPCR that chromatin occupancy of Spt5 does not change dramatically over the time course detailed in Fig. 1a (Additional file 4: Figure S4a). In order to gain mechanistic insight into the significance of the interaction between Pho and Spt5, we sought to identify the amino acid residues in Pho that mediate the protein–protein interaction with Spt5. Pho exhibits a modular structure with its protein–protein interaction domains localised in the N-terminus and its DNA-binding property ascribed to the Zn-finger domains in the C-terminus [5, 29, 30]. To identify putative Spt5-interacting regions of Pho, we searched for evolutionary conserved functional domain(s) by performing a multiple protein alignment between Pho and YY1, the mouse and human homologue of Pho. The Zn-finger domains in Pho and YY1 showed a high degree of conservation in the multiple protein alignment (Fig. 3a). However, a second stretch of amino acids, located in the N-terminus of Pho, also showed conservation with YY1 in the multiple protein alignment (Fig. 3a). This region mapped to a domain known as the Pho spacer region [5],

and its homologous region within YY1 is known as the Recruitment of Polycomb (REPO) as it was shown to be necessary and sufficient for PcG-mediated repression in *Drosophila* [31].

We hypothesised that the REPO domain of Pho could be involved in mediating its protein–protein interaction with Spt5. To test this hypothesis, we generated two mutant constructs with the aim of disrupting the structure of the REPO domain [5] and analysed the effect of the mutation on the protein–protein interactions of Pho by co-immunoprecipitation (co-IP). The first construct, PhoV164D, contained a point mutation resulting in a valine (a hydrophobic amino acid residue) to aspartic acid (a hydrophilic amino acid residue) substitution to interfere with its ability to contact hydrophobic surfaces, while the second construct, ΔREPO, carried a 66-nucleotide deletion in the REPO domain [5] (Fig. 3b).

In agreement with previous findings, both mutant Pho constructs were incapable of interacting with dSfmbt [5] (Fig. 3c). However, both mutant Pho constructs also significantly impaired the interaction between Pho and Spt5, indicating that the REPO domain represents an interaction surface with multiple protein interaction partners (Fig. 3d). Next, we sought to explain the residual interaction between Pho and Spt5 in the mutant background using different N-terminal and C-terminal Pho constructs (1-351a.a N-Pho, 1-200a.a N-Pho, 173-351a.a N-Pho, 351-520a.a C-Pho). Our analysis revealed that in addition to the REPO domain containing N-terminal fragments of Pho, the C-terminus of Pho could also

(See figure on next page.)
**Fig. 3** Pho interacts dynamically with Spt5 and dSfmbt. **a** Multiple protein alignment between Pho and YY1, the mouse and human homologue of Pho. The REPO domain is surrounded by a blue rectangle, while the Zn-finger domains are surrounded by an orange rectangle. The asterisk mark on the bottom line 'consensus' denotes the amino acid residues that are fully conserved between the three species. **b** Representation of the N- and C-termini of Pho and the mutant constructs generated for this study. **c, d** co-IP assays of S2 DRSC cells transiently transfected with plasmids expressing FLAG-tagged dSfmbt or Spt5 (FLAG-dSfmbt, FLAG-Spt5) and HA-tagged Pho, PhoV164D, Pho ΔREPO, EGFP (HA-Pho, HA-PhoV164D, HA-Pho REPO del, HA-EGFP) as indicated. Cell lysates were used for pull-downs using an anti-FLAG antibody and were later probed by Western blot using an anti-HA antibody. MW = molecular weight in kDa. **e** co-IP assays of S2 DRSC cells transiently transfected with plasmids expressing FLAG-tagged dSfmbt or Spt5 (FLAG-dSfmbt, FLAG-Spt5) and HA-tagged Pho or Spt4 (HA-Pho, HA-Spt4) as indicated. S2 DRSC cells were either maintained at normal growth temperature (25 °C) or heat shocked at 37 °C for 15 min, cell lysates were used for pull-downs using an anti-FLAG antibody and were later probed by Western blot using an anti-HA antibody. MW = molecular weight in kDa

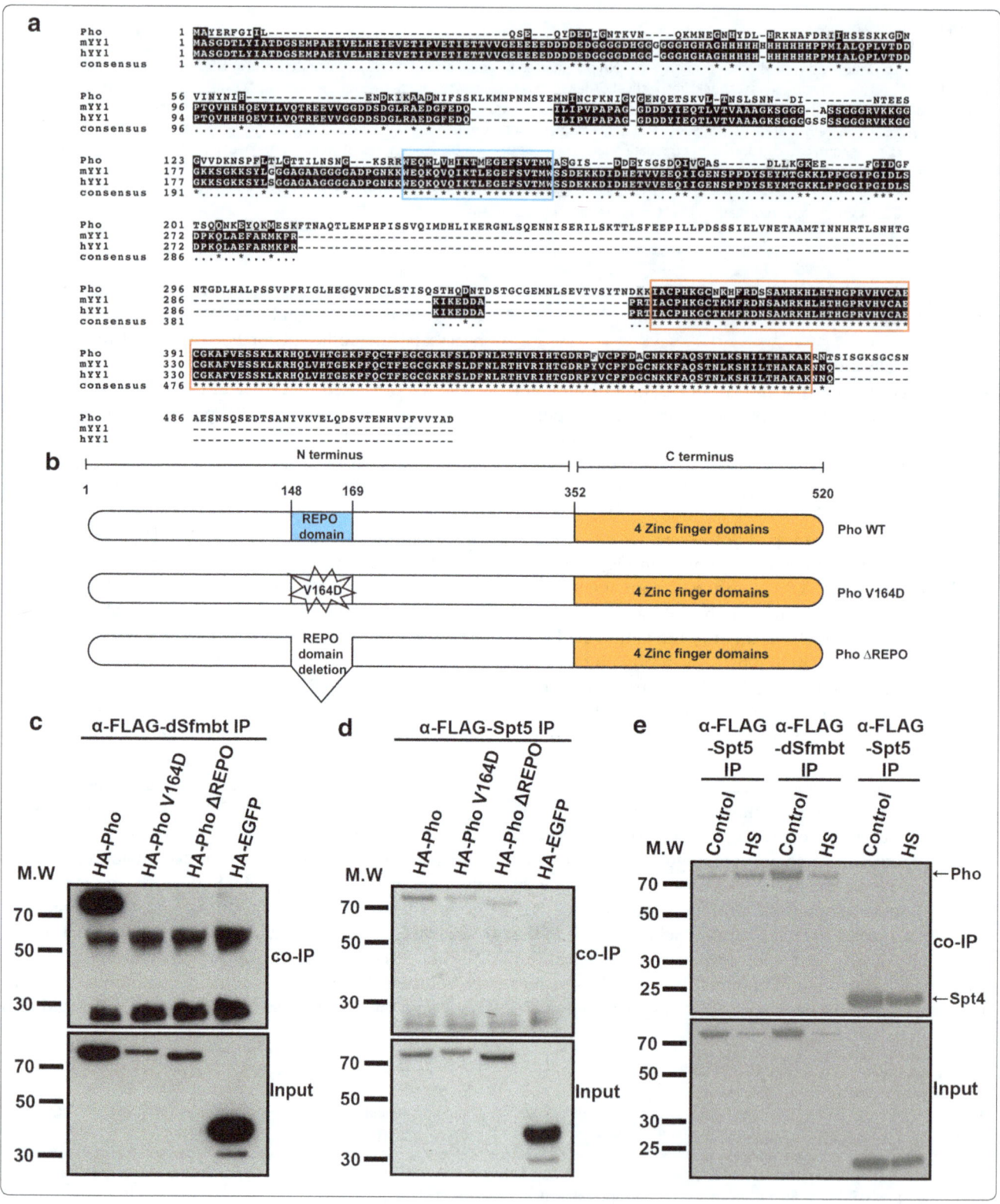

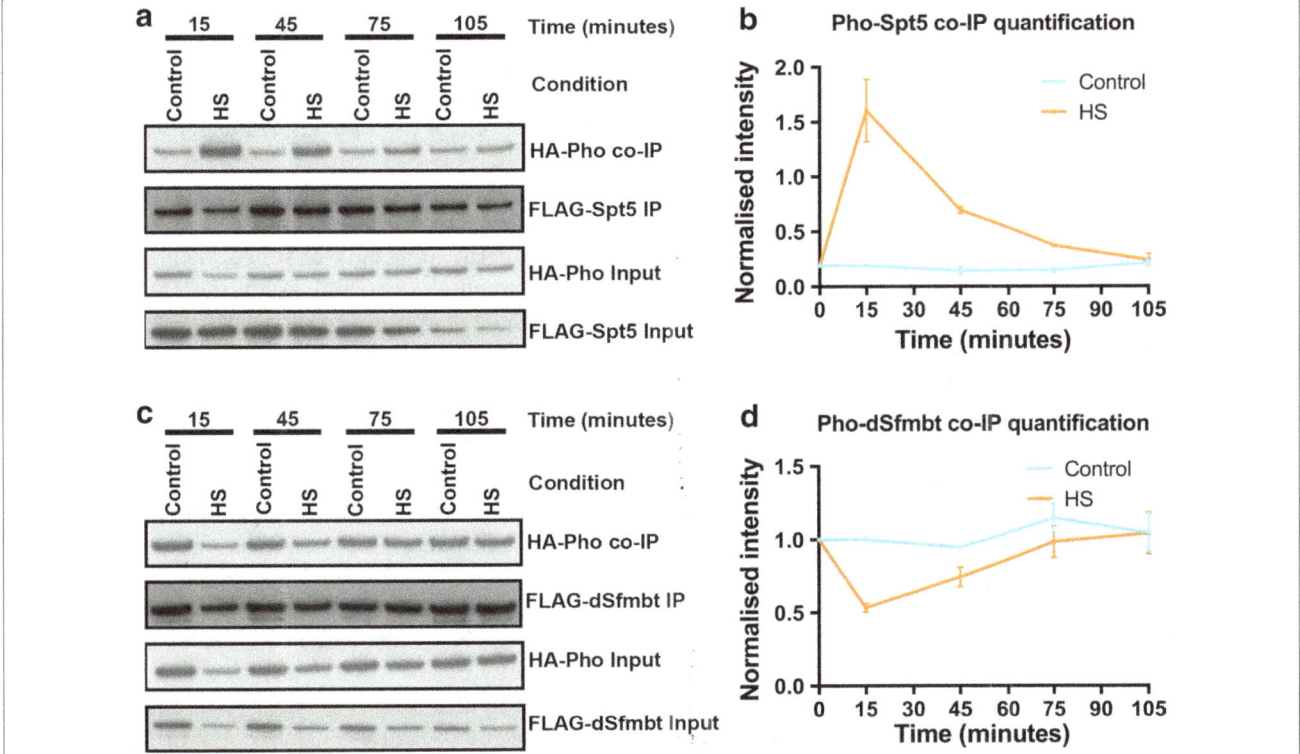

**Fig. 4** Pho preferentially interacts with Spt5 and dSfmbt during the active and silencing phase, respectively, of the heat shock response. **a, c** co-IP assays of S2 DRSC cells transiently transfected with plasmids expressing FLAG-tagged dSfmbt or Spt5 (FLAG-dSfmbt, FLAG-Spt5) and HA-tagged Pho (HA-Pho). The S2 DRSC cells were subjected to a heat shock and then allowed to recover from the heat shock according to the experimental scheme depicted in Fig. 1a. At the indicated time points in the experimental scheme, cell lysates were prepared and were used for pull-downs using an anti-FLAG antibody. They were later probed by Western blot using an anti-HA antibody. **b, d** Quantification of the co-IP assays in A and C. The control line represents cells that were maintained at 25 °C for the entire duration of the time course detailed in Fig. 1a. Details of the procedure used for quantification can be found in Materials and Methods. Data information: In **b** and **d**, data are presented as mean ± SEM for n = 2

interact with Spt5, thus explaining the residual interaction in the absence of a functional REPO domain (Additional file 4: Figure S4c). In contrast, the Pho–dSfmbt interaction was solely dependent on the REPO domain (Additional file 4: Figure S4b). In conclusion, the REPO domain facilitated the interaction of Pho with dSfmbt (a PcG protein) as well as with Spt5 (a transcription elongation factor). These data led us to hypothesise that the protein–protein interactions of Pho might be dynamic under ON/OFF transcriptional states.

To test this hypothesis, we performed co-IPs of Pho–Spt5 and Pho–dSfmbt under control and heat shock conditions. Upon heat shock, we observed a considerable increase in the interaction between Pho and Spt5 and a concomitant reduction in the interaction between Pho and dSfmbt, which was stronger in non-stimulated cells (Fig. 3e and Additional file 4: Figure S4d). As a negative control, we tested whether the interaction between Spt4 and Spt5 is also dynamic upon heat shock. However, the interaction between Spt4 and Spt5 appeared stable under control and heat shock conditions (Fig. 3e). Thus, our

co-IP experiments support the view that Pho can dynamically interact with Spt5 and dSfmbt depending upon the transcriptional state of the *hsp70* locus.

Next, we tested whether the dynamic protein–protein interactions of Pho recapitulated the dynamic chromatin binding patterns of activators and repressors during the heat shock response. To this end, we analysed Pho–Spt5 and Pho–dSfmbt interactions during the heat shock response. Interestingly, the Pho–Spt5 interaction followed the Pol II and HSF (activator) binding dynamic. In contrast, the interaction between Pho and dSfmbt closely resembled the dynamic exhibited by Ph (PRC1, repressor) (Fig. 4a–d). This result suggested the possibility of a relationship between the chromatin binding dynamic of HSF and the dynamic protein–protein interaction of Pho with an activator (Spt5).

### HSF recruitment to chromatin is necessary and sufficient for the dynamic interaction between Pho and Spt5
Since our data suggested that HSF binding kinetics at the *hsp70* locus could influence the protein–protein

interaction between Pho and Spt5, we asked whether HSF recruitment to chromatin would be sufficient and necessary for this dynamic protein–protein interaction. We, thus, relied on heat shock-independent ways to recruit HSF to chromatin.

HSF is canonically chaperoned by Hsp90, which prevents it from trimerising and assuming its active conformation [32]. However, Hsp90 inhibition by radicicol triggers HSF to acquire its active confirmation through trimerisation, thus binding to its cognate site in the *hsp70* locus and causing transcriptional activation. We confirmed both recruitment of HSF to the *hsp70* locus by ChIP-qPCR and transcriptional activation of the locus upon radicicol treatment (Fig. 5a, b). Interestingly, the interaction between Pho and Spt5 was enhanced by Hsp90 inhibition (Fig. 5c, d). Therefore, we set out to decouple this increase in interaction from the transcriptional activation of the *hsp70* locus by recruiting HSF under conditions that caused little or no transcription from the *hsp70* locus.

Upon flavopiridol treatment followed by heat shock (FP + HS), HSF was recruited to the *hsp70* locus, but inhibition of P-TEFb led to a significant decrease in *hsp70* transcript levels (Fig. 5a, b). However, under FP + HS conditions, the interaction between Pho and Spt5 was still enhanced (Fig. 5c, d). Treatment with sodium salicylate has also been shown to recruit HSF to chromatin without affecting transcription from the *hsp70* locus [33]. While HSF was robustly recruited to chromatin upon sodium salicylate treatment, *hsp70* transcript levels remained unchanged when compared to the control condition (Fig. 5a, b). Upon sodium salicylate treatment, the interaction between Pho and Spt5 was, also, increased (Fig. 5c, d). We could, thus, uncouple the increase in

interaction between Pho and Spt5 upon HSF recruitment from transcriptional activation. Next, we tested whether HSF binding was also necessary for this interaction switch between Pho and Spt5 by RNAi-mediated knockdown of HSF. By knocking down HSF, the dynamic nature of the Pho–Spt5 protein–protein interaction was undetectable when compared to the control LacZ RNAi condition (Fig. 5e, f).

Thus, by performing gain-of-function and loss-of-function experiments, we could establish that recruitment of HSF to chromatin is necessary and sufficient for the dynamic protein–protein interaction switch between Pho and Spt5.

In order to confirm that the protein–protein interaction between Pho and Spt5 was independent of the DNA-binding ability of Pho, we performed a co-IP between Pho and Spt5 under *DNase I* conditions. Indeed, we could confirm that the interaction as well as the dynamic switch between Pho and Spt5 was independent of DNA (Additional file 5: Figure S5a). We further went on to confirm the DNA-independent nature of the Pho–Spt5 interaction by performing a co-IP between N-Pho (lacking the DNA-binding Zn-finger domains) and Spt5 (Additional file 5: Figure S5b). However, we did observe dependence for the dynamic interaction switch between Pho and Spt5 on RNA binding (Additional file 5: Figure S5a). This observation is in line with our conclusion that HSF is necessary and sufficient for the dynamic interaction switch between Pho and Spt5 since HSF binding can trigger release of paused RNA Pol II leading to the production of nascent RNA transcripts [23]. Furthermore, the KOW4-5 domain of Spt5 has been shown to be capable of binding RNA and thereby influence the position of paused RNA Pol II [34].

(See figure on next page.)
**Fig. 5** HSF recruitment to chromatin is necessary and sufficient for the dynamic protein–protein interaction between Pho and Spt5. **a** ChIP-qPCR measurements for the occupancy levels of HSF at the *hsp70* locus in S2 DRSC cells upon HS (heat shock), radicicol, FP + HS (flavopiridol + heat shock) and Na Sal (Sodium Salicylate) treatment conditions, respectively. In case of radicicol treatment, the control condition was DMSO treatment. For FP + HS, the control condition was FP control and for Na Sal, the control condition was nuclease-free water. Distance of the location of the primers used from the *hsp70* TSS is − 85 bp and has been depicted in the form of a cartoon below the data figure. **b** qRT-PCR measurements of *hsp70* transcript levels in S2 DRSC cells under HS, radicicol, FP + HS and Na Sal treatment conditions, respectively. Distance of the location of the primers used from the *hsp70* TSS is + 1492 bp and has been depicted in the form of a cartoon below the data figure. **c** co-IP assays of S2 DRSC cells transiently transfected with plasmids expressing FLAG-tagged Spt5 (FLAG-Spt5) and HA-tagged Pho (HA-Pho). The S2 DRSC cells were subjected to HS, radicicol, FP + HS and Na Sal, respectively; cell lysates were used for pull-downs with an anti-FLAG antibody and later probed by Western blot with an anti-HA antibody. **d** Quantification of the co-IP assays in **c**. Details of the procedure used for quantification can be found in Materials and Methods. **e** co-IP assays of S2 DRSC cells transiently transfected with plasmids expressing FLAG-tagged Spt5 (FLAG-Spt5) and HA-tagged Pho (HA-Pho). Cells were either treated with HSF or LacZ RNAi for 4 days, then either maintained at control conditions or heat shocked, cell lysates were prepared and used for pull-downs using an anti-FLAG antibody and later probed by Western blot using an anti-HA antibody. **f** qRT-PCR measurements of *hsp70* transcript levels in S2 DRSC cells under HSF and LacZ RNAi conditions. Distance of the location of the primers used from the *hsp70* TSS is + 1492 bp and has been depicted in the form of a cartoon below the data figure. Data information: In **a** and **b**, data are presented as mean ± SEM for n = 2, while in **e**, data are presented as mean ± SEM for n = 3

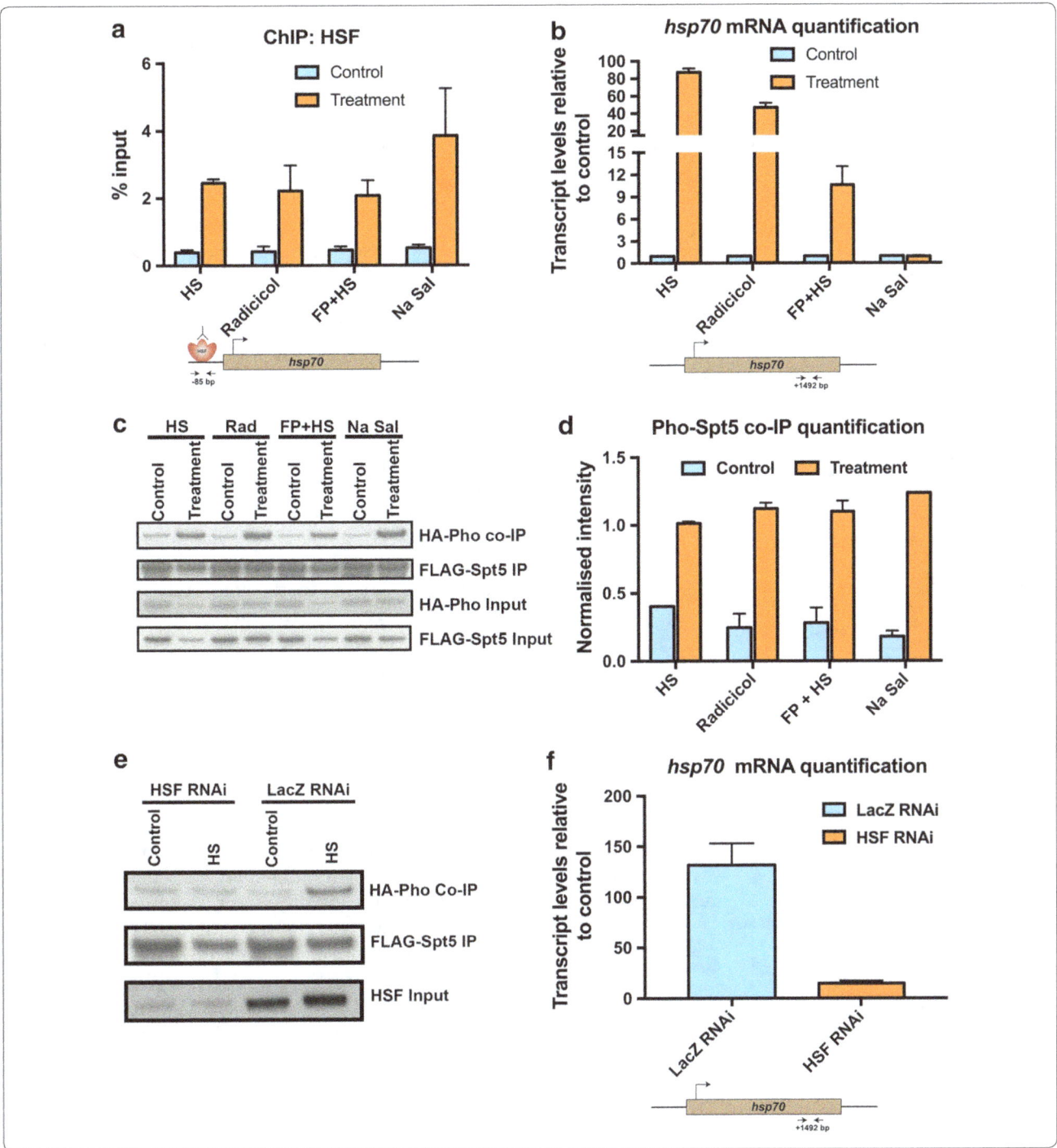

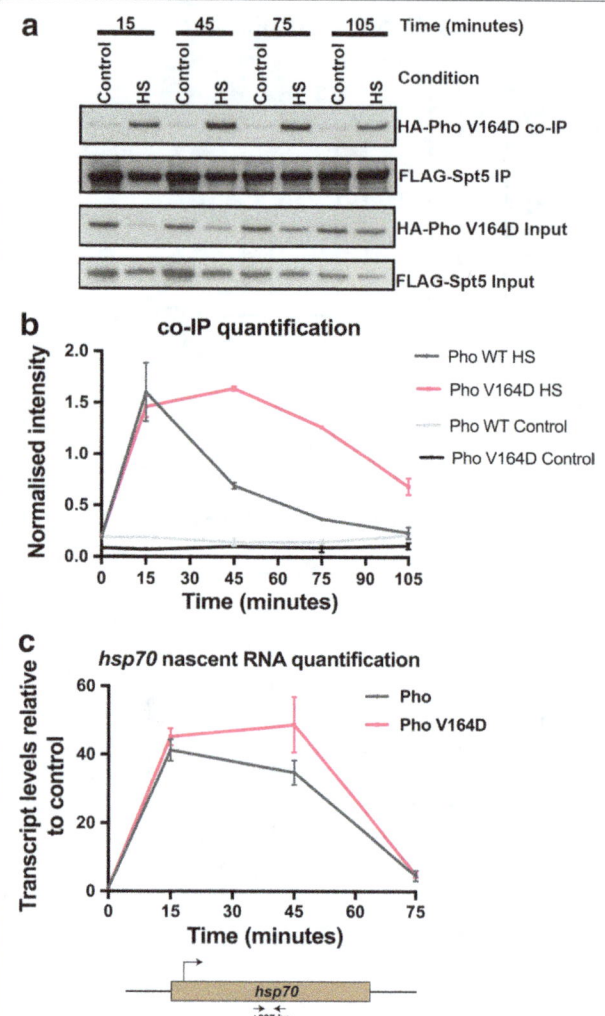

**Fig. 6** Absence of a functional REPO domain disrupts the transcriptional kinetics of the *hsp70* locus during the heat shock response. **a** co-IP assays of S2 DRSC cells transiently transfected with plasmids expressing FLAG-tagged Spt5 (FLAG-Spt5) and HA-tagged PhoV164D (HA-PhoV164D). The S2 DRSC cells were subjected to a heat shock and then allowed to recover from the heat shock according to the experimental scheme depicted in Fig. 1a. At the indicated time points in the experimental scheme, cell lysates were prepared and were used for pull-downs using an anti-FLAG antibody. They were later probed by Western blot using an anti-HA antibody. **b** Quantification of the co-IP assay in **a** and its comparison with the quantification of the Pho–Spt5 co-IP assay across the same time course. The control lines represent cells that were maintained at 25 °C for the entire duration of the time course detailed in Fig. 1a. Details of the procedure used for quantification can be found in Materials and Methods. **c** qRT-PCR measurements of nascent *hsp70* transcripts produced over the course of the heat shock response from S2 DRSC cells upon over-expression of WT Pho and mutant Pho (PhoV164D), respectively. Distance of location of the primers from the *hsp70* TSS is + 687 bp and has been depicted in the form of a cartoon below the data figure. Data information: In **b**, data are presented as mean ± SEM for *n* = 2, while in **c**, data are presented as mean ± SEM for *n* = 3

## Absence of a functional REPO domain disrupts the transcriptional kinetics of the *hsp70* locus during the heat shock response

We next tested the effect of mutating the REPO domain on the dynamic protein–protein interaction between Pho and Spt5 during the heat shock response. To this end, we performed a co-IP for Pho V164D–Spt5 over the time course detailed in Fig. 1a and compared it to the Pho WT–Spt5 protein–protein interaction dynamic.

To our surprise, mutant and wild-type Pho exhibited comparable interaction strengths with Spt5 upon heat shock, indicating that the dynamic interaction switch upon heat shock is not affected by the V164D point mutation in the REPO domain (Fig. 6a, b). However, while the interaction between Pho WT–Spt5 diminished as the *hsp70* locus was repressed in the absence of the activating stimulus, the increase in interaction between Pho V164D–Spt5 remained sustained during the recovery phase (Fig. 6a, b). This could be explained by the fact that in the absence of a functional REPO domain, dSfmbt could no longer compete with Spt5 for interaction with Pho during the recovery phase of the heat shock response (Fig. 6a, b). Similar to the Pho WT–Spt5 interaction dynamic, the Pho V164D–Spt5 interaction switch was independent of DNA but dependent upon RNA (Additional file 6: Figure S6a).

To directly assess the effect of the PhoV164D mutant on transcription from the *hsp70* locus upon heat shock, we expressed the PhoV164D mutant by placing it under the control of the copper-inducible metallothionein promoter. As a control, PhoWT was expressed under the control of the same promoter. The expression levels of the copper-inducible constructs were compared to endogenous Pho after 72 h of induction (Additional file 6: Figure S6b). After expression of the WT and mutant constructs for 72 h, respectively, the transcriptional kinetics of the heat shock response was assessed by measuring nascent RNA production from the *hsp70* locus. The expression of the PhoV164D mutant did not affect the up-regulation of the *hsp70* locus in response to a heat shock as compared to the control (Fig. 6c). However, upon withdrawing the heat shock stimulus, while nascent RNA production in the control cells sharply decreased, this was not observed in the case of PhoV164D mutant (Fig. 6c). Thus, the absence of a functional REPO domain negatively affected the recovery kinetics of the *hsp70* locus during the heat shock response.

## Discussion

Our study aimed at gaining mechanistic understanding about how PcG proteins regulate their target genes, which are transiently activated by an extracellular stimulus. To this end, we used the heat shock-responsive *hsp70* gene in *Drosophila* as it exhibits a dynamic pattern of expression during the well-characterised heat shock response and it is also a target for PRC1. Furthermore, mechanistic analysis supports a role for PcG proteins in the re-establishment of silencing at the *hsp70* locus in the absence of the heat shock stimulus.

Our results highlight antagonism between HSF and PRC1 for binding at the *hsp70* locus, suggesting a role for PRC1 in repressing the locus in the absence of the heat shock stimulus. Furthermore, we could uncouple the eviction of PRC1 from the *hsp70* locus upon heat shock from production of mRNA by treating cells with fla-vopiridol, a P-TEFb inhibitor. We demonstrated that Pho could preferentially interact with Spt5 and dSfmbt under ON/OFF conditions, respectively, at the *hsp70* locus. Our data also conclusively showed that the recruitment of HSF to chromatin is sufficient and necessary for the dynamic interaction between Pho and Spt5. The REPO domain of Pho is central for the interaction switch, since a mutation in the REPO domain not only disrupted the kinetics of the protein–protein interaction between Pho and Spt5 but also disrupted the kinetics of nascent RNA production from the *hsp70* locus during the heat shock response.

Several reports have described a role for Pho in both the activation and the repression of its target genes [19, 35, 36]. In order to dissect the relationship between binding patterns of PRC1, Pho and Trx and their respec-tive effect on transcriptional activity, ChIP-chip was employed to identify their binding sites in the ANT-C and BX-C gene clusters in two different *Drosophila* cell lines. At silent loci, the Pho signal localised at PREs. However, in case of active loci, the Pho signal spread over the gene body [19]. Another study demonstrated the requirement of a Pho-binding site and a functional Pho protein for a pairing-sensitive silencing element in the *Drosophila even skipped* (*eve*) locus for propagation of the silenced as well as the activated state in different cellular contexts, respectively [35]. Yet another reported example is the Ubx gene in *Ubx*[ON] and *Ubx*[OFF] cells in the developing *Drosophila* larvae; the levels of Pho remained unchanged at the *bx* PRE in *Ubx*[ON] cells, while the occu-pancy of PRC1 and PRC2 was reduced by twofold [36].

Our results demonstrate the ability of Pho to switch its protein interaction partners in a HSF-dependent manner. Our current observations at the *hsp70* locus lead us to propose that, in the absence of transcription factor binding, Pho interacts with dSfmbt, forming a recruitment platform for PRC1 and PRC2, and stably maintains the silent state [6, 9, 37]. However, upon acti-vation, TF binding could lead to an increase in interac-tion affinity between Pho and Spt5, thereby relieving

Polycomb-mediated silencing and enabling a switch to the activated state. It is thus plausible that dynamic pro-tein–protein interaction switches between Polycomb group members and Spt5 could potentially play a role in enabling rapid switches in expression states at Pho target genes.

Our study also assessed the contribution of the dynamic switch in Pho protein–protein interactions on the transcriptional output of the *hsp70* locus during the heat shock response. It was notable that Pho prefer-entially interacts with Spt5 and dSfmbt when the *hsp70* locus is active and silent, respectively. A point mutation in the REPO domain of Pho resulted in ablation of its interaction with dSfmbt. While the PhoV164D mutant was capable of strongly interacting with Spt5, immedi-ately upon heat shock, it was incapable of switching back to its interaction with dSfmbt in the absence of the heat shock stimulus. This result underscores the importance of a functional REPO domain in mediating the switch in the protein interaction partners of Pho depending upon the binding kinetics of HSF at the *hsp70* locus. Nascent RNA analysis also suggested that the dynamic protein–protein interactions of Pho could contribute to the tran-scriptional regulation of the *hsp70* locus. Expression of PhoV164D mutant negatively affected the transcriptional kinetics of the *hsp70* locus in response to a heat shock when compared to the control. We speculate that in the absence of a functional REPO domain, the dynamic nature of the protein–protein interactions of Pho is lost, thereby preventing cells from efficiently switching between ON/OFF states in the presence/absence of the activating stimulus.

Taken together, our study highlights how dynamic pro-tein interaction switches between PcG members and Spt5 could facilitate the transcriptional regulation of a Poly-comb-regulated inducible gene like *hsp70*. Additional open questions remain regarding the complex relation-ship between PcG proteins and paused RNA polymerase II. Nonetheless, this study provides a new perspective on how the plasticity of Polycomb gene silencing is medi-ated by dynamic protein–protein interaction switches between activators and repressors.

## Conclusions

In this study, we sought to understand the mechanistic basis of dynamic gene regulation by Polycomb group pro-teins at stress-inducible loci like *hsp70*. We identified a role for the dynamic protein–protein interaction switch between Pho and Spt5 in the transcriptional regulation of the *hsp70* locus. Crucially, we show that the recruitment of HSF to chromatin and a functional REPO domain in Pho is necessary for the dynamic interaction switch between Pho and Spt5. Thus, our study reveals how a dynamic interaction switch between PcG proteins and an elongation factor could enable inducible genes regulated by Polycomb to switch between active and repressed transcriptional states.

## Materials and methods

### Cell culture

S2 DRSC cells (DGRC, stock #181) were cultured in Schneider's medium (Sigma-Aldrich) + 10% heat-inactivated FBS (PAN Biotech GmbH) at 25 °C.

Cells were instantaneously heat shocked by adding Schneider's medium, which was pre-warmed to 48 °C, and then, the temperature was maintained at 37 °C for 15 min by incubating the cells in a water bath as previously described [18]. The cells were then placed into a water bath at 25 °C to enable recovery from the heat shock. In case of treatment with 40 μM radicicol, 500 nm flavopiridol and 10 μM sodium salicylate, the treatment was performed for 1 h, 40 and 10 min, respectively, at 25 °C.

### Chemicals and antibodies

The following chemicals were used in this study: flavopiridol (F3055, Sigma-Aldrich), radicicol (R-370, Alomone), sodium salicylate (S3007, Sigma-Aldrich), copper sulphate (102790, Merck).

The following primary antibodies were used in this study: α-mouse RNA polymerase II CTD repeat (Abcam, ab817), α-mouse RNA polymerase II CTD repeat phosphor S5 (Abcam, ab5408), α-rabbit RNA polymerase II CTD repeat phospho S2 (Abcam, ab5095), α-rat RNA polymerase II CTD repeat phosphor S2 (Active Motif, Cat#61083), α-rabbit HSF (gift from John Lis), α-rabbit Polyhomeotic (Paro lab stock), α-rabbit Pleiohomeotic (gift from Judith Kassis), α-rabbit Pleiohomeotic (gift from Jurg Muller), α-mouse FLAG (Sigma, F1804), α-rat HA (Roche, 11 867 423 001).

The following secondary antibodies were used in this study: α-mouse IgG HRP-linked whole antibody (GE Healthcare, NA931), α-rabbit IgG HRP-linked whole antibody (GE Healthcare, NA934V), α-rat IgG HRP-linked whole antibody (GE Healthcare, NA935V), Goat α-mouse IgG Alexa Fluor 488 (Thermo Fisher Scientific, A-11001), Goat α-rabbit IgG Alexa Fluor 568 (Thermo Fisher Scientific, A-11011).

### Gene cloning

cDNA clones for Spt5 (LD10265) and dSfmbt (LD14884) were obtained from the *Drosophila* Genome Resource Centre (DGRC). cDNA for Pho and Spt4 was prepared from S2 DRSC cells using the First Strand cDNA synthesis kit (Thermo Fisher Scientific, K1612).

In order to express the constructs in S2 DRSC cells, the cDNA clones were amplified using appropriate primers and cloned in pENTR/D-TOPO vector using the pENTR™/D-TOPO™ cloning kit (Thermo Fisher Scientific, K240020). The entry clones were subsequently cloned into pAFW, pAHW, pAWH, pMT/FHW (*Drosophila* Gateway Vector Collection by DGRC) by using the Gateway™ LR clonase™ II enzyme mix (Thermo Fisher Scientific, 11791020). All plasmids were sequenced by Sanger Sequencing (Microsynth AG). All primer sequences used in this study are listed in Additional file 7: Table S1.

### Cell Transfection

In case of transient co-transfection for co-IP experiments, 1 μg of each plasmid was used for transfection, and procedure was performed according to the Effectene transfection reagent protocol (Qiagen, 301425).

In case of stable transfection for generation of cell lines, 2 μg of pMT/FHW HygRes [38] was transfected into cells according to the Effectene transfection reagent protocol and cells were selected with Hygromycin B (Thermo Fisher Scientific, 10687010) 3 days post-transfection until the emergence of drug-resistant colonies was observed.

### RNAi

Sequences used for the production of double-stranded RNA against HSF and LacZ have been previously described [23]. The procedure of dsRNA production was performed according to the manufacturer's protocol (MEGAscript™ RNAi kit, Thermo Fisher Scientific, AM1626).

Cells were treated with dsRNA according to the protocol by DRSC (Cell RNAi, 6-well) with the following exceptions. One million cells were plated in 1 ml of serum-free medium onto 6-well plates. Ten micrograms of dsRNA was added to the well, and this was followed by a 45-min incubation period. Thereafter, 1 ml of Schneider's medium + 20% heat-inactivated FBS was added to the wells, and the cells were incubated for a period of 96 h.

### Chromatin immunoprecipitation

Approximately $25 \times 10^6$ cells were resuspended in PBS and fixed with 1% formaldehyde (Sigma, F8775) for 10 min. The reaction was quenched with 125 mM glycine for 5 min. Cells were washed with ice-cold PBS and centrifuged at 500*g* for 5 min. The cells were resuspended in buffer A (100 mM Tris ph8, 10 mM DTT) and incubated at 4 °C on a rotor for 15 min. Thereafter, the cells were incubated on a thermomixer at 30 °C at 300 rpm. The cells were briefly vortexed for 5 s and centrifuged at 500*g* for 5 min. The cells were resuspended in buffer B (10 mM HEPES ph8, 10 mM EDTA, 10 mM EGTA, 0.25% Triton X-100) and incubated at 4 °C on a rotor for 5 min. The cells were centrifuged at 500g for 5 min. The cells were resuspended in buffer C (10 mM HEPES ph8, 10 mM EDTA, 0.5 mM EGTA, 200 mM NaCl) and incubated at 4 °C on a rotor for 5 min. The cells were centrifuged at 500g for 5 min. The cells were resuspended in buffer D (50 mM HEPES ph8, 10 mM EDTA, 0.5 mM EGTA, 0.1% SDS) and sonicated using S220 ultrasonicator (Covaris) to obtain chromatin fragments approximately 500 bp long.

The chromatin was adjusted to RIPA conditions and centrifuged at 4 °C for 10 min at 13,000 rpm. One hundred microlitres of the chromatin was aliquoted as an input fraction. Appropriate amount of the antibody (amount of each antibody used per IP is stated in Additional file 8: Table S2) was added to the chromatin solution and incubated overnight on a rotor at 4 °C.

In order to prepare the Dynabeads™ Protein A/G (Thermo Fisher Scientific) for purification of the immunocomplexes, 40 μl of the magnetic beads was washed with 1 ml RIPA. The ChIP sample was centrifuged at 4 °C for 5 min at 13,000 rpm, transferred to the tube containing the magnetic beads and incubated at 4 °C on a rotor for 2–3 h. The immunocomplexes along with the magnetic beads were washed five times with RIPA buffer (10 mM Tris ph8, 1 mM EDTA, 1% Triton X-100, 0.1% SDS, 0.1% Na-deoxycholate), once with LiCL buffer (10 mM Tris ph8, 250 mM LiCl, 1 mM EDTA, 0.5% NP40, 0.5% Na-deoxycholate) and twice with TE (10 mM Tris ph8, 1 mM EDTA). The immunocomplexes were eluted thrice with the elution buffer (1% SDS, 0.1 M NaHCO$_3$) at 65 °C for 15 min with gentle agitation (800 rpm). The ChIP sample was reverse cross-linked with 200 mM NaCl and incubated at 65 °C overnight. The input and ChIP samples were treated with RNase (Roche, 11 119 915 001) and incubated at 37 °C for 90 min. This was followed by treatment with Proteinase K (Axon Lab AG, A4392.0001) and incubated at 55 °C for 90 min. The DNA was purified using QIAquick PCR purification kit (QIAGEN, 28106). All buffers were supplemented with protease inhibitors (cOmplete EDTA-free, Roche, 4693132001).

## Polytene immunostaining

*Drosophila* salivary glands were dissected from third-instar larvae, permeabilised in solution 1 (1xPBS, 1% Triton X-100) and then fixed for 10 min (50% acetic acid, 3.7% p-formaldehyde). The squashed salivary glands were prepared as described previously [39]. The slides were washed with PBS and blocked for 60 min with the blocking solution (3% BSA, 0.2% NP40, 0.2% Tween 20, 10% non-fat dry milk powder). Dilutions of the primary antibody (Pho, 1:50 and RNA pol II, 1:100) were made with the blocking solution, and the slides were incubated with the primary antibody overnight in a humid chamber at 4 °C. The slides were washed with PBS, dilutions of the secondary antibody (1:200) were made with the blocking solution and the slides were incubated with the secondary antibodies for 60 min in a humid chamber at room temperature. The slides were washed with wash buffer '300' and wash buffer '400' (1xPBS, 300/400 mM NaCl, 0.2% NP40, 0.2% Tween 20). The slides were then stained with 1 μg/ml DAPI (Roche, 236 276), mounted using 40 μl Fluoromount G (Southern Biotech, 0100-01) and analysed using a Leica DMI6000 B fluorescent microscope.

## Native ChIP (N-ChIP)

The protocol for N-ChIP had been previously described in [27]. Approximately $2 \times 10^8$ cells were harvested by scraping, washed with ice-cold PBS and centrifuged at 1200g at 4 °C for 4 min. The cells were resuspended in TM2+ buffer (10 mM Tris pH 7.5, 2 mM MgCl$_2$, 0.5 mM PMSF). Nuclei were released by adding 0.6% NP-40 to the cell suspension and incubating for 3 min and 30 s on ice. During the incubation, the cell suspension was gently vortexed at intervals of 1 min. The nuclei were centrifuged at 4 °C for 10 min at 1000g and washed once with TM2+. The nuclei were resuspended in TM2+ with protease inhibitors (cOmplete EDTA-free, Roche, 4693132001). MNase treatment was performed as follows: the nuclei were preheated at 37 °C for 3 min. 1 mM CaCl$_2$ was added to the nuclei followed by addition of 80U MNase (M0247S, NEB). The digestion with MNase was carried out for precisely 6 min, and the reaction was stopped by addition of 2 mM EGTA. The nuclei were incubated on ice for 2 min after the MNase treatment and centrifuged at 4 °C for 10 min at 1000g. Nuclei were washed with TM2+ containing protease inhibitors and 2 mM EGTA and centrifuged at 4 °C for 10 min at 1000g. The nuclei were resuspended in 80T buffer (70 mM NaCl, 10 mM Tris pH 7.5, 2 mM MgCl$_2$, 2 mM EGTA, 0.1% Triton X-100, 0.5 mM PMSF, protease inhibitors), cavitated 10 times through a 26G × ½" needle (4710004512, Henke Sass Wolf), and chromatin was centrifuged at 4 °C for 10 min at 1000g. Ten percentage of the chromatin was aliquoted as the input fraction. Anti-Pho antibody (rabbit polyclonal, a gift from Jurg Muller) was used for each IP at 1:100 dilution and incubated at 4 °C on a rotating wheel overnight. Forty microlitres of Dynabeads™ Protein G magnetic beads was added to the IP and incubated at 4 °C for 2 h. Beads were washed twice with 80T buffer, and beads were resuspended in 80T buffer. Following treatment with *RNase A* (Thermo Fisher Scientific, EN0531) and Proteinase K (Axon Lab AG, A4392.0001),

DNA was extracted by phenol–chloroform and ethanol precipitation.

## Co-immunoprecipitation (co-IP)

The co-IP protocol was adapted from [40]. Cells were harvested, washed with ice-cold PBS and centrifuged at 4 °C for 5 min at 500$g$. The cells were resuspended and lysed in co-IP buffer (50 mM Tris ph7.5, 150 mM NaCl, 1% NP40, 1 mM EGTA, 20 mM NaF) for 30 min at 4 °C. The cells were centrifuged at 4 °C for 20 min at 12,000 rpm. Prior to treating the cell lysate with nucleases, 0.4 mM $MgCl_2$ was supplemented to the cell lysate. The cell lysate was incubated with *RNase A* (Thermo Fisher Scientific, EN0531) or *DNase I* (Thermo Fisher Scientific, AM2222) for 10 min at room temperature. The reaction with the nuclease was terminated by addition of 5 mM EDTA to the cell lysate. The supernatant was transferred to a siliconised tube. Ten percentage was aliquoted as the input fraction. 2.5 µg of α-mouse FLAG was added to the IP fraction and incubated at 4 °C for 2–3 h. Forty microlitres of Dynabeads™ Protein G magnetic beads was added to the IP fraction and incubated at 4 °C for 2–3 h. The immunocomplexes were washed thrice with co-IP wash buffer (10 mM Tris ph7.5, 1 mM EDTA, 1 mM EGTA, 150 mM NaCl, 1% Triton X-100). The immunocomplexes were eluted with co-IP buffer, denatured with a mixture containing NuPage® 4x LDS sample buffer (Thermo Fisher Scientific, NP0008) and 200 mM DTT upon heating to 70 °C with mild agitation at 800 rpm for 15 min. Western blots were visualised using Amersham ECL Western Blotting Detection reagent (GE Healthcare, RPN2106) and Amersham Hyperfilm ECL (GE Healthcare, 28906836). All buffers were supplemented with protease inhibitors (cOmplete EDTA-free, Roche, 4693132001) and phosphatase inhibitors (PhosSTOP phosphatase inhibitor cocktail tablets, Roche, 4906837001). For the quantification of HA-Pho co-IP, the intensity of the co-IP band was normalised against the intensity of the corresponding FLAG IP band using ImageJ software.

## Native elongating transcript isolation and quantification

The protocol used for the isolation of native elongating transcripts had been previously described [41]. Cells were grown in medium containing 0.5 mM $CuSO_4$ for 72 h to induce the expression of stably integrated constructs. Cells were harvested and centrifuged at 2000 rpm for 5 min at 4 °C. The cells were resuspended in PBS and centrifuged at 2000 rpm for 5 min at 4 °C. The cells were resuspended in cytoplasmic buffer (10 mM Tris pH 7, 150 mM NaCl, 0.15% NP-40) and incubated on ice for 5 min. The cell lysate was layered on top of NET-seq sucrose buffer (10 mM Tris pH 7, 150 mM NaCl,

0.34 M sucrose) and centrifuged at 16,000$g$ for 10 min at 4 °C. Nuclei were washed twice with nuclei wash buffer (1 mM EDTA, 0.05% Triton X-100, PBS) and centrifuged at 2000$g$ for 5 min at 4 °C. The nuclei were resuspended in glycerol buffer (20 mM Tris pH 8, 75 mM NaCl, 0.5 mM EDTA, 50% glycerol, 0.85 mM DTT) and harsh nuclei lysis buffer (20 mM HEPES ph 8, 300 mM NaCl, 1% NP-40, 1 M urea, 0.2 mM EDTA, 1 mM DTT). This was followed by 4 rounds of vortexing for 2 s each with an incubation period of 1 min on ice between each round of vortexing. Samples were incubated on ice for 2 min and centrifuged at 18,500$g$ for 2 min at 4 °C. The supernatant represented the nucleoplasmic fraction while the pellet contained the chromatin fraction. The chromatin fraction was washed once using a mix of the glycerol and harsh nuclei lysis buffer. The pellet was resuspended in 500 µl of TRIzol. All the buffers were supplemented with protease inhibitors (cOmplete EDTA-free, Roche, 4693132001) and actinomycin D (Sigma, A1410).

Nascent RNA was isolated from the chromatin fraction following manufacturer's protocol (Thermo Fisher Scientific, 15596018). The RNA was precipitated overnight using ethanol. Mature RNAs were depleted from the sample by performing two rounds of Oligo dT-based depletion (Dynabeads™ Oligo(dT)$_{25}$, Thermo Fischer Scientific, 61002). The RNA was DNase-treated using Turbo DNA-free™ kit (Thermo Fisher Scientific, AM1907) and reverse transcribed using First Strand cDNA synthesis kit (Thermo Fisher Scientific, K1612). cDNA was diluted (1:50), mixed with FastStart Essential DNA Green Master (Roche, 06402712001) and assayed using Light Cycler® 96 (Roche). $\Delta\Delta C^T$ method was used to measure transcript levels relative to control.

## Quantitative real-time PCR

For measurement of mRNA levels using qRT-PCR, cells were lysed using TRIzol™ reagent (Thermo Fisher Scientific, 15596018), and RNA was extracted using Direct-zol™ RNA MiniPrep (Zymo Research, R2052). The RNA was DNase-treated using Turbo DNA-free™ kit (Thermo Fisher Scientific, AM1907) and reverse transcribed using First Strand cDNA synthesis kit (Thermo Fisher Scientific, K1612). cDNA was diluted (1:50), mixed with FastStart Essential DNA Green Master (Roche, 06402712001) and assayed using Light Cycler® 96 (Roche). $\Delta\Delta C^T$ method was used to measure transcript levels relative to control.

For measurement of % input values for the ChIP and N-ChIP samples, the input DNA samples were diluted in fivefold dilution series (1:10 up to 1:1250) and the ChIP sample was diluted (1:5). The standard curve obtained from the input DNA dilutions was used to determine the amount of DNA in the ChIP, which was calculated in terms of % input.

## Additional files

**Additional file 1: Figure S1.** Schematic representation of the proteins involved in the silencing, activation and re-silencing of the *hsp70* locus. a At optimal growth temperature, Pho, a DNA-binding PcG member, binds to promoter region of the *hsp70* locus. Pho interacts with dSfmbt, which together form a recruitment platform for PRC1 to the *hsp70* locus. In addition, RNA polymerase II is maintained in the paused state by NELF and Spt5, which act as pausing factors. Upon heat shock, HSF, along with P-TEFb, is recruited to chromatin and releases RNA polymerase II from the paused state. P-TEFb modifies Spt5 and converts into an elongation factor. It also modifies the CTD of RNA polymerase II to enable productive elongation. However, upon removal of the heat shock stimulus, the locus should eventually return to its paused state. Thus, *hsp70* is an ideal model gene to study the eviction and recruitment of PRC1 upon activation and re-silencing, respectively. The colour code for the protein names is as follows: silencing in red, pausing factors in orange and activators in green.

**Additional file 2: Figure S2.** RNA polymerase II binding dynamics at *Act42A* locus during the heat shock response. a, b ChIP-qPCR measurements of occupancy levels of RNA polymerase II CTD and S2P form of RNA polymerase II at the *Act42A* locus during the heat shock response detailed in Fig. 1a. The control line represents cells that were maintained at 25 °C for the entire duration of the time course detailed in Fig. 1a. Distance of the location of the primers used from the *Act42A* TSS is as follows: for a and b (+ 104 bp) and has been depicted in the form of a cartoon below the data figure. Data information: In (a, b), data are presented as mean ± SEM for $n = 2$.

**Additional file 3: Figure S3.** Effect of flavopiridol on RNA polymerase II occupancy at the *hsp70* locus upon heat shock. a N-ChIP-qPCR measurement of the occupancy level of Pho at the *hsp70* locus in S2 DRSC cells during the heat shock response detailed in Fig. 1a. Distance of the location of the primers used from the *hsp70* TSS is − 85 bp and has been depicted in the form of a cartoon below the data figure. b, c ChIP-qPCR measurements of occupancy levels of RNA polymerase II CTD and S2P form of RNA polymerase II, respectively, at the *hsp70* locus in S2 DRSC cells. Distances of the location of the primers used from the *hsp70* TSS are as follows: for a (+ 78 bp) and for b (+ 802 bp) and has been depicted in the form of a cartoon below the data figure. d qRT-PCR measurements of *hsp70* transcript levels in third-instar *Drosophila* larvae under the conditions used for double polytene immunostaining. Distance of the location of the primers used from the *hsp70* TSS is + 687 bp and has been depicted in the form of a cartoon below the data figure. For b–d, the control bar represents cells that were maintained at 25 °C for 15 min, the HS bar represents cells that were maintained at 37 °C for 15 min, the FP control bar represents cells that were treated with 500 nM flavopiridol for 40 min and then maintained at 25 °C for 15 min and the FP HS bar represents cells that were treated with 500 nM flavopiridol for 40 min and then maintained at 37 °C for 15 min. Data information: In (a–d), data are presented as mean ± SEM ($n = 2$).

**Additional file 4: Figure S4.** Chromatin binding dynamics of Spt5 and the dissection of the protein–protein interaction domains of Pho. a ChIP-qPCR measurements of occupancy levels of FLAG-Spt5 at the *hsp70* locus over the time course detailed in Fig. 1a. The control line represents cells that were maintained at 25 °C for the entire duration of the time course detailed in Fig. 1a. The cartoon at the bottom of the figure represents the distance of the location of the primers used from the *hsp70* TSS. b–c co-IP assays of S2 DRSC cells transiently transfected with plasmids expressing FLAG-tagged dSfmbt or Spt5 (FLAG-dSfmbt, FLAG-Spt5) and HA-tagged Pho, N-Pho (a.a 1-351), 1-200N-Pho (a.a 1-200), 173-351N-Pho (a.a 173-351), C-Pho (a.a 352-520). Cell lysates were used for pull-downs using an anti-FLAG antibody and were later probed by Western blot using an anti-HA antibody. MW = molecular weight in kDa. d co-IP assays for S2 DRSC cells transiently transfected with plasmids expressing FLAG-tagged Pho and HA-tagged Spt5 or HA-tagged dSfmbt. S2 DRSC cells were either maintained at normal growth temperature (25 °C) or heat shocked at 37 °C for 15 min. Cell lysates were used for pull-downs using an anti-FLAG

antibody and were later probed by Western blot using an anti-HA antibody. MW = molecular weight in kDa.

**Additional file 5: Figure S5.** The dynamic interaction switch between Pho and Spt5 is independent of DNA but dependent upon RNA. a co-IP assays of S2 DRSC cells transiently transfected with plasmids expressing FLAG-tagged Spt5 (FLAG-Spt5) and HA-tagged Pho (HA-Pho). S2 DRSC cells were either maintained at normal growth temperature (25 °C) or heat shocked at 37 °C for 15 min. Prior to performing the pull-down, the cell lysates were treated with either *RNase A* or *DNase I*. Thereafter, the treated cell lysates were used for pull-downs using an anti-FLAG antibody and later probed by Western blot using an anti-HA antibody. b co-IP assays of S2 DRSC cells transiently transfected with plasmids expressing FLAG-tagged Spt5 (FLAG-Spt5) and HA-tagged N-Pho (HA-N-Pho). S2 DRSC cells were either maintained at normal growth temperature (25 °C) or heat shocked at 37 °C for 15 min. Cell lysates were prepared and were used for pull-downs using an anti-FLAG antibody. They were later probed by Western blot using an anti-HA antibody.

**Additional file 6: Figure S6.** The dynamic interaction switch between Pho V164D–Spt5 is independent of DNA but dependent upon RNA. a co-IP assays of S2 DRSC cells transiently transfected with plasmids expressing FLAG-tagged Spt5 (FLAG-Spt5) and HA-tagged Pho V164D (HA-Pho V164D). S2 DRSC cells were either maintained at normal growth temperature (25 °C) or heat shocked at 37 °C for 15 min. Prior to performing the pull-down, the cell lysates were treated with either *RNase A* or *DNase I*. Thereafter, the treated cell lysates were used for pull-downs using an anti-FLAG antibody and later probed by Western blot using an anti-HA antibody. b A western blot depicting the levels of the copper-inducible constructs (Pho WT/Pho V164D) after 72 h of induction to the endogenous Pho. MW = molecular weight in kDa.

**Additional file 7: Table S1.** List of the primer sequences used in the study

**Additional file 8: Table S2.** Amount of antibody used per IP in ChIP.

## Abbreviations
PRC1: Polycomb repressive complex 1; *hsp70*: Heat shock protein 70; HSF: Heat shock factor; Pho: Pleiohomeotic; PRE: Polycomb response element; DSIF: DRB sensitivity-inducing factor; NELF: Negative elongation factor; P-TEFb: Positive transcription elongation factor; dSfmbt: *Drosophila* Scm-related gene containing four MBT domains; Spt5: Suppressor of TY's; CTD: C-terminal domain; S2P: Serine 2 phosphorylated; ChIP: Chromatin immunoprecipitation; N-ChIP: Native chromatin immunoprecipitation; co-IP: Co-immunoprecipitation; YY1: Ying yang 1; REPO: Recruitment of Polycomb domain.

## Authors' contributions
The study was conceived by AP and RP. All the experiments and analyses were performed by AP. The results of the experiments were interpreted by AP and RP. The manuscript was written by AP and RP. Both authors read and approved the final manuscript.

## Author details
Department of Biosystems Science and Engineering, ETH Zurich, 4058 Basel, Switzerland. ² Faculty of Sciences, University of Basel, 4056 Basel, Switzerland.

## Acknowledgements
We would like to thank Jorge Beira, Federico Comoglio and Cem Sievers for their critical comments on the manuscript. We would also like to acknowledge

Fabian Rudolf for the discussion and input during the generation of the Pho mutants and Christian Beisel for the discussion and input on performing N-ChIP. We would like to thank John Lis, Judith Kassis and Jurg Muller for sharing antibodies. Allwyn Pereira is a member of the Molecular Life Sciences Programme, Life Science Graduate School Zurich.

**Competing interests**
The authors declare that they have no competing interests.

**Funding**
This work was supported by a grant from the Swiss National Science Foundation to Renato Paro and the ETH Zurich.

**References**
1. Prezioso C, Orlando V. Polycomb proteins in mammalian cell differentiation and plasticity. FEBS Lett. 2011;585:2067–77.
2. Schwartz YB, Pirrotta V. A new world of Polycombs: unexpected partnerships and emerging functions. Nat Rev Genet. 2013;14:853–64.
3. Brown JL, Mucci D, Whiteley M, Dirksen ML, Kassis JA. The *Drosophila* Polycomb group gene pleiohomeotic encodes a DNA binding protein with homology to the transcription factor YY1. Mol Cell. 1998;1:1057–64.
4. Müller J, Kassis JA. Polycomb response elements and targeting of Polycomb group proteins in *Drosophila*. Curr Opin Genet Dev. 2006;16:476–84.
5. Alfieri C, Gambetta MC, Matos R, Glatt S, Sehr P, Fraterman S, et al. Structural basis for targeting the chromatin repressor Sfmbt to Polycomb response elements. Genes Dev. 2013;27:2367–79.
6. Frey F, Sheahan T, Finkl K, Stoehr G, Mann M, Benda C, et al. Molecular basis of PRC1 targeting to Polycomb response elements by PhoRC. Genes Dev. 2016;30:1116–27.
7. Francis NJ, Kingston RE, Woodcock CL. Chromatin compaction by a polycomb group protein complex. Science. 2004;306:1574–7.
8. Pengelly AR, Copur Ö, Jäckle H, Herzig A, Müller J. A histone mutant reproduces the phenotype caused by loss of histone-modifying factor Polycomb. Science. 2013;339:698–9.
9. Laprell F, Finkl K, Müller J. Propagation of Polycomb-repressed chromatin requires sequence-specific recruitment to DNA. Science. 2017;356:85–8.
10. Coleman RT, Struhl G. Causal role for inheritance of H3K27me3 in maintaining the OFF state of a *Drosophila* HOX gene. Science. 2017;356:eaai8236.
11. Enderle D, Beisel C, Stadler MB, Gerstung M, Athri P, Paro R. Polycomb preferentially targets stalled promoters of coding and noncoding transcripts. Genome Res. 2011;21:216–26.
12. Rougvie AE, Lis JT. The RNA polymerase II molecule at the 5′ end of the uninduced *hsp70* gene of D. melanogaster is transcriptionally engaged. Cell. 1988;54:795–804.
13. Missra A, Gilmour DS. Interactions between DSIF (DRB sensitivity inducing factor), NELF (negative elongation factor), and the *Drosophila* RNA polymerase II transcription elongation. Proc Natl Acad Sci USA. 2010;107:11301–6.
14. Wu CH, Yamaguchi Y, Benjamin LR. NELF and DSIF cause promoter proximal pausing on the *hsp70* promoter in *Drosophila*. Genes Dev. 2003;17:1402–14.
15. Westwood JT, Clos J, Wu C. Stress-induced oligomerization and chromosomal relocation of heat-shock factor. Nature. 1991;353:822.
16. Lis JT, Mason P, Peng J, Price DH, Werner J. P-TEFb kinase recruitment and function at heat shock loci. Genes Dev. 2000;14:792–803.
17. Ivanov D, Kwak YT, Guo J, Gaynor RB. Domains in the SPT5 protein that modulate its transcriptional regulatory properties. Mol Cell Biol. 2000;20:2970–83.
18. Boehm AK, Saunders A, Werner J, Lis JT. Transcription factor and polymerase recruitment, modification, and movement on *dhsp70* in vivo in the minutes following heat shock. Mol Cell Biol. 2003;23:7628–37.
19. Beisel C, Buness A, Roustan-Espinosa IM, Koch B, Schmitt S, Haas SA, et al. Comparing active and repressed expression states of genes controlled by the Polycomb/Trithorax group proteins. Proc Natl Acad Sci USA. 2007;104:16615–20.
20. Harvey R, Schuster E, Jennings BH. Pleiohomeotic interacts with the core transcription elongation factor Spt5 to regulate gene expression in *Drosophila*. PLoS ONE. 2013;8:e70184.
21. Andrulis ED, Guzmán E, Döring P, Werner J, Lis JT. High-resolution localization of *Drosophila* Spt5 and Spt6 at heat shock genes in vivo: roles in promoter proximal pausing and transcription elongation. Genes Dev. 2000;14:2635–49.
22. Teves SS, Henikoff S. Heat shock reduces stalled RNA polymerase II and nucleosome turnover genome-wide. Genes Dev. 2011;25:2387–97.
23. Duarte FM, Fuda NJ, Mahat DB, Core LJ, Guertin MJ, Lis JT. Transcription factors GAF and HSF act at distinct regulatory steps to modulate stress-induced gene activation. Genes Dev. 2016;30:1731–46.
24. Boyer LA, Plath K, Zeitlinger J, Brambrink T, Medeiros LA, Lee TI, et al. Polycomb complexes repress developmental regulators in murine embryonic stem cells. Nature. 2006;441:349–53.
25. Gaertner B, Johnston J, Chen K, Wallaschek N, Paulson A, Garruss AS, et al. Poised RNA polymerase II changes over developmental time and prepares genes for future expression. Cell Rep. 2012;2:1670–83.
26. Chao SH. Flavopiridol inactivates P-TEFb and blocks most RNA polymerase II transcription in vivo. J Biol Chem. 2001;276:31793–9.
27. Kasinathan S, Orsi GA, Zentner GE, Ahmad K, Henikoff S. High-resolution mapping of transcription factor binding sites on native chromatin. Nat Methods. 2013;11:203–9.
28. Ni Z, Schwartz BE, Werner J, Suarez J-R, Lis JT. Coordination of transcription, RNA processing, and surveillance by P-TEFb kinase on heat shock genes. Mol Cell. 2004;13:55–65.
29. Mohd-Sarip A, Venturini F, Chalkley GE, Verrijzer CP. Pleiohomeotic can link polycomb to DNA and mediate transcriptional repression. Mol Cell Biol. 2002;22:7473–83.
30. Klymenko T, Papp B, Fischle W, Köcher T, Schelder M, Fritsch C, et al. A Polycomb group protein complex with sequence-specific DNA-binding and selective methyl-lysine-binding activities. Genes Dev. 2006;20:1110–22.
31. Wilkinson FH, Park K, Atchison ML. Polycomb recruitment to DNA in vivo by the YY1 REPO domain. Proc Natl Acad Sci USA. 2006;103:19296–301.
32. Zou J, Guo Y, Guettouche T, Smith DF, Voellmy R. Repression of heat shock transcription factor HSF1 activation by HSP90 (HSP90 complex) that forms a stress-sensitive complex with HSF1. Cell. 1998;94:471–80.
33. Winegarden NA, Wong KS, Sopta M, Westwood JT. Sodium salicylate decreases intracellular ATP, induces both heat shock factor binding and chromosomal puffing, but does not induce hsp 70 gene transcription in *Drosophila*. J Biol Chem. 1996;271:26971–80.
34. Qiu Y, Gilmour DS. Identification of regions in the Spt5 subunit of DRB sensitivity-inducing factor (DSIF) that are involved in promoter-proximal pausing. J Biol Chem. 2017;292:5555–70.
35. Fujioka M, Yusibova GL, Zhou J, Jaynes JB. The DNA-binding Polycomb-group protein Pleiohomeotic maintains both active and repressed transcriptional states through a single site. Development. 2008;135:4131–9.
36. Papp B. Histone trimethylation and the maintenance of transcriptional ON and OFF states by trxG and PcG proteins. Genes Dev. 2006;20:2041–54.
37. Wang L, Brown JL, Cao R, Zhang Y, Kassis JA, Jones RS. Hierarchical recruitment of polycomb group silencing complexes. Mol Cell. 2004;14:637–46.
38. Kockmann T, Gerstung M, Schlumpf T, Xhinzhou Z, Hess D, Beerenwinkel N, et al. The BET protein FSH functionally interacts with ASH1 to orchestrate global gene activity in *Drosophila*. Genome Biol. 2013;14:R18.
39. Paro R. Mapping protein distributions on polytene chromosomes by immunostaining. Cold Spring Harbor Protoc. 2008;2008:pdb.prot4714-4.
40. Genevet A, Wehr MC, Brain R, Thompson BJ, Tapon N. Kibra is a regulator of the Salvador/Warts/Hippo signaling network. Dev Cell. 2010;18:300–8.
41. Mayer A, Churchman LS. Genome-wide profiling of RNA polymerasetranscription at nucleotide resolution in humancells with native elongating transcript sequencing. Nat Protoc. 2016;11:813–33.

# 8

# Treatment with a DNA methyltransferase inhibitor feminizes zebrafish and induces long-term expression changes in the gonads

Laia Ribas[1], Konstantinos Vanezis[2], Marco Antonio Imués[3] and Francesc Piferrer[1*] ⓘ

## Abstract

**Background:** The role of epigenetic modifications such as DNA methylation during vertebrate sexual development is far from being clear. Using the zebrafish model, we tested the effects of one of the most common DNA methyltransferase (dnmt) inhibitor, 5-aza-2′-deoxycytidine (5-aza-dC), which is approved for the treatment of acute myeloid leukaemia and is under active investigation for the treatment of solid tumours. Several dose–response experiments were carried out during two periods, including not only the very first days of development (0–6 days post-fertilization, dpf), as done in previous studies, but also, and as a novelty, the period of gonadal development (10–30 dpf).

**Results:** Early treatment with 5-aza-dC altered embryonic development, delayed hatching and increased teratology and mortality, as expected. The most striking result, however, was an increase in the number of females, suggesting that alterations induced by 5-aza-dC treatment can affect sexual development as well. Results were confirmed when treatment coincided with gonadal development. In addition, we also found that the adult gonadal transcriptome of 5-aza-dC-exposed females included significant changes in the expression of key reproduction-related genes (e.g. cyp11a1, esr2b and figla), and that several pro-female-related pathways such as the Fanconi anaemia or the Wnt signalling pathways were downregulated. Furthermore, an overall inhibition of genes implicated in epigenetic regulatory mechanisms (e.g. dnmt1, dicer, cbx4) was also observed.

**Conclusions:** Taken together, our results indicate that treatment with a DNA methylation inhibitor can also alter the sexual development in zebrafish, with permanent alterations of the adult gonadal transcriptome, at least in females. Our results show the importance of DNA methylation for proper control of sexual development, open new avenues for the potential control of sex ratios in fish (aquaculture, population control) and call attention to possibly hidden long-term effects of dnmt therapy when used, for example, in the treatment of prepuberal children affected by some types of cancer.

**Keywords:** 5-Aza-dC, Methylation, dnmt, Sex ratio, Epigenetic, Reproduction, Zebrafish

## Background

DNA methylation is one of the main epigenetic modifications involved in gene expression regulation. In vertebrates, it consists in the addition of a methyl group to the 5′ position of cytosine followed by a guanine (CpG) [1, 2]. Proper control of DNA methylation is essential for many phenomena, including X-chromosome inactivation [3], genomic imprinting [4] or ageing [5]. DNA methylation is carried out by enzymes named DNA methyltransferases (dnmts) [6]. In mammals, the main dnmts include one responsible for DNA methylation maintenance (*dnmt1*) and another two for de novo DNA methylation (*dnmt3a/b*) [7, 8]. Addition of methyl groups to CpGs by dnmts can prevent transcription factor binding and hence gene expression [6, 9]. Many studies have focused on the consequences of DNA methylation alterations by using dnmt inhibitor agents

*Correspondence: piferrer@icm.csic.es
[1] Institut de Ciències del Mar, Consejo Superior de Investigaciones Científicas (CSIC), Passeig Marítim, 37–45, 08003 Barcelona, Spain
Full list of author information is available at the end of the article

to control the expression of genes involved in the onset of cancer [10]. Furthermore, many tumour cells have hypermethylation in the promoters of tumour suppressor genes [11, 12] and thus research has also contemplated the effects of demethylating agents to regain the expression of these silenced genes [13, 14].

The most popular demethylation agents are 5-azacytidine (5-aza-CR), 5-aza-2′-deoxycytidine (5-aza-dC), commonly named as decitabine, and zebularine. 5-Aza-dC is more potent than 5-aza-CR, but both are more toxic and unstable than zebularine [14, 15]. These agents block DNA methylation when incorporated in the DNA as cytidine nucleoside analogues [16], forming a covalent bond in which dnmts become removed from the active nuclear pool and the genome results hypomethylated [17, 18]. However, despite many studies on the underlying biochemical reactions taking place in cells exposed to these agents, their exact in vivo mechanism still remains unclear [18, 19].

In recent years, zebrafish (*Danio rerio*) has become widely accepted as a model for the study of epigenetic regulatory mechanisms, which are generally conserved with respect to those of mammals [20, 21]. Thus, for example, epigenetic alterations that occur during germ cell development are common between mice and zebrafish [22, 23]. The paternal zebrafish methylome is inherited through the sperm. After fertilization, the maternal zebrafish methylome is reprogrammed to match the paternal methylome [22, 24]. Subsequently, during earlier development stages, about 80% of the CpGs in the zebrafish genome are methylated with some fluctuations along development, i.e. blastula and gastrula stages [25, 26].

Few studies have investigated the effects of dnmt inhibitors in fish models. In zebrafish, demethylation agents have been used to better understand the role of DNA methylation during early development, where lack of proper DNA methylation resulted in different types of malformations [27]. Cranial deformities were also observed in another fish model, the Japanese rice fish (*Oryzias latipes*), after early exposure to 5-aza-CR [28]. As it occurs in mammalian cells [17, 29], treatment of zebrafish with 5-aza-CR results in global hypomethylation in embryonic cells [30] as well as in adult hepatocytes [31]. In female zebrafish fed by 5-aza-dC during 32 days a decrease in global DNA methylation was observed [32], likewise, in larvae treated with 10 or 25 μM during 0–6 dpf [33]. However, transgenerational effects up to the F2 generation were only observed in the latter study [33]. Few data are found in parental imprinting in fish gametes (reviewed in [34]), a process responsible for the heritance of the DNA methylome. In zebrafish, it has been documented that dynamic changes in DNA methylation occur during imprinting [25] and that the

DNA methylome is inherited through the sperm, but no through the oocyte [22].

During the last few years, the importance of epigenetic regulatory mechanisms for sexual development has been realized, particularly in organisms where sex is the result of the interplay between genetic and environment (reviewed in [35]). Thus, in fish the methylation levels of the promoter of gonadal aromatase (*cyp19a1a*)—the enzyme that converts androgens to oestrogens—in the European sea bass, *Dicentrarchus labrax*, were positively correlated with temperature during early development [36]. In the olive flounder, *Paralichthys olivaceus* [37] and in zebrafish [38] *cyp19a1a* methylation levels during ovarian development have been studied, showing different methylation patterns during folliculogenesis. Whole-genome approaches have revealed global hypermethylation in various chromosomes in the gonads of Nile tilapia, *Oreochromis niloticus*, exposed to elevated temperatures when compared to control fish [39]. Also, in the half-smooth tongue sole, *Cynoglossus semilaevis*, genome-wide DNA methylation analysis revealed the existence of an epigenetic regulatory mechanism on the suppression of the female-specific W chromosomal genes in high-temperature masculinized fish [40]. However, the role of DNA methylation during gonadal development is far from being clear.

The zebrafish is also increasingly becoming a useful model for aquaculture-related research, where, for example, the control of sex ratios is pursued due to the frequent sexual dimorphism in growth [41]. Domesticated zebrafish have a polygenetic sex-determining system in which genetic factors in combination with environmental factors determine the sexual phenotype [42, 43]. In contrast, wild zebrafish has a chromosomal (WZ/ZZ) sex determination system [44]. Thus, domesticated zebrafish is a well-suited model for studying the effects of environmental perturbations on its development, particularly sexual development. After preliminary trials, in this study we report the establishment of the appropriate conditions for treatment of zebrafish with the most common dnmt inhibitor, 5-aza-dC. Importantly, treatments were not limited to embryonic development, as done earlier, but included treatments covering the period of gonadal development. We report the effects of dose, timing and duration of treatment with 5-aza-dC in terms of resulting survival, deformities and growth. Interestingly, we show that treatment with 5-aza-dC consistently results in an increase in the number of females after different treatments, opening the possibility for a new approach to study the epigenetic regulation of sex and its control, and provide a detailed description of the effects on the gonadal transcriptome as a result of 5-aza-dC treatment. We also raise the possibility that some of the novel effects found in zebrafish ovaries could also be happening in other vertebrates, including humans, particularly prepuberal children affected by some types of cancer, where treatment with DNA-demethylating agents is clinically used.

## Results

### 5-Aza-dC decreases survival and induces teratologies when administered during zebrafish early development

#### Early development experiments

Treatment of zebrafish eggs with 5-aza-dC at 0, 5, 15 or 25 µM added to the embryo medium from 0 to 6 days post-fertilization (dpf) resulted in a progressive decrease in survival at the end of the treatment: 75.3, 62, 66 and 44%, respectively. At 30 dpf, these survival values had further dropped to 34, 42, 44 and 26%, respectively.

Treatment of eggs with 5-aza-dC at 75 µM from 0 to 2 dpf resulted in significantly ($P < 0.05$) lower survival at 8 dpf but not before (Fig. 1a). A non-significant delay

on development, as assessed by hatching rate, was also observed (Fig. 1b). Teratologies were already observed at 2 dpf in two out of seven tested families, but in all of them teratology was observed between 3 and 4 dpf onwards (Fig. 2); however, significant differences ($P < 0.05$) were not found until 4 days and onwards, with ~ 75% of the surviving treated fish affected at 8 dpf ($P < 0.05$; Fig. 1c). Teratologies included three major types: body curvature, reduced yolk-sac reabsorption and overall body deformation (Fig. 2).

#### Gonadal development experiments

Treatment of larvae with 5-aza-dC at 25 µM at different periods during gonadal development resulted in significantly ($P < 0.05$) lower survival, particularly if the treatment started at 10 dpf and lasted until 30 dpf (Fig. 1d). Differences in survival persisted until the end of the experiment at 90 dpf only in the fish treated from 10 to 30 dpf. Survival of fish treated only between 10 and 20 dpf

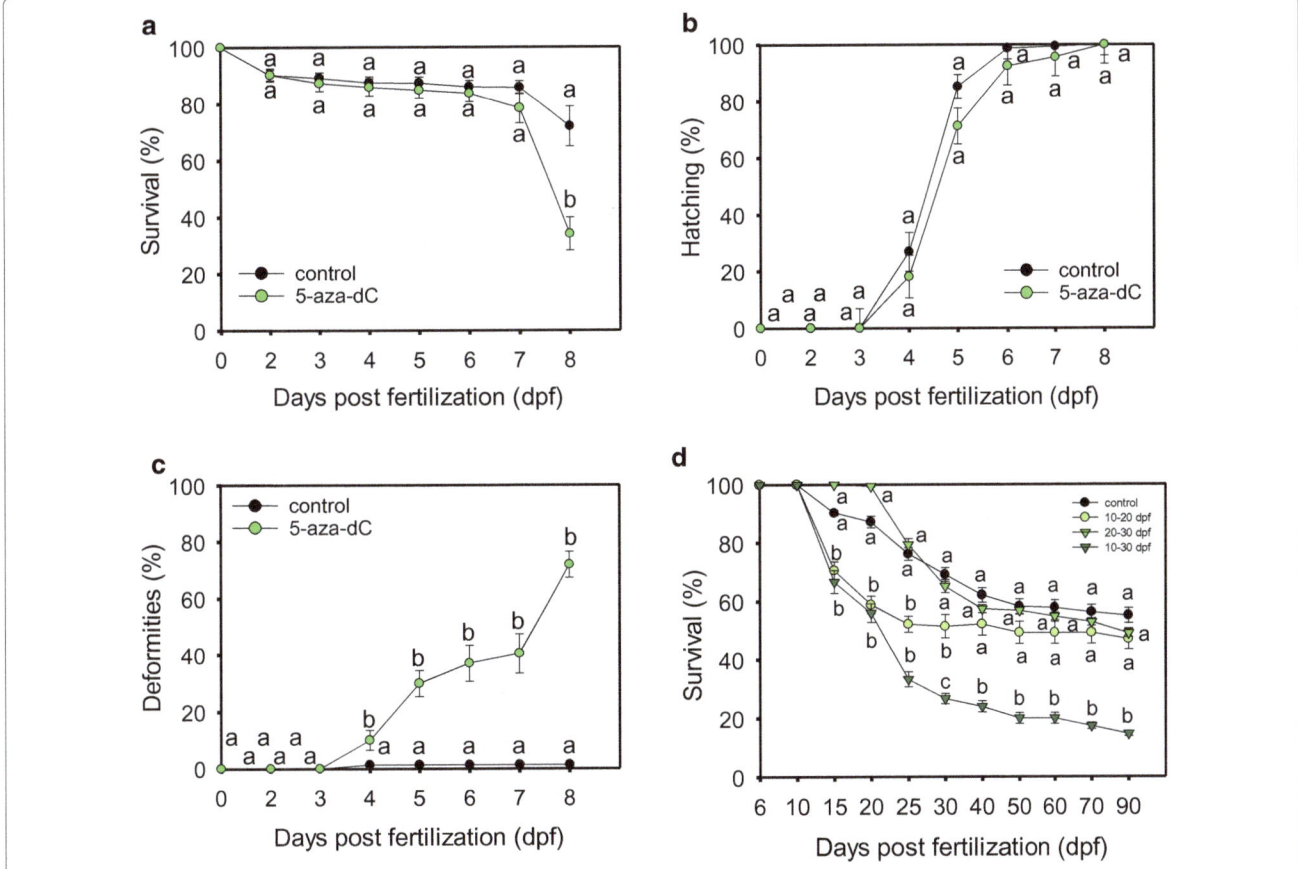

**Fig. 1** Zebrafish treated with 5-aza-dC. **a** Survival, **b** hatching rate and **c** teratology of zebrafish embryos treated with 5-aza-dC at 75 µM from 0 to 2 dpf. Each shown datapoint is the mean ± s.e.m. of seven independent experiments. Within each experiment, each datapoint is the mean of three technical replicates. Significant differences $P < 0.05$ (in **a**) or $P < 0.01$ (in **c**) among groups at a given age are indicated by different letters and were examined by Student's *t* test. **d** Survival of treated zebrafish with 25 µM of 5-aza-dC during gonadal development. Each datapoint is the mean ± s.e.m. of two independent experiments. Within each experiment, each datapoint is the mean of 2–4 technical replicates, originated from five breeding pairs. Significant differences ($P < 0.05$) among groups at a given sampling age were tested by one-way ANOVA and are indicated by different letters

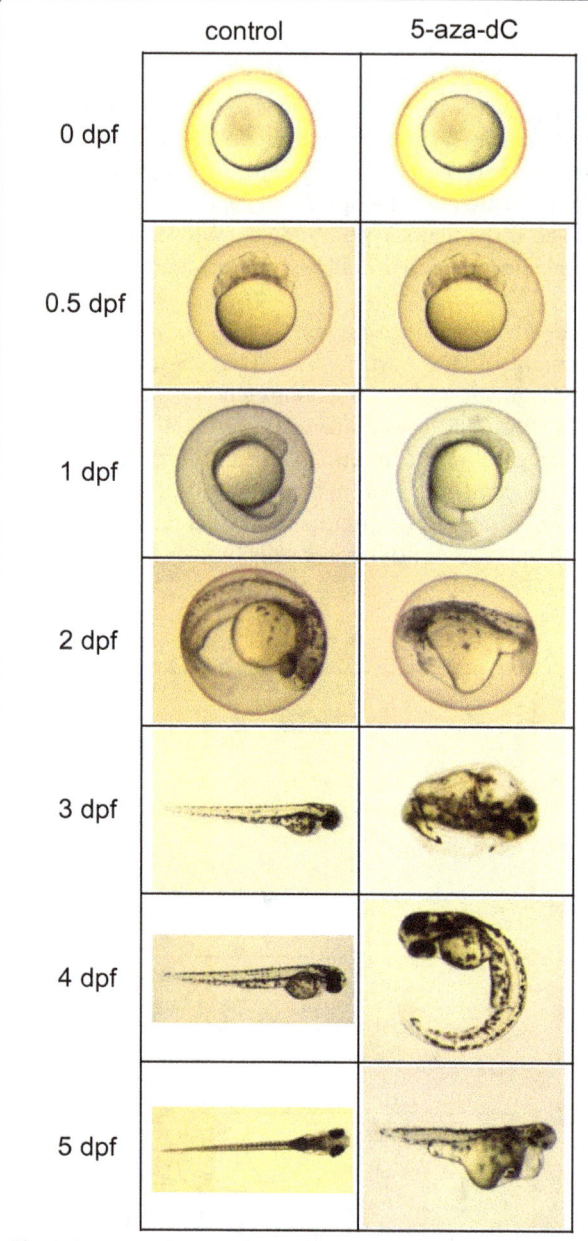

**Fig. 2** Representative teratology observed from zebrafish embryos treated with 5-aza-dC at 75 μM from 0 to 2 dpf. Teratology was observed from 2 dpf in two out of seven families, and between 3 and 4 dpf, it was observed in all tested families. Different types of teratologies were observed; body curvature, problems with yolk-sac reabsorption and body deformation

was also significantly ($P < 0.05$) reduced from 15 to 30 dpf but stabilized at 40 dpf, without differences with the controls. No differences in fish survival were found in the 20–30-dpf period when compared to controls (Fig. 1d).

The survival of fish treated with 5-aza-dC at the highest concentration (75 μM) from 10 to 30 dpf was reduced to 56% one week after starting the treatment (17 dpf) and to 42.6% at the end of the treatment (30 dpf). At 90 dpf, survival of treated fish was only 13.3%. Furthermore, surviving fish were smaller than the controls (Additional file 1: Fig. S1A) both in standard length (SL) in males ($P < 0.05$) and in females ($P < 0.01$) and in body weight (BW) in males ($P < 0.05$) and females ($P = 0.08$) (Additional file 1: Fig. S1 B, C).

### 5-Aza-dC treatment consistently alters the sex ratio

In this study, the number of males in the control groups of the different experiments was in the range of 60–75%, a typical value for domestic zebrafish (AB strain). Treatment with 5-aza-dC at 75 μM from 0 to 2 dpf significantly ($P < 0.05$) reduced the number of males at 90 dpf (Fig. 3a). Furthermore, a clear dose–response effect was elicited when treatment was carried out from 0 to 6 dpf (Fig. 3b), with significant differences ($P < 0.05$) in the number of males with respect to the untreated controls observed with the 15- and 25-μM doses. Consistent with these results, the number of males also decreased in fish treated with 5-aza-dC at 25 μM when the treatment included the 20–30- or 10–30-dpf periods; however, significant differences ($P < 0.05$) could be recorded only in the 20–30 dpf due to the lack of replication in the 10–30-dpf period. In contrast, no significant differences in sex ratio were observed when treatment took place during the 10–20-dpf period. As stated in the previous section, larvae treated with 5-aza-dC fish at 75 μM from 10 to 30 dpf had very low survival. Therefore, in this group sex ratios could not be assessed accurately due to the low number of fish available for statistics. Taken together, the data shown above indicate that treatment with 5-aza-dC is able to alter sexual development in zebrafish.

### Long-term effects of 5-aza-dC treatment on the expression of dnmt1 and dnmt3b

Treatment at 75 μM from 0 to 2 dpf did not affect the expression of the *dnmt1* (Fig. 4a) and *dnmt3* (specifically, *dnmt3bb.2*, which will be referred to as *dnmt3b* in the rest of the paper for simplicity) (Fig. 4b) in 4-dpf larvae. In addition, 5-aza-dC did not alter the expression of *dnmt1* in the gonads of 90-dpf adults (Fig. 4c), but significantly ($P < 0.05$) decreased the expression of *dnmt3b* in testes (Fig. 4d). No effects were observed in 30-dpf juveniles after treatment at 25 μM from 0 to 6 dpf (Fig. 4e, f). However, the same dose administered between 20 and 30 dpf significantly ($P < 0.05$) increased *dnmt1* expression at 30 dpf (Fig. 4g), while *dnmt3b* expression was not affected (Fig. 4h).

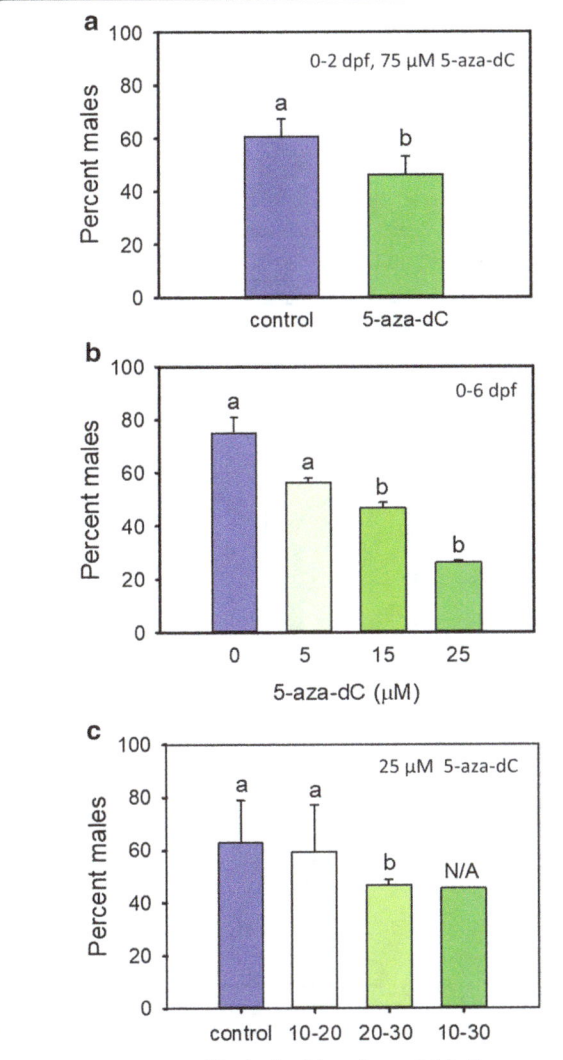

**Fig. 3** Effects of treatment with 5-aza-dC on the resulting sex ratio at 90 dpf. **a** Sex ratio of fish treated with 5-aza-dC at 75 µM from 0 to 2 dpf. Each datapoint is the mean ± s.e.m., corresponding to five independent experiments using five different breeding pairs. Numbers of fish are 162 and 81 in control and 5-aza-dC, respectively. Significant differences among groups (P < 0.05) were analysed by the Chi-squared test with Yate's correction. **b** Sex ratio of zebrafish larvae (0–6 dpf) treated with different 5-aza-dC doses (0, 5, 15, 25 µM n = 13, 27, 31, 19, respectively). Each datapoint is the mean ± s.e.m. of two independent experiments with two technical replicates each. **c** Sex ratio of zebrafish treated with 25 µM of 5-aza-dC during different periods during gonadal development. Data shown as mean ± s.e.m. of three independent experiments with 1–3 biological replicates (n = 133, 59, 72 and 11 in control, 10–20 dpf, 20–30 dpf and 10–30 dpf groups, respectively). Significant differences among groups were analysed by the Chi-squared test with Yate's correction. Significant differences (P < 0.05) are indicated by different letters. N/A, variation not assessed because there was only one experiment one technical replicate

**Effects of 5-aza-dC on the ovarian transcriptome of treated**

**females include downregulation of reproduction-related signalling pathways and a repression of genes related to epigenetic regulatory mechanisms**

Since treatment with 5-aza-dC affected sex ratios by increasing the number of females, we wanted to examine the ovarian transcriptome of females resulting from exposure to 5-aza-dC. To do this, we compared females treated with 5-aza-dC at 75 µM from 10 to 30 dpf during the gonadal development period (Fig. 3c) with untreated females from the control group (n = 4 fish per group) as that was the group that showed the highest differences in growth in comparison with the lowest concentration (25 µm), suggestive of clear treatment effects. Expression profiles using a zebrafish homologous microarray (see materials and methods) were subjected to principal component analysis (PCA), which classified the samples into two clusters corresponding to control and treated fish. The PCA component 1 alone explained 64.0% of the variance, while component 2 explained an additional 10.6% (Fig. 5a). Between the two groups, there were a total of 998 differentially expressed genes (DEG), with 298 up- and 700 downregulated genes with a fold change (FC) ≥ 1.2 including both upregulation and downregulation and a $P$ value < 0.01 (Fig. 5b). Likewise, the number of up- and downregulated DEG with a FC ≥ 2 was 74 and 30, respectively (Fig. 5b). Validation by quantitative (q) PCR using DEG between the two groups, and primarily related to reproduction, showed that the results obtained matched those obtained with the microarray ($R^2 = 0.963$, $P < 0.0001$), thus validating the microarray data (Additional file 2: Fig. S2).

In all gene ontology (GO) terms identified by analysing level 3, a larger number of downregulated GO terms were found (76) in comparison with the upregulated terms (43) (Additional file 3: Table S1, Additional file 4: Fig. S3). In the upregulated terms in the Biological process category (Additional file 4: Fig. S3A), the most enriched subcategories were: multicellular organism development (GO:0007275), system development (GO:0048731) and animal organ development (GO:0048513). The most downregulated subcategories (Additional file 4: Fig. S3D) were: cellular nitrogen compound metabolic process (GO:0034641), cellular macromolecule metabolic process (GO:0044260) and nucleobase-containing compound metabolic process (GO:0006139). We also found a downregulation of a GO term directly related to methylation: methylation-dependent chromatin silencing (GO:0006346) (Additional file 4: Fig. S3D). In the Cellular component category (Additional file 4: Fig. S3B, E) only seven GO terms were characterized in the upregulated subcategory. Likewise, we identified 20 downregulated subcategories, most of them with a high significant enriched $p$ value ($p = 1\text{E}^{-23}\text{–E}^{-4}$),

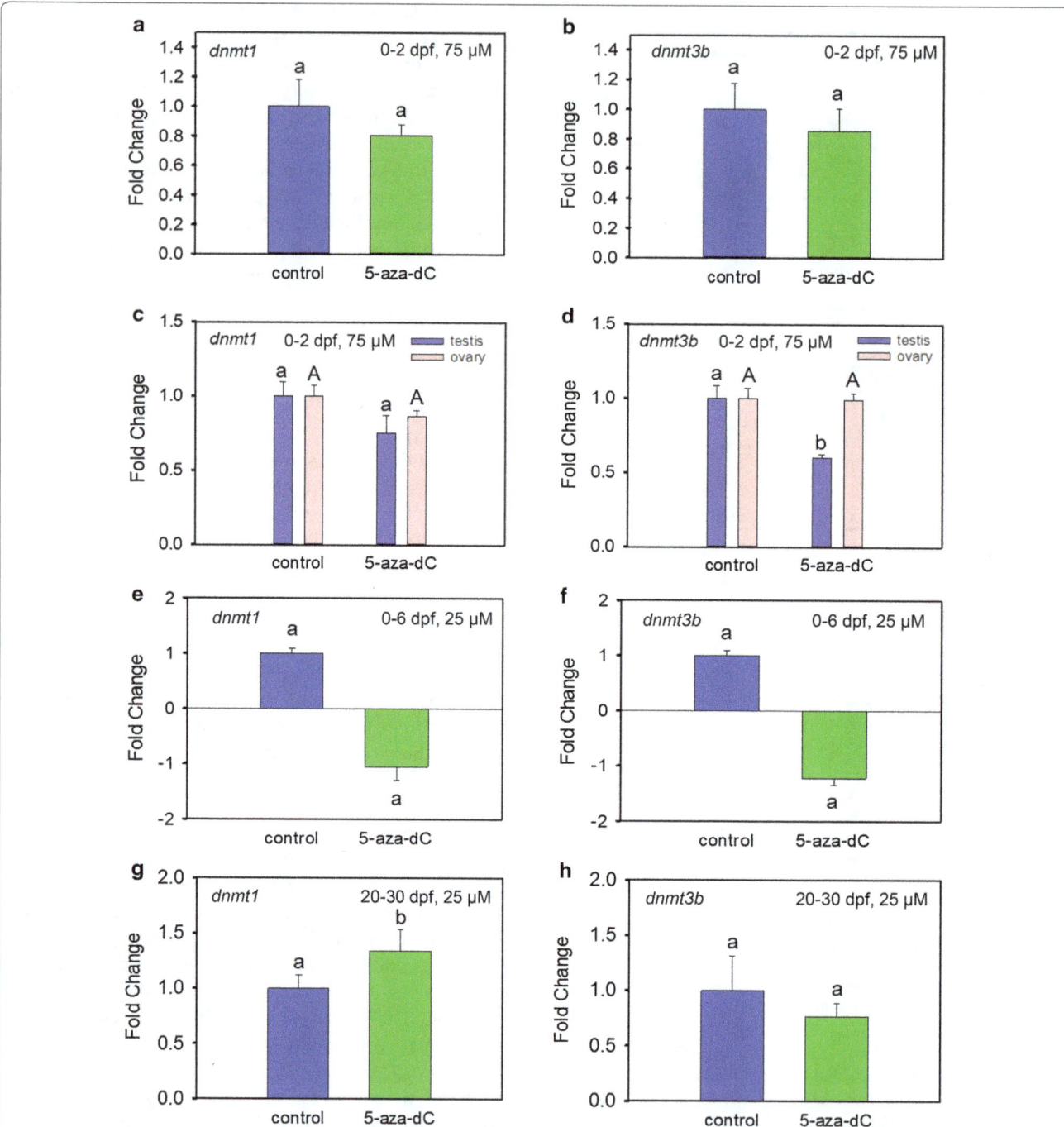

**Fig. 4** Gene expression profiles of DNA methyltransferases 1 and 3 (*dnmt1* and *dnmt3b*) in zebrafish treated with 5-aza-dC. **a**, **b** 0–2-dpf treatment, short-term effects. Gene expression of *dnmt1* (**a**) and *dnmt3b* (**b**) in larvae at 4 dpf previously treated with 5-aza-dC at 75 μM from 0 to 2 dpf. Each datapoint is the mean ± s.e.m., corresponding to 5 pools of larvae each, in turn, made of a pool of ~ 20 larvae from two independent experiments. **b**, **c** 0–2-dpf treatment, long-term effects. Gene expression of *dnmt1* (**c**) and *dnmt3b* (**d**) in zebrafish gonads at 90 dpf after treatment with 5-aza-dC at 75 μM from 0 to 2 dpf. Data shown as mean ± s.e.m. of fold change using control values set at 1. Sample size *n* = 7–9 gonads per sex and treatment. Within the same sex, different letters indicate significant differences (*P* < 0.01) between treated and control fish analysed by Student's *t* test. **e**, **f** 0–6-dpf treatment, mid-term effects. Gene expression of *dnmt1* (**e**) and *dnmt3b* (**f**) in juvenile fish at 30 dpf treated with 5-aza-dC at 25 μM in the period 0–6 dpf. Each datapoint is the mean ± s.e.m. with *n* = 7 individual larvae corresponding to 3 technical replicates. The same letter between groups indicates no significant differences (*P* > 0.05) among groups were tested by Student's *t* test. **g**, **h** 20–30-dpf treatment, short-term effect. Gene expression of *dnmt1* **g** and *dnmt3b* **h** in body trunks of juvenile zebrafish at 30 dpf after treatment from 20 to 30 dpf with 25 μM 5-aza-dC. Data shown as mean ± s.e.m. fold change of *n* = 12 samples per group using control values set at 1. Significant differences (*P* < 0.05) are indicated by different letters

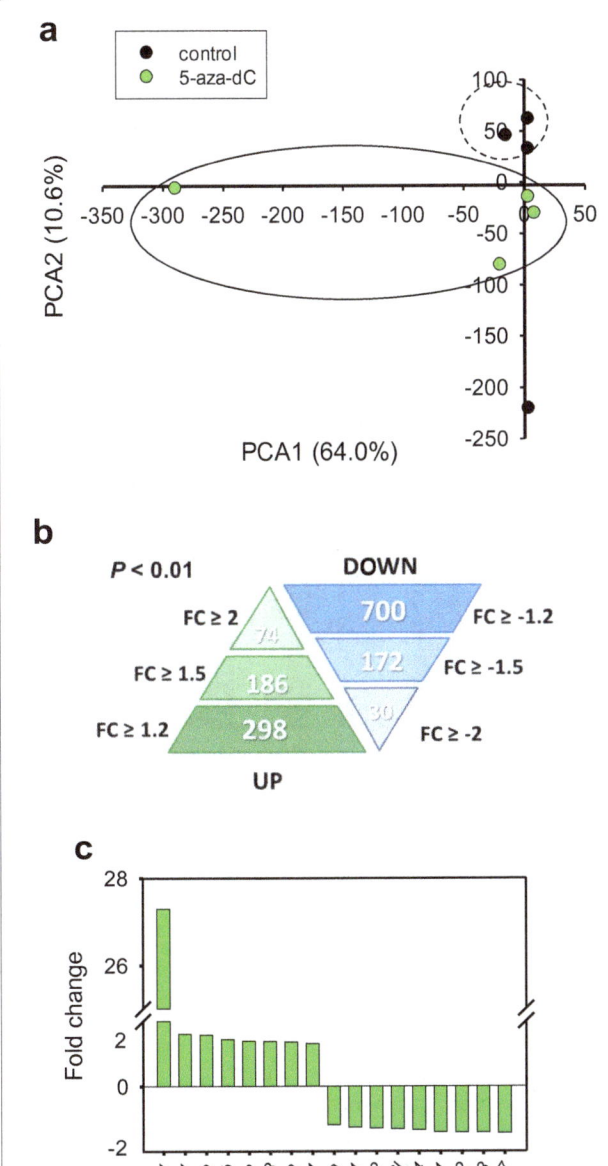

**Fig. 5** Transcriptomic effects in ovaries treated with 75 μM of 5-aza-dC during gonadal development (10–30 dpf). **a** Clustering of fish based on their ovarian transcriptomic profiles at 90 dpf as assessed by principal component analysis (PCA). Per cent values refer to variance (total variance explained ~ 75%). **b** Description of the number of differentially expressed genes (white numbers) according to different fold change values with a P < 0.01. **c** Gene expression at 90 dpf in zebrafish ovaries of some reproduction-related genes present in the microarray. Full gene names are listed in Additional file 5: Table S2

downregulated GO term was related to epigenetics: chromatin binding (GO:0003682).

We identified a total of 24 DEG involved in reproduction-related functions [43, 45–47] (Additional file 5: Table S2) of which 17 are shown in Fig. 5c. The most upregulated gene (FC = 27.30) was cytochrome P450 family (cyp) 11, subfamily a, member 1 (*cyp11a1*), which is involved in the glucocorticoid and steroid pathways catalysing the conversion of cholesterol to pregnenolone during gonad formation [48]. We found other gene members of the cytochrome P450 superfamily also upregulated: 17β-hydroxysteroid dehydrogenase 3 (*hsd17b3*), which is predominantly expressed in the testis, catalysing the conversion of androstenedione to testosterone [49] and cytochrome P450 family 26 subfamily B member 1 (*cyp26b1*), which degrades retinoic acid [50]. In contrast, *hsd11b1*, which catalyses the conversion of the stress hormone cortisol to the inactive metabolite cortisone [51], was downregulated, together with other male-related genes such as spermatogenesis-associated (spata) 6 like *spata6 l*, *spata4* and azoospermia-associated protein 1 (*dazap1*). The oestrogen receptor (esr) 2a, *esr2b* and folliculogenesis-specific BHLH transcription factor (*figl*) were upregulated, whereas zona pellucida glycoprotein 2 (*zp2l2*) and ovarian tumour suppressor candidate 2 (*ovca2*) genes were downregulated.

We identified three enriched Kegg pathways that were upregulated: focal adhesion (dre04510), oxidative phosphorylation (dre00190) and regulation of actin cytoskeleton (dre04810) as well as ten downregulated Kegg pathways (Additional file 6: Table S3). Among them, the most enriched was the mTOR signalling pathway. We also found one reproduction-related pathway, progesterone-mediated oocyte maturation (dre04914), which was also inhibited. Next, we looked at four pathways typically associated with female development in zebrafish [43] (I: Fanconi anaemia, II: Wnt signalling, III: oocyte meiosis and IV: progesterone-mediated oocyte maturation), as well as four pathways typically associated with male development (V: PPAR signalling, VI: p53 signalling, VII: cytokine–cytokine interaction and VIII: cardiac muscle contraction) [43]. The fold change in the expression of genes belonging to these eight signalling pathways in the ovaries of 5-aza-dC-treated females compared to control females ranged approximately from − 2 to + 3 (Fig. 6a, b and Additional file 7: Table S4). Thus, in each pathway there were upregulated and downregulated genes. However, in the four pro-female pathways most genes were downregulated and these accounted for 78–90% of the total number of genes in each pathway (Fig. 6c). In contrast, the number of up- and downregulated genes was similar in two of the pro-male pathways (PPAR and p53

i.e. intracellular (GO:0005622) and intracellular part (GO:0044424). For the Molecular process (MP) category (Additional file 4: Fig S3 C, F) the most enriched

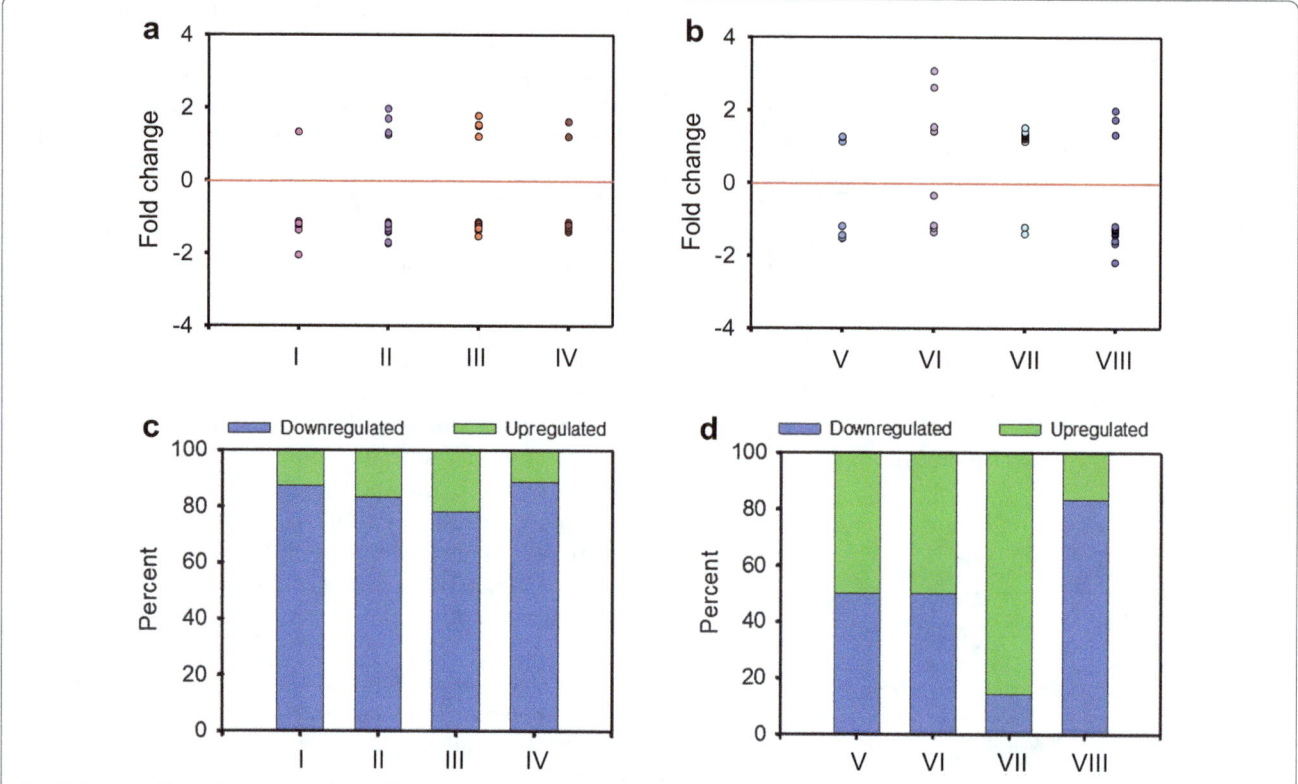

**Fig. 6** Lasting effects of treatment of zebrafish with 5-aza-dC at 75 µM between 10 and 30 dpf during gonadal development on the signalling pathways observed in 90-dpf adult ovaries. **a** Gene expression fold change differences in pro-female pathways. I: Fanconi anaemia (*n* = 8 genes), II: Wnt signalling (*n* = 30 genes), III: oocyte meiosis (*n* = 23 genes) and IV: progesterone-mediated oocyte maturation (*n* = 18 genes). **b** Gene expression fold change differences in pro-male pathways. V: PPAR signalling (*n* = 10 genes), VI: cardiac muscle contraction (*n* = 10 genes), VII: cytokine–cytokine interaction (*n* = 14 genes) and VIII: p53 signalling (*n* = 18 genes). **c, d** Percentage of up- or downregulated genes in the pro-female and pro-male pathways observed in **a** and **b**, respectively

signalling), while in the cytokine–cytokine interaction pathway the number of upregulated genes clearly predominated, whereas in the cardiac muscle contraction pathway it was the opposite (Fig. 6b, d).

Finally, we investigated genes with a known epigenetic-related function and we found a total of 40 DEG which were classified according to their regulatory mechanisms [35, 52] (Fig. 7 and Additional file 8: Table S5): chromatin-related, e.g. chromatin assembly factor 1 subunit A (*chaf1a*); CpG binding domain, e.g. methyl–CpG binding domain protein 3a (*mbd3a*); demethylases, e.g. lysine (K)-specific demethylase 5Bb (*kdm5bb*); dicer, e.g. ribonuclease type III (*dicer1*); histone-related, e.g. histone cluster 1 H4c (*hist1h4c*), methyltransferases, e.g. calmodulin–lysine N–methyltransferase (*camkmt*), and polycomb-associated proteins, e.g. chromobox homologue 4 (*cbx4*). Regardless of their mechanism, 35 out of 40 of these genes were downregulated in the ovaries of treated females (Fig. 7a). These included all but one methyltransferase and all but three chromatin-related genes (Fig. 7b).

## Discussion

Changes in the methylome are part of normal development and occur throughout life in all vertebrates [53]. These changes can be artificially induced by DNA methylation inhibitors, the development of which has been fuelled for their promise in the treatment of some types of cancer. In fish such as medaka, *Oryzias latipes* [54], zebrafish [27] or goldfish, *Carasius auratus* [55], the effects of early treatment with DNA methylation inhibitors have been studied. Here we report the effects of 5-aza-dC treatment applied during either early development or gonadal development in zebrafish. We show for the first time that this DNA inhibitor is able to affect not only the process of sexual development but also the gonadal transcriptome of adults.

Regarding early developmental treatments, these were conducted within the first 2 h after fertilization, since a previous study in zebrafish already showed that such treatments indeed result in a global DNA hypomethylation [27]. In our study, early treatment with 5-aza-dC resulted in lower survival, which was evidenced eight

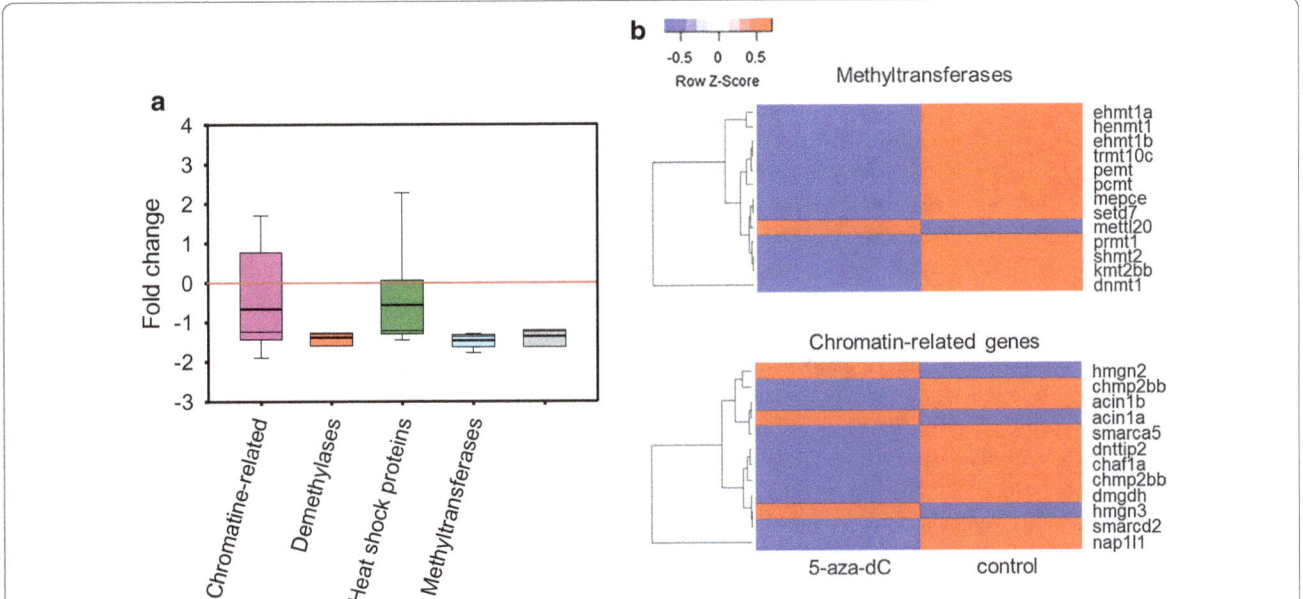

**Fig. 7** Expression of genes involved in different types of epigenetic regulatory mechanisms in the ovaries of 90-dpf adult zebrafish treated with 75 μM of 5-aza-dC during gonadal development (10–30 dpf). **a** Fold change with respect to control values set at 0. In the boxplot, the solid and thick lines indicate the median and mean, respectively; the lower and upper edges indicate the 25th and 75th percentiles, respectively; the lower and upper whiskers indicate the 5th and 95th percentiles, respectively. **b** Heatmap analysis of methyltransferases and chromatin-related genes in treated ovaries. Full gene names are listed in Additional file 8: Table S5

days later. In fact, the toxicity of 5-aza-dC, first reported in human cells, prompted the development of less toxic and more stable agents such as zebularine [19, 56]. Cytotoxic effects have also been reported in medaka [28] and are thought to be due to induced DNA hypomethylation, emphasizing the importance of DNA methylation for proper development [57]. When the 5-aza-dC dose was 75 μM and exposure included most of the gonadal development (10–30 dpf) not only survival but also growth was reduced. This reduction in growth is in accordance with a previous study in which zebrafish was fed with 5-aza-dC at 10 mg/kg for 32 days and where adult females had reduced growth [32].

One of the most striking results found in this study is that 5-aza-dC treatment consistently resulted in a sex ratio bias towards females in all the experiments performed. This is in line with the results obtained with an hermaphroditic fish, the ricefield eel, *Monopterus albus*, in which 5-aza-dC treatment drove the natural sex reversal towards female development [58]. Our study shows that 5-aza-dC at 75 μM applied during 48 hpf was capable to feminize zebrafish. When the exposure time was longer (6 dpf) but the drug was administered in lower concentrations (15 and 25 μM) similar results were obtained. Further, when exposure lasted longer and included the gonadal development period (20–30 dpf) the number of females was also significantly increased.

In contrast, in a recent study in which zebrafish was treated with 5-aza-dC during 0–6 dpf, no differences in sex ratios were found in the F0, but a shift towards males was observed in the F1 [33]. However, in that study group spawnings were used and it is thus difficult to ascertain the genetic contribution of parents, which can heavily influence the results [43, 46]. Our results are unlikely due to differential mortality because the early treatment experiments ended at 6 dpf at the latest, well before the differentiation of the gonads in zebrafish. Adult stickleback, *Gasterosteus aculeatus*, fed with 5-aza-dC (5 μg/g feed) had global DNA changes in both testes and ovaries, with significant changes in methylation levels particularly in the testes, indicating that 5-aza-dC is also able to affect differentiated gonads [31]. Interestingly, treatment with 5-aza-dC of children suffering from acute leukaemia resulted in hypogonadism, indicating that 5-aza-dC can affect the reproductive tissues in humans as well [59]. Taken together, our results with zebrafish along with the results in the other fish species mentioned above suggested that epigenetic mechanisms involving DNA methylation—for example, by decreasing methylation of the *cyp19a1a* promoter in ricefield eel [58]—suggest the possibility of testing DNA methylation inhibitors as a new option to control fish sex ratios as done, for example, in fish farming or population control. Nevertheless, further research should be carried out in additional species.

Transcriptomic data from the ovaries of treated zebrafish with 75 µM of 5-aza-dC during gonadal development showed changes in the expression of genes related to reproduction and sex differentiation. We identified three affected genes of the P450 cytochrome family. The cyp11a1, coding for a key enzyme implicated in steroid biosynthesis, was the highest upregulated reproduction-related gene in ovaries of treated zebrafish, although normally higher expression levels are found in testes [43]. This gene is mostly expressed in the brain, head, kidney and testis of adult fish, but it has also been involved in ovarian formation [48, 60]. In humans, the expression levels of cyp11a1 also increased in placental cells treated with 5-aza-dC [61]. The cyp26b1 gene, which was also upregulated, is involved in the sexually dimorphic entry into meiosis in mammals (downregulation in ovaries and upregulation in testis) [62, 63]. In zebrafish this role seems to be carried on by cyp26a1 [45], although higher levels of cyp26b1 were found in testis of adult heat-treated males when compared to control female ovaries [43]. The cyp17a1 gene was upregulated in 5-aza-dC-treated lymphocytes of infertile men [64], but it was downregulated in our study. We also identified an upregulation of two oestrogen receptors (esr2a and esr2b), similar to was observed in breast cancer cultured cells treated with 5-aza-dC [65]. In addition, we identified a downregulation of ovarian tumour suppressor candidate 2 (ovca2) gene similar to that occurred in 5-aza-dC-treated lung cancer cell lines [66].

Zebrafish is a gonochoristic species in which all individuals initially differentiate female-like gonads. Then, the gonads of about half of the fish enter apoptosis and are transformed into testis [67]. Zebrafish treated with 5-aza-dC showed an increase of the number of females; thus, the transformation into testis might have been interrupted due to the treatment. Treatment also resulted in an inhibition, persisting in adults, of four signalling pathways associated with female development, while one out of four pro-male pathways was clearly upregulated.

This is in agreement with the observation that some male-related genes and male-related pathways were found upregulated in ovaries of exposed fish, indicating that permanent but hidden effects of the treatment during sex differentiation lasted until adulthood. When exposing zebrafish to elevated temperature some females remain as such and others become masculinized. However, recently we discovered that some of the heat-resistant zebrafish females have a male transcriptome [43]. Thus, regardless of exposure, either to heat or to a demethylation agent, some zebrafish females exhibit resistant ovaries in terms of form, although they present a transcriptome similar to that of male gonads.

Epigenetic regulatory mechanisms are implicated in sex determination/differentiation in a wide variety of organisms, including plants and animals (reviewed in [35]). Thus, we also examined genes such demethylases, chromatin-related and histone-related genes. The expression of most of these genes was inhibited long after the end of treatment. This resembles the situation after cancer therapy, where demethylation agents are administered during 3–6 months [68, 69], but side effects such as hematologic and renal toxicities can persist 9–13 months later [70]. The effects of 5-aza-dC are thought to be limited to euchromatin, suggesting some sort of genome selectivity [71]. However, how this favours the development and maintenance of the female phenotype is not known.

The observed upregulation of some genes related to the epigenetic machinery (e.g. dicer1, ehmt2, hdac11, mettl22) is consistent with observations in European sea bass juveniles after early exposure to heat [52]. We also found a repression of 12 out of 13 methyltransferases (e.g. mepce, kmt2ab, ehmt1a, prmt1 and dnmt1). These enzymes catalyse the transfer of methyl groups on histones [72]. Some methyltransferases have been identified in the gonads of fish, for example prmt5, which is implicated in oogenesis and spermatogenesis [73], and ehmt2, the transgenerational regulation of which was recently reported in the testis [74, 75]. Due to its important role in DNA methylation of the genome, dnmts are the most well-studied methyltransferases, not only in mammals, but also in fish. 5-Aza-dC treatment results in a hypomethylation of the genome because the complex DNA-5-aza-dC depletes the activity of the dnmts by the proteasome pathway and activates gene expression [18, 76, 77]. We found inhibition of dnmt3b in adult testes of fish treated for 48 h with 75 µM of 5-aza-dC, but gene expression of the two major dnmts in ovaries was not affected. However, when treatment included the gonadal development (10–30 dpf) period, a significant inhibition of dnmt1 expression in the ovaries at 90 dpf was observed (Additional file 8: Table S5) despite that just at the end of the treatment it expression was increased (Fig. 4g). These results are in agreement with observations made in Japanese ricefield eel embryos, where dnmt1 and dnmt3b expression was enhanced after 5-aza-CR treatment during 2 dpf [28]. Dnmt3aa and dnmt3bb.1 were also upregulated in hatched embryos of Solea senegalensis after 24 h of 5-aza-CR treatment, while dnmt1 was downregulated [78]; in zebrafish larvae treated with 5-aza-dC during 0–6 dpf, where dnmt3bb.2 was upregulated, whereas dnmt1 was not [33]. Thus, the responsiveness to 5-aza-dC treatment can be variable as this drug has multiple in vivo targets [79].

Dicer1 is a key enzyme required for miRNA formation, and so this gene is involved in transcriptional repression functions [80]. Dicer1 is crucial for oocyte maturation in invertebrates [81], and its depletion renders sterile females in insects [82]. In zebrafish, *dicer1* has no role in oocyte maturation as primordial germ cells proliferate and remain pluripotent to form the adult germ line in the *dicer1* mutant [83]. However, *dicer1* is essential for zebrafish development [84]. In this study, we found *dicer1* downregulated in the ovaries after 5-aza-dC treatment during sex differentiation process together with other epigenetic-related genes. We also found downregulation of genes of the polycomb group, which are also repressors of the gene transcription machinery, in particular, the polycomb homologue 1 (*epc1*) and two chromobox genes (*cbx4* and *cbx5*). In mammals, the role of the *cbx2* in sex determination and differentiation has been shown [85, 86], whereas in Nile tilapia it was demonstrated that the expression of *cbx1b*, *cbx3a* and *cbx5* was sexually dimorphic in the gonads [87].

## Conclusions

We studied the effects of DNA methylation on vertebrate sexual development in a well-established model, the zebrafish. We report that the demethylating agent 5-aza-dC results in a sex ratio bias towards females in this species. The scarce data in other species point also to this direction, but whether this is a truly generalized effect is at present unknown. Thus, our results show the importance of DNA methylation for proper control of sexual development and open new avenues for the potential control of sex ratios in fish (aquaculture, population control). We also show that gene expression patterns of reproduction and epigenetic-related genes are affected by 5-aza-dC treatment in gonads, suggesting underlying DNA methylation changes that should be further studied. The long-term effects of treatment with 5-aza-dC at the time when the gonads are still not differentiated on the resulting adult gonadal transcriptome should be considered and explored in other situations. This could include, for example, prepuberal children treated with DNA-demethylating drugs as part of cancer therapy, given the fact that hypogonadism resulting from these treatments has already been reported.

## Methods
### Animal rearing conditions

Domesticated zebrafish (AB strain) were housed in 2.8-l plastic tanks (mod. ZT280) placed in a close-circuit system (Aquaneering, San Diego, CA, USA) inside a thermoregulated walk-in chamber at the aquarium facility of our institute. Inside the chamber the conditions were as follows: 12-h light/12-h dark constant photoperiod, air temperature of $26 \pm 1$ °C and humidity of $50 \pm 3\%$. The water quality parameters were monitored daily

(temperature: $28 \pm 0.2$ °C; pH: $7.2 \pm 0.5$; conductivity: 750–900 µS; dissolved oxygen: 6.5–7.0 mg l$^{-1}$). Ammonium, nitrite, nitrate, silicate and phosphates were checked 2–3 times monthly by the water analysis service of our institute to ensure they remained in the appropriate ranges [41]. Fish were fed ad libitum three times a day with a commercial food (AquaSchwarz, Göttingen, Germany) according to their developmental stages and supplemented with live *Artemia* nauplii (AF48, INVE Aquaculture, Dendermonde, Belgium). Fertilization always followed natural spawning involving single-pair crossings. Batch size and fertilization rates were determined for each clutch to ensure values within accepted range for this species [41]. Likewise, it was ensured that post-hatch survival in the untreated groups was in accordance with the OECD's guidelines for the Fish Sexual Development Test [88]. In order to avoid unwanted masculinization due to elevated rearing density, the number of fish per tank was kept in the range 25–35, based on our previous study of effects of density on zebrafish sex ratios [89].

### Ethics statement

Fish were kept in agreement with the European regulations of animal welfare (ETS N8 123, 01/01/91). Our fish facilities are approved for animal experimentation by the Ministry of Agriculture and Fisheries (certificate number 08039–46–A) in accordance with the Spanish law (R.D. 223 of March 1988). The experimental protocol was approved by the Spanish National Research Council (CSIC) Ethics Committee within the project AGL2013–41047–R.

### 5-Aza-dC treatments
#### Early development experiments

Fertilized eggs were reared at $26 \pm 1$ °C in 6-well Petri dishes (Thermo Fisher Scientific, Waltham, MA, USA) at 30 eggs/well filled with embryo medium (EM, pH $7.2 \pm 0.5$) supplemented with 0.1% methylene blue (Sigma-Aldrich, Madrid, Spain). Starting within the first 2 h post-fertilization (hpf), when zebrafish is sensitive to 5-aza-dC treatments [27], embryos were treated with 5-aza-dC (A3656, Sigma-Aldrich, Saint Louis, USA) added to the EM at a final concentration of 0 (control), 5, 15, 25 or 75 µM (Additional file 9: Fig. S4).

For the 5-, 15- and 25-µM concentrations, treatment lasted until 6 dpf. At 2 and 4 dpf, 50% of the EM volume was replaced with EM containing fresh 5-aza-dC at the appropriate concentration. At 6 dpf, fish were counted and thoroughly rinsed in EM and then housed in the 2.8-l tanks described above. At 30 dpf fish were counted again and a random sample ($n = 7$) of fish were killed and a cross section of the body trunk was cut and flash-frozen in liquid nitrogen until analysis. The remaining fish were left alive until 90 dpf. The experiment was replicated twice using eggs originating from the same breeding pair and involved a total of ~ 600 fertilized eggs.

For the 75-µM concentration, treatment lasted only from 0 to 2 dpf, since preliminary trials showed that at this concentration survival was unacceptably low with longer durations. At 1 dpf, 50% of the EM volume was replaced with EM containing fresh 5-aza-dC. At 2 dpf embryos were thoroughly rinsed in EM and reared in untreated EM. At 4 dpf, 5 pools of 20 larvae each from two technical replicates were collected from the 0- and 75-µM group. Larvae were flash-frozen in liquid nitrogen and kept at − 80 °C for further analysis. At 6 dpf, the remaining larvae were housed in the 2.8-l tanks described above. The experiment was repeated seven times using seven different breeding pairs, involving a total of ~ 1260 fertilized eggs.

The effects of all tested concentrations on survival, hatching rate and teratology were monitored daily until 8 dpf. At 90 dpf, fish were euthanized on iced water followed by severing the spinal cord. Survival, growth and sex ratios were recorded. To assess sex in adults, we used visual inspection of the gonad under a dissecting microscope as previously described [53]. Fish sex was determined individually after dissection, and the sex ratio was calculated for each biological replicate. Gonads were carefully dissected and flash-frozen in liquid nitrogen and stored at − 80 °C until further analysis (Additional file 9: Fig. S4).

### Gonadal development experiments

This experiment targeted the period of gonadal sexual development in zebrafish [38, 40]. The effects of 5-aza-dC at 25 µM were studied in three different periods: 10–20, 20–30 and 10–30 dpf. In the latter period, the concentration of 75 µM was also tested (Additional file 9: Fig. S4). For each period, 2–4 technical replicates were used and the whole experiment was repeated twice. For these experiments, larvae obtained from pooled eggs of five different broodstock pairs were used. To carry out this experiment, 10-dpf larvae were randomly assigned to 2.8-l tanks. Treatments were carried out with static bath at different nominal 5-aza-dC concentrations (0, 25 or 75 µM). The tanks were placed inside a large thermoregulated tub to ensure constant temperature. Three times week during the treatment period, 50% of the water in each tank was replaced with water containing fresh 5-aza-dC at the appropriate concentration. Once all treatments with 5-aza-dC were finished at 30 dpf, the water of all tanks was replaced and fish were still maintained in the tanks inside the tub for an additional

2 weeks to ensure complete clearance of the drug before being returned to the commercial rack.

Survival was recorded periodically every 5–10 days during the course of this experiment. At 30 dpf 12 juvenile fish (whole body) per group were flash-frozen individually and kept at − 80 °C for further gene expression analysis. At 90 dpf all remaining fish were killed and sampled as described above.

### Gene expression analysis

Tissues were homogenized with 0.5 ml of TRIzol (Sigma), and total RNA was extracted with chloroform, precipitated with isopropanol and washed with 75% ethanol. Pellets were suspended in 25 µl DEPC–water and stored at − 80 °C. Total RNA concentration was determined by spectrometry (ND-1000 spectrophotometer, NanoDrop Technologies), and quality was checked on a 1% agarose/formaldehyde gel. RNA (200 ng) was treated with DNAse I, Amplification Grade (Thermo Fisher Scientific Inc., Wilmington, DE, USA H) and retrotranscribed to cDNA using SuperScript III RNase Transcriptase (Invitrogen, Spain) and Random hexamer (Invitrogen, Spain) following the manufacturer's instructions. Quantitative PCR (qPCR) was carried out with the SYBR Green chemistry (Power SYBR Green PCR Master Mix; Applied Biosystems). All qPCRs were run in triplicate in optically clear 384-well plates. Cycling parameters were: 50 °C for 2 min, 95 °C for 10 min, followed by 40 cycles of 95 °C for 15 s and 60 °C for 1 min. Finally, a temperature-determining dissociation step was performed at 95 °C for 15 s, 60 °C for 15 s and 95 °C for 15 s at the end of the amplification phase. qPCR data were collected by SDS 2.3 and RQ Manager 1.2 software, and relative quantity (RQ) values were calculated by the $2\Delta\Delta Ct$ method [90, 91]. Specificity for each primer pair was also confirmed by dissociation step, primers efficiency curves and PCR product sequencing. Primer sequences used for gene expression study are shown in Additional file 10: Table S6.

### Microarray hybridization and analysis

For microarray analysis, RNA samples ($n = 4$ for control and $n = 4$ for 5-aza-dC groups) from ovaries at 90 dpf of fish subjected to 75 µM of 5-aza-dC during 10–30 dpf were used. RNA integrity was measured by a Bioanalyzer 2100 (RNA 6000 Nano LabChip kit Agilent, Spain). Samples with a RNA integrity number (RIN) > 8.5 were used for microarray hybridizations. Briefly, 50 ng of total RNA was labelled using the Low Input Quick Amp Labeling Kit, One-Color (Cy3; Agilent Technologies). Samples were hybridized individually in a 4 × 44 K Agilent platform (G2519F) at the Barcelona Biomedical Research Park (PRBB). cRNA was prepared for overnight hybridization with the corresponding buffers during 17 h at 65 °C and washed on the following day. Hybridized slides were scanned using an Agilent G2565B microarray scanner (Agilent Technologies, USA). Agilent software was used to avoid saturation, and feature extraction generated the raw data for further preprocessing. Statistical analyses were carried out with the statistical language R (2.13.1 version). Array normalization was implemented using the Quantile method in the Limma package in R (http://www.R-project.org/). $P$ value < 0.01 threshold was applied to identify genes that showed statistically significant differences in gene expression from comparisons of interest. Microarray analysis software Multiple Experiment Viewer (MeV) version 4.8.1 was used to analyse microarray data and visualized samples by PCA. For the heatmaps statistical language R (3.3.2 version) was used with the gplot package. The log2 transformation of the fluorescence values was used for the statistical analysis. DAVID Bioinformatic Resources 6.8 and REVIGO software [92, 93] was used to analyse and study the enriched gene ontology (GO) terms in the DEG between groups. For GO terms analysis, a Fisher exact test ($P < 0.05$) false discovery rate (FDR) corrected for multiple testing was performed using all genes in our microarray as background and the DEG of each comparison as query. Microarray data were submitted to NCBI's Gene Expression Omnibus (GEO) [94] and are accessible through GEO Series accession number GSE93367 (https://www.ncbi.nlm.nih.gov/geo/query/acc.cgi?acc=GSE93367).

Microarray was validated by quantifying the gene expression of 16 genes by qPCR (Additional file 2: Fig. S2D). The genes and the primers used are listed in Additional file 10: Table S6. The RNAs of the same individuals used for microarrays were retrotranscribed, and qPCR was performed as previously described. Genes were selected based on their importance to reproduction (Additional file 5: Table S2) and epigenetics (Additional file 8: Table S5) in fish sex determination and differentiation [35, 43, 45–47, 52].

### Statistical analysis

Data normality and the homoscedasticity of variances were checked with the Kolmogorov–Smirnov's and Levene's tests, respectively. One-way analysis of variance (ANOVA) was used to detect possible differences among groups in survival, BW and SL. Post hoc multiple comparisons were made with the Tukey's test. The Student's $t$ test was used to detect differences in gene expression analysis by using $2\Delta Ct$ values [91]. For sex ratio analysis, Chi-squared test with Yate's correction was used [95]. All data analyses were performed with Stat Graphics software (version 17). Data were expressed as mean ± s.e.m. In all tests, differences were accepted as significant when $P < 0.05$.

### Additional files

**Additional file 1: Fig. S1.** Effects of zebrafish treatment with 5-aza-dC at 75 µM during the period of gonadal development (10–30 dpf). (A) External differences between control and 5-aza-dC-treated adult zebrafish females at 90 dpf. Scale in cm. (B) Body weight and (C) standard length of adults at 90 dpf. Data shown as mean ± s.e.m. ($n = 8$ and 6 males, and 11 and 4 females in control and 5-aza-dC groups, respectively). Within each sex, significant differences ($P < 0.05$ for males and $P < 0.01$ for females) in growth were determined by the Student's $t$ test and are indicated by different letters.

**Additional file 2: Fig. S2.** Microarray analysis. (A) qPCR validation of microarray results using 16 genes. (Only 15 datapoints can be seen due to overlap.) See Additional file 6: Table S3 for further primer information.

**Additional file 3: Table S1.** List of enriched GO terms (level 3) found in the ovaries of fish treated with 75 µM of 5-aza-dC between 10 and 30 dpf.

**Additional file 4: Fig. S3.** Third level of gene ontology terms of differentially expressed genes found by microarray analysis of ovaries of fish subjected to 75 µm of 5-aza-dC between 10 and 30 dpf during gonadal development. (A, B, C) show the upregulated GO terms, (D, E, F) show the downregulated GO terms, (A, D) biological process, (B, E) cellular component and (C, F) molecular function.

**Additional file 5: Table S2.** List of reproduction-related genes differentially expressed ($P < 0.01$) obtained by microarray analysis in ovaries of zebrafish treated with 75 µM of 5-aza-dC between 10 and 30 dpf.

**Additional file 6: Table S3.** List of enriched Kegg pathways found in the ovaries of fish treated with 75 µM of 5-aza-dC between 10 and 30 dpf.

**Additional file 7: Table S4.** List of genes found differentially expressed in eight pathways associated with female or male development, in the ovaries of fish treated with 75 µM of 5-aza-dC between 10 and 30 dpf.

**Additional file 8: Table S5.** List of epigenetic-related genes differentially expressed ($P < 0.01$) obtained by microarray analysis in ovaries of zebrafish treated with 75 µM of 5-aza-dC between 10 and 30 dpf.

**Additional file 9: Fig. S4.** Experimental design to study the effects of 5-aza-dC treatment on zebrafish development and survival, growth, sex ratio and gene expression. (A) Experiments performed during the early stages of development, with treatments either from 0 to 2 or 0–6 days post-fertilization. (B) Experiments performed during the gonadal development period. In the different experiments, 2–7 biological replicates per treatment were used, with 2–3 technical replicates each.

**Additional file 10: Table S6.** Gene symbols, names, Refseq IDs and primer sequences for all genes used in qPCR (in alphabetical order) in this study.

### Abbreviations
5-aza-dC: 5-aza-2′-deoxycytidine; dnmt: DNA methyltransferase; dpf: days post-fertilization; FC: fold change; GO: gene ontology; PCA: principal component analysis.

### Authors' contributions
FP and LR designed the study and collected data. LR, KV and MAI conducted the experiments and analysed the data. LR drafted the initial manuscript. LR, KV, MAI and FP wrote the manuscript. All authors read and approved the final manuscript.

## Author details

[1] Institut de Ciències del Mar, Consejo Superior de Investigaciones Científicas (CSIC), Passeig Marítim, 37–45, 08003 Barcelona, Spain. [2] Imperial Centre for Translational and Experimental Medicine, Hammersmith Hospital, Du Cane Road, London W12 0NN, UK. [3] Departamento de Recursos Hidrobiológicos, Universidad de Nariño, Torobajo, Pasto, Colombia.

## Acknowledgements

We thank Lia Sarrà, Mónica Genestar and Sergi Ruíz for technical assistance.

## Competing interests

The authors declare that they have no competing interests.

## Funding

Work supported by Ministry of Economy and Competitiveness (MEC) grants from Spanish government ("EpiFarm": AGL2013–41047–R) and Aquagenomics (CDS2007-0002) to FP LR was supported by an Epifarm contract.

## References

1. Gardiner-Garden M, Frommer M. CpG Islands in vertebrate genomes. J Mol Biol. 1987;196:261–82.

2. Bird AP. CpG-rich islands and the function of DNA methylation. Nature. 1986;321:209–13.

3. Ross MT, Grafham DV, Coffey AJ, Scherer S, McLay K, Muzny D, Platzer M, Howell GR, Burrows C, Bird CP, et al. The DNA sequence of the human X chromosome. Nature. 2005;434:325–37.

4. Reik W, Walter J. Genomic imprinting: parental influence on the genome. Nat Rev Genet. 2001;2:21–32.

5. Lindahl T. Instability and decay of the primary structure of DNA. Nature. 1993;362:709–15.

6. Turek-Plewa J, Jagodzinski PP. The role of mammalian DNA methyltransferases in the regulation of gene expression. Cell Mol Biol Lett. 2005;10:631–47.

7. Pradhan S, Bacolla A, Wells RD, Roberts RJ. Recombinant human DNA (cytosine-5) methyltransferase I. Expression, purification, and comparison of de novo and maintenance methylation. J Biol Chem. 1999;274:33002–10.

8. Okano M, Bell DW, Haber DA, Li E. DNA methyltransferases Dnmt3a and Dnmt3b are essential for de novo methylation and mammalian development. Cell. 1999;99:247–57.

9. Razin A, Riggs AD. DNA methylation and gene-function. Science. 1980;210:604–10.

10. Plumb JA, Strathdee G, Sludden J, Kaye SB, Brown R. Reversal of drug resistance in human tumor xenografts by 2′-deoxy-5-azacytidine-induced demethylation of the hMLH1 gene promoter. Cancer Res. 2000;60:6039–44.

11. Esteller M, Corn PG, Baylin SB, Herman JG. A gene hypermethylation profile of human cancer. Cancer Res. 2001;61:3225–9.

12. Yan PS, Chen CM, Shi HD, Rahmatpanah F, Wei SH, Caldwell CW, Huang THM. Dissecting complex epigenetic alterations in breast cancer using CpG island microarrays. Cancer Res. 2001;61:8375–80.

13. Egger G, Liang GN, Aparicio A, Jones PA. Epigenetics in human disease and prospects for epigenetic therapy. Nature. 2004;429:457–63.

14. Lyko F, Brown R. DNA methyltransferase inhibitors and the development of epigenetic cancer therapies. J Natl Cancer Inst. 2005;97:1498–506.

15. Cheng JC, Matsen CB, Gonzales FA, Ye W, Greer S, Marquez VE, Jones PA, Selker EU. Inhibition of DNA methylation and reactivation of silenced genes by zebularine. J Natl Cancer Inst. 2003;95:399–409.

16. Jones PA, Taylor SM. Cellular-differentiation, cytidine analogs and DNA methylation. Cell. 1980;20:85–93.

17. Juttermann R, Li E, Jaenisch R. Toxicity of 5-aza-2′-deoxycytidine to mammalian-cells is mediated primarily by covalent trapping of dna methyltransferase rather than DNA demethylation. PNAS. 1994;91:11797–801.

18. Stresemann C, Lyko F. Modes of action of the DNA methyltransferase inhibitors azacytidine and decitabine. Int J Cancer. 2008;123:8–13.

19. Saleh MH, Wang L, Goldberg MS. Improving cancer immunotherapy with DNA methyltransferase inhibitors. Cancer Immunol Immun. 2016;65:787–96.

20. White R, Rose K, Zon L. Zebrafish cancer: the state of the art and the path forward. Nat Rev Cancer. 2013;13:624–36.

21. Cavalieri V, Spinelli G. Environmental epigenetics in zebrafish. Epigenetics Chromatin. 2017;10:11.

22. Jiang L, Zhang J, Wang J-J, Wang L, Zhang L, Li G, Yang X, Ma X, Sun X, Cai J, et al. Sperm, but not oocyte, dna methylome is inherited by zebrafish early embryos. Cell. 2013;153:773–84.

23. Potok ME, Nix DA, Parnell TJ, Cairns BR. Reprogramming the maternal zebrafish genome after fertilization to match the paternal methylation pattern. Cell. 2013;153:759–72.

24. Potok ME, Nix DA, Parnell TJ, Cairns BR. Reprogramming the maternal zebrafish genome after fertilization to match the paternal methylation pattern. Cell. 2013;153:759–72.

25. Mhanni AA, McGowan RA. Global changes in genomic methylation levels during early development of the zebrafish embryo. Dev Genes Evol. 2004;214:412–7.

26. Wu SF, Zhang H, Hammoud SS, Potok M, Nix DA, Jones DA, Cairns BR. DNA methylation profiling in zebrafish. Methods Cell Biol. 2011;104:327–39.

27. Martin CC, Laforest L, Akimenko MA, Ekker M. A role for DNA methylation in gastrulation and somite patterning. Dev Biol. 1999;206:189–205.

28. Dasmahapatra AK, Khan IA. DNA methyltransferase expressions in Japanese rice fish (Oryzias latipes) embryogenesis is developmentally regulated and modulated by ethanol and 5-azacytidine. Comput Biochem Physiol C: Toxicol Pharmacol. 2015;176–177:1–9.

29. Ferguson AT, Vertino PM, Spitzner JR, Baylin SB, Muller MT, Davidson NE. Role of estrogen receptor gene demethylation and DNA methyltransferase DNA adduct formation in 5-aza-2′-deoxycytidine-induced cytotoxicity in human breast cancer cells. J Biol Chem. 1997;272:32260–6.

30. Bouwmeester MC, Ruiter S, Lommelaars T, Sippel J, Hodemaekers HM, van den Brandhof EJ, Pennings JLA, Kamstra JH, Jelinek J, Issa JPJ, et al. Zebrafish embryos as a screen for DNA methylation modifications after compound exposure. Toxicol Appl Pharmacol. 2016;291:84–96.

31. Aniagu SO, Williams TD, Allen Y, Katsiadaki I, Chipman JK. Global genomic methylation levels in the liver and gonads of the three-spine stickleback (Gasterosteus aculeatus) after exposure to hexabromocyclododecane and 17-beta oestradiol. Environ Int. 2008;34:310–7.

32. Olsvik PA, Williams TD, Tung HS, Mirbahai L, Sanden M, Skjaerven KH, Ellingsen S. Impacts of TCDD and MeHg on DNA methylation in zebrafish (Danio rerio) across two generations. Comp Biochem Physiol Toxicol Pharmacol: CBP. 2014;165:17–27.

33. Kamstra JH, Sales LB, Alestrom P, Legler J. Differential DNA methylation at conserved non-genic elements and evidence for transgenerational inheritance following developmental exposure to mono(2-ethylhexyl) phthalate and 5-azacytidine in the zebrafish. Epigenetics Chromatin. 2017;10:20.

34. Labbé C, Robles V, Herraez MP. Epigenetics in fish gametes and early embryo. Aquaculture. 2017;472:93–106.

35. Piferrer F. Epigenetics of sex determination and gonadogenesis. Dev Dyn. 2013;242:360–70.

36. Navarro-Martín L, Viñas J, Ribas L, Díaz N, Gutiérrez A, Di Croce L, Piferrer F. DNA methylation of the gonadal aromatase (cyp19a) promoter is involved in temperature-dependent sex ratio shifts in the European sea bass. PLoS Genet. 2011;7:1002447.

37. Si Y, Ding YX, He F, Wen HS, Li JF, Zhao JL, Huang ZJ. DNA methylation level of cyp19a1a and foxl2 gene related to their expression patterns and reproduction traits during ovary development stages of Japanese flounder (Paralichthys olivaceus). Gene. 2016;575:321–30.

38. Bai J, Gong W, Wang C, Gao Y, Hong W, Chen SX. Dynamic methylation pattern of cyp19a1a core promoter during zebrafish ovarian folliculogenesis. Fish Physiol Biochem. 2016;42:947–54.

39. Sun LX, Wang YY, Zhao Y, Wang H, Li N, Ji XS. Global DNA methylation changes in Nile tilapia gonads during high temperature-induced masculinization. PLoS One. 2016;11:e0158483.

40. Shao CW, Li QY, Chen SL, Zhang P, Lian JM, Hu QM, Sun B, Jin LJ, Liu SS, Wang ZJ, et al. Epigenetic modification and inheritance in sexual reversal of fish. Genome Res. 2014;24:604–15.

41. Ribas L, Piferrer F. The zebrafish (Danio rerio) as a model organism, with emphasis on applications for finfish aquaculture research. Rev Aquac. 2014;6:209–40.

42. Liew WC, Bartfai R, Lim Z, Sreenivasan R, Siegfried KR, Orban L. Polygenic sex determination system in zebrafish. PLoS One. 2012;7:e34397.

43. Ribas L, Liew WC, Díaz N, Sreenivasan R, Orbán L, Piferrer F. Heat-induced masculinization in domesticated zebrafish is family-specific and yields a set of gonadal transcriptomes. PNAS. 2017;114:E941–50.

44. Wilson CA, High SK, McCluskey BM, Amores A, Yan YL, Titus TA, Anderson JL, Batzel P, Carvan MJ III, Schartl M, Postlethwait JH. Wild sex in zebrafish: loss of the natural sex determinant in domesticated strains. Genetics. 2014;198:1291–308.

45. Rodriguez-Mari A, Canestro C, BreMiller RA, Catchen JM, Yan YL, Postlethwait JH. Retinoic acid metabolic genes, meiosis, and gonadal sex differentiation in zebrafish. PLoS One. 2013;8:e73951.

46. Liew WC, Orban L. Zebrafish sex: a complicated affair. Brief Funct Genom. 2014;13:172–87.

47. Ribas L, Robledo D, Gómez-Tato A, Viñas A, Martínez P, Piferrer F. Comprehensive transcriptomic analysis of the process of gonadal sex differentiation in the turbot (Scophthalmus maximus). Mol Cell Endocrinol. 2016;422:132–49.

48. Vizziano D, Randuineau G, Baron D, Cauty C, Guiguen Y. Characterization of early molecular sex differentiation in rainbow trout, Oncorhynchus mykiss. Dev Dyn. 2007;236:2198–206.

49. Andersson S, Geissler WM, Wu L, Davis DL, Grumbach MM, New MJ, Schwarz HP, Blethen SL, Mendonca BB, Bloise W, et al. Molecular genetics and pathophysiology of 17 beta-hydroxysteroid dehydrogenase 3 deficiency. J Clin Endocrinol Metab. 1996;81:130–6.

50. Pennimpede T, Cameron DA, MacLean GA, Li H, Abu-Abed S, Petkovich M. The role of cyp26 enzymes in defining appropriate retinoic acid exposure during embryogenesis. Birth Defects Res Part a-Clin Mol Teratol. 2010;88:883–94.

51. Tomlinson JW, Walker EA, Bujalska IJ, Draper N, Lavery GG, Cooper MS, Hewison M, Stewart PM. 11 beta-hydroxysteroid dehydrogenase type 1: a tissue-specific regulator of glucocorticoid response. Endocr Rev. 2004;25:831–66.

52. Diaz N, Piferrer F. Lasting effects of early exposure to temperature on the gonadal transcriptome at the time of sex differentiation in the European sea bass, a fish with mixed genetic and environmental sex determination. BMC Genom. 2015;16:679.

53. Bird A. DNA methylation patterns and epigenetic memory. Genes Dev. 2002;16:6–21.

54. Iida A, Shimada A, Shima A, Takamatsu N, Hori H, Takeuchi K, Koga A. Targeted reduction of the DNA methylation level with 5-azacytidine promotes excision of the medaka fish Tol2 transposable element. Genet Res. 2006;87:187–93.

55. Zhang X, Li H, Qiu Q, Qi Y, Huang D, Zhang Y. 2,4-Dichlorophenol induces global DNA hypermethylation through the increase of S-adenosylmethionine and the upregulation of dnmts mRNA in the liver of goldfish Carassius auratus. Comp Biochem Physiol C: Toxicol Pharmacol. 2014;160:54–9.

56. Yoo CB, Cheng JC, Jones PA. Zebularine: a new drug for epigenetic therapy. Biochem Soc Trans. 2004;32:910–2.

57. Andersen IS, Lindeman LC, Reiner AH, Ostrup O, Aanes H, Alestrom P, Collas P. Epigenetic marking of the zebrafish developmental program. Curr Top Dev Biol. 2013;104:85–112.

58. Zhang Y, Zhang S, Liu Z, Zhang L, Zhang W. Epigenetic modifications during sex change repress gonadotropin stimulation of cyp19a1a in a teleost ricefield eel (Monopterus albus). Endocrinology. 2013;154:2881–90.

59. Rivard GE, Momparler RL, Demers J, Benoit P, Raymond R, Lin KT, Momparler LF. Phase-I study on 5-aza-2'-deoxycytidine in children with acute-leukemia. Leuk Res. 1981;5:453–62.

60. Blasco M, Fernandino JI, Guilgur LG, Strussmann CA, Somoza GM, Vizziano-Cantonnet D. Molecular characterization of cyp11a1 and cyp11b1 and their gene expression profile in pejerrey (Odontesthes bonariensis) during early gonadal development. Comp Biochem Physiol A: Mol Integr Physiol. 2010;156:110–8.

61. Hogg K, Robinson WP, Beristain AG. Activation of endocrine-related gene expression in placental choriocarcinoma cell lines following DNA methylation knock-down. Mol Hum Reprod. 2014;20:677–89.

62. Bowles J, Knight D, Smith C, Wilhelm D, Richman J, Mamiya S, Yashiro K, Chawengsaksophak K, Wilson MJ, Rossant J, et al. Retinoid signaling determines germ cell fate in mice. Science. 2006;312:596–600.

63. Koubova J, Menke DB, Zhou Q, Capel B, Griswold MD, Page DC. Retinoic acid regulates sex-specific timing of meiotic initiation in mice. PNAS. 2006;103:2474–9.

64. Park JH, Lee J, Kim CH, Lee S. The polymorphism (−600 C > A) of CpG methylation site at the promoter region of CYP17A1 and its association of male infertility and testosterone levels. Gene. 2014;534:107–12.

65. Bovenzi V, Momparler RL. Antineoplastic action of 5-aza-2'-deoxycytidine and histone deacetylase inhibitor and their effect on the expression of retinoic acid receptor beta and estrogen receptor alpha genes in breast carcinoma cells. Cancer Chemother Pharmacol. 2001;48:71–6.

66. Prowse AH, Vanderveer L, Milling SWF, Pan ZZ, Dunbrack RL, Xu XX, Godwin AK. OVCA2 is downregulated and degraded during retinoid-induced apoptosis. Int J Cancer. 2002;99:185–92.

67. Orban L, Sreenivasan R, Olsson PE. Long and winding roads: testis differentiation in zebrafish. Mol Cell Endocrinol. 2009;312:35–41.

68. Kantarjian H, Issa JPJ, Rosenfeld CS, Bennett JM, Albitar M, DiPersio J, Klimek V, Slack J, de Castro C, Ravandi F, et al. Decitabine improves patient outcomes in myelodysplastic syndromes—results of a Phase III randomized study. Cancer. 2006;106:1794–803.

69. Derissen EJB, Beijnen JH, Schellens JHM. Concise drug review: azacitidine and decitabine. Oncologist. 2013;18:619–24.

70. Silverman LR, McKenzie DR, Peterson BL, Holland JF, Backstrom JT, Beach CL, Larson RA. Further analysis of trials with azacitidine in patients with myelodysplastic syndrome: studies 8421, 8921, and 9221 by the Cancer and Leukemia Group B. J Clin Oncol. 2006;24:3895–903.

71. Ramos MP, Wijetunga NA, McLellan AS, Suzuki M, Greally JM. DNA demethylation by 5-aza-2'-deoxycytidine is imprinted, targeted to euchromatin, and has limited transcriptional consequences. Epigenetics Chromatin. 2015;8:11.

72. Nishioka K, Rice JC, Sarma K, Erdjument-Bromage H, Werner J, Wang YM, Chuikov S, Valenzuela P, Tempst P, Steward R, et al. PR-Set7 is a nucleosome-specific methyltransferase that modifies lysine 20 of histone H4 and is associated with silent chromatin. Mol Cell. 2002;9:1201–13.

73. Chen W, Cao M, Yang Y, Nagahama Y, Zhao H. Expression pattern of prmt5 in adult fish and embryos of medaka, Oryzias latipes. Fish Physiol Biochem. 2009;35:325–32.

74. Tse AC-K, Li J-W, Wang SY, Chan T-F, Lai KP, Wu RS-S. Hypoxia alters testicular functions of marine medaka through microRNAs regulation. Aquat Toxicol. 2016;180:266–73.

75. Wang SY, Lau K, Lai K-P, Zhang J-W, Tse AC-K, Li J-W, Tong Y, Chan T-F, Wong CK-C, Chiu JM-Y, et al. Hypoxia causes transgenerational impairments in reproduction of fish. Nature Communications. 2016;7:12114.

76. Chuang JC, Yoo CB, Kwan JM, Li TW, Liang G, Yang AS, Jones PA. Comparison of biological effects of non-nucleoside DNA methylation inhibitors versus 5-aza-2'-deoxycytidine. Mol Cancer Ther. 2005;4:1515–20.

77. Ghoshal K, Datta J, Majumder S, Bai SM, Kutay H, Motiwala T, Jacob ST. 5-Aza-deoxycytidine induces selective degradation of DNA methyltransferase 1 by a proteasomal pathway that requires the KEN box, bromo-adjacent homology domain, and nuclear localization signal. Mol Cell Biol. 2005;25:4727–41.

78. Firmino J, Carballo C, Armesto P, Campinho MA, Power DM, Manchado M. Phylogeny, expression patterns and regulation of DNA Methyltransferases in early development of the flatfish, Solea senegalensis. BMC Dev Biol. 2017;17:11.

79. Liu K, Wang YF, Cantemir C, Muller MT. Endogenous assays of dna methyltransferases: evidence for differential activities of dnmt1, dnmt2, and dnmt3 in mammalian cells in vivo. Mol Cell Biol. 2003;23:2709–19.

80. Hutvagner G, McLachlan J, Pasquinelli AE, Balint E, Tuschl T, Zamore PD. A cellular function for the RNA-interference enzyme Dicer in the maturation of the let-7 small temporal RNA. Science. 2001;293:834–8.

81. Zhang XY, Lu K, Zhou Q. Dicer1 is crucial for the oocyte maturation of telotrophic ovary in *Nilaparvata lugens* (stal) (Hemiptera: geometroidea). Arch Insect Biochem Physiol. 2013;84:194–208.

82. Tanaka ED, Piulachs MD. Dicer-1 is a key enzyme in the regulation of oogenesis in panoistic ovaries. Biol Cell. 2012;104:452–61.

83. Giraldez AJ, Cinalli RM, Glasner ME, Enright AJ, Thomson JM, Baskerville S, Hammond SM, Bartel DP, Schier AF. MicroRNAs regulate brain morphogenesis in zebrafish. Science. 2005;308:833–8.

84. Wienholds E, Koudijs MJ, van Eeden FJM, Cuppen E, Plasterk RHA. The microRNA-producing enzyme Dicer1 is essential for zebrafish development. Nat Genet. 2003;35:217–8.

85. Katoh-Fukui Y, Tsuchiya R, Shiroishi T, Nakahara Y, Hashimoto N, Noguchi K, Higashinakagawa T. Male-to-female sex reversal in M33 mutant mice. Nature. 1998;393:688–92.

86. Katoh-Fukui Y, Miyabayashi K, Komatsu T, Owaki A, Baba T, Shima Y, Kidokoro T, Kanai Y, Schedl A, Wilhelm D, et al. Cbx2, a polycomb group gene, is required for sry gene expression in mice. Endocrinology. 2012;153:913–24.

87. Liu XY, Zhang XB, Li MH, Zheng SQ, Liu ZL, Cheng YY, Wang DS. Genome-wide identification, evolution of chromobox family genes and their expression in Nile tilapia. Comp Biochem Physiol B: Biochem Mol Biol. 2016;203:25–34.

88. OECD. Test No. 234: Fish sexual development test OECD Publishing, Paris; 2011.

89. Ribas L, Valdivieso A, Diaz N, Piferrer F. Appropriate rearing density in domesticated zebrafish to avoid masculinization: links with the stress response. J Exp Biol. 2017;220:1056–64.

90. Livak KJ, Schmittgen TD. Analysis of relative gene expression data using real-time quantitative PCR and the 2(-Delta Delta C(T)) Method. Methods. 2001;25:402–8.

91. Schmittgen TD, Livak KJ. Analyzing real-time PCR data by the comparative CT method. Nat Protoc. 2008;3:1101–8.

92. Huang DW, Sherman BT, Lempicki RA. Systematic and integrative analysis of large gene lists using DAVID bioinformatics resources. Nat Protoc. 2009;4:44–57.

93. Supek F, Bosnjak M, Skunca N, Smuc T. REVIGO summarizes and visualizes long lists of gene ontology terms. PLoS One. 2011;6:e21800.

94. Edgar R, Domrachev M, Lash AE. Gene expression omnibus: NCBI gene expression and hybridization array data repository. Nucl Acids Res. 2002;30:207–10.

95. Fowler J, Cohen L, Jarvis P. Practical statistics for field biology. Chichester: Wiley; 2008.

# 9

# Differential DNA methylation at conserved non-genic elements and evidence for transgenerational inheritance following developmental exposure to mono(2-ethylhexyl) phthalate and 5-azacytidine in zebrafish

Jorke H. Kamstra[1], Liana Bastos Sales[2], Peter Aleström[1] and Juliette Legler[2,3]* (iD)

## Abstract

**Background:** Exposure to environmental stressors during development may lead to latent and transgenerational adverse health effects. To understand the role of DNA methylation in these effects, we used zebrafish as a vertebrate model to investigate heritable changes in DNA methylation following chemical-induced stress during early development. We exposed zebrafish embryos to non-embryotoxic concentrations of the biologically active phthalate metabolite mono(2-ethylhexyl) phthalate (MEHP, 30 μM) and the DNA methyltransferase 1 inhibitor 5-azacytidine (5AC, 10 μM). Direct, latent and transgenerational effects on DNA methylation were assessed using global, genome-wide and locus-specific DNA methylation analyses.

**Results:** Following direct exposure in zebrafish embryos from 0 to 6 days post-fertilization, genome-wide analysis revealed a multitude of differentially methylated regions, strongly enriched at conserved non-genic elements for both compounds. Pathways involved in adipogenesis were enriched with the putative obesogenic compound MEHP. Exposure to 5AC resulted in enrichment of pathways involved in embryonic development and transgenerational effects on larval body length. Locus-specific methylation analysis of 10 differentially methylated sites revealed six of these loci differentially methylated in sperm sampled from adult zebrafish exposed during development to 5AC, and in first and second generation larvae. With MEHP, consistent changes were found at 2 specific loci in first and second generation larvae.

**Conclusions:** Our results suggest a functional role for DNA methylation on cis-regulatory conserved elements following developmental exposure to compounds. Effects on these regions are potentially transferred to subsequent generations.

**Keywords:** Phthalate, 5-Azacytidine, Epigenetics, DNA methylation, Transgenerational, Zebrafish, Toxicology, Environmental stress

*Correspondence: juliette.legler@brunel.ac.uk
[3] Institute for Environment, Health and Societies, College of Health and Life Sciences, Brunel University London, Uxbridge, UK
Full list of author information is available at the end of the article

## Background

Exposure to environmental stressors early in life, such as malnutrition, stress and chemical compounds has been hypothesized to play a role in the latent onset of diseases and adverse effects that may be transferred to subsequent generations [1]. In agreement with this 'developmental origins of health and disease' paradigm [2], a plurality of epidemiological and animal studies during the last decade have reported latent and transgenerational effects of developmental exposure to environmental stressors (reviewed in [3–5]). These latent and heritable effects may not be attributed to genetic variation and are suggested to be of an epigenetic nature [6]. DNA methylation and chemical modifications on histone tails are both considered epigenetic marks with high potential to be inherited and could therefore act as the drivers behind latent and transgenerational effects [7].

DNA methylation, by cytosine (mC), is dynamically regulated throughout life, particularly during early development. During mitosis, hemimethylated DNA in daughter cells is remethylated to the state of the mother cell, by maintenance of DNA methyltransferase 1 (DNMT1) [8]. Genome-wide reprogramming of DNA methylation takes place during both early zygotic development and the development of the gametes [8, 9]. From zygote to blastula stage, a wave of genome-wide demethylation ensures a totipotent cell state, followed by remethylation mediated by de novo DNMTs. A second wave of reprogramming follows during primordial germ cell (PGC) development, to ensure a gender-specific methylation state in gametes [10]. Recently, dynamic enhancer methylation during early developmental stages has been observed in vertebrates, linked to many developmental genes [11, 12]. Clearly, during these dynamic periods of epigenetic regulation, environmental stress targeted to the epigenome could potentially affect early embryonic development.

In this study, we used zebrafish as an alternative model to study transgenerational epigenetic inheritance. Zebrafish are a suitable vertebrate model in epigenetic studies, as they harbor similar methylation patterns compared to mammals and show conservation of DNA methylation and other epigenetic pathways [9]. However, there are distinct differences between the methylome of zebrafish and mammals. It is suggested that the paternal genome during zebrafish zygotic development is relatively resistant to demethylation [13], whereas recent research in mice suggests that the paternal genome is actively demethylated [14]. Also, in zebrafish, active developmental enhancers are hypermethylated, which has not been observed in other species [15]. Furthermore, the second wave of DNA methylation reprogramming in PGCs has not been confirmed in zebrafish. However,

the methylome of sperm and oocytes in zebrafish differs significantly [16], which suggests that DNA methylation reprogramming events occur in zebrafish PGCs as well. Compared to mammalian models, zebrafish has the advantage that external exposures of eggs directly after fertilization is possible, thereby enabling the inclusion of both reprogramming events during exposures. Exposure of zebrafish embryos directly after fertilization means that the F0 generation is directly exposed, as well as the developing primordial germ cells which will ultimately become the F1. The F2 generation is the first completely unexposed progeny, as opposed to the F3 in mammalian studies [7].

Here, we examined the direct, latent and transgenerational effects of two model compounds with different modes of action on DNA methylation in zebrafish. We used mono(2-ethylhexyl)phthalate (MEHP), a major metabolite of di-2-(ethylhexyl) phthalate (DEHP), a high-volume production plasticizer ubiquitously present in the environment [17]. Developmental DEHP exposure has been associated with many health effects, such as reproductive toxicity, obesity and dyslipidemia [18–20], and its use in food contact materials, baby products and toys has been restricted in the European Union, although it is still allowed in electronic devices and medical equipment [21]. DEHP is rapidly metabolized to monophthalates, such as MEHP, and the toxicity of DEHP is considered to be mediated by MEHP rather than the parent compound [20]. Several studies have shown latent and transgenerational effects on DNA methylation in different tissues following in utero DEHP exposures in different rodent models, as a pure compound or in a mixture with other plasticizers [22–26]. One of these studies has shown an obese phenotype in rat offspring and linked differentially methylated regions (DMRs) to an obesity-related gene network [23]. In our study, we exposed zebrafish during early life stages (0–6 dpf) to the active metabolite MEHP, as gut and liver mediated metabolism of DEHP during these early stages may be limited. Furthermore, we exposed embryos to 5-azacytidine (5AC), a DNA methylation inhibitor used in the treatment of myelodysplastic syndrome and acute myeloid leukemia [27]. During the cell cycle, 5AC incorporates into DNA, irreversibly binds to and inactivates DNMT1, resulting in genome-wide hypomethylation [28]. As the hypomethylating properties of 5AC have been previously observed in zebrafish [29], we used this compound as a positive control, and as a proof of principle of transgenerational epigenetic inheritance following chemical exposure. To our knowledge, no previous studies assessing the transgenerational effects of early exposures to 5AC on the methylome have been carried out.

In this study, we assessed direct, latent and transgenerational effects of MEHP and 5AC using three different approaches. We analyzed global 5-hydroxymethyl-2'-deoxycytidine and 5-methyl-2'-deoxycytidine (hmC and mC) levels, genome-wide, and loci-specific mC levels, with liquid chromatography–tandem mass spectrometry (LC/MS), reduced representation bisulfite sequencing (RRBS) and amplicon bisulfite high-throughput sequencing (BisPCR2), respectively. Our data indicate genome-wide effects on DNA methylation following developmental exposure to both MEHP and 5AC on a multitude of loci, which were associated with specific biological pathways and enriched at conserved non-genic elements. A subset of DMRs were transgenerationally inherited in F2 larvae following both MEHP and 5AC exposures.

## Results

### Quality control

To assess whether developmental exposures to MEHP and 5AC altered mC and hmC at global levels, we used LC/MS analysis based on a method we recently developed [29]. Using internal standards to account for inter and intra experimental variation, we observed excellent reproducibility between experiments. Quality control DNA showed a relative standard deviation of 1.27 and 0.81% for hmC and mC, respectively (data not shown).

For RRBS analysis, we used a mapping pipeline specifically designed for RRBS data, developed by the Babraham Institute (Trim_galore and Bismark [30]). Using this pipeline, we were able to map around 59% of the sequences, generating an average of 444,855 analyzed Cs in CpG context per replicate with at least 10 reads (Additional file 1: Table 1). This is comparable to previously reported RRBS analysis in zebrafish brain and liver [31]. methylKit analysis estimates the bisulfite conversion efficiency using Cs in non-CpG context, which was very consistent between the samples, at an average of 99.2% (Additional file 1: Table 1).

Specific analysis of differentially methylated CpG sites (DMCs) was performed with a recently developed method, BisPCR2 [32]. To account for unforeseen biases, the method was thoroughly validated using a bisulfite converted standard curve of 0–100% methylated DNA to check for PCR bias and two samples that were analyzed in duplicate for technical variation. We were able to map >90% of the reads to the 10 loci covering a total of 103 CpG sites, with high accuracy between technical replicates (Additional file 1: Figure 1). Linear relationships were found at all loci analyzed, except for 4 CpG sites at Chr2:32025720, Chr2:32025757, Chr25:36706591 and Chr25:36706627, and these were excluded from the analyses (Additional file 1: Figure 2).

Finally, we validated our RRBS results against the BisPCR2 method. We were able to assess 49 mutually analyzed CpG sites between the two methods, which showed a high correlation (Spearman $r = 0.889$, $P < 0.0001$, Additional file 1: Figure 3), indicating that the results were consistent between the two methods.

### 5AC exhibits transgenerational phenotypic effects

We exposed zebrafish embryos from 0 to 6 dpf and followed them up until adulthood. In-cross F1 and F2 generations were established, in which F2 is the first unexposed progeny (Fig. 1). We used non-embryotoxic concentrations of MEHP and 5AC (30 and 10 μM, respectively) that did not cause observable effects on developmental endpoints as defined by the standard zebrafish embryo toxicity assay [33]. In addition to 10 μM, embryos exposed to 25 μM 5AC from 0 to 6 dpf were included as a positive control for global hypomethylation [29]. Significant effects were observed on F0 larval length with both MEHP and 5AC exposure at 3 and 6 dpf (Fig. 2a, b). A transgenerational effect was observed on larval length in F1 and F2 exclusively for the 5AC exposure, which was most pronounced at 6 dpf (Fig. 2b).

A clear effect on swim bladder inflation and abnormal intestinal development was observed in 5AC treated F0 fish at 15 dpf (Fig. 2c and Additional file 1: Figure 4). However, no larval lethality was found and fish grew to adulthood without apparent effects. Additionally, the effects on intestine and swim bladder were not observed in F1 (data not shown). We observed a significant shift in gender toward males in the F1 generation with 5AC, but not after MEHP exposures (Fig. 2d, e).

### 5AC and MEHP affect *dnmt* gene expression and global mC and hmC levels

We assessed DNA-methyltransferase (*dnmt*) gene expression of all 3 *dnmt* orthologues and their respective paralogues in F0 larvae at 6 dpf (Fig. 3a). Dnmt1 is mainly involved in maintenance of DNA methylation during cell replication, whereas the other 6 genes encode de novo Dnmts, which are suggested to have both tissue- and promoter-specific functions [9]. Significant upregulation of *dnmt1* was observed with MEHP exposure, but not with 5AC (Fig. 3a). Both exposures show very similar differential expression profiles for the de novo *dnmt3* orthologues, where *dnmt3aa* and *dnmt3ab* paralogues are downregulated and *dnmt3bb.1* and *dnmt3bb.2* paralogues are upregulated. The differential expression profiles of the *dnmts* indicate interference in DNA methylation pathways with both exposures.

We observed a significant decrease in global mC levels at 25 μM 5AC in 6 dpf F0 larvae DNA (Additional file 1: Figure 5). Additionally, a decrease in global mC levels

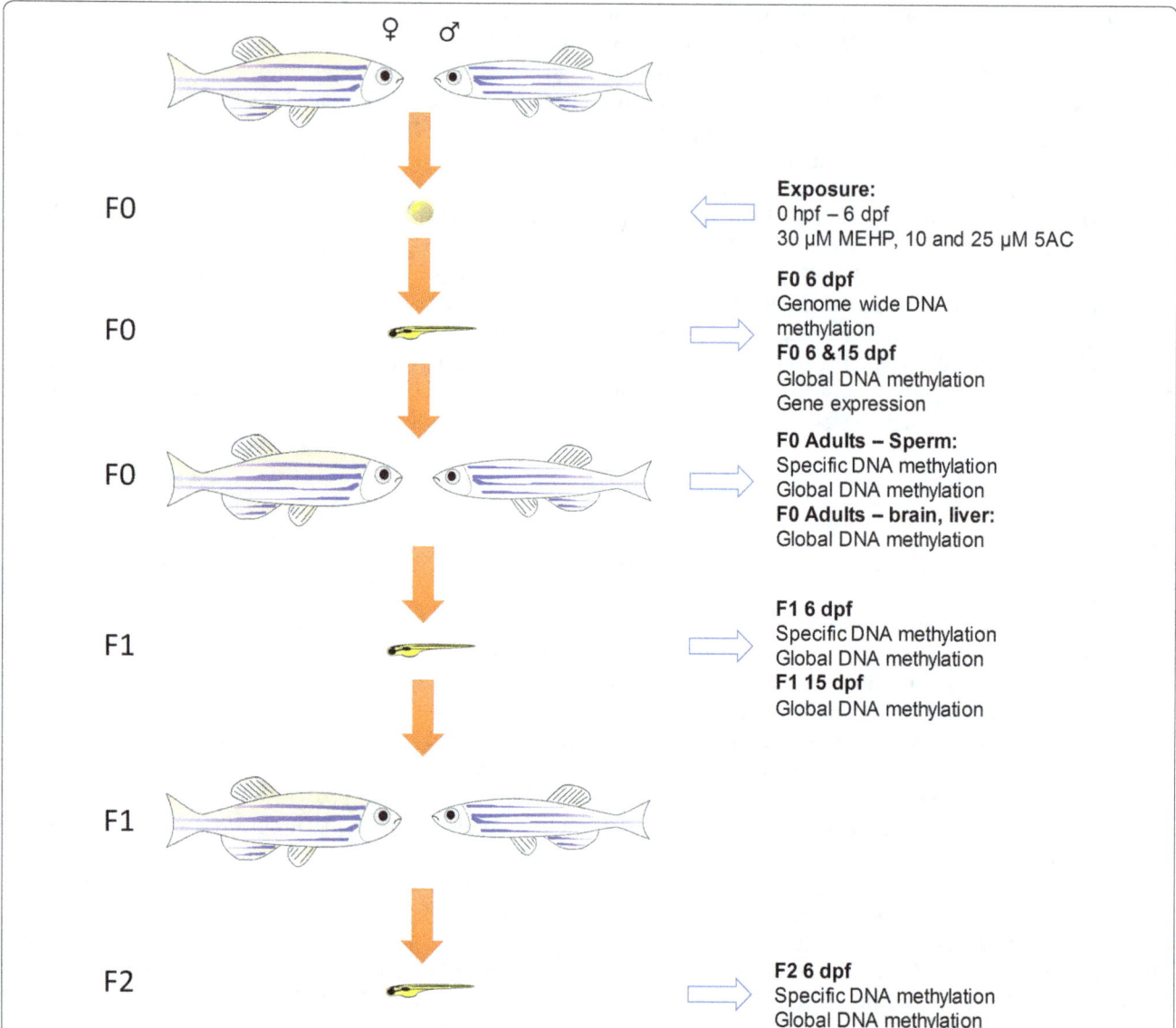

**Fig. 1** Setup of the transgenerational exposure experiment. Fish were exposed to mono(2-ethylhexyl) phthalate and 5-azacytidine (5AC) directly after fertilization up to 6 dpf. Adult F0 were in-crossed to generate F1 and F2. Analysis was performed at the different stages as indicated in Materials and methods. *hpf* hours post-fertilization, *dpf* days post-fertilization)

was observed at 15 dpf F0 following 10 μM 5AC exposures (Fig. 3b). Global demethylation was also observed in livers from adult females exposed to both MEHP and 5AC (Fig. 3b). Decreased hmC levels following MEHP exposure were observed at 6 dpf F0 and in brain tissue of male fish, whereas no changes were observed on mC levels (Fig. 3b).

### RRBS reveals enrichment of DMRs on conserved non-genic elements

With RRBS, we were able to analyze around 200,000 mutually measured Cs (read depth >10) and over 60,000 mutually measured 300-bp tiles (Fig. 4a). Global

methylation changes by RRBS in features as promoter regions (2000-bp upstream of transcriptional start sites) (TSSs), CpG islands (CGis) and shores and gene bodies were assessed using 300-bp tiles (Fig. 4a). A sharp decline in methylation around both TSSs and CpG islands was observed, with no apparent difference between the treatments (Fig. 4b, c). Overlap of tiles to a computationally derived list of conserved non-genic elements in zebrafish (zfCNEs) [33] also showed a general decline in methylation, but a distinct difference was found with MEHP exposures, which showed hypermethylation compared to control and 5AC samples (Fig. 4d). Calculation of the average methylation relative to the control over the

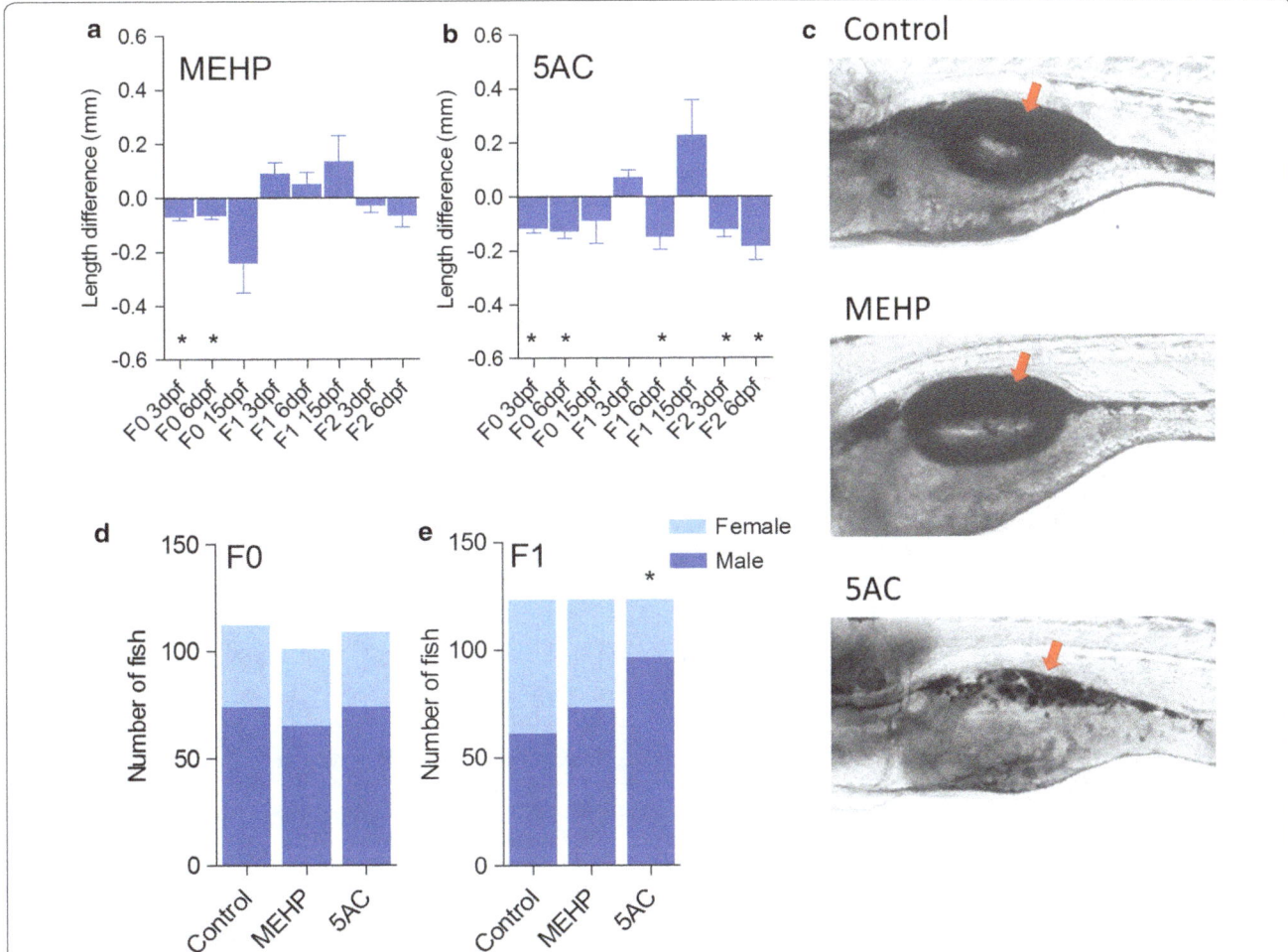

**Fig. 2** Phenotypic effects after developmental exposure of zebrafish (0-6 days post-fertilization) to mono(2-ethylhexyl) phthalate and 5-azacytidine (5AC). **a, b** Absolute difference in length of larvae compared to control (*error bars* represent SEM, *$P < 0.05$, two-tailed *t* test, with Bonferroni multiple comparison correction). **c** Representative images of effects on swim bladder inflation and intestinal development with 5AC exposures at 15 dpf. *Arrows* indicate swim bladder. **d, e** Sex ratios in F0 and F1 generations (*$P < 0.001$, Chi-square)

different features revealed a significant decrease in methylation between control and 5AC at promoter regions and a significant increase between control and MEHP at zfCNEs (Fig. 4e).

methylKit analysis revealed 410 DMRs in F0 6 dpf larvae following exposure to MEHP, with a cutoff of 10% methylation difference and a *Q* value <0.01 (Fig. 4f). From these DMRs, more hypermethylated regions (70%) were observed than hypomethylated (Fig. 4g). When we mapped the MEHP DMRs to different genomic features, we observed an enrichment of DMRs at zfCNEs (Fig. 4h and Additional file 1: Table 2A, $P = 1.3E-31$). Additionally, limited overlap was found on gene bodies which indicates that DMRs are predominantly located outside gene bodies (Fig. 4h and Additional file 1: Table 2A, $P = 7.5E-4$). With 5AC, 580 DMRs were found with equal numbers of hyper and hypomethylated regions (Fig. 4i, j). As with

MEHP, enrichment was found on zfCNEs ($P = 5.2E-23$) and limited overlap on gene bodies ($P = 5.1E-05$) (Fig. 4k and Additional file 1: Table 2B). Additionally, 5AC-specific DMRs had limited overlap at CGis ($P = 3.2E-54$) (Fig. 4k and Additional file 1: Table 2B). For both 5AC and MEHP exposures, we calculated that 44% of the DMRs overlapped with the 23% zfCNEs that are conserved with human and mice. We also analyzed developmental enhancer regions, as indicated by histone H3K4me1 and H3K27Ac marks, but could not find any enrichment of DMRs on these specific sites (data not shown) [34].

### DMR-associated genes are involved in several developmental and disease-related pathways

Next, we were interested to find out whether these DMRs are associated with specific biological pathways. Therefore, the Genomic Regions Enrichment of Annotations

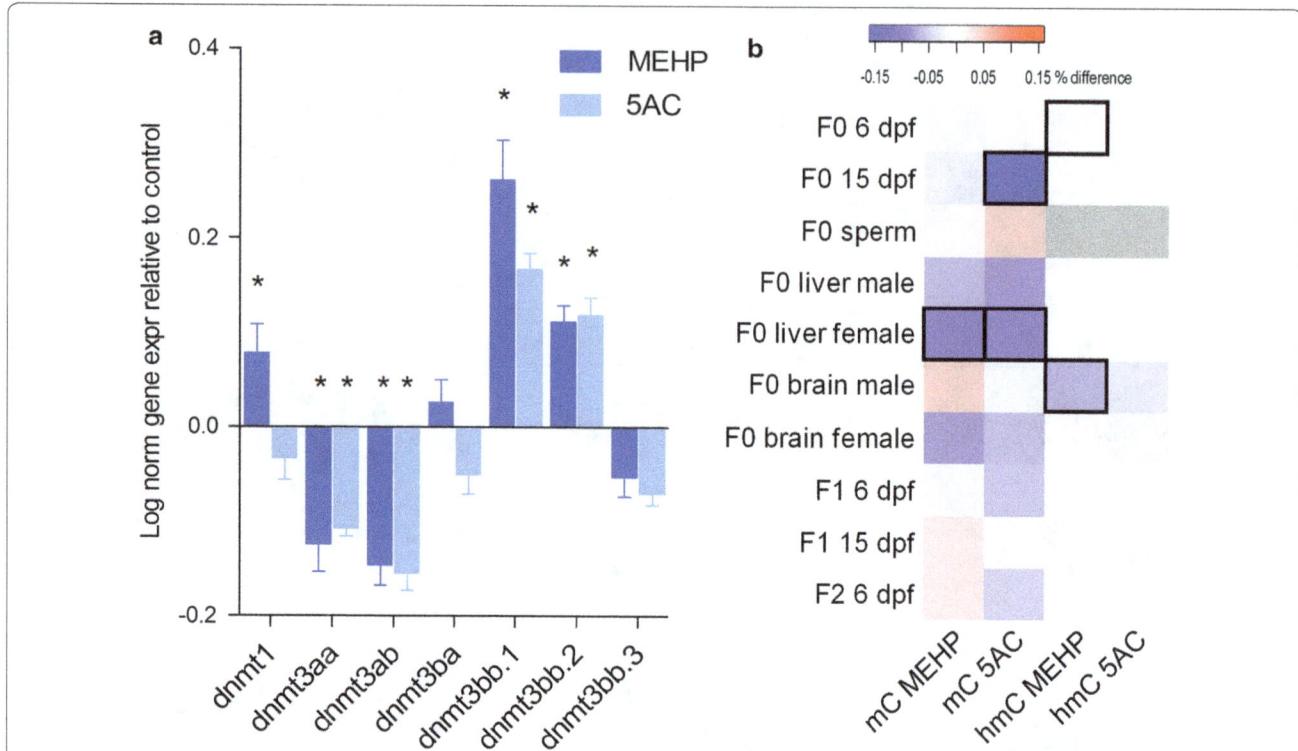

**Fig. 3** Gene expression analysis of *dnmt* variants and global methylation. **a** Log-normalized gene expression relative to control in F0 6 days post-fertilization larvae (*Error bars* represent SEM, *P < 0.05, ANOVA with Dunnets post hoc test). **b** Heat map of absolute differences in global cytosine methylation and cytosine hydroxymethylation compared to control. *Black squares* indicate significant changes relative to control (P < 0.05, two-tailed *t* test, using Bonferroni multiple comparisons correction). *Gray* indicates non-detectable levels

Tool (GREAT) was used to associate the DMRs with genes, which we subsequently imported into Ingenuity Pathway Analysis (IPA). We analyzed all DMRs, hypermethylated DMRs, hypomethylated DMRs and zfCNE-specific DMRs. With both exposures, pathways involved in transcriptional and developmental processes were enriched, such as pathways involved in pluripotency, TGF-β, Gα 12/13, Wnt/β-catenin (Table 1). Upstream regulators involved in development, such as SSH, TGF-β1, SOX2, POU5F1, were also enriched with both exposures, which could be a specific response to the toxicological stress of the compounds (Additional files 3 and 4).

Figure 5 presents heat maps of the compound-specific top 20 lists of canonical pathways, upstream regulators and toxicological lists, supplemented with a custom imported list (adipose tissue development). For MEHP, pathways and upstream regulators involved in adipogenesis and neuronal development were the most prominently enriched pathways (Fig. 5a). Involvement of the adipogenesis pathway was especially enriched in hypomethylated DMRs. Other upstream regulators, such as PPARG, PPARα/RXRα, TGFB1 and WNT7a, together with the custom adipogenesis list, also predicted

involvement of MEHP in adipogenesis. Processes involving axonal guidance signaling, together with upstream regulators involved in neuronal development (ASCL1, SOX2, PAX6 and GLI1) of which some were specifically enriched at zfCNEs and hypermethylated DMRs, point toward disruption in nervous system development (Fig. 5a).

5AC-specific predicted pathways were involved in development and expression control of genes, such as KLF4, POU4F1, SHH, SOX3 (Fig. 5b). Also, as with MEHP, neuronal development may be impaired, as indicated by the enrichment of axonal guidance signaling pathways. Upstream regulators involved in (sensory) neuronal development were enriched (SOX2, SOX3, POU4F1) (Fig. 5b). Notably, POU4F1 was specifically enriched at zfCNEs (Fig. 5b). Furthermore, IPA analysis predicted effects on gastrointestinal diseases (P values of 6.25E−4–4.64E−13) and showed a strong enrichment in the upstream regulator HNF4a (P = 2.49E−6), a transcription factor known to be involved in gastrointestinal development (Additional file 4). Notably, development of body axis was among the most significant enriched lists in diseases and bio functions (P = 7.35E−17, Additional file 4), which is consistent with the effects found on body

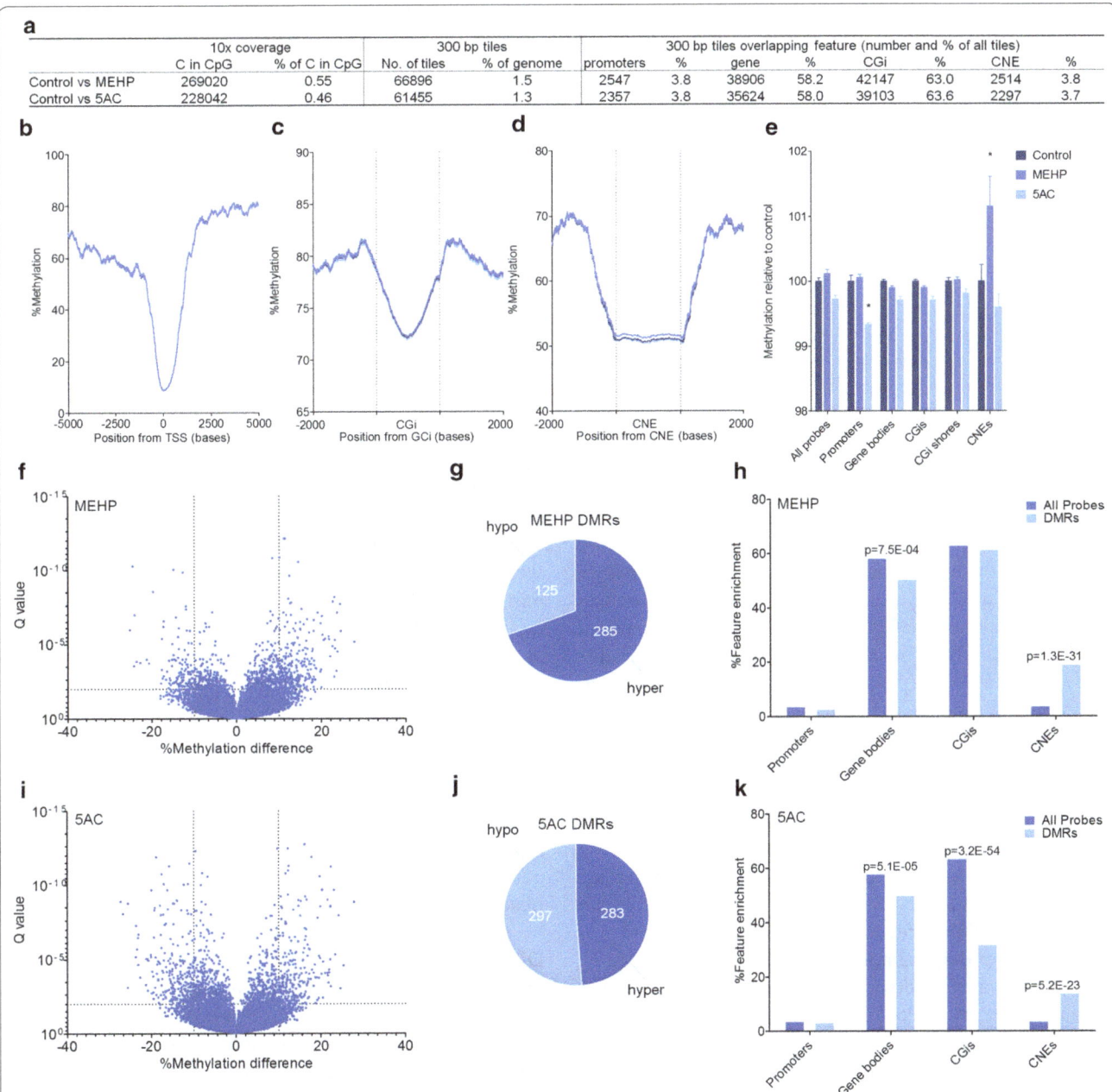

**Fig. 4** Reduced representation bisulfite sequencing (RRBS) results following developmental exposure to mono(2-ethylhexyl) phthalate (MEHP) and 5-azacytidine (5AC). **a** General statistics of RRBS analyses, showing mutual analyzed Cs in CpG context at a read depth of 10, number of tiles analyzed with regional analysis and the number of tiles overlapping different features. **b–d** Methylation profile spanning transcriptional start sites (TSSs), CpG islands and zebrafish conserved non-genic elements (zfCNEs). **e** Methylation levels relative to control overlapping different features. *Error bars* represent SEM; significance was calculated with two-way ANOVA, using Bonferroni multiple comparisons correction ($P < 0.05$). **f** Volcano plot of the methylation difference between MEHP and control. **g** Number of MEHP hypo and hyper differentially methylated regions (DMRs). **h** Enrichment plot of all tiles and DMRs overlapping different features. Significant enrichment calculated using a hypergeometric test. **i** Volcano plot of the methylation difference between 5AC and control. **j** Number of 5AC hypo and hyper DMRs. **k** Enrichment plot of all probes and DMRs overlapping different features. Significant enrichment calculated using a hypergeometric test

length following exposure to 5AC. 5AC did not seem to affect general toxicological pathways, since only weak enrichments were found in toxicological lists.

**Transgenerational effects observed with BisPCR2**

Regions around ten specific DMCs were selected from the RRBS analysis that exhibited a difference in

**Table 1 Top canonical pathways after ingenuity pathway analysis of all differentially methylated region-associated genes in zebrafish larvae exposed from 0-6 days post-fertilization to mono(2-ethylhexyl) phthalate (MEHP, 30 μM) and 5-azacytidine (5AC, 10 μM)**

| Canonical pathway | MEHP (*P* value) | 5AC (*P* value) |
|---|---|---|
| Human embryonic stem cell pluripotency | 7.36E−05 | 1.92E−05 |
| Factors promoting cardiogenesis in vertebrates | 2.32E−03 | 1.32E−06 |
| TGF-β signaling | 8.43E−06 | 6.28E−04 |
| Wnt/β-catenin signaling | 7.10E−06 | 8.00E−04 |
| Axonal guidance signaling | 2.68E−03 | 1.05E−05 |
| Regulation of the epithelial–mesenchymal transition pathway | 1.87E−05 | 4.91E−03 |
| Gα12/13 signaling | 3.59E−06 | 4.21E−02 |
| Adipogenesis pathway | 5.23E−03 | 1.17E−03 |
| Role of NANOG in mammalian embryonic stem cell pluripotency | 6.56E−03 | 9.54E−04 |
| Epithelial adherent junction signaling | 6.66E−04 | 1.93E−02 |

methylation larger than 20%. As nomenclature for the analyzed loci, we used the nearest annotated genes (Table 2). We analyzed 6 dpf F0, F1 and F2 larvae, as well as samples from 15 dpf F0 larvae and sperm from F0 fish, each with their respective control. Additionally, we included 25 μM 5AC exposures (6 dpf) in this analysis in order to investigate CpG site-specific dose-dependent relationships. These relationships were clearly visible following cluster analysis, with clusters of high methylation showing a decrease in methylation with the higher concentration and clusters of low methylation increasing with higher-concentration 5AC (Additional file 1: Figure 6). A number of these sites showed significant effects with both concentrations, with some showing a clear dose response (Fig. 6a).

Hierarchical clustering of methylation levels at specific loci, as determined by BisPCR2 analysis, revealed that samples from F0 sperm and 15 dpf larvae clustered distinctively from 6 dpf samples (Additional file 1: Figure 7). This age-specific difference in methylation pattern was also apparent when looking at methylation after compound exposure, where methylation differences compared to control for MEHP at 15 dpf and sperm as well as 5AC at 15 dpf clustered distinctively from the 6 dpf samples (Fig. 6b). When focused on the difference between controls and exposed samples at all stages and F0 sperm, methylation patterns for MEHP at F0, F1 and F2 cluster together, as well as for 5AC at F0 and F2 (Fig. 6b), indicating that methylation changes caused by developmental exposure to these compounds in F0 are persistent from one generation to the next.

While the largest effects on methylation after BisPCR2 analysis were generally observed with 5AC exposures, we found significant effects over two generations following exposure to both compounds, in 2 and 6 out of 10 loci for MEHP and 5AC, respectively (Fig. 6c). With MEHP, a transgenerational increase in methylation was observed

at the entire *cbfa2t2* locus, averaging from 6.8% in F1 to 11.6 and 10.7% in F1 and F2, respectively. Interestingly opposite effects were observed at 15 dpf and sperm, where methylation was decreased. MEHP-specific effects were observed at the *CT583728.4* locus up to F2 at CpG1 and CpG9, and F0 and in F1 at all CpGs at the *cps1* locus, but not F2. For 5AC, a strong transgenerational effect at the *cbfa2t2* locus was found, with an average regional increase in methylation up to 25% in F2 compared to control. Furthermore, transgenerational effects were observed at specific CpG sites at *nrp1b*-CpG2 (hyper), *si:ch211-245b21.1*-CpG3 (hypo), *si:ch211-245b21.1*-CpG4 (hyper), *CT583728.4*-CpG1 (hypo), *si:dkey-234i14.6*-CpG14 (hyper). At the *cps1* locus, a regional hypomethylating effect is present at F0, but not propagated to F1 and F2; however, a transgenerational effect is observed at CpG11.

## Discussion

In this study, we used next-generation sequencing to analyze DNA methylation on a genome-wide scale using zebrafish as an alternative model, in order to detect regional and site-specific changes following exposures to MEHP and 5AC. Our *dnmt* gene expression data and global methylation approach confirmed that both compounds interfered with DNA methylation pathways. RRBS analysis allowed us to link DMRs to specific pathways and aided in the prediction of adverse effects of these compounds. With the use of loci-specific bisulfite sequencing, we detected differentially methylated sites that persisted over two generations with both MEHP and 5AC exposures. We show that the combination of genome-wide analysis, followed by loci-specific analysis of newly discovered DMRs in subsequent generations, provides important insights in DNA methylation changes involved in transgenerational effects of developmental exposure to xenobiotic compounds.

**Fig. 5** Ingenuity pathway analysis of mono(2-ethylhexyl) phthalate (MEHP) and 5-azacytidine (5AC)-specific differentially methylation regions (DMRs). Heat maps of −log(P) values for DMR-associated genes of **a** MEHP and **b** 5AC exposures. Heat maps show top 20 lists of predicted canonical pathways, upstream regulators, toxicology-related gene lists (Tox) of all (All), hypermethylated (hyper) and hypomethylated (hypo) DMRs, and DMR-specific zebrafish conserved non-genic elements (zfCNE). A custom imported list was also added (adipose tissue development). Extended lists are found in Additional file 3 and 4

To our knowledge, we show for the first time that developmental exposure to compounds specifically targets DNA methylation at conserved non-genic regions. We used a computationally derived list of zfCNEs which contained over 54,000 regions [35]. CNEs are generally located outside genic regions and can have cis-regulatory functions

**Table 2 Overview of the 10 F0 loci analyzed with BisPCR2**

| Location | Nearest gene | CpGs analyzed |
|---|---|---|
| chr2:32025472–32025772 | *mycb* | 7 |
| chr2:43611512–43611846 | *nrp1b* | 4 |
| chr3:48276549–48276951 | *si:ch211–245b21.1* | 10 |
| chr4:53831094–53831415 | *CT583728.4* | 14 |
| chr6:57641533–57641533 | *cbfa2t2* | 14 |
| chr9:39356640–39356640 | *cps1* | 11 |
| chr20:43514527–43514862 | *si:dkey–14a7.2* | 10 |
| chr12:29105587–29105587 | *gabrz* | 12 |
| chr21:20158921–20158921 | *si:dkey–247m21.3* | 7 |
| chr25:36706556–36706894 | *si:dkey–234i14.6* | 14 |

such as enhancers or silencers [36]. The zfCNEs are conserved in many species (at least three species per region), and have a 22% overlap with mice and human CNEs [35]. Furthermore, these regions show about 23% overlap with empirically derived developmental enhancer regions (H3K4me1 and H3K27Ac [34]), indicating a significant role for zfCNEs in early development. However, no enrichment was observed on developmental enhancer regions in this study, suggesting a regulatory function for zfCNE-specific DMRs outside developmental enhancers. Notably, DMRs were generally found outside gene bodies and outside CpG islands (specifically for 5AC), and were not enriched at promoter regions which indicates that DMRs are located at distal regulatory sites. A recent study summarizing DMRs derived from several transgenerational studies in rats exposed to different classes of compounds found an overrepresentation of DMRs at low CpG content areas, from both somatic tissues as sperm-specific DMRs [37]. Although we did not observe such effects with MEHP, 5AC-specific DMRs were over represented outside CpG islands. Further research to elucidate the functional and phenotypical significance of differential DNA methylation on these conserved elements is warranted. The application of novel methods using clustered regularly interspaced short palindromic repeat-CAS9 (Crispr/CAS9) engineered with de novo methylation and demethylation catalytic domains would be useful to target these specific regions and shed light on the functional significance of induced changes in DNA methylation [38, 39].

5-Azacytidine belongs to the family of azanucleosides, which are known cytotoxic and teratogenic agents, of which 5AC is the least toxic derivate [40]. We exposed embryos to 5AC at concentrations that were below the effect concentration in zebrafish embryo toxicity tests based on our own results and others [29, 41, 42]. Nevertheless, developmental exposure of zebrafish embryos to 5AC resulted in transgenerational effects on larval body length as well as direct effects on gastrointestinal development in F0 larvae at 15 dpf. IPA analysis revealed enrichments of genes involved in gastrointestinal diseases as well as HNF4a regulation. HNF4a is known to be involved in liver and intestinal development and is in combination with CDX2 crucial for columnar cell formation [43]. Interestingly, loss of columnar cells formation after *dnmt1* knockdown in zebrafish has been observed previously [44]. Additionally, we found direct and transgenerational effects of 5AC exposure on the *cbfa2t2* gene body. Effects on intestinal development and secretory cells formation in the small intestine have been reported for CBFA2T2$^{-/-}$ mice, which also exhibit smaller phenotypes [45]. Although further research is necessary, our results point to a role for DNA methylation in intestinal development via Hnf4a signaling and suggest that the regional change in *cbfa2t2* methylation could be an interesting target.

MEHP-specific DMRs could be linked to genes that are involved in diseases known to be associated with this compound, in particular pathways related to obesity, mostly found on hypomethylated DMRs. MEHP is known to exert its adipogenic action via peroxisome proliferator-activated receptors (PPARs) [18]. IPA analysis revealed significant enrichment in upstream regulation of both PPARγ and PPARα and enrichments in upstream regulators TGFβ and WNT7a, all involved in adipogenic processes. The significant enrichment of the adipose tissue development gene list together with the enriched prediction molecules, cyclic AMP and dexamethasone and isobutylmethylxanthine, essential factors in the stimulation of adipogenic differentiation, implies a strong role of MEHP in adipogenesis. However, no effects were found on adipocyte differentiation in vivo (Bastos-Sales et al., unpublished results), suggesting that DMRs related to adipogenesis after direct exposure do not persist

(See figure on next page.)

**Fig. 6** Locus-specific methylation analysis of different larval stages and sperm over generation following exposure to mono(2-ethylhexyl) phthalate (MEHP) and 5-azacytidine (5AC). **a** Loci showing significant effects with both 10 and 25 μM 5AC exposures compared to control at F0 6 dpf. *Error bars* represent SEM. **b** Hierarchical clustering of generational effects on DNA methylation differences of MEHP and 5AC exposures at 6 and 15 dpf and in sperm from F0 compared to their respective generational or tissue-specific controls (ward clustering). **c** Heat map showing the methylation difference compared to control of all CpG sites over generations (F0, F1 and F2) and 15 dpf (15) and sperm (sp) with at least one significant differentially methylated CpG, exhibiting a methylation difference of more than 10% as indicated by the *black squares*. Each stage or tissue-specific sample was compared with the controls from the same stage

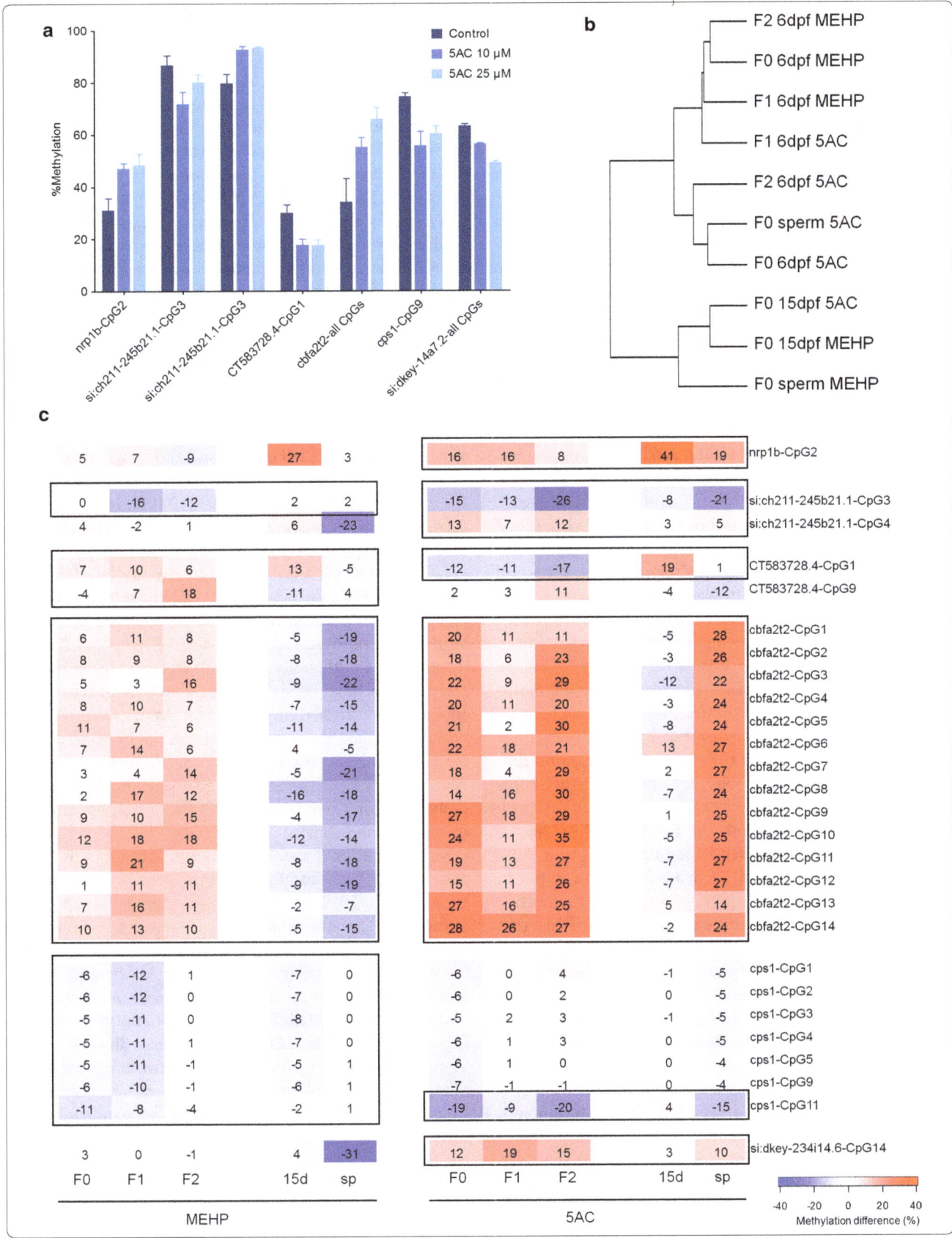

in adulthood. Furthermore, enrichments of upstream regulators as SOX2 and ASCL1 point to neurodevelopmental effects following MEHP exposure. Some of these upstream regulators are specifically enriched at zfCNEs, suggesting a regulating role of these conserved elements on neuronal development. Interestingly, the increased mRNA expression of *dnmt3bb.1* and 2, known to be involved in brain development [46], coincides with hypermethylation of these genes at zfCNE DMRs.

Transgenerational effects with persistent differential DNA methylation of specific loci following developmental exposure to both compounds were observed up to the F2 generation. We identified 2 and 6 out of 10 loci that showed transgenerational effects for MEHP and 5AC, respectively. From the 6 loci that were transgenerationally inherited following 5AC exposures, 5 exhibited the same effect on methylation that was found in sperm samples of F0 fish. Although studies have shown altered methylation patterns in progeny from exposed ancestors, site-specific transgenerational effects that are persistent over generations, as we observe, are rarely reported [47]. For example, in an exposure study of zebrafish adults to either 5AdC, 2,3,7,8-tetrachlorodibenzo-p-dioxin or methylmercury, effects on DNA methylation were observed from F0 livers to F1 zebrafish larvae in methylmercury exposures and no transgenerational effects were observed with any of the exposures [48]. In adult male mice exposed to 5AdC, direct effects on the sperm epigenome were found, but in subsequent generations no effects on DNA methylation were seen [49]. Both studies, however, assessed methylation differences after exposure to adults, which could disrupt methylation in gametes, but misses the sensitive windows of reprogramming events during early development. Furthermore, the mode of action of 5AdC compared to 5AC might differ, since only 10–20% of 5AC will be actively incorporated into the DNA [28]. 5AC shows persistent locus-specific effects spanning generations and suggests that specifically by 5AC, a methylation state can be inherited in zebrafish from one generation to another. Interestingly, the transgenerational effects on methylation of specific loci in F0, F1 and F2 larvae were not consistently reflected in the methylation status of sperm DNA from F0 adults for MEHP. For example, developmental MEHP exposure showed transgenerational hypermethylation of the *cbfa2t2* locus, while F0 sperm showed hypomethylation. This raises the question whether only the paternal methylome acts as a template for DNA methylation, as suggested by others [13, 16]. Alternatively, differential methylation in somatic tissue does not necessarily reflect the methylation status observed in sperm [37]. In contrast to 5AC, we found moderate effects at the 10 analyzed loci following MEHP exposure; however, a few loci exhibited differences in DNA methylation

over generations. Early life exposure to DEHP, the parent compound of MEHP, is known to exhibit effects later in life in rodents [19], and transgenerational effects have been found, related to behavior, obesity, early onset of puberty and effects on reproductive organs [24, 50–52]. Furthermore, similar effects on reproductive endpoints have been found in zebrafish after chronic exposure to MEHP, indicating similar modes of action between mammals and zebrafish [53]. Our finding of transgenerational DNA methylation changes in zebrafish exposed to MEHP is also corroborated by other studies in rodents with the DEHP. Transgenerational effects on the sperm methylome have been observed in F3 progeny of F0 pregnant rats exposed to a mixture of plasticizers (DEHP, bisphenol A and dibutylphthalate) [23]. Specifically, over 190 DMRs were found in sperm of F3 progeny; however, no analysis was performed in F2 or F1, which could link these specific DMRs toward transgenerational epigenetically inherited DMRs. A recent study observed moderate intragenerational effects (F0–F1) and no transgenerational effects (F3) on imprinted genes in mice prospermatogonia exposed to 300 mg/kg/day DEHP [25]. However, both studies focused on primordial germ cell reprogramming, since this reprogramming event specifically establishes gender-specific gene imprints in mammals [54]. However, the first wave of reprogramming could be equally essential in epigenetic inheritance as is emphasized by a study in which mice exposed to low doses of DEHP (40 μg/kg/day) directly after fertilization showed effects on DNA methylation at imprinted genes in oocytes up to F2 progeny [26]. In this study, both reprogramming events were included and may explain the strong effects on the imprinted genes up to F2 progeny. These results indicate that not only the reprogramming of primordial germ cells but also the first demethylation wave is of importance in transgenerational epigenetic inheritance.

Many studies have shown the strength of using zebrafish as a vertebrate model to assess DNA methylation [41, 48, 55, 56]. The use of whole zebrafish larvae allows for the assessment of an advanced functional organism for changes in DNA methylation. The drawback of using whole larvae is the complexity of many different cell lineages, each with their own transcriptome and methylome. Small differences in phenotype following exposures may reflect DNA methylation differences due to cell type composition, rather than a compound-specific change in DNA methylation. We cannot completely exclude the possibility that compound exposures may have affected specific cell populations, leading to non-specific effects on DNA methylation. However, to avoid this, we measured DNA methylation in embryos exposed to concentrations that were below the no effect concentration for embryo toxicity. For MEHP, the persistent effects on DNA methylation in subsequent generations were not accompanied by any effects on length. Furthermore, pathways described in previous studies and known to be affected by these compounds were confirmed in our study. Therefore, our results indicate that effects on methylation were directly related to the mode of action of the compound, and not to non-specific effects.

## Conclusions

Developmental exposure of zebrafish embryos to MEHP and 5AC resulted in differential DNA methylation specifically at zfCNEs, suggesting a functional role of DNA methylation on these sequences, conserved throughout evolution. A number of loci that were differentially methylated directly after exposure were also differentially methylated in subsequent generations, indicating transgenerational effects on DNA methylation after developmental exposure to 5AC and MEHP. In the case of 5AC, phenotypic changes were observed in embryo morphology in F1 and F2 progeny. Further research is needed to demonstrate the functional significance of methylation changes in the specific loci studied, as well as genome-wide DMR characterization in F1 and F2 progeny to further map transgenerational effects. Additionally, it is important to link DMRs to histone modifications, and expression of mRNAs and non-coding RNAs to get a view of possible interactions in epigenetic landscapes.

## Methods
### Chemicals

5-Azacytidine (5AC, >98%) was obtained from Sigma (Germany). Mono(2-ethylhexyl) phthalate (MEHP, >99%) was purchased from AccuStandard (USA). Dimethylsulfoxide was purchased from Acros Chemicals (Belgium). The standards 5-methyl-2'-deoxycytidine (mC) and 2'-deoxyguanosine (G) and 5-hydroxymethyl-2'-deoxycytidine (hmC) were purchased at MP biomedicals (the Netherlands), Sigma (Germany) and Carbosynth (UK), respectively. The internal standards for LC/MS analysis (2'-deoxyguanosine-[13]C10,[15]N5 (G-[13]C10[15]N5), 5-methyl-2'-deoxycytidine-d3 (mC-d3) and 5-hydroxymethyl-2'-deoxycytidine-d3 (hmC-d3)) were obtained from Toronto research chemicals (TRC, Canada).

### Zebrafish husbandry

This study was performed in accordance with European Directive 2010/63/EU implemented in the Dutch Act on Animal Experiments. The protocol was approved by The Committee on the Ethics of Animal Experiments of the VU University of Amsterdam under permit number DEC IVM 11-01. All efforts were made to minimize suffering.

Wild-type AB adult zebrafish (*Danio rerio*) were maintained in a recirculation system under a light regimen of 14 h light/10 h dark at a density of 8 fish/L. The parameters of the recirculation system water were: temperature, $26 \pm 0.5$ °C, pH $7.4 \pm 0.2$ and conductivity, $525 \pm 50$ µS. Embryos were obtained by natural mating for an hour of 2 family crosses (6 males:6 females per tank) in 2-L tanks.

### Exposures and transgenerational design of experiment

Fresh stock solutions of 5AC and MEHP were prepared daily in DMSO at concentrations of 10 and 25 mM for 5AC, and 30 mM for MEHP. Embryos (F0) were collected directly after fertilization and immediately transferred to petri dishes containing the chemicals at the final concentration of 0.01% DMSO (1.4 µM), 10 µM 5AC, 25 µM 5AC or 30 µM MEHP in medium (294 mg/L $CaCl_2$, 123 mg/L $MgSO_4$, 64.7 mg/L $NaHCO_3$, 5.7 mg/L KCl), allowing exposures directly after fertilization. Fertilized and good quality eggs were selected under a stereomicroscope (M7.5 Leica, Eindhoven, the Netherlands), and randomly transferred to a 24-well suspension culture plate in pools of 12 embryos/well containing 2 mL of exposure medium, and maintained in an incubator on a 14-h light/10-h dark cycle at 26 °C. Each day, embryos (from 1 to 6 dpf) were checked for developmental malformations and 90% exposure medium was refreshed with freshly made exposure medium. After 6 days, the exposure was terminated, and pools of zebrafish larvae were collected for RNA and DNA analysis (Table 3) and a total of 120 larvae were equally divided over 4 tanks containing 300 mL medium per treatment. Feeding of 2.3 mg powdered baby food (Sera micron) and 2 mL of *Tetrahymena* suspension twice a day was started at this time point. At 15 dpf, one tank of fish was used to for DNA and RNA analysis (Table 3). The fish from the other three tanks were transferred to a 2-L tank and maintained in the recirculation system. Fish were fed twice with 6 mg powdered feed (Tetraminbaby/Tetrapro) and 5 mL of *Tetrahymena* suspension and once with 5 droplets of Artemia suspension a day per tank. From 20 dpf on, fish were fed with 6 mg powdered feed and 5 mL Artemia suspension. After 23 weeks, fish were able to produce enough eggs for subsequent analyses and the establishment of F1 generation. From each tank, 8 males and 12 females were crossed in 2 separate groups, and eggs were collected, and after 24 h pooled similarly as described above. From each tank, replicate pools of larvae were taken for RNA and DNA analysis at 6 dpf (Table 3) and from each tank 80 larvae were equally divided over $2 \times 300$ mL tanks. At 15 dpf one tank was used for DNA analysis (Table 3) and the fish from the other tanks were transferred to 2-L tanks. Similar as above, F1 was raised and mated at 23 weeks of age to generate F2. F2 was raised until 6 dpf, when larvae were collected for further analysis. See Fig. 1 for details.

## Length analysis

Images from hatched zebrafish (3, 6 and 15 dpf) were analyzed by ImageJ. From each time point at each generation, the difference in length was calculated, and a two-tailed $t$ test was performed with a Bonferroni correction for multiple comparisons per exposure, comparing the different generations.

## Gene expression analysis

RNA was purified using the Nucleospin total RNAII extraction kit (Macherey–Nagel, Germany). RNA was extracted from pools of 10 or 2 larvae (6 or 15 dpf, respectively) (Table 3). Larvae were collected in tubes with ceramic beads and snap-frozen in liquid nitrogen. RA1 lysismix (Macherey–Nagel, Germany) was added to the tubes and samples were homogenized using Precellys homogenization (Precellys, USA), followed by RNA extraction according the manufacturers' protocol. Equal amounts of RNA were reverse transcribed with the high-capacity cDNA RT kit (Applied Biosystems, Grand Island, NY), followed by a 10 times dilution of the cDNA reaction with Milli-Q water. QPCRs on the diluted cDNA were performed in 10 µL, containing 5 µL Lightcycler 480 SYBR Green I Master mix (Roche, Norway), 250 nM of forward and reverse primers, 2 µL diluted cDNA and Milli-Q water, in technical duplicates. PCR was performed on a Roche Lightcycler 96 (Roche, Norway), with 5-min denaturation at 95 °C, followed by 40 cycles of 15 s at 95 °C and 45 s at 60 °C. After the run a melting curve was generated from 60 to 90 °C. Primers for reference genes were developed using the Primer-BLAST software from NCBI (http://www.ncbi.nlm.nih.gov/) (Additional file 1: Table 3). *Dnmt* primer sequences

were obtained from a recently published study [57] (Additional file 1: Table 3). All primers were validated for specificity by melting curve analysis and gel electrophoresis. Efficiency of primers was determined against a dilution curve of pooled zebrafish cDNA. Cq values were determined using linreg [58]. Five reference genes were measured (*hprt1*, *ef1a*, *beta-actin*, *hmbs* and *rps18*), of which *ef1a* and *hprt1* were most stable and were used for relative gene expression calculations, using the geometric average of two reference genes as described earlier [59]. Statistical analysis was performed in GraphPad 5.04 (GraphPad Software Inc., La Jolla, CA). For statistics, relative gene expression was log2-normalized and ANOVA was performed per gene, with Dunnets' post hoc tests for multiple comparisons.

## DNA purification

Precipitation and purification of genomic DNA was performed with the Gentra puregene tissue DNA extraction kit (Qiagen, Germany). Pools of zebrafish (Table 3) were snap-frozen in liquid nitrogen. Prior to DNA extraction the pools of zebrafish larvae were disrupted in lysis buffer with a 20G needle. DNA was extracted as described earlier [29]. Quality of DNA was assessed by gel electrophoreses for fragmentation and RNA contamination, and quantity and purity by NanoDrop (ND-1000, Thermo Scientific, Germany).

## Global methylation analysis

Analysis of mC and hmC relative against G was analyzed with liquid chromatography–tandem mass spectrometry (LC/MS, Agilent 6490) as described previously [29], with modifications. To account for fluctuations between LC runs, we adapted the protocol with the inclusion of labeled internal standards for each of the analytes. Genomic DNA (200 ng in 10 µL E buffer) was digested

## Table 3 Number of biological replicates analyzed per exposure

| Samples per exposure group | Gen | Length | DNA[a] | RNA[a] | Global methylation[b] | RRBS[b] | BisPCR2[b] |
|---|---|---|---|---|---|---|---|
| 3 dpf Larvae | F0 | 68–76 | | | | | |
| 6 dpf Larvae | F0 | 50–60 | 6 | 5 | 6 | 5 | 6 |
| 15 dpf Larvae | F0 | 10 | 5 | | 5 | | 5 |
| Brain per gender | F0 | | 15 | | 15 | | |
| Liver per gender | F0 | | 15 | | 15 | | |
| Sperm | F0 | | 15 | | 15 | | 6 |
| 6 dpf Larvae | F1 | 30 | 9 | | 9 | | 5 |
| 15 dpf Larvae | F1 | 30 | 9 | | 9 | | |
| 6 dpf Larvae | F2 | 30 | 9 | | 9 | | 5 |

[a] Pool of 10 or 2 larvae for 6 or 15 dpf, respectively

[b] Number of biological replicates used in specific methylation analysis

by adding 10 µL of a mixture of benzonase, phosphodiesterase and alkaline phosphatase (50 U/mL, 60 mU/mL and 40 U/mL, respectively (Sigma, Germany) in buffer (20 mM TRIS, 100 mM NaCl and 20 mM $MgCl_2$, pH 7.9). A standard curve was included with percentages that were expected in zebrafish (0–12% mC and 0–1% hmC), based on molarity of mC and hmC relative to G (345 nM in 200 µL). Following incubation of 6 h at 37 °C, a mix of internal standards was added to the samples and standards, with final concentrations of 345, 20 and 0.69 nM for G-$^{13}C10^{15}N5$, mC-d3 and hmC-d3, respectively, in a final volume of 200 µL. For LC/MS analysis, the same mass transitions were used as described earlier [29], but with extra mass transitions of 283.3/167.2 for $^{13}C10^{15}N5$-G, 261.1/145.1 for hmC-d3 and 245.1/129.1 for mC-d3. Ionization-specific parameters were similar for all compounds; dwell time 50 ms, collision energy 2 V, fragmentor 380 V, cell accelerator voltage 4. QQQ-specific parameters were a gas temperature of 200 °C, gas flow of 14 L/min, nebulizer gas at 45 psi, sheath gas temperature of 350 °C, sheath gas flow of 7 L/min, capillary voltage of 3000 V (positive mode), nozzle voltage of 500 an iFunnel parameters of high-pressure RF of 150 psi and low-pressure RF of 60 psi. For all samples and calibration curves, we first calculated the ratio in peak areas of mC/ mC-d3 (ratio mC), hmC/hmC-d3 (ratio hmC) and G/G-$^{13}C10^{15}N5$ (ratio G), followed by (ratio mC/ratio G) and (ratio hmC/ratio G). For the standard curve, these ratios were plotted against the percentage (hydroxy)methylation (0–12% for mC and 0–1% for hmC). We added a quality control (QC) sample to every series to calculate the deviation between experiments. This QC sample consisted of a pool of 48 hpf zebrafish DNA. All samples were interpolated in the calibration curves. Statistical significance was performed with ANOVA, using Dunnets' multiple comparison post hoc tests within Graph-Pad software 5.04 (Inc., La Jolla, USA).

## Reduced representation bisulfite sequencing (RRBS)

Genomic DNA was measured with the Quant-it Picogreen dsDNA assay (Thermo Fischer) and 1 µg total genomic DNA was digested overnight with MspI at 37 °C (NEB, USA). Digestions were terminated by adding 0.5 M EDTA, and digested DNA was purified on a GeneJET PCR purification column (Thermo Fischer). A library was made using the NEBNext Ultra DNA library preparation kit for Illumina, including methylated index adapters. Adapter ligated fragments were bisulfite converted using the Zymo EZ DNA Methylation Gold kit (Zymo Research, USA), followed by 14 cycles of PCR. PCR products were purified using AMPure XP beads and quality was assessed on a Bioanalyzer (Agilent, Belgium), using a high-sensitivity DNA chip (Agilent,

Belgium). Concentration of the library was measured by QPCR. Samples were pooled in equal concentrations and sequencing was performed on an Illumina HiSeq 2500 in paired end (2 × 100 bp), according to the manufacturers' recommendations for RRBS sequencing. Data analysis was performed using a bisulfite analysis pipeline, developed by Babraham institute (UK). Detailed procedures are provided in supplementary materials and methods (Additional file 2). In short, after sequencing, Fasta files were first adapter and quality trimmed with Trim_galore (version 0.4.0, Babraham bioinformatics), followed by Bismark alignment (version 0.14.5, Babraham bioinformatics, [30]) to the recently released zebrafish genome assembly GRCz10. Downstream analysis was performed with the methylKit package in R (version 3.2.2), using logistic regression analysis with a sliding linear model to correct for multiple comparisons, using Benjamini–Hochberg false discovery rate (FDR) correction (Q-value) [60]. Within methylKit, our approach was to detect differentially methylated regions (DMRs) by dividing the genome in 300-bp tiles containing at least 4 mutually covered Cs in CpG context, with at least 5 reads per C, resulting in ≥20 observations in each replicate, while controlling the FDR at 0.01 and using a methylation difference cutoff of 10%. The use of 5 pooled replica's per treatment, a sufficient read depth over multiple CpG sites, and the use of logistic regression, combined with an FDR approach should account for sampling bias. A second approach was to identify specific differentially methylated CpG sites (DMCs) for downstream locus-specific analysis. We used logistic regression analysis on Cs in CpG context with at least 10 reads, with FDR controlled at 5% and a difference cutoff of 20%.

We used the DMRs in the Seqmonk genome browser (version 0.32) to investigate enrichment on different features, gene promoters, gene bodies, CpG islands and shores. Gene promoters were defined as 2000-bp upstream of a transcriptional start site (TSS). CpG islands were calculated according to the Takai and Jones algorithm [61]. We adjusted the parameters because of the different GC content and observed/expected ratio of CpG sites in zebrafish compared to mammals (CG > 0.45, o/e > 0.65), and defined shores as 2000-bp regions surrounding the CpG islands. Furthermore, we used a list of zebrafish conserved non-genic elements (zfCNEs) [35] and regions with developmental enhancer marks [12] to calculate enrichment on these specific elements.

## Pathway analysis

Associated genes from DMRs were predicted using Genomic Regions Enrichment of Annotations Tool (GREAT) [62]. This tool predicts gene functions of cis-regulatory elements in a genome. We took the standard parameters used in GREAT to predict functional genes (5000-bp upstream and 1000-bp downstream of a TSS, with an extension of max 1 Mb to the next regulatory domain of the nearest gene). The resulting gene list was used in Ingenuity Pathway Analysis (IPA, QIAGEN Redwood City, www.qiagen.com/ingenuity). IPA uses canonical pathways based on human, mice and rat data, but is able to import and analyze zebrafish gene homologs, based on mammalian knowledge bases. Since there is no apparent relationship between genes expressed and the methylation state of a regulatory region, we did not provide the methylation difference as an extra parameter in IPA. We used IPA to predict upstream regulatory features and to search for enrichments in toxicology and disease gene lists. IPA analysis was performed using Fischer's exact test and P values <0.05 were considered significant.

## BisPCR2

For validation of RRBS data and analysis of DMCs in F0 15 dpf, sperm and subsequent 6 dpf generations (F1 and F2), we used a recently developed method by Bernstein et al. [32]. Detailed procedures are presented in supplementary materials and methods (Additional file 2). BisPCR2 uses two rounds of PCR on bisulfite converted DNA. One round is used for amplifying bisulfite converted DNA using specific primers with Illumina adapter overhangs. After PCR#1, all amplicons of one sample are pooled in equal amounts and subjected to a second PCR using the standard Illumina library and index primers as reported by Bernstein et al. [32]. From the single CpG analysis from the RRBS results, we initially selected 10 Cs in loci which were differentially methylated in either 5AC or MEHP exposures (Additional file 1: Table 4). We validated this method using a calibration curve of control DNA from 0 to 100% methylated DNA for efficiency assessment of the PCRs and technical variation using two samples that were assessed in duplicate. Sequencing was performed on an Illumina MiSeq sequencer (Illumina Inc., USA) as described by Bernstein et al. Mapping of sequences and statistics was performed similarly as with RRBS analysis.

## Additional files

**Additional file 1.** A Microsoft Word document that contains the additional Tables and Figures as cited in the main text.

**Additional file 2.** A Microsoft Word document that contains detailed descriptions of used bioinformatics and the BisPCR2 method.

**Additional file 3.** A Microsoft Excel spreadsheet that contains all the DMRs and IPA output for MEHP.

**Additional file 4.** A Microsoft Excel spreadsheet that contains all the DMRs and IPA output for 5AC.

## Abbreviations

MEHP: mono(2-ethylhexyl) phthalate; 5AC: 5-azacytidine; dpf: days post-fertilization; hpf: hours post-fertilization; mC: methyl cytosine; DNMT: DNA methyltransferase; PGC: primordial germ cell; DEHP: di-2-(ethylhexyl) phthalate; DMR: differentially methylated region; hmC: hydroxymethyl cytosine; LC/MS: liquid chromatography–tandem mass spectrometry; BisPCR2: multiplex loci-specific bisulfite sequencing; DMC: differentially methylated CpG; TSS: transcriptional start site; CGi: CpG island; zfCNE: zebrafish conserved non-genic elements; GREAT: Genomic Regions Enrichment of Annotations Tool; IPA: ingenuity pathway analysis; TGF: transforming growth factor; SSH: slingshot protein phosphatase; SOX: SRY (sex determining region Y)-box; POU: POU domain; PPAR: peroxisome proliferator-activated receptor; RXR: retinoic X receptor; ASCL: achaete-scute family bHLH transcription factor; PAX: paired box; GLI: GLI family zinc finger; KLF: kruppel-like factor; HNF: hepatocyte nuclear factor 4; RRBS: reduced representation bisulfite sequencing; myc: v-myc avian myelocytomatosis viral oncogene homolog; nrp: neuropilin; cbfa: core-binding factor, runt domain; cps: carbamoyl-phosphate synthase; gabr: gamma-aminobutyric acid (GABA) A receptor.

## Authors' contributions

JL, LBS and JK developed the transgenerational setup. LBS performed the exposure studies with zebrafish and related analysis, including imaging, measurements, dissections and generation of F1 and F2. JK and LBS performed DNA and RNA extractions. Methylation analyses (except RRBS sample prep and sequencing), QPCR and bioinformatics and statistics were performed by JK. Manuscript was written by JK, with substantial input from JL, PA and LBS. All authors read and approved the final manuscript.

## Author details

[1] Faculty of Veterinary Medicine, Department of Basic Sciences and Aquatic Medicine, CoE CERAD, Norwegian University of Life Sciences, P.O. Box 8146 Dep., 0033 Oslo, Norway. [2] Institute for Environmental Studies, VU University Amsterdam, Amsterdam, The Netherlands. [3] Institute for Environment, Health and Societies, College of Health and Life Sciences, Brunel University London, Uxbridge, UK.

## Acknowledgements

RRBS sequencing was carried out by NXT-DX, Gent, Belgium. Frode Lingaas and Ole Guttersrud (both BaSAM, NMBU) are acknowledged for their assistance with Illumina MiSeq sequencing. Helene Thorsen Rønning (Matinf, NMBU) is acknowledged her support with LC/MS analysis. The support of Marjo den Broeder and Peter Cenijn, IVM/VU University, in the zebrafish experiments is much appreciated.

## Competing interests

The authors declare that they have no competing interests.

## Funding

JK, LBS and JL were supported by the Netherlands Organization for Scientific Research (NWO), project number VIDI/864.09.005. JK also received support from the Research Council of Norway through its Centre of Excellence funding scheme, Project No. 223268/F50, Centre for Environmental Radioactivity (CERAD) 2013–2022.

## References

1. Godfrey KM, Lillycrop KA, Burdge GC, Gluckman PD, Hanson MA. Epigenetic mechanisms and the mismatch concept of the developmental origins of health and disease. Pediatr Res. 2007;61:31–6.
2. Barker DJ, Clark PM. Fetal undernutrition and disease in later life. Rev Reprod. 1997;2:105–12.
3. Jiménez-Chillarón JC, Nijland MJ, Ascensão AA, Sardão VA, Magalhães J, Hitchler MJ, et al. Back to the future: transgenerational transmission of xenobiotic-induced epigenetic remodeling. Epigenetics. 2015;10:259–73.
4. Vickers MH. Early life nutrition, epigenetics and programming of later life disease. Nutrients. 2014;6:2165–78.
5. Marczylo EL, Jacobs MN, Gant TW. Environmentally induced epigenetic toxicity: potential public health concerns. Crit Rev Toxicol. 2016;8444:1–25.
6. Jirtle RL, Skinner MK. Environmental epigenomics and disease susceptibility. Nat Rev Genet. 2007;8:253–62.
7. Nilsson EE, Skinner MK. Environmentally induced epigenetic transgenerational inheritance of disease susceptibility. Transl Res. 2015;165:12–7.
8. Wu H, Zhang Y. Reversing DNA methylation: mechanisms, genomics, and biological functions. Cell. 2014;156:45–68.
9. Kamstra JH, Aleström P, Kooter JM, Legler J. Zebrafish as a model to study the role of DNA methylation in environmental toxicology. Environ Sci Pollut Res. 2015;22:16262–76.
10. Kobayashi H, Sakurai T, Miura F, Imai M, Mochiduki K, Yanagisawa E, et al. High-resolution DNA methylome analysis of primordial germ cells identifies gender-specific reprogramming in mice. Genome Res. 2013;23:616–27.
11. Lee HJ, Lowdon RF, Maricque B, Zhang B, Stevens M, Li D, et al. Developmental enhancers revealed by extensive DNA methylome maps of zebrafish early embryos. Nat Commun. 2015;6:6315.
12. Bogdanović O, Smits AH, de la Calle MustienesE, Tena JJ, Ford E, Williams R, et al. Active DNA demethylation at enhancers during the vertebrate phylotypic period. Nat Genet. 2016;48:417–26.
13. Jiang L, Zhang J, Wang J-J, Wang L, Zhang L, Li G, et al. Sperm, but not oocyte, DNA methylome is inherited by zebrafish early embryos. Cell. 2013;153:773–84.
14. Amouroux R, Nashun B, Shirane K, Nakagawa S, Hill PWS, D'Souza Z, et al. De novo DNA methylation drives 5hmC accumulation in mouse zygotes. Nat Cell Biol. 2016;18:225–33.
15. Kaaij LJT, Mokry M, Zhou M, Musheev M, Geeven G, Melquiond ASJ, et al. Enhancers reside in a unique epigenetic environment during early zebrafish development. Genome Biol. 2016;17:146.
16. Potok ME, Nix DA, Parnell TJ, Cairns BR. Reprogramming the maternal zebrafish genome after fertilization to match the paternal methylation pattern. Cell. 2013;153:759–72.
17. Johns LE, Cooper GS, Galizia A, Meeker JD. Exposure assessment issues in epidemiology studies of phthalates. Environ Int. 2015;85:27–39.
18. Feige JN, Gelman L, Rossi D, Zoete V, Métivier R, Tudor C, et al. The endocrine disruptor monoethyl-hexyl-phthalate is a selective peroxisome proliferator-activated receptor γ modulator that promotes adipogenesis. J Biol Chem. 2007;282:19152–66.
19. Martinez-Arguelles DB, Papadopoulos V. Prenatal phthalate exposure: epigenetic changes leading to lifelong impact on steroid formation. Andrology. 2016;4:573–84.
20. ECB. European Union Risk Assessment Report: Bis(2-ethylhexyl)phthalate (DEHP). Eur. Comm.—Jt. Res. Centre. Luxemb 2008;80:588.
21. European Chemical Agency. Review of new available information for Bis (2-Ethylhexyl) Phthalate (DEHP). Evaluation of new scientific evidence concerning the restrictions contained in Annex XVII to Regulation (Ec) No 1907/2006 (Reach). 2010;2006:1–24.
22. Prados J, Stenz L, Somm E, Stouder C, Dayer A, Paoloni-Giacobino A. Prenatal exposure to DEHP affects spermatogenesis and sperm DNA methylation in a strain-dependent manner. PLoS ONE. 2015;10:1–27.
23. Manikkam M, Tracey R, Guerrero-Bosagna C, Skinner MK. Plastics derived endocrine disruptors (BPA, DEHP and DBP) induce epigenetic transgenerational inheritance of obesity, reproductive disease and sperm epimutations. PLoS ONE. 2013;8:e55387.
24. Manikkam M, Guerrero-Bosagna C, Tracey R, Haque MM, Skinner MK. Transgenerational actions of environmental compounds on reproductive disease and identification of epigenetic biomarkers of ancestral exposures. PLoS ONE. 2012;7:e31901.
25. Iqbal K, Tran DA, Li AX, Warden C, Bai AY, Singh P, et al. Deleterious effects of endocrine disruptors are corrected in the mammalian germline by epigenome reprogramming. Genome Biol. 2015;16:59.
26. Li L, Zhang T, Qin XS, Ge W, Ma HG, Sun LL, et al. Exposure to diethylhexyl phthalate (DEHP) results in a heritable modification of imprint genes DNA methylation in mouse oocytes. Mol Biol Rep. 2014;41:1227–35.
27. Estey EH. Epigenetics in clinical practice: the examples of azacitidine and decitabine in myelodysplasia and acute myeloid leukemia. Leukemia. 2013;27:1803–12.
28. Stresemann C, Lyko F. Modes of action of the DNA methyltransferase inhibitors azacytidine and decitabine. Int J Cancer. 2008;123:8–13.
29. Kamstra JH, Løken M, Aleström P, Legler J. Dynamics of DNA hydroxymethylation in zebrafish. Zebrafish. 2015;12:230–7.
30. Krueger F, Andrews SR. Bismark: a flexible aligner and methylation caller for Bisulfite-Seq applications. Bioinformatics. 2011;27:1571–2.
31. Chatterjee A, Ozaki Y, Stockwell PA, Horsfield JA, Morison IM, Nakagawa S. Mapping the zebrafish brain methylome using reduced representation bisulfite sequencing. Epigenetics. 2013;8:979–89.
32. Bernstein DL, Kameswaran V, Le Lay JE, Sheaffer KL, Kaestner KH. The BisPCR2 method for targeted bisulfite sequencing. Epigenetics Chromatin. 2015;8:27.
33. OECD. Test No. 236: Fish Embryo Acute Toxicity (FET) Test. OECD Guidel. Test. Chem. Sect. 2, OECD Publ. 2013;1–22.
34. Bogdanovic O, Fernandez-Minan A, Tena JJ, de la Calle-Mustienes E, Hidalgo C, van Kruysbergen I, et al. Dynamics of enhancer chromatin signatures mark the transition from pluripotency to cell specification during embryogenesis. Genome Res. 2012;22:2043–53.
35. Hiller M, Agarwal S, Notwell JH, Parikh R, Guturu H, Wenger AM, et al. Computational methods to detect conserved non-genic elements in phylogenetically isolated genomes: application to zebrafish. Nucleic Acids Res. 2013;41:e151.
36. Nelson AC, Wardle FC. Conserved non-coding elements and cis regulation: actions speak louder than words. Development. 2013;140:1385–95.
37. Skinner M, Guerrero-Bosagna C. Role of CpG deserts in the epigenetic transgenerational inheritance of differential DNA methylation regions. BMC Genom. 2014;15:692.
38. Xu X, Tao Y, Gao X, Zhang L, Li X, Zou W, et al. A CRISPR-based approach for targeted DNA demethylation. Cell Discov. 2016;2:16009.
39. Vojta A, Dobrinić P, Tadić V, Bočkor L, Korać P, Julg B, et al. Repurposing the CRISPR-Cas9 system for targeted DNA methylation. Nucleic Acids Res. 2016;44:1–14.
40. Christman JK. 5-Azacytidine and 5-aza-2'-deoxycytidine as inhibitors of DNA methylation: mechanistic studies and their implications for cancer therapy. Oncogene. 2002;21:5483–95.
41. Bouwmeester MC, Ruiter S, Lommelaars T, Sippel J, Hodemaekers HM, van den Brandhof E-J, et al. Zebrafish embryos as a screen for DNA methylation modifications after compound exposure. Toxicol Appl Pharmacol. 2015;291:84–96.
42. Martin CC, Laforest L, Akimenko MA, Ekker M. A role for DNA methylation in gastrulation and somite patterning. Dev Biol. 1999;206:189–205.
43. San Roman AK, Aronson BE, Krasinski SD, Shivdasani RA, Verzi MP. Transcription factors GATA4 and HNF4A control distinct aspects of intestinal homeostasis in conjunction with the transcription factor CDX2. J Biol Chem. 2014;290:1850–60.
44. Rai K, Nadauld LD, Chidester S, Manos EJ, James SR, Karpf AR, et al. Zebra fish Dnmt1 and Suv39h1 regulate organ-specific terminal differentiation during development. Mol Cell Biol. 2006;26:7077–85.
45. Amann JM, Chyla BJI, Ellis TC, Martinez A, Moore AC, Franklin JL, et al. Mtgr1 is a transcriptional corepressor that is required for maintenance of the secretory cell lineage in the small intestine. Mol Cell Biol. 2005;25:9576–85.
46. Rai K, Jafri IF, Chidester S, James SR, Karpf AR, Cairns BR, et al. Dnmt3 and G9a cooperate for tissue-specific development in zebrafish. J Biol Chem. 2010;285:4110–21.

47. Tillo D, Mukherjee S, Vinson C. Inheritance of cytosine methylation. J Cell Physiol. 2016;234:1–7.

48. Olsvik PA, Williams TD, Tung H, Mirbahai L, Sanden M, Skjaerven KH, et al. Impacts of TCDD and MeHg on DNA methylation in zebrafish (Danio rerio) across two generations. Comp Biochem Physiol C: Toxicol Pharmacol. 2014;165:17–27.

49. Kläver R, Sánchez V, Damm OS, Redmann K, Lahrmann E, Sandhowe-Klaverkamp R, et al. Direct but no transgenerational effects of decitabine and vorinostat on male fertility. PLoS ONE. 2015;10:e0117839.

50. Quinnies KM, Doyle TJ, Kim KH, Rissman EF. Transgenerational effects of Di-(2-Ethylhexyl) phthalate (DEHP) on stress hormones and behavior. Endocrinology. 2015;156:3077–83.

51. Doyle TJ, Bowman JL, Windell VL, McLean DJ, Kim KH. Transgenerational effects of di-(2-ethylhexyl) phthalate on testicular germ cell associations and spermatogonial stem cells in mice. Biol Reprod. 2013;88:112.

52. Quinnies KM, Harris EP, Snyder RW, Sumner SS, Rissman EF. Direct and transgenerational effects of low doses of perinatal di-(2-ethylhexyl) phthalate (DEHP) on social behaviors in mice. PLoS ONE. 2017;12:e0171977.

53. Zhu Y, Hua R, Zhou Y, Li H, Quan S, Yu Y. Chronic exposure to mono-(2-ethylhexyl)-phthalate causes endocrine disruption and reproductive dysfunction in zebrafish. Environ Toxicol Chem. 2016;35:2117–24.

54. Smallwood SA, Kelsey G. De novo DNA methylation: a germ cell perspective. Trends Genet. 2012;28:33–42.

55. Corrales J, Fang X, Thornton C, Mei W, Barbazuk WB, Duke M, et al. Effects on specific promoter DNA methylation in zebrafish embryos and larvae following benzo[a]pyrene exposure. Comp Biochem Physiol C: Toxicol Pharmacol. 2014;163:37–46.

56. Fang X, Thornton C, Scheffler BE, Willett KL. Benzo[a]pyrene decreases global and gene specific DNA methylation during zebrafish development. Environ Toxicol Pharmacol. 2013;36:40–50.

57. Santangeli S, Maradonna F, Gioacchini G, Cobellis G, Piccinetti CC, Dalla Valle L, et al. BPA-induced deregulation of epigenetic patterns: effects on female zebrafish reproduction. Sci Rep. 2016;6:21982.

58. Ramakers C, Ruijter JM, Lekanne Deprez RH, Moorman AFM. Assumption-free analysis of quantitative real-time polymerase chain reaction (PCR) data. Neurosci Lett. 2003;339:62–6.

59. Vandesompele J, De Preter K, Pattyn F, Poppe B, Van Roy N, De Paepe A, et al. Accurate normalization of real-time quantitative RT-PCR data by geometric averaging of multiple internal control genes. Genome Biol. 2002;3:RESEARCH0034.

60. Akalin A, Kormaksson M, Li S, Garrett-Bakelman FE, Figueroa ME, Melnick A, et al. MethylKit: a comprehensive R package for the analysis of genome-wide DNA methylation profiles. Genome Biol. 2012;13:R87.

61. Takai D, Jones PA. Comprehensive analysis of CpG islands in human chromosomes 21 and 22. Proc Natl Acad Sci. 2002;99:3740–5.

62. McLean CY, Bristor D, Hiller M, Clarke SL, Schaar BT, Lowe CB, et al. GREAT improves functional interpretation of cis-regulatory regions. Nat Biotechnol. 2010;28:495–501.

# 10

# Well-positioned nucleosomes punctuate polycistronic pol II transcription units and flank silent *VSG* gene arrays in *Trypanosoma brucei*

Johannes Petrus Maree[1], Megan Lindsay Povelones[2], David Johannes Clark[3], Gloria Rudenko[4] and Hugh-George Patterton[1*] (iD)

## Abstract

**Background:** The compaction of DNA in chromatin in eukaryotes allowed the expansion of genome size and coincided with significant evolutionary diversification. However, chromatin generally represses DNA function, and mechanisms coevolved to regulate chromatin structure and its impact on DNA. This included the selection of specific nucleosome positions to modulate accessibility to the DNA molecule. *Trypanosoma brucei*, a member of the Excavates supergroup, falls in an ancient evolutionary branch of eukaryotes and provides valuable insight into the organization of chromatin in early genomes.

**Results:** We have mapped nucleosome positions in *T. brucei* and identified important differences compared to other eukaryotes: The RNA polymerase II initiation regions in *T. brucei* do not exhibit pronounced nucleosome depletion, and show little evidence for defined −1 and +1 nucleosomes. In contrast, a well-positioned nucleosome is present directly on the splice acceptor sites within the polycistronic transcription units. The RNA polyadenylation sites were depleted of nucleosomes, with a single well-positioned nucleosome present immediately downstream of the predicted sites. The regions flanking the silent variant surface glycoprotein (VSG) gene cassettes showed extensive arrays of well-positioned nucleosomes, which may repress cryptic transcription initiation. The silent VSG genes themselves exhibited a less regular nucleosomal pattern in both bloodstream and procyclic form trypanosomes. The DNA replication origins, when present within silent VSG gene cassettes, displayed a defined nucleosomal organization compared with replication origins in other chromosomal core regions.

**Conclusions:** Our results indicate that some organizational features of chromatin are evolutionarily ancient, and may already have been present in the last eukaryotic common ancestor.

**Keywords:** Genome-wide nucleosome positions, *Trypanosoma brucei*, Polycistronic transcription units, Silent variant surface glycoproteins, *VSG*, Bloodstream expression sites, Evolution, MNase-seq

## Background

African trypanosomes lie in an ancient evolutionary branch of the Excavates supergroup, which split from the last eukaryotic common ancestor (LECA) along with the SAR, Archaeplastida, and Unikonta supergroups some 2 billion years ago [1, 2]. Despite this early divergence,

*Trypanosoma brucei* encodes an extensive repertoire of proteins associated with chromatin structure, modification, and functional regulation [3–6]. The presence of an epigenome in trypanosomes is perhaps expected, given the evolutionary origin of functional core histones in the ancestral Archaea [7], and the presence of linker histone homologs in evolutionarily distant bacteria [8]. Although Archaea lack multi-domain chromatin remodelers [9], the SNF2 domain, which has DNA-dependent ATPase activity and is present in a broad range of chromatin remodelers [10], is identifiable in bacterial helicases.

*Correspondence: hpatterton@sun.ac.za
[1] Department of Biochemistry, Stellenbosch University, Matieland 7602, South Africa
Full list of author information is available at the end of the article

The occurrence of histone modification enzymes, and a functional role for modified DNA packaging proteins, was also demonstrated in Archaea [11], suggesting that the regulation of chromatin structure and the epigenetic definition of different functional states of chromatin predates LECA.

Nucleosomes can influence diverse DNA functions [12–14], and the precise nucleosome positions present in regulatory elements are functionally crucial [15]. Nucleosomes can also assume a similarly defined distribution around RNA polymerase II (pol II) transcription start sites (TSSs) in diverse eukaryotes from the Unikonta supergroup, including *Dictyostelium discoideum*, *Saccharomyces cerevisiae*, and *Homo sapiens*. Incredibly, this nucleosomal arrangement appears evolutionarily ancient, as a similar nucleosome-depleted region bracketed by positioned −1 and +1 "nucleosomes" was observed upstream of genes in the archaeal *Haloferax volcanii*. This is despite the fact that the structural chromatin unit in this archaeal cell is composed of a tetramer instead of an octamer of histones [16].

There is currently a lack of insight into the genome-wide nucleosomal organization of genomes from organisms that are evolutionarily far removed from the Unikonta supergroup of eukaryotes, yet encode conventional nucleosomes composed of a conserved octamer of histones. In this regard, *T. brucei* represents a very intriguing subject. Here, all four core histones are present, and the classic conservation benchmark, canonical H4, is 79% similar to that of *H. sapiens*. The most recent ancestor shared between *T. brucei* and the Unikonta supergroup is therefore, in all likelihood, LECA.

*Trypanosoma brucei* is a unicellular parasite that is transmitted to humans by one of several *Glossina* fly species, and causes human African trypanosomiasis (HAT) [17]. Upon initial human infection, *T. brucei* invades interstitial spaces, the lymph system, and the bloodstream. With prolonged infection, the parasite crosses the blood–brain barrier and invades the central nervous system [18]. Without treatment HAT is often fatal and although the number of cases is declining, more than 1.8 million people are still thought to be at high risk of the disease [19]. As the parasite cycles between the mammalian host and the insect vector, it differentiates into different life cycle stages including the bloodstream form (BF) in the mammal or the procyclic form (PF) in the midgut of the tsetse fly [20].

In the bloodstream of the mammalian host, *T. brucei* escapes clearance by the immune system by periodically switching a mono-allelically expressed variant surface glycoprotein (*VSG*), an abundant cell surface protein that masks invariant cell surface proteins [21, 22]. The active *VSG* is expressed from a single pol I-transcribed

subtelomeric *VSG* expression site (ES) [23]. The expressed *VSG* gene can be switched through multiple mechanisms [24]. First of all, a transcriptional switch can result in silencing of the active ES and the activation of one of approximately 15 other silent ESs. Alternatively, DNA recombination can be involved. Gene conversion can result in all or part of the active *VSG* gene being swapped with sequences from a different silent *VSG* cassette, present on a variety of types of chromosomes. *T. brucei* contains 11 megabase chromosomes (>1 Mb), ~5 intermediate chromosomes (200–900 kb), and ~100 mini-chromosomes (30–150 kb), and all of these contain silent *VSGs* [25]. Lastly, *VSGs* can be switched through telomere exchange with another *VSG*-containing telomere.

ESs are telomeric transcription units. There is a relatively localized telomeric silencing gradient extending up to 10 kb from the telomere end, although this is not implicated in the ES regulation involved in antigenic variation [26–29]. The telomeric repression observed in the immediate vicinity of the telomeres of the silent ESs appears superficially reminiscent of that observed in yeast and *Drosophila* in that it requires RAP1, among other factors [26]. However, SIR2, which plays an important role in telomere position effect in eukaryotes, appears to also have unrelated functions in *T. brucei* [29]. Additional repressive mechanisms appear to operate on the ES promoter itself. These is about 40–60-kb upstream from the chromosome end and is effectively silenced, even though distance-wise it would be expected to escape the effects of typical telomere position effect [30]. A number of proteins including the chromatin remodelers ISWI [31, 32], ORC1 [33], FACT [34], and HDAC3 [35] among others, play a role in ES promoter silencing. In addition, *T. brucei* histone H1, similar to the C-terminal tail of the H1 from metazoans, renders chromatin in the BF stage more resistant to nucleases, presumably due to a more closed chromatin conformation. H1 is also required for full transcriptional repression of silent ESs [36, 37]. Another unusual feature of ESs is that they are transcribed in a mono-allelic fashion by pol I, which in eukaryotes normally exclusively transcribes ribosomal DNA (rRNA) [38].

Unusually for a eukaryote, protein-coding genes in *T. brucei* are arranged in extensive, polycistronic transcription units (PTUs). These are constitutively transcribed by pol II from poorly defined promoters and can span up to several hundred kilobases [39]. There is no transcriptional regulation of pol II in *T. brucei*, and virtually all regulation of mRNA and protein levels appears to occur posttranscriptionally [40]. Pol II transcription initiation typically occurs in the strand switch regions (SSRs) between two divergently transcribed PTUs, but pol II start sites have also been identified between head-to-tail aligned PTUs [41]. Various epigenetic marks are located at SSRs and are likely to functionally define pol II regulatory regions [42]. The pol II transcription start sites are enriched for the H2A.V and H2B.V histone variants, as well as for modified H3K4me3 and H4K10ac [42, 43]. Termination of pol II transcription also occurs in the SSRs, for example, between two convergent PTUs. These termination regions are enriched for the H3.V and H4.V histone variants and modified H3K76me1/2 [42, 44]. In addition, the modified thymidine base, β-D-glucosyl-hydroxymethyluracil (base J), is also located at termination sites. Base J was shown to contribute to transcriptional termination in both *T. brucei* [45, 46] and *Leishmania tarentolae* [47], and the knockdown of base J and H3.V in *T. brucei* results in transcriptional read-through, and the appearance of downstream anti-sense RNA [45]. In addition, pol II termination is often also associated with a tRNA gene [42].

In this study, we mapped the genome-wide nucleosome positions in two different life cycle stages (BF and PF) of *T. brucei* 427 using MNase-seq. We report the nucleosomal organization at pol I and pol III promoters, as well as in regions flanking pol II-transcribed PTUs, including the adjacent transcription start and termination regions. We find that the pol II PTUs are punctuated internally by well-positioned nucleosomes at regions involved in RNA processing and analyzed the possible contribution of DNA sequence elements in the nucleosomal positions that were observed. In addition, we find that the silent *VSG* gene arrays are flanked by regions of well-positioned nucleosomes. This could play a role in suppressing fortuitous initiation by pol II, and ensuring mono-allelic expression of a single *VSG* from the active ES.

## Results
### Distribution of positioned nucleosomes in the *T. brucei* genome

In eukaryotic chromatin, approximately 147 nucleotides (nt) of DNA are wrapped around each histone octamer. Digestion of this chromatin with micrococcal nuclease (MNase) results in release of these 147-nt fragments, which if sequenced using high-throughput methods, allows nucleosome positioning over the entire genome. We therefore mapped nucleosomes at the whole genome level in the *T. brucei* 427 BF or PF life cycle stages using MNase-seq. This involved mapping the paired-end reads of isolated nucleosomal fragments of ~147 bp to the *T. brucei* 427 reference genome. Nucleosome dyad positions were assumed to correspond to the center of the mapped fragments. To gain insight into the distribution of nucleosomes in the genome of *T. brucei*, we performed a binning analysis, which makes it easier to visualize nucleosome density and positioning in different genomic regions. Here we summed the number of times that specific dyad frequencies were observed in the genome, using bin values from 1 to the maximum observed dyad frequency (Fig. 1a). The bin size represents the number of co-aligned nucleosome dyads, reflecting nucleosome positioning strength, and the number of members in each bin represents the number of times this degree of positioning was observed in the tested genomic region. Nucleosomes that aligned with AT-rich repeats, which are also present in intermediate- and mini-chromosomes not included in the current *T. brucei* reference genome, were also masked to avoid copy-number effects (Fig. 1b, c).

There was not a significant difference in nucleosome density between the BF and PF life cycle stages, and coding sequences contained 0.22 and 0.21 dyads/bp in both BF and PF *T. brucei*, respectively (Fig. 1d). The intergenic regions (SSRs as well as noncoding regions between individual genes in a PTU) had a nucleosomal dyad density of 0.22 dyads/bp in both the BF and PF life cycle stages (Fig. 1e). A slight extension of the bin population to higher value bins (or more well-positioned nucleosomes at a given genomic region) was visible in the intergenic, compared to the coding regions (compare Fig. 1d, e). This increase is statistically significant (Whitney–Mann *U* test; $p < 0.001$), showing that highly positioned nucleosomes were present more often in intergenic regions. No statistically significant difference could be detected between the distribution of nucleosomes in BF and PF forms in either the coding sequences or intergenic regions (Whitney–Mann *U* test; $p > 0.05$). This result suggests that the bulk nucleosome density and proportion of well-positioned nucleosomes were comparable in BF and PF *T. brucei*.

In the case of the predominantly pol III-transcribed *tRNA* genes, an average dyad density of 0.1 dyads/bp was found, with a normalized bins distribution as shown in Fig. 1f. The absence of high-value dyad bins indicates that nucleosomes are generally weakly positioned on the *tRNA* genes, which may be related to the transcriptional activity of *tRNA* genes and the size of the pol III transcription complex compared to the size of the *tRNA* gene itself. In yeast, *tRNA* genes are occupied by the TFIIIB–TFIIIC complex, displaying a distinctive occupancy pattern termed a "bootprint," and are generally nucleosome free [48]. A TFIIIC ortholog has not been identified in *T. brucei*, and it is not clear whether any transcription-related protein complex binds to the putative A-box [49]. *rRNA* genes had a dyad density of 0.34 dyads/bp and accommodated more well-positioned nucleosomes (Fig. 1g) compared to the genome average (see Fig. 1c; indicated by the gray diagonal in Fig. 1g). This may indicate the presence of a subpopulation of inactive *rRNA*

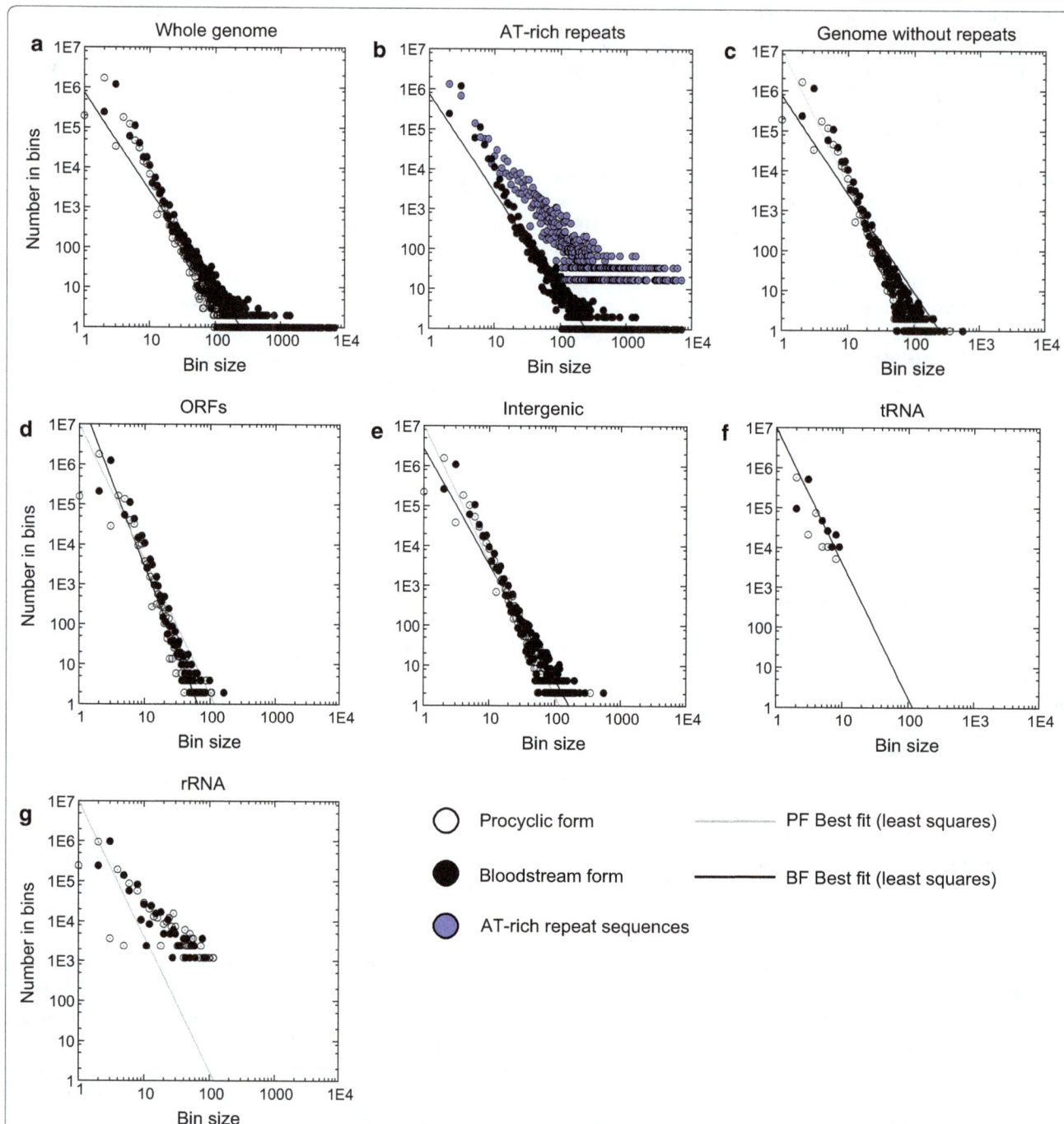

**Fig. 1** Binning analysis of nucleosome dyad co-alignment in the *T. brucei* genome. Binning analysis allows the visualization of nucleosome density and positioning in different genomic regions. The bin size (*x axes*) indicates the number of co-aligned nucleosome dyads reflecting nucleosome positioning strength, and the number of members in each bin (*y axes*) represents the number of times this degree of positioning was observed in the tested genomic region. **a** These analyses were performed on the entire *T. brucei* 427 genome using chromatin from either BF or PF. **b** Nucleosome distribution in the whole genome compared with AT-rich repeat sequences in BF chromatin. **c** Comparison of chromatin from BF and PF genomes excluding AT-rich repeat sequences. **d** Nucleosome distribution at the coding regions in BF and PF chromatin. **e** Comparison of nucleosome positioning at intergenic regions (SSRs as well as noncoding regions between individual genes in a PTU) in BF and PF *T. brucei*. **f** Nucleosomes at *tRNA* genes in BF and PF. **g** Nucleosomal distributions at the *5S* subunit *rRNA* genes in BF and PF. The *lines* represent the best fit (*least squares*) line to the BF (*black line*) or PF (*gray line*) datasets, respectively

genes, which is a common feature of the rDNA transcription units in other eukaryotes [50], assuming the number of *rRNA* genes annotated in the reference genome is accurate. The bulk nucleosome repeat length, or average nucleosome dyad to dyad distance, was determined at 194 bp in both the BF and PF life stages (data not shown).

### 5S rRNA genes

The pol III-transcribed *5S rRNA* genes are arranged as a single cluster from position 454,171–460,512 on chromosome 8 with a spacing of approximately 620 bp between each 118-bp *5S* gene. When the nucleosomal distribution of each *5S* gene was aligned relative to the transcript start position (Fig. 2a), defined as the first nucleotide in the *5S rRNA* gene sequence, a very clear nucleosomal organization emerged. Three well-positioned nucleosomes were seen on the *5S* transcription unit, with the first nucleosome centered on the transcript start. The downstream two nucleosomes were positioned directly downstream of the transcript end (represented by the ellipses in Fig. 2a). Nucleosome I overlapped with the *5S rRNA* gene up to approximately position 80 of the 5S transcript, which would include the putative A box (involved in pol III transcription initiation) located at position 51–62 of the gene (Additional file 1: Fig. S1).

A nucleosome-depleted region (NDR) was observed upstream of the *5S rRNA* gene, preceded by a well-positioned nucleosome at approximately position −400, which was equivalent to the second downstream nucleosome in the repeating 5S unit (labeled III' and III, respectively, in Fig. 2a). The eight *5S rRNA* genes in the *T. brucei* 427 reference genome encode transcripts that are 100% identical, and the 620-bp intergenic regions are 98% identical at nucleotide level. The alignment of sequence reads to the reference genome was therefore averaged between the individual genes by the mapping procedure. The distribution of the 3 identified nucleosomes was at sterically allowed distances (nucleosome I–II, 185 bp, and nucleosome II–III, 160 bp). There was no detectable difference in the nucleosomal organization of the *5S* genes in the

BF and PF life cycle stages (Fig. 2a; Additional file 1: Fig. S2A).

*Trypanosoma brucei* 427 contains two larger *rRNA* gene clusters with 24% identity at the nucleotide level located on chromosomes 2 and 3, and six smaller clusters of *rRNA* genes on chromosomes 6, 7, 8, and 9. The small number of annotated *rRNA* gene clusters precludes an informative alignment of the nucleosomal dyads at these loci.

### tRNA genes

We next aligned the 64 *tRNA* encoding genes at the transcript start positions. A clear NDR was again discernible directly upstream of the *tRNA* genes (Fig. 2b). This region was bracketed by groups of nucleosomes that appear to be located in multiple, overlapping frames. Three groups of nucleosomes were typically visible downstream of the *tRNA* gene and may represent three abutting nucleosomes in multiple phases. Upstream of the *tRNA* gene another group of overlapping nucleosome positions was typically visible. The nucleosomal arrangement at the *tRNA* genes appears functionally important, since the NDR, and the single upstream and three downstream nucleosomes were independently visible on the Watson and Crick strands (Additional file 1: Fig. S2B; Kendall's tau correlation; $\tau = 0.56, p < 0.001$), suggesting that they were arranged relative to the direction of transcription of the *tRNA* gene. However, the nucleosomal dyad distributions shown in Fig. 2b represent the contribution from both active and inactive genes, and the first downstream nucleosome may be unstable or disrupted in active *tRNA* genes, thereby not contributing substantially to the dyad distribution.

### Nucleosomal organization of pol II PTU transcription initiation regions

In other eukaryotes, a general picture has emerged for nucleosomal organization associated with the upstream regions of pol II-transcribed genes, where well-positioned nucleosomes flank an NDR that largely overlaps

(See figure on next page.)
**Fig. 2** Alignment of nucleosomal dyads at pol I-, II- and III-transcribed loci in BF trypanosomes. The number of nucleosome dyads assigned to each nucleotide was summed for all aligned features. Genes were aligned relative to the TSS, SAS or PAS, indicated by the *vertical black line* at dyad position 0 in each panel. **a** The average, cumulative nucleosomal dyad distribution is shown for *5S rRNA* genes and **b** for *tRNA* genes. The extent of the 119-bp *5S rRNA* (**a**) and 74-bp *tRNA* (**b**) genes are shown by the *black, rectangular arrows*. The ellipses **a**, **b** indicate the span of 160-bp nucleosomes, with the ellipse centers (nucleosomal dyads) aligned to the center of the corresponding, major peaks. The nucleosomes in **a** are labeled I–III, and the III' and III nucleosomes indicate identical nucleosomes in the repeating *5S* array. **c** Alignment relative to the SAS of the first coding sequence of all PTUs. **d** Separate alignment of the Watson and Crick strand data relative to the SAS of the first coding sequence of all PTUs. **e** Alignment relative to the SAS of all coding sequences within all PTUs. **f** Separate alignment of the Watson and Crick strand data relative to the SAS of all coding sequence within all PTUs. **g** Dyad axes aligned relative to the PAS of all genes in all PTUs. **h** Dyad axes aligned to the Watson and Crick strand data relative to the PAS of all genes within all PTUs

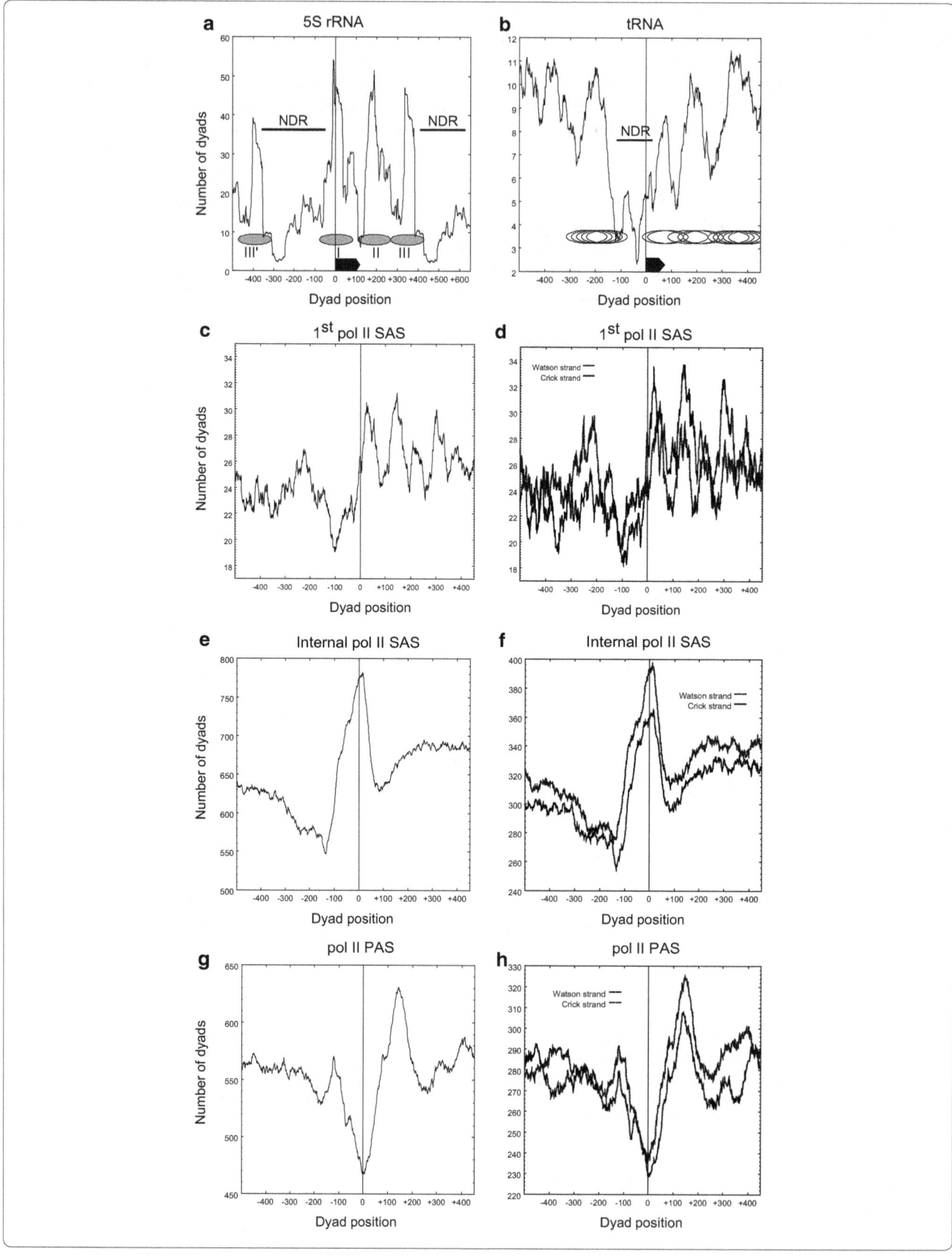

with the TSS [14]. This NDR is thought to allow assembly of the pol II pre-initiation complex in the region of the TSS [14]. However, in *T. brucei* pol II-mediated transcription is unusual in the sense that conventional pol II promoters are absent or undefined, and transcription commences at divergent SSRs and at a few other unique non-SSR locations [51]. Sequence analysis has failed to identify conventional TATA boxes or initiator sequences in the pol II transcription start regions [52], and it is thought that *T. brucei* lacks the many transcriptional activators, basal factors and *cis*-elements associated with pol II gene expression in eukaryotes from the Unikonta supergroup [53]. Although the pol II transcription start region is enriched for the H2A.V and H2B.V histone variants, and epigenetic marks associated with transcriptional activation in model eukaryotes [42], it is not understood how these histone variants and modifications are targeted to these precise chromatin regions. We were, therefore, interested in the nucleosomal organization of these pol II transcription start regions, and the idea that the local nucleosomal landscape may define functional pol II TSSs in *T. brucei*.

As the pol II TSSs remain unmapped in *T. brucei*, we aligned the nucleosomal dyads of PTUs relative to the most upstream splice acceptor site (SAS) which was identified in a *T. brucei* transcriptomic study [51]. The SAS is located upstream of each coding sequence in a PTU and is where the 39-nt sequence spliced leader RNA is *trans*-spliced onto the nascent RNA. The SAS is functionally related to the 3′ end of introns at intron–exon junctions in metazoans [54]. In the case of the SAS of the first gene of the PTU, this would be the genomic feature that is closest to the pol II transcription initiation region. When aligned relative to the first SAS (Fig. 2c), a region that was weakly depleted of nucleosome dyads was visible. This region was depleted of nucleosomal dyads in an approximately 100-bp region upstream of the SAS of the first gene of the PTU on both the Watson and the Crick strand (Fig. 2d), suggesting that this nucleosomal organization was sensitive to the direction of transcription (Kendall's tau correlation $p < 0.001$).

Although evidence of positioned nucleosomes downstream of the NDR was visible, these nucleosomes did not form a single phased array relative to a defined genomic feature, such as a +1 nucleosome, as is seen in other eukaryotes [14, 55, 56]. The observed peaks were less than the allowed nucleosome-to-nucleosome distance, indicating the absence of a regularly spaced nucleosome array, in both BF and PF cell lines (Additional file 1: Figs. S3 and S4). Although this nucleosomal arrangement is reminiscent of nucleosomes that were mapped downstream of the pol II TSSs in *S. cerevisiae* [55] and *Drosophila* [56], we note that the pol II PTUs in

*T. brucei* are constitutively expressed, and a well-defined nucleosomal array in this region would be unexpected. A similar pattern of nucleosomal distribution surrounding pol II TSS has been observed in a related kinetoplastid, *Leishmania major* [57]. A high nucleosome occupancy (defined as the average nucleosome dyad density per base pair in a specified region), but little positioning (defined as the number of co-aligned dyads at a specific base pair), was observed across the constitutively transcribed pol II PTUs [57]. The nucleosomes downstream of the first SAS appeared less organized in the PF life cycle stage (Additional file 1: Fig. S4A–D). In our experiments, we did not partially crosslink the nucleosomes chemically. However, the less-ordered nucleosomal arrangement in these regions is unlikely to be due to nucleosome sliding, since we do observe stable and strongly positioned nucleosomes at the *5S rRNA* genes.

These results showed that a region displaying a weak nucleosomal depletion was present upstream of the first SAS of pol II PTUs in *T. brucei*. This NDR was not as well defined as that seen at pol II TSSs in other eukaryotes such as in *S. cerevisiae* [14], and its functional association with the TSS, which remains unmapped in *T. brucei*, is also uncertain. If this NDR is indeed linked to transcription initiation, it suggests that this organizational feature, probably inherited from an Archaeal ancestor, was not maintained to the same extent in *T. brucei* compared with other eukaryotes. A clear enrichment of the H2A.Z histone variant was seen in nucleosome +1 bordering the NDR at the pol II TSSs in eukaryotic model organisms from the Unikonta supergroup [14]. The structurally less stable H2A.Z-containing nucleosome was proposed to facilitate elongation by pol II [58]. We therefore mapped H2A.V ChIP-seq data from *T. brucei* (kindly made available by N. Siegel) to our MNase-seq data. Although a clear enrichment in H2A.V-containing nucleosomes was seen in the upstream regions of PTUs and overlapping with the first gene, as was previously reported [42], this broad enrichment of nucleosomes did not map to any single, well-positioned nucleosome (data not shown).

### The NDR is not a feature of all SAS regions

We next assessed whether the nucleosomal organization observed for the region surrounding the SAS of the first gene in pol II PTUs was specific to this region, or simply reflected the nucleosomal organization of all upstream SAS regions for all genes within a PTU. We therefore repeated the dyad alignment analysis for all genes with assigned SAS sites, excluding that of the first gene in a PTU (3178 genes). The result is shown in Fig. 2e. In stark contrast to the arrangement seen at the SAS upstream of the first gene in a PTU, a striking enrichment of nucleosomal dyad axes that were aligned with the mapped SAS was seen. This was observed in the biological duplicates in both cell lines of the BF and PF life cycle stages (Additional file 1: Fig. S5) and was also independently seen (Kendall's tau correlation $p < 0.001$) on the assigned SAS of both the Watson (1545 genes) and Crick (1633 genes) strands (Fig. 2f). This suggests that there is a well-positioned nucleosome preferentially aligned with each internal SAS in *T. brucei*, contrary to the NDR observed at internal SASs in *L. major* [57]. When studying the nucleosomal arrangement at the SAS at the level of individual genes, the consistent, tight positioning of a nucleosome in the direct vicinity of the SAS is often observed, with a more random distribution of nucleosomes in the regions adjacent to the SAS (Additional file 1: Fig. S6). In contrast, bordering nucleosomes appeared disorganized.

### Nucleosomal organization at pol II termination regions

In addition to the SAS, the individual genes in a PTU typically contain a 3′ UTR with an average length of 676 bp, terminating at the polyadenylation site (PAS) [51]. The genomic region downstream of the PAS of the last gene of the PTU does not appear to contain defined transcription termination sequences [59]. The pol II transcription termination regions are enriched for the histone variants H3.V and H4.V [42], and both H3.V and base J deposition in this region appear to act synergistically to mediate efficient pol II termination [46, 60]. We were, therefore, interested in possible specialized nucleosomal arrangements in the regions of pol II transcription termination.

We first aligned the nucleosomal dyad axes relative to the PAS elements of all genes within PTUs [51] to get an overview of the general PAS structure. In contrast to the average nucleosomal organization around the internal SAS element, the PAS was clearly depleted of nucleosomes, with a single well-positioned nucleosome present immediately downstream of the PAS (Fig. 2g). Again, similar nucleosomal arrangements were independently observed on the Watson (1364 genes) and Crick (1448 genes) strands (Kendall's tau correlation; $\tau = 0.46, p < 0.001$, Fig. 2h), suggesting that the nucleosomal arrangement was important to a directional process on the DNA molecule (presumably transcription). This nucleosomal arrangement is very similar to that seen in human and yeast genomes, where nucleosomes are also depleted on polyadenylation sites [61, 62]. A similar organization was also seen in *L. major*, although the observed NDR was present immediately upstream of the PAS [57]. However, the relatively small number of terminal PAS elements ($n = 35$) precluded any statistically meaningful analysis of the average nucleosomal organization at this genomic position.

### Genomic distribution of nucleosome refractory sequences

It has previously been shown that oligo-d(A·T) runs are generally excluded from central locations of isolated chicken nucleosome cores [63], and it was suggested that this depletion is due to the inherently rigid structure of A·T tracts due to a series of bifurcated hydrogen bonds [64]. In *T. brucei*, runs of oligo-d(A·T) of 7 bp and longer are present in nucleosomes at approximately 70% of that expected for a random distribution (Additional file 1: Fig. S7A). Interestingly, runs of oligo-d(G·C) of up to 4 nucleotides were present in nucleosomes more often or the same as that expected from a random distribution, with a striking absence of runs longer than 7 bp (Additional file 1: Fig. S7B). Oligo-d(A-T) and oligo-d(T-A) are markedly depleted in *T. brucei* nucleosomes (Additional file 1: Fig. S8), even though these sequence runs occur at very high frequencies in the genome (Additional file 1: Table S1). This depletion might possibly be due to the destruction of nucleosomal fragments containing these sequences during nucleosome core preparation and subsequent rarefaction in the sequencing sample. It therefore appears that runs of oligo-d(A·T) and oligo-d(G·C) contribute to the relative absence of nucleosomes in specific regions of the *T. brucei* genome.

The polycistronic nature of the PTUs in *T. brucei* requires a SAS upstream of each open reading frame (ORF) to allow the *trans*-splicing of the 39-nt spliced leader (SL) RNA. The 3′ splice acceptor site contains the AG dinucleotide and a polypyrimidine tract (PPT) which is typically 10–40-nt upstream of the SAS, and is recognized by a U2AF35 and U2AF65 heterodimer of the spliceosome in the pre-mRNA [65]. As we have found that oligo-d(A·T) and oligo-d(G·C) runs are underrepresented in nucleosomes, we wondered whether the PPTs upstream of the SAS of each gene, and in particular the first gene of a PTU, was involved in the structural organization of nucleosomes in these regions. The preference for oligo-dT, as opposed to oligo-dA on the coding strand, is a requirement of the splicing mechanism [66]. Looking at the distribution of T runs of 7 bp and longer, a clear concentration of such runs is visible upstream of the first SAS (Fig. 3a) as well as all internal SASs (Fig. 3b). A second region enriched for oligo-dT appears downstream of the first SAS in a region that mostly falls within the 5′ UTR of the RNA transcripts. This second region of oligo-dT enrichment is absent in the average distribution of T runs at internal SASs (Fig. 3a, b).

The nucleosomal organization in the region of the first SAS of a PTU compared to the average organization of all PTUs differed significantly (see Fig. 2c, e). A region weakly depleted of nucleosomes was observed upstream of the first SAS (Fig. 2c, d), and a well-positioned nucleosome was observed at all internal SAS sites (Fig. 2e, f).

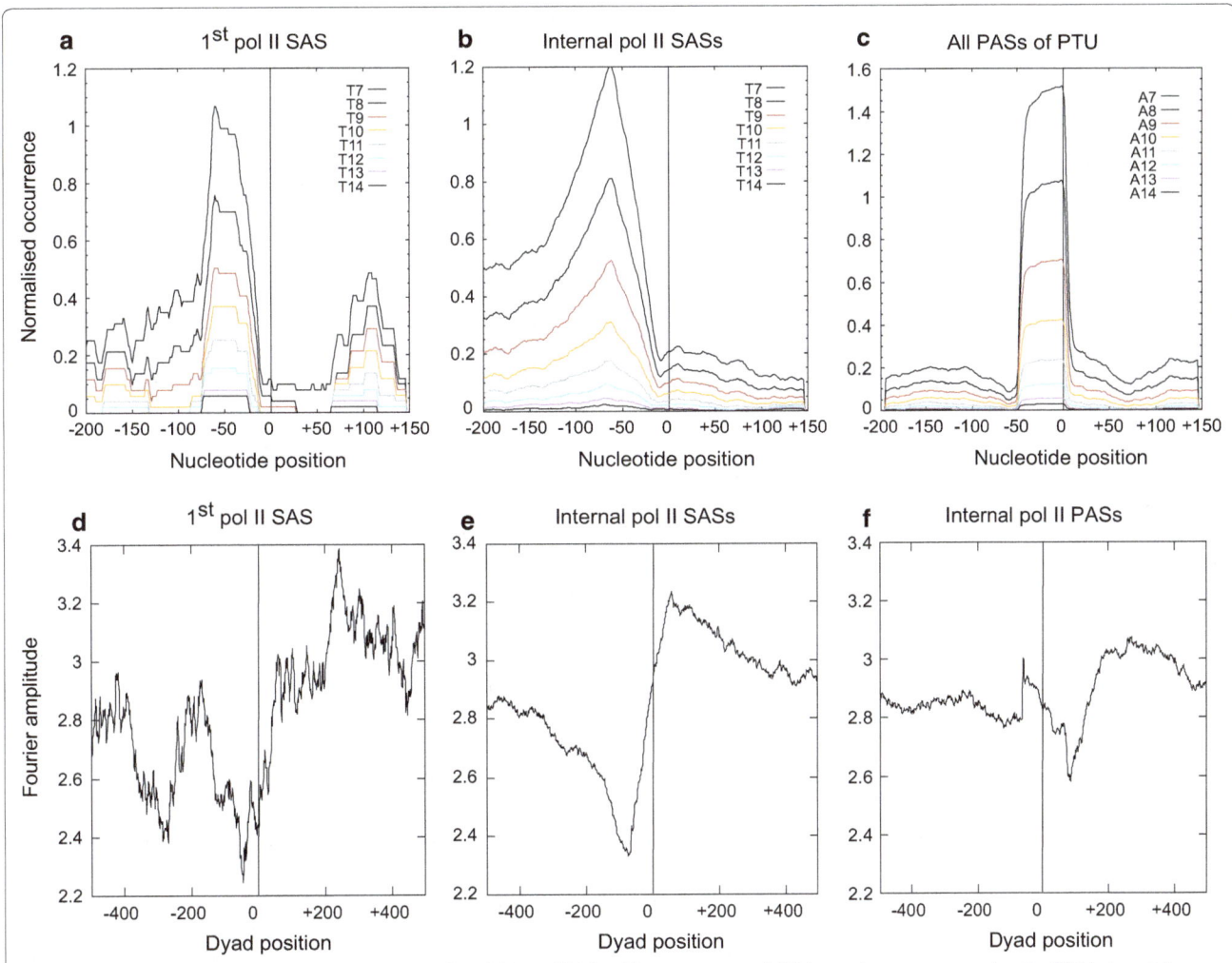

**Fig. 3** Distribution of nucleosome positioning signals at SASs and PASs. **a** The sequences of 400-bp regions encompassing the SAS (where it has been annotated [51]) were retrieved for the first gene in all PTUs, **b** for all internal SASs, and **c** for all PASs in all PTUs. The number of oligo-dT or oligo-dA runs (7–14 bp) was determined for the sequences aligned at the annotated SASs and PASs and is shown as a value normalized to the number of sequences. The Fourier amplitude of the distribution of A–A dinucleotides at a 10-bp periodicity was determined in a sliding 128-bp window, and normalized to the number of sequences in the window. **d** The cumulative Fourier amplitude is shown in a range from −500 to +500 bp relative to the first SAS upstream of all PTUs, **e** relative to internal SASs in all PTUs, and **f** relative to the internal PAS in all PTUs. The location of the SAS or PAS is indicated by the *vertical black line*

The region of nucleosome depletion at position −50 to −150 does not precisely align with the region enriched for T·A runs at position −70 to −20. However, the overlap in these regions, and the observation that oligo A runs in excess of 7 bp are generally depleted of nucleosomes, makes it highly likely that the presence of the oligo-d(A·T) runs upstream of the SAS contributes to the appearance of an NDR. However, the positional mismatch between the oligo-d(A·T) run and the NDR suggests the involvement of additional factors in establishing the NDR. The second region of oligo-dA enrichment at position +100 relative to the SAS of the first gene may serve to position the first nucleosome of the first gene

of a PTU (Fig. 2c, d). The absence of this second oligo-dA-enriched area at the remaining genes in a PTU may contribute to the absence of positioned nucleosomes downstream of the nucleosome positioned on the SAS (Fig. 2e, f).

A striking enrichment of oligo-dA runs is present directly upstream of the PASs (Fig. 3c), partially overlapping with the NDR observed in this region (Fig. 2g). In higher eukaryotes, a highly conserved AATAAA hexanucleotide that signals polyadenylation is located upstream of the PAS and may be refractory to nucleosomes [62]. However, this *cis*-acting element seems to be absent in *T. brucei* [67] and the role of the observed oligo-dA run abutting the PASs is unclear, but might contribute to the observed NFR at PASs.

## Nucleosome positioning signals at the initial SAS, internal SASs and internal PASs

Travers and colleagues showed that the rotational position of isolated nucleosome cores is defined by a distribution of dinucleotides at a periodicity equal to that of the DNA helix [63]. This was interpreted in terms of the structural constraints imposed on the rotational freedom of specific dinucleotide steps, and the ability to accommodate a narrowed or expanded minor groove. It was subsequently shown in genome-wide studies that positioned nucleosomes were often associated with di- and trinucleotide distributions equal to the DNA periodicity [55]. We therefore investigated whether the positioned nucleosomes upstream of the first gene in a *T. brucei* PTU, as well as those present at internal SASs and at the PASs, was positioned by underlying sequence periodicity.

We analyzed the distribution of the A–A dinucleotide by plotting the Fourier amplitude in a 128-nt window at consecutive settings across sites of interest in the genome, showing the strength of a periodic distribution of a given nucleotide. In the case of the first gene of a PTU, the highest Fourier amplitude is seen at approximately position +200 bp (Fig. 3d), indicating the likely presence of a strong nucleosome positioning sequence in this region. This aligns with the nucleosome immediately downstream of the NDR (see Fig. 2c) of the first SAS in a PTU. Interestingly, this nucleosome fits into the region forming a saddle between the two peaks of enrichment for oligo-dA (Fig. 3a), and the NDR (Fig. 2c), and partially overlaps with the peak at −50 in the distribution of oligo-dA (Fig. 3a). Therefore, the nucleosome distribution in the region of the SAS upstream of the first gene of a PTU can be explained in terms of the enrichment and depletion of specific sequence elements in this region and the known preference of nucleosome cores for such elements.

When looking at the internal SASs, a region generally enriched for 10-bp A–A periodicities is seen downstream of the SAS (Fig. 3e). Upstream of the SAS, centered at approximately position −80 bp, a region depleted of 10-bp A–A periodicities is evident. This aligns almost perfectly with the region enriched for oligo-dA tracts (Fig. 3b). The average nucleosomal structure in the region showed a nucleosome positioned at approximately position 0 (Fig. 2e, f). Thus, this nucleosome position is also consistent with the distribution of 10-bp A–A periodicities and oligo-dA tracts in these regions.

Looking at the Fourier amplitude of all annotated PASs (Fig. 3f), a region slightly enriched for a 10-bp

AA periodicity is seen downstream of position +150. Comparing this distribution to that of oligo-dA runs (Fig. 3c), a strong peak of oligo-dA runs is typically seen directly upstream of the PAS. Thus, there appears to be a sequence arrangement that discourages nucleosome formation from position 0 of the PAS and is more facilitative of nucleosome deposition downstream of the PAS. This is exactly what was seen in the average nucleosomal organization surrounding the PAS elements (Fig. 2g, h), where position 0 was depleted of nucleosomes, and a strongly positioned nucleosome was visible centered at approximately position +150. Again, the nucleosomal organization can be explained in terms of the sequence elements present at the PAS.

It was previously shown that nucleosome positions in vivo are directed by sequences as well as DNA binding proteins that may initiate "statistical positioning," as well as by chromatin remodelers [68]. Our results do not exclude the contribution from agents other than sequence, and the NDR upstream of the first SAS of a PTU cannot be fully attributed only to the polypyrimidine tract, which overlaps only partially with the NDR, thus implying the involvement of other factors.

## Nucleosome organization in BF and PF *T. brucei* life cycle stages is highly comparable

There is little evidence for pol II transcriptional control in *T. brucei*, and the life-cycle-specific control of most genes occurs posttranscriptionally [40, 69, 70]. However, some pol I-transcribed loci are differentially transcribed in the different *T. brucei* life cycle stages. For example, only one of about 15 VSG expression sites is active in BF *T. brucei*, whereas all ESs are repressed in the procyclic form. In contrast, the procyclin genes are repressed in the BF and active in the PF stage. We were therefore interested in establishing whether there were any clear differences in the nucleosomal organization of the *T. brucei* genome in the BF or PF life cycle stages that could explain this differential expression.

The average density of nucleosomal dyads in the *T. brucei* genome was determined, and the ratio of the number of dyads in a 1000-nt window compared with the genome average was plotted on a $\log_2$ scale. Chromosome 6 was chosen as a representative (Fig. 4a; all chromosomes are shown in Additional file 1: Fig. S9). A superficial inspection did not reveal any striking differences in the nucleosome occupancy traces on any chromosomes when comparing BF versus PF *T. brucei* (Fig. 4a, Additional file 1: Fig. S9). However, to quantitate possible subtle differences, we performed a Whitney–Mann $U$ correlation analysis. Statistically significant ($p < 0.01$) differences between the two life cycle stages are shown as the lower red trace for each chromosome (Fig. 4a). The five

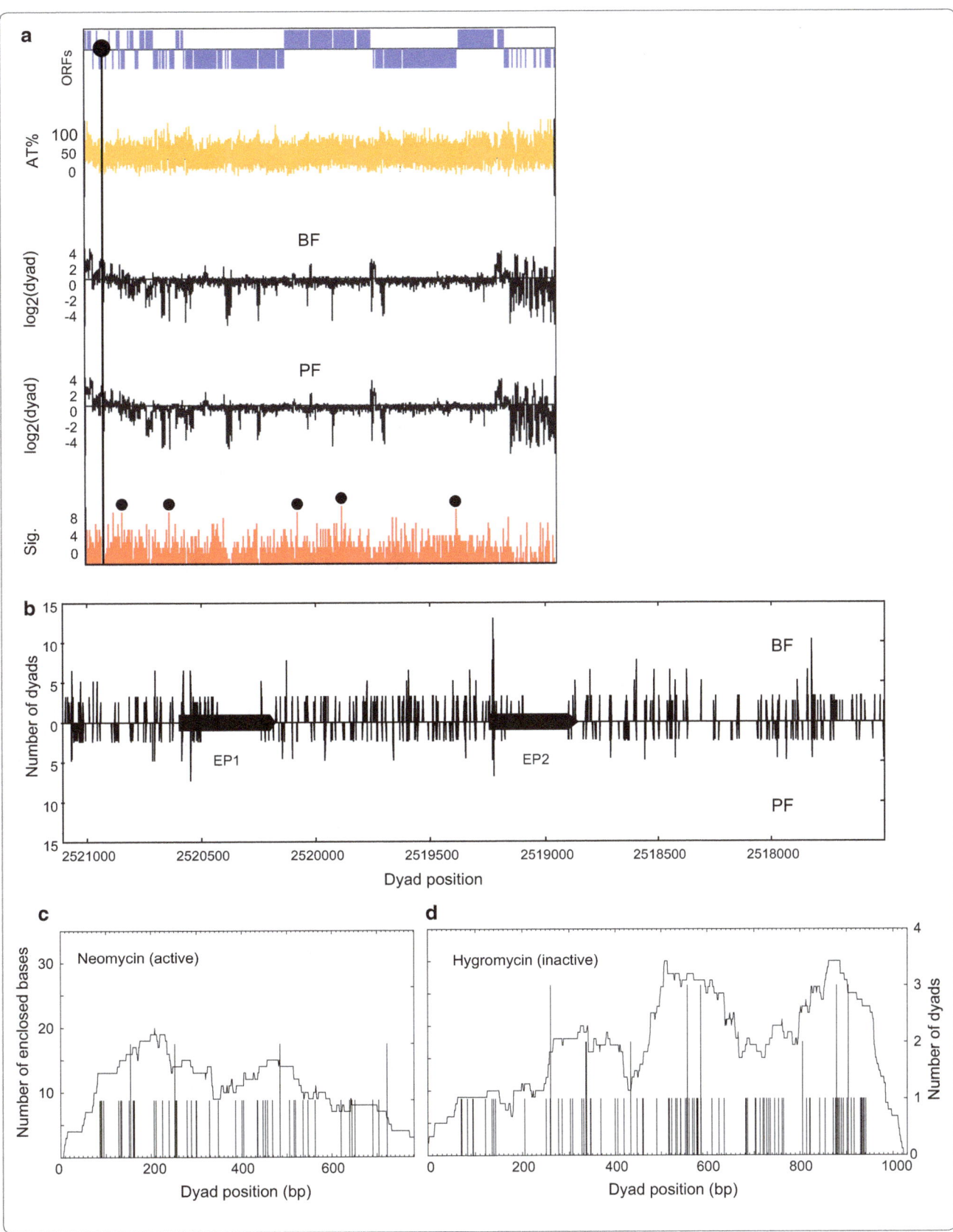

(See figure on previous page.)

**Fig. 4** Genome-wide nucleosomal organization in BF and PF *T. brucei*. **a** Chromosome 6 is selected as example, and statistically significant differences between BF and PF stages are shown. The *top line* indicates the position of assigned genes on the Watson strand (*above horizontal axis*) and the Crick strand (*below horizontal axis*). The location of the centromere is indicated by the *filled circle*. The *yellow line* shows the A/T % at each setting of a 100-bp scanning window. The *black traces below* show the traces of the $\log_2$ ratio of the average number of assigned dyad axes in a 1000-bp scanning window to the genome average in either BF or PF. The correlation between the relative nucleosome density in BF and PF chromatin was calculated by the Whitney–Mann *U* test at 50-bp intervals in a 500-bp sliding window. The number of statistically significant ($p < 0.01$) samples in four biological replicates was added and is shown as a "significance" value in the *red traces*. The *small black circles* identify peaks of high significance where the nucleosomal organization of the corresponding regions was individually assessed. **b** The distribution of dyads at the *EP1–EP2* procyclin locus (*black bars*) in a 3500-bp region. In each case, the *red* and *blue impulse* plots represent two biological replicates of BF (*top panel*) and PF (*bottom panel*) chromatin. **c, d** Alignment of nucleosomal dyads in chromatin from BF HNI_VO2 cell line. Coding sequences of the neomycin and hygromycin single copy resistance marker genes, present immediately downstream of the pol I ES promoters, were analyzed. Lower levels of well-positioned nucleosomes were present on the active neomycin gene (**c**), with regions of dyad enrichment observed for the transcriptionally inactive hygromycin gene (**d**). Y-axes scales are mirrored, with the *left-hand y-axes* indicating number of enclosed bases (*black trace*), and the *right-hand axes* indicating number of dyads (*blue*)

most significant peaks for each chromosome were chosen, and the corresponding region in the genome was further investigated. The regions with different nucleosome occupancy did not map to any predominant functional feature. Identified differences were located within and between PTUs, on coding and intergenic regions, and were not correlated with classes of genes that were relevant to a specific life cycle stage. This suggests that the identified differences in nucleosomal occupancy were probably not meaningful.

We next investigated the chromatin structure of transcription units which are expressed in a life-cycle-specific fashion. The *EP1* and *EP2* procyclin genes are not expressed in BF *T. brucei*. A larger number of nucleosomal dyads were present at these loci in BF compared with the PF *T. brucei* (Fig. 4b) and extended further downstream into the procyclin coding sequence in the BF cells. This indicated that the *EP1* and *EP2* procyclin genes have structurally more compact chromatin in BF cells, possibly impeding transcription of these procyclin loci in BF, compared with the PF cells. ESs are only transcribed at a high rate in BF *T. brucei*. Unfortunately, the high degree of sequence identity of the 15 ESs precluded unique read mapping and the analysis of nucleosome organization at active and inactive ESs. However, the BF HNI_VO2 cell line contains unique sequences in the form of neomycin and hygromycin resistance genes in the active and inactive ESs, respectively. Looking at the nucleosomal distributions on these single copy, unique sequence markers, fewer well-positioned nucleosomes were present on the active drug resistance gene compared with the inactive one (Fig. 4c, d). In fact, 0.07 (53 dyads per 782-bp gene) dyads were assigned per base pair in the case of the transcriptionally active neomycin gene, as opposed to 0.11 (109 dyads per 1031-bp gene) dyads per base pair in the case of the inactive hygromycin gene. Although this is in agreement with previous observations [71, 72], it is not clear to what extent sequence differences contributed to

this difference in nucleosome occupancy. In summary, subtle differences in the nucleosomal organization of life-cycle-specific genes were observed, which could be an effect of active transcription. However, we saw no difference in the organization of chromatin over extensive genomic regions between the two *T. brucei* life cycle stages (Fig. 4a; Additional file 1: Fig. S9).

### Nucleosomal organization at the silent *VSG* arrays

*Trypanosoma brucei* encodes a large number of transcriptionally silent *VSG* genes and pseudogenes present in large tandem arrays on chromosomes 5, 9 and 11. In addition, silent *VSGs* are present immediately at the telomeres of all classes of chromosomes [22]. We investigated whether the silent *VSG* arrays have a defined nucleosomal organization, with specific bordering structures. This could represent possible silencing elements, as seen at the silent mating-type loci in *S. cerevisiae* [73]. We therefore investigated the chromatin structure around these silent *VSG* arrays. The nucleosomal dyads were aligned in 10-kb regions relative to the first nucleotide on the most upstream gene of a block of tandem *VSG* genes ("start"), or the last nucleotide of the most downstream gene in a block of tandem *VSGs* ("end"), and are shown in Fig. 5. It is immediately clear that the silent *VSG* arrays on chromosome 9 are enclosed by regions of well-positioned nucleosomes, both upstream of the "start" (Fig. 5a) and downstream of the "end" (Fig. 5b). The silent *VSG* arrays themselves are clearly packaged into nucleosomes in both BF and PF *T. brucei*, although the positions are less defined compared with the flanking regions. A representative organization of one of these *VSG* arrays in BF cells is shown in Fig. 5. Similar clusters of well-positioned nucleosomes were seen bordering silent *VSG* arrays on the left-hand side of chromosome 9 as well as on chromosomes 5 and 11.

The observed well-positioned nucleosome structures flanking the silent *VSG* arrays could result in repressing

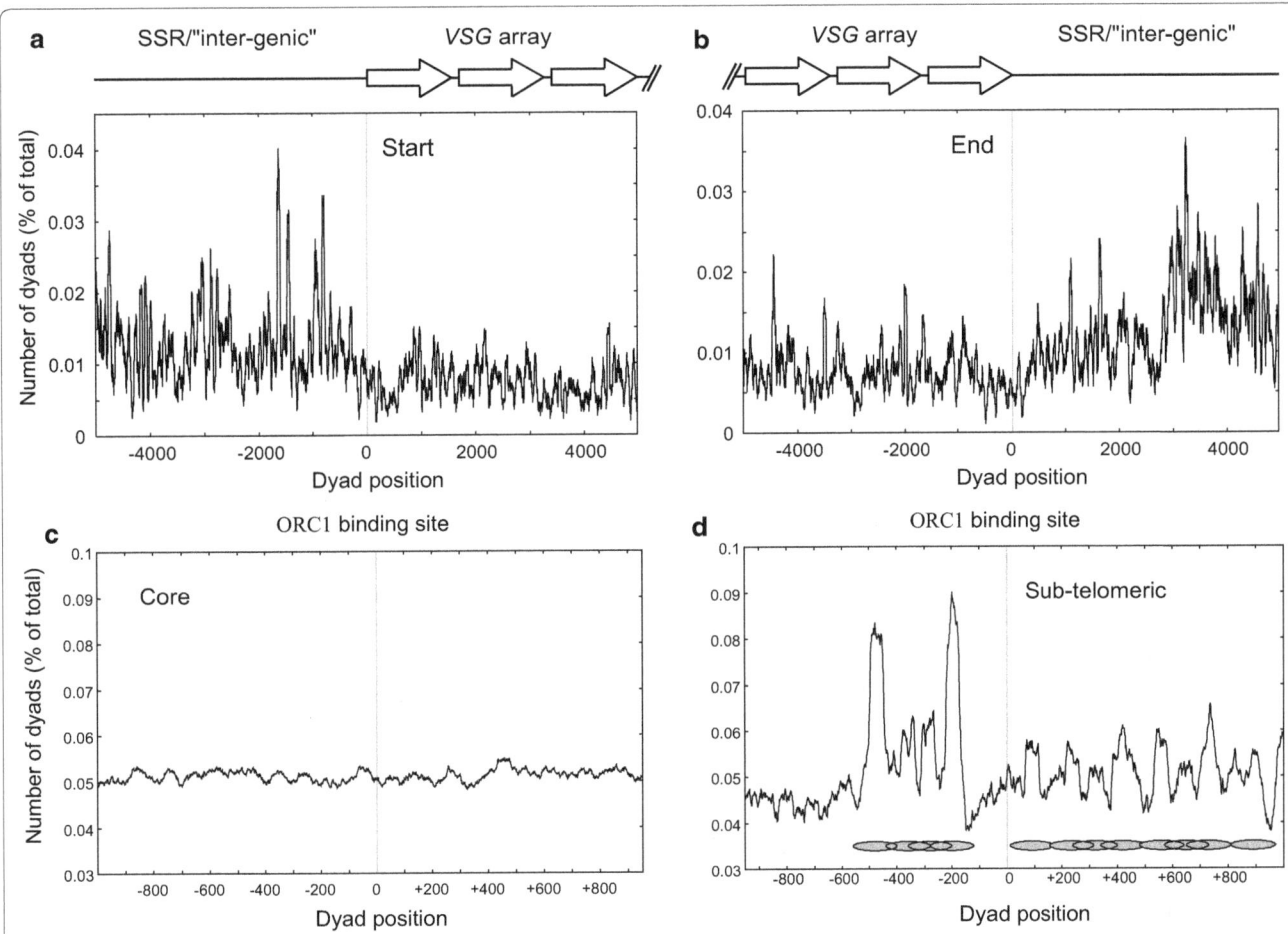

**Fig. 5** Nucleosomal organization around the silent *VSG* gene arrays and DNA origins of replication. Nucleosome dyads were analyzed in a 10-kb region at the beginning or end of arrays of silent *VSG* genes located at the right-hand side of chromosome 9. **a** These dyads were analyzed within a 10-kb region at the beginning of VSG arrays (18,449 dyads), or **b** at the ends of these arrays of silent *VSG* genes (21,824 dyads). These beginning and end points are indicated by the *vertical gray lines* in **a**, **b**. The alignments of the regions between adjacent blocks of co-aligned *VSG* genes (SSR/"intergenic") and the *VSG* genes and pseudogenes (*VSG* array) are schematically shown above *each panel*. **c** The distribution of nucleosomal dyads located in the chromosomal core regions of the *T. brucei* genome (434,752 dyads). The *gray line* represents the position of the center of the ORC1 binding site. The number of nucleosomal dyads present at each nucleotide position is expressed as a percentage of the total number of dyads in the analysis. **d** The distribution of nucleosomal dyads in the subtelomeric region (72,616 dyads) was aligned relative to the center of mapped ORC1 binding sites [74]. The alignment of nucleosomes with major nucleosomal dyad peaks is shown as *gray ovals*, with the telomere end located at the *right of the panel*

fortuitous transcription initiation of these silent *VSGs*, thereby maintaining mono-allelic expression of the active *VSG*. The more poorly positioned nucleosomes covering the *VSG* arrays may represent an open chromatin structure more amenable to DNA recombination events, or may be due to a repressive chromatin structure resulting in decreased MNase cleavage.

### TbORC1 and DNA replication origins

It had previously been shown that ORC1, a component of the origin recognition complex, binds to numerous regions in the *T. brucei* genome, many of which act as origins of DNA replication [74]. It has also previously been shown in *S. cerevisiae* that ORC1, apart from its role in DNA

replication, is also a component of silencing complexes assembled at silencing elements such as the E-element of the Mat alpha silent mating-type locus. Consequently, we were interested in establishing whether the *T. brucei* ORC1 was similarly involved in a specialized nucleosomal organization which could be implicated in gene silencing.

We mapped ORC1 sites identified in *T. brucei* 927 [74] to the equivalent sequences in the genome of *T. brucei* 427. The ORC1 binding sequences originally identified by Tiengwe and colleagues [74] ranged from 65 bp to 3 kb. This mapping was therefore at a low resolution, below that of single nucleosomes. Nevertheless, we utilized this dataset to investigate the nucleosomal organization in the vicinity of assigned ORC1 binding sites.

We first investigated the DNA region surrounding ORC1 sites in the core region of chromosomes (994 sites), which contain the constitutively expressed housekeeping genes. Here, very little nucleosomal organization relative to the center of the assigned DNA replication origin was discernible (Fig. 5c). We next investigated the subtelomeric regions (119 sites), which are defined as regions adjacent to telomeric ends. These contain the silent *VSG* gene and pseudogene arrays, expression site-associated genes (*ESAGs*), and the highly repetitive retrotransposon hotspot proteins (*RHS*) gene family. Here, a very clear nucleosomal organization was evident (Fig. 5d). The differences in organization of nucleosomes within the coding and flanking regions of the silent *VSG* arrays are consistent with the presence of specialized bordering silencing complexes, as also suggested by McCulloch and colleagues [72]. Intriguingly, of the 15 high confidence (false discovery rate (FDR) <0.05; [74]) ORC1 binding sites identified in the subtelomeric region of chromosome 9, 14 were present in the silent *VSG* array. This result strongly suggests that ORC1 binding sites, and presumably ORC1 itself are involved in the demarcation of specialized chromatin domains associated with silencing of the *VSG* gene and pseudogene arrays, as is the case in *S. cerevisiae* silent mating-type loci [73]. This suggests that the role of ORC1 in arranging surrounding chromatin structure and repressing regional transcription is evolutionarily ancient, and was likely present in LECA.

## Discussion

*Trypanosoma brucei* regulates the expression of most of its protein-coding genes at the posttranscriptional level. However, the *T. brucei* genome encodes homologs for putative chromatin writers, readers and erasers, as well as chromatin remodeling enzymes and histone variants [3–6]. It has been shown that specific histone variants demarcate the borders of PTUs [42]. In addition, various epigenetic players were shown to be important in the regulation of the mono-allelic expression of the active *VSG* ES, and the concomitant repression of the approximately 14 silent ESs [28, 33, 36, 37, 42, 44, 71, 74–76].

In this study, we present a whole genome analysis of nucleosomal positioning in *T. brucei*. We provide clear evidence for locally organized nucleosomal structures in both BF and PF *T. brucei*. A general feature of the transcription initiation region of pol II-transcribed genes in model organisms from the Unikonta supergroup includes an NDR region overlapping the TSS. Although a weak NDR was observed upstream of the first gene of pol II-transcribed PTUs in *T. brucei*, the pol II TSS remains unmapped. It is therefore unclear whether this NDR is functionally related to the transcription process or is due to a polypyrimidine tract that is required by the splicing mechanism, but is refractory to nucleosomes.

A detailed analysis of the DNA sequence in these regions showed that the NDR overlapped with a region of low distribution of A–A dinucleotides at a 10-bp periodicity, as well as with a region enriched for oligo-dA (see Fig. 6a). The distribution of A–A dinucleotides at a

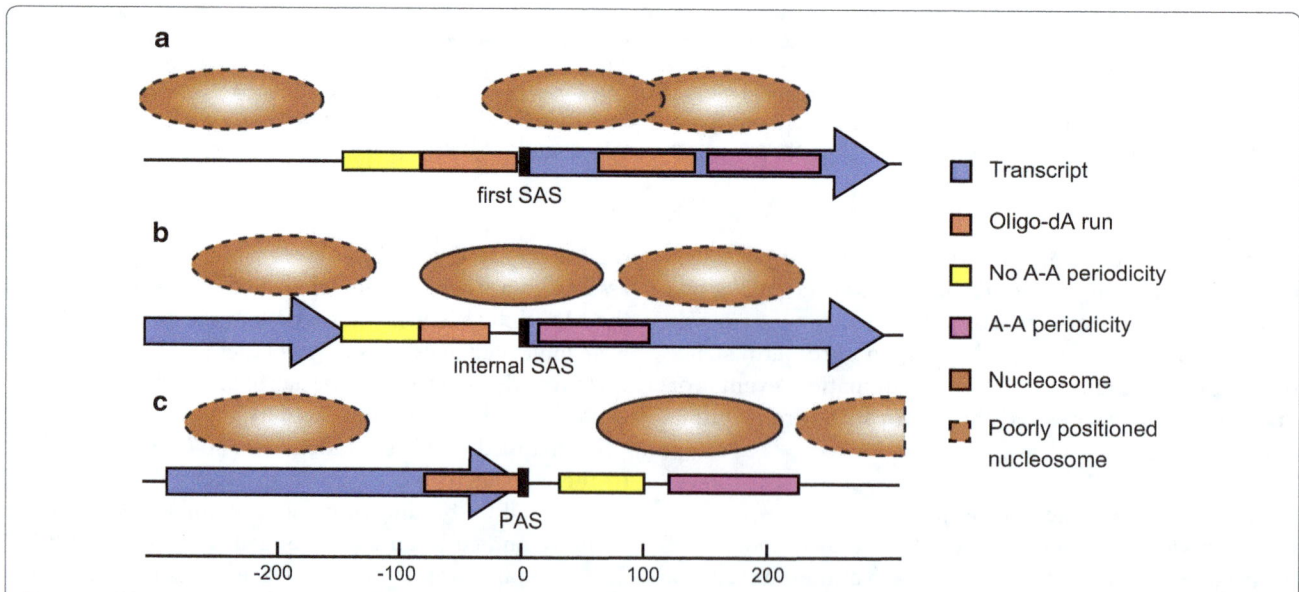

**Fig. 6** Model of nucleosomal organization at pol II-transcribed PTUs in *T. brucei*. Schematic representation of the average nucleosomal architecture and underlying DNA sequence positioning signals of *T. brucei* in a 400-bp window centered on (*a*) the first SAS of a PTU, representing the nucleosomal organization around the putative pol II TSS, (*b*) the internal SASs, and (*c*) the polyadenylation site

10-bp periodicity is generally associated with anisotropically flexible DNA, amenable to tight spooling onto the histone octamers [63]. Oligo-dA, however, is flexurally rigid due to the presence of bifurcated hydrogen bonds and is thus not compatible with the bent path of nucleosomal DNA [64]. Oligo-dG is also averse to bending, due to the stacking of consecutive guanine bases [77]. It is unclear to what extent basal transcription factors and activators contribute to the observed NDR. Although the *T. brucei* genome does appear to encode transcription activators, none have yet been mapped to pol II PTU initiation regions. *T. brucei* does contain a TBP-related factor, TRF4. However, the protein motifs shown to interact with the DNA in *S. cerevisiae* TBP are absent in TRF4 [78]. TATA box elements have also not been identified in the *T. brucei* genome. Within the PTUs themselves, the internal SASs showed a single well-positioned nucleosome covering the SAS element. This nucleosome was similar to that seen at the 3′ end of a metazoan intron at the intron–exon boundary [54, 79], a splicing feature mechanistically related to a trypanosome SAS. This SAS nucleosome appears to incorporate the 5′ AG acceptor, where the nucleosome position may be directed by the oligo-dA run, present at a larger average distance upstream of the internal SAS compared to the first SAS (Fig. 6b).

In the case of the PASs, the oligo-dA abutting the end of the PAS would in theory be refractory for nucleosomes, whereas the presence of a weak A–A distribution at a 10-bp periodicity would accommodate the well-positioned nucleosome in its observed location (Fig. 6c). The nucleosomal organization seen at metazoan intron–exon boundaries and at polyadenylation sites, compared to that seen at *T. brucei* SASs and PASs, is thus highly similar. This could be a consequence of the evolutionarily conserved sequence elements that direct local nucleosome placement. Alternatively, this could be a nucleosome arrangement that is required for the respective genetic mechanisms involving these elements. The nucleosome organization at the SASs and PASs therefore appears evolutionarily ancient and was probably already established before the evolutionary divergence of the Excavata from the other eukaryotic supergroups [1, 80]. The aforementioned is supported by recent findings where comparable nucleosomal patterns were observed in *L. major*, a related kinetoplastid with genomic synteny to *T. brucei* [57]. The apparent organization of nucleosomes relative to these SAS and PAS sequence elements is intriguing, as these sequence elements, and notably the oligo-dA runs, are functionally relevant for RNA processing rather than transcription. Although the nucleosomal organization in the region of the first SAS is related to that seen at TSSs in other eukaryotes, this could be a consequence of the role of these sequences in RNA processing. It is possible that these sequences have dual functions in contributing both to the organization of the TSS region, as well as to spliceosome binding on the nascent RNA.

The functional relevance of the nucleosomal organization seen at internal SASs is less clear. One possibility is that the concerted placement of a nucleosome on internal SASs would slow the RNA polymerase at this position, ensuring time for splicing at the SAS and subsequent splicing and polyadenylation at the upstream PAS [81]. The juxtaposition of a poly-d(A·T) tract upstream of the nucleosome-associated internal SASs makes this an intriguing possibility, as the polypyrimidine tract is known to affect both *trans*-splicing and polyadenylation of adjacent genes [82]. It is also theoretically possible that factors involved in RNA processing are preloaded onto the DNA and then transferred to the growing, nascent RNA by the elongating pol II. In fact, a physical interaction between the SF3a60 spliceosome factor and the largest subunit of pol II was shown using a two-hybrid approach [83], suggesting a possible link between the elongating pol II and RNA splicing in *T. brucei*. In addition, TbRRM1, an RNA-binding SR nucleoprotein, has been shown to directly interact with histones and modulate chromatin structure, maintaining permissive chromatin to facilitate transcription and RNA processing, and may also be involved in splicing commitment for a subset of transcripts [84].

Alternatively, the nucleosomal presence at internal SASs (or the 3′ acceptor region of a metazoan intron) could limit DNA recombination in this area [85, 86]. This would protect exon units over evolutionary time, or limit mutation of the SAS, and thus conserve a mechanistically functional SAS. In support of this, the rate of C → T hydrolytic deamination was reported to be reduced by twofold in nucleosomally wrapped DNA in *S. cerevisiae*, *C. elegans*, and *Oryzias* [87]. Given that a single point mutation in the 3′ AG acceptor sequence can functionally destroy the entire downstream gene, it is likely that protection of such sites would be evolutionarily advantageous. The PAS sites themselves showed an NDR and a well-positioned nucleosome downstream, similar to the organization seen in other organisms [14]. This arrangement may also reflect a requirement for transient pausing of pol II transcription elongation.

Chromatin can play an important role in silencing areas of eukaryotic genomes. The regions flanking the silent *VSG* arrays showed a unique nucleosomal organization, with an extensive array of well-positioned nucleosomes. The nucleosomes on the silent *VSG* coding regions themselves appeared less well-positioned. We propose that the bordering chromatin structure of well-positioned nucleosomes serves to limit cryptic pol II initiation. For antigenic variation to work, it is imperative for *T. brucei* to maintain the silent *VSG* arrays in a transcriptionally repressed state. Promiscuous expression of these silent *VSGs* would allow the host immune system to develop an immune response to a wide variety of VSGs and therefore facilitate immune clearance of the parasite. This repressive nucleosomal arrangement found at the silent *VSG* arrays is reminiscent to that found at the silent mating-type loci *HML* and *HMR* in *S. cerevisiae*. In contrast, the less defined nucleosomal structure on the silent *VSG* genes themselves could be more permissive for gene conversion events copying the silent *VSG* to the active ES. In this way, the chromatin structure at these silent *VSG* arrays could facilitate antigenic variation in African trypanosomes.

In *T. brucei*, ORC1 and RAP1 were shown to be required for full repression of the silent ESs, and ORC1 was shown to bind at positions bordering many silent *VSG* cassettes. The similarity between silencing at the *S. cerevisiae* HM loci and telomeres, and silencing at the *T. brucei* ESs and the *VSG* gene arrays is striking. However, no Sir-related protein other than SIR2RP1 has been identified in *T. brucei*, and although SIR2RP1 is involved in silencing the immediate regions of the telomeres [29], it does not appear to play a direct role in ES silencing. In *S. cerevisiae*, *ORC1* is thought to be the ancestral gene of *SIR3*, which arose after a genome duplication event [88]. However, unlike yeast ORC1, the *T. brucei* ORC1 does not contain a BAH domain. It is therefore unclear whether it can substitute for SIR3 in *T. brucei* and therefore contribute to the propagation of a repressive heterochromatic structure at the telomeres and silent *VSG* arrays. Of all the mapped *T. brucei* ORC1 binding sites, 38% were found in the chromosomal cores, localizing to transcription boundaries, with the remaining 62% localizing to subtelomeric and silent *VSG* arrays. All active DNA origins of replication (ORI) were found in the chromosomal cores with no evidence of ORIs originating from subtelomeric sites or silent *VSG* arrays [74]. It is possible that these non-replicative ORC1 binding sites have a repressive function at transcriptionally silent *VSG*-containing subtelomeric regions in *T. brucei*. A gradient of repression extends up to 10 kb from the telomeric ends in *T. brucei* [30], implying the propagation of a repressive chromatin structure. However, the proteins that participate in the establishment of this repressive structure remain unknown.

## Conclusions

In summary, our genome-wide nucleosomal analysis revealed striking correlations, as well as stark differences, in the nucleosomal architecture of *T. brucei* compared to other model eukaryotes studied to date. These similarities

suggest that some chromatin features, like the weak NDR upstream of PTUs, the strong positioning of nucleosomes on the internal SASs, the depletion of nucleosomes from PASs, and the nucleosomal organization at silent *VSG* arrays and ORC1 binding sites, were already established in LECA, before the divergence of the eukaryotic super groups. These findings, in conjunction with the co-localization of histone variants, histone and DNA modifications, and chromatin modulators, indicate the presence and importance of a functional epigenome in *T. brucei*, possibly providing a regulatory interface to genome regulation.

## Methods

### Trypanosome strains and culturing

Bloodstream form *T. brucei* Lister 427 was cultured in HMI-9 medium as previously described [89] supplemented with 15% fetal calf serum (FCS) and appropriate drugs at 37 °C under 5% $CO_2$. Procyclic form trypanosomes were cultured in SDM-79 medium supplemented with 10% (v/v) FCS, 5 µg/ml hemin, and appropriate drugs at 27 °C [90]. Two cell lines were chosen for both the BF and the PF life cycle stages to account for possible cell line-specific differences. These are the BF (HNI_VO2 and RYT3) and PF (Amsterdam wild-type and 221BsrDsRed) cell lines. HNI_VO2 cells have a hygromycin resistance gene in the silent VSG221 ES and a neomycin resistance gene in the active VO2 ES, providing single copy sequences which allow the differentiation of the silent and active VSG ESs [91]. The RYT3 cell line has a blasticidin resistance gene in the active VSGT3 ES and an eGFP gene and a puromycin resistance gene in the silent VSG221 ES [32]. The PF 221BsrDsRed cell line has a blasticidin resistance gene and a DsRed gene in the silent 221 ES [32].

### Core particle preparation

MNase digestion of chromatin was performed as described [71]. Briefly, $5 \times 10^7$ cells/sample were harvested by centrifugation (1200g, 5 min) and permeabilized with 400 µM digitonin for 5 min at room temperature (Sigma-Aldrich). Chromatin was digested with 1–32 units MNase (Worthington Biochemicals) for 5 min at 37 °C, followed by phenol–chloroform extraction and ethanol precipitation. Recovered DNA was resolved on a 2% (w/v) agarose gel. DNA fragments of ~147 bp were extracted from the gel using the freeze-and-squeeze technique (Bio-Rad laboratories, Hercules) and 50 nt from each end was sequenced by paired-end methodology (Illumina) as previously described [92].

### Alignment to reference genome

The FASTQ sequence files were aligned to version 4.2 of the *T. brucei* 427 genome using *Bowtie 2* [93] allowing no mismatches, limiting alignments to 7 ambiguous (N) bases in the reference genome, and filtering for convergent primer pairs separated by between 145 and 155 bp, end-to-end. The detail of the alignment output is shown in Additional file 1: Table S2.

### Dyad value files

The file of dyad values, listing the number of dyads assigned to each nucleotide position of each chromosome, was normalized to chromosome length to allow inter-chromosome comparisons, as well as normalized between different experiments, to allow analysis between different experimental conditions. We assumed that the median nucleosome density was the same between experimental conditions and chromosomes. The genomic positions of tandem repeats, previously identified with the *TRF* program [94], were downloaded from TriTrypDB, and the dyad values of all nucleotides that fell within tandem repeat sequences were set to "−1" with the program *strip_tandem_repeats*, and ignored in the calculation of average dyad densities. To mask the contribution of unmatched, N-rich sequences to the generation of spurious NDRs, all possible dyad positions of fragments between 145 and 155 bp, where either of the aligned 50-bp paired-end fragments exceeded the maximum cutoff for base ambiguity, were set to "−2" in the dyad value files, and ignored in subsequent calculations.

### Bin analysis

A binning analysis was performed with the program *dyad_bins* using the normalized dyad values. The lowest and highest numbers of dyad axes co-localized on single nucleotides, representing the minimum and maximum bin values, were identified. Dyad values associated with AT-rich repeats were masked as indicated. Bins intermediate to the minimum and maximum bins were defined at an increment of 1, and the number of occurrences of dyad values equal to each bin value identified in each file of dyad values. The number of members in each bin was normalized to the genome size to allow direct comparison between regions of different sizes. The number of samples in each bin was plotted with *Gnuplot* version 4.6.3.

### Normalized dyad densities

The number of dyads in a 10-bp scanning window was normalized with the program *dyads_genome_wide* to the number of nucleotides (excluding the "−1" and "−2" values), and expressed as the $\log_2$ of the ratio to the average number of dyads in the genome, as shown by the following equation:

$$\log_2 \frac{\frac{1}{N}\sum_{i=1}^{N} d_i}{\frac{1}{M}\sum_{j=1}^{M} d_j}$$

where $N$ is the width of the scanning window, $M$ is the genome size, and $d_i$ and $d_j$ represent the number of assigned dyads at positions $i$ and $j$, respectively.

### Differences in the nucleosomal organization in BF and PF *T. brucei*

The statistical significance of differences between the nucleosomal occupancy of each chromosome in the BF and PF life stage of *T. brucei* was determined with the Whitney–Mann $U$ test. The autocorrelation function:

$$R_k = \frac{\sum_{i=1}^{N-k}\left(Y_i - \bar{Y}\right)\left(Y_{i+k} - \bar{Y}\right)}{\sum_{i=1}^{N}\left(Y_i - \bar{Y}\right)^2}$$

where $N$ is the number of data points, $k$ the offset, $Y$ the average value for the dataset, and $Y_i$ and $Y_{i+k}$ the values at positions $i$ and $i + k$, and it was applied to the $\log_2$ dyad values and smoothed with a 10-bp running average. The autocorrelation plot showed a significant flattening of the autocorrelation coefficient at values beyond $k = 10$ (Additional file 1: Fig. S10), as expected. We therefore chose values 50 nt apart to ensure independence of data points, as is required by the statistical test. The nonparametric Whitney–Mann $U$ test was performed in a scanning 500-bp window at consecutive settings on each chromosome, using data points at 50-bp intervals. Statistical significance was determined by using a Whitney–Mann $U$ value $p < 0.01$, and the presence of a 500-bp setting that was statistically significant between the BF and PF stages, recorded at the start of each window setting. To increase the rigor of the significance test, we further summed the number of statistically significant settings in a 10-bp window for the two biological replicates as well as the two different *T. brucei* cell lines, to correct for possible cell line related differences. Only cumulative statistical significance peaks in the top 10th percentile of the range were tagged for further investigation.

### Alignment of dyads

The alignment of dyads relative to specific genomic positions was performed with the program *align_dyads*, which aligned the dyads in an orientation depending on whether the feature was on the Watson or the Crick strand, and reported the average value in a 50-bp running window.

### Statistical test

The Kendall's tau correlation and Pearson product moment correlation test were performed with R scripts.

### Mapping of *T. brucei* 927 genes to *T. brucei* 427

The list of translated coding sequences of *T. brucei* strain 927 (Tb927) was downloaded from TriTypDB and BLASTed against a database of *T. brucei* 427 protein sequences (Tb427). This had been generated with version 7 of the fastA format translated CDS files downloaded from tritrypdb.org, using the *makeblastdb*. The homologous sequences were listed, and entries with $E$ values $<10^{-10}$ selected. Where Tb927 genes were mapped to multiple Tb427 genes, only the match with the smallest $E$ value was selected. Where a Tb927 gene had more than one splice acceptor site, the site furthest upstream from the coding sequence was chosen as it was likely to be closest to the site of transcription initiation. The position of the splice acceptor sites in the genome sequence of the Tb427 was mapped using the program *927_to_427_map*.

## Mapping of ORC1 binding sites from Tb927 to Tb427

The ORC1 binding sites assigned with a signal ratio >1.0 and a FDR <0.05 were selected from the list of ORC1 sites (kindly provided by R. McCulloch) [74], and the corresponding sequences in Tb927 retrieved from TriTrypDB. The retrieved sequences were BLASTed against version 4.2 of the *T. brucei* Lister 427 genome, and the single, best hit for each query sequence recovered. Only target sequences that were >95% identical and with $E$ values $<10^{-10}$ were retained (1114 sites, Additional file 2: Table S3). The ORC1 site was defined as the center of each such recovered sequence.

## Definition of chromosome regions

Subtelomeric regions were identified as the terminal regions of chromosomes enriched for retrotransposon hotspot proteins (*RHS*), expression site-associated genes (*ESAGs*), *VSG* genes and pseudogenes, and leucine-rich repeat protein (*LRRP*) genes. Subtelomeric regions were further divided into telomere–proximal ($\leq$10% of the chromosomal length from the telomere) and chromosome internal/core (>10% of the chromosomal length away from telomere) subtelomeric regions.

## Alignment to resistance markers

The hygromycin resistance gene (Hygromycin-B 4-O-kinase; accession number V01499) as well as the neomycin resistance marker (aminoglycoside 3'-phosphotransferase; accession number P00551) was indexed using Bowtie 2 with default settings, and the sequenced pairs for the BF HNI_VO2 aligned to the indexed sequences, allowing accordant alignments of sequence pairs resulting in fragment lengths of between 145 and 155 bp.

## Probability of sequence repeats

Nucleotide repeat sequences were identified in the *T. brucei* genome with *polyA_genome_distribution*. The occurrence of runs between specific genomic positions,

such as within transcription start regions, was identified with the program *in_or_out*. The probability of specific sequence repeats occurring in the genome was calculated as a hypergeometric distribution. Where the sample size $n \ll N$, the population size, the hypergeometric distribution can be approximated as:

$$\frac{1}{f^n}$$

where $n$ is the oligonucleotide sequence length and $f$ is the fractional occurrence of the given nucleotide. The fractional occurrence of A·T and G·C in the genome is 0.2651 and 0.2349, respectively. The probability of occurrence of each oligonucleotide decamer with a given dinucleotide composition was calculated and is shown in Additional file 1: Table S1.

## Probability of finding an A10, G10, and AT10 repeat

Specific sequences were recovered from the genome of *T. brucei* using the program *get_sequences*. The enrichment of specific nucleotide oligomers was determined in aligned sequences using the program *polyA_enrich*. The number of sequence runs was normalized to the number of sequences analyzed, smoothed with a 50-nt running average, and expressed as a percentage of the smoothing window size. The probability of finding 10-bp homo- or hetero-oligomer motifs in the coding and noncoding regions of *T. brucei* 427 was calculated with a hypergeometric distribution function using the coding frequencies shown in Additional file 1: Table S4.

The transcription start regions identified in Tb927 were mapped to the equivalent genomic positions in Tb427 using the correlation list derived from the BLAST analysis explained above. The presence of upstream *t-*, *r-* or *snRNA* genes was verified for internal start regions. The list of mapped transcription start regions is shown in Additional file 3: Table S5.

## Discrete Fourier analysis of dinucleotide frequencies

The presence of strong 10-bp A–A dinucleotide periodicities was assessed from the Fourier magnitude of the distribution in specific regions using the program *hp_fftw* that utilizes the *FFTW* library (www.fftw.org).

## Software

All software was written in C++ (ISO/IEC 14882:2011) and compiled with the g++ version 4.8.2 64-bit compiler (gcc.gnu.org) using the mingw-w64 version 4.8.2 (mingw-w64.sourceforge.net) toolchain on a Windows version 8.1 operating system platform. The program *hp_fftw* was compiled with g++ version 4.8 on Linux openSUSE version 12.3 (www.opensuse.org). All plots were prepared with scripts using gnuplot version 4.6 (www.gnuplot.info). The source code for software developed and used in this study is freely available (sourceforge.net/projects/nucpos/).

## Additional files

**Additional file 1.** Supplementary nucleosome dyad profiles, oligonucleotide runs, (di)nucleotide occurrence and frequencies, megabase chromosome panels, and auto-correlation of nucleosome positions

**Additional file 2: Table S3.** Genomic positions of equivalent sequences mapped as ORC1 sites in *T. brucei* strain 927 in *T. brucei* strain 427.

**Additional file 3: Table S5.** Mapped transcription start regions from Tb927 to Tb427.

## Abbreviations

LECA: last eukaryotic common ancestor; pol II: RNA polymerase II; TSS: transcription start site; HAT: human African trypanosomiasis; BF: bloodstream form; PF: procyclic form; VSG: variant surface glycoprotein; ES: expression site; PTU: polycistronic transcription unit; SSR: strand switch region; Base J: β-D-glucosyl-hydroxymethyluracil; MNase: micrococcal nuclease; NDR: nucleosome-depleted region; SAS: splice acceptor site; PAS: polyadenylation site; SL RNA: spliced leader RNA; PPT: polypyrimidine tract; RHS: retrotransposon hotspot protein; ESAG: expression site-associated gene; LRRP: leucine-rich repeat protein; FDR: false discovery rate; ORI: origins of replication; FCS: fetal calf serum.

## Authors' contributions

HGP conceived and coordinated the study; JPM performed the experimental work; MLP performed experimental work; GR provided the strains and facilities for the experimental work; DJC performed the DNA repair and paired-end sequencing; HGP wrote the software; HGP and JPM performed the analysis and drafted the manuscript. All authors read and approved the final manuscript.

## Author details

[1] Department of Biochemistry, Stellenbosch University, Matieland 7602, South Africa. [2] Department of Biology, Pennsylvania State University (Brandywine Campus), Media, PA 19063, USA. [3] Division of Developmental Biology, Eunice Kennedy Shriver National Institute for Child Health and Human Development, National Institutes of Health, Bethesda, MD, USA. [4] Department of Life Sciences, Imperial College London, South Kensington, London SW7 2AZ, UK.

## Acknowledgements

The authors thank Robert Schall for his advice on statistical analyses, Razvan Chereji for his assistance with the analysis of the nucleosome organization at TSSs, the NHLBI Sequencing Core Facility (Yan Luo, Poching Lu, Yoshi Wakabayashi and Jun Zhu), Nicolai Siegel for making his H2A.V ChIP-seq data available to us before publication, and Richard McCulloch for providing ORC1 data.
 Johannes Petrus Maree and Hugh-George Patterton are members of the H3Africa Consortium.

## Competing interests

The authors declare that they have no competing interests.

## Funding

This work was supported by the H3Africa program of the National Institutes of Health [Grant 1U01HG007465, to HGP] and in part by the Intramural Research Program of the National Institutes of Health [to DJC]. G.R. is a Wellcome Senior Research Fellow funded by the Wellcome Trust. The funding bodies did not contribute to the design of the study, collection, analysis, and interpretation of data, or to writing the manuscript.

## References

1. He D, Fiz-Palacios O, Fu C-J, Fehling J, Tsai C-C, Baldauf SL. An alternative root for the eukaryote tree of life. Curr Biol. 2014;24:465–70.
2. Cavalier-Smith T. Kingdoms Protozoa and Chromista and the eozoan root of the eukaryotic tree. Biol Lett. 2010;6:342–5.
3. Figueiredo LM, Cross GAM, Janzen CJ. Epigenetic regulation in African trypanosomes: a new kid on the block. Nat Rev Microbiol. 2009;7:504–13.
4. Maree JP, Patterton HG. The epigenome of *Trypanosoma brucei*: a regulatory interface to an unconventional transcriptional machine. Biochim Biophys Acta. 2014;1839:743–50.
5. Duraisingh MT, Horn D. Epigenetic regulation of virulence gene expression in parasitic protozoa. Cell Host Microbe. 2016;19:629–40.
6. Günzl A, Kirkham JK, Nguyen TN, Badjatia N, Park SH. Mono-allelic VSG expression by RNA polymerase I in *Trypanosoma brucei*: expression site control from both ends? Gene. 2015;556:68–73.
7. Reeve JN, Sandman K, Daniels CJ. Archaeal histones, nucleosomes, and transcription initiation. Cell. 1997;89:999–1002.
8. Kasinsky HE, Lewis JD, Dacks JB, Ausio J. Origin of H1 linker histones. FASEB J. 2001;15:34–42.
9. Sandman K, Reeve JN. Archaeal chromatin proteins: different structures but common function? Curr Opin Microbiol. 2005;8:656–61.
10. Peterson CL, Workman JL. Promoter targeting and chromatin remodeling by the SWI/SNF complex. Curr Opin Genet Dev. 2000;10:187–92.
11. Marsh VL, Peak-Chew SY, Bell SD. Sir2 and the acetyltransferase, Pat, regulate the archaeal chromatin protein, Alba. J Biol Chem. 2005;280:21122–8.
12. Simpson RT. Nucleosome positioning can affect the function of a *cis*-acting DNA element in vivo. Nature. 1990;343:387–9.
13. Hinz JM, Czaja W. Facilitation of base excision repair by chromatin remodeling. DNA Repair. 2015;36:91–7.
14. Jiang C, Pugh BF. Nucleosome positioning and gene regulation: advances through genomics. Nat Rev Genet. 2009;10:161–72.
15. Bai L, Morozov AV. Gene regulation by nucleosome positioning. Trends Genet. 2010;26:476–83.
16. Ammar R, Torti D, Tsui K, Gebbia M, Durbic T, Bader GD, et al. Chromatin is an ancient innovation conserved between Archaea and Eukarya. eLife. 2012;1:e00078.
17. Fèvre EM, Wissmann BV, Welburn SC, Lutumba P. The burden of human African trypanosomiasis. In: Brooker S, editor. PLoS Negl Trop Dis. 2008;2:e333.
18. Mogk S, Meiwes A, Boßelmann CM, Wolburg H, Duszenko M. The lane to the brain: how African trypanosomes invade the CNS. Trends Parasitol. 2014;30:470–7.
19. Simarro PP, Cecchi G, Franco JR, Paone M, Diarra A, Priotto G, et al. Monitoring the progress towards the elimination of gambiense human African trypanosomiasis. PLoS Negl Trop Dis. 2015;9:e0003785.
20. Fenn K, Matthews KR. The cell biology of *Trypanosoma brucei* differentiation. Curr Opin Microbiol. 2007;10:539–46.
21. Horn D. Antigenic variation in African trypanosomes. Mol Biochem Parasitol. 2014;195:123–9.
22. Taylor JE, Rudenko G. Switching trypanosome coats: what's in the wardrobe? Trends Genet. 2006;22:614–20.
23. Hertz-Fowler C, Figueiredo LM, Quail MA, Becker M, Jackson A, Bason N, et al. Telomeric expression sites are highly conserved in *Trypanosoma brucei*. PLoS ONE. 2008;3:e3527.
24. Vink C, Rudenko G, Seifert HS. Microbial antigenic variation mediated by homologous DNA recombination. FEMS Microbiol Rev. 2012;36:917–48.
25. Daniels J-P, Gull K, Wickstead B. Cell biology of the trypanosome genome. Microbiol Mol Biol Rev. 2010;74:552–69.
26. Perrod S, Gasser SM. Long-range silencing and position effects at telomeres and centromeres: parallels and differences. Cell Mol Life Sci. 2003;60:2303–18.
27. Renauld H, Aparicio OM, Zierath PD, Billington BL, Chhablani SK, Gottschling DE. Silent domains are assembled continuously from the telomere and are defined by promoter distance and strength, and by SIR3 dosage. Genes Dev. 1993;7:1133–45.

28. Yang X, Figueiredo LM, Espinal A, Okubo E, Li B. RAP1 is essential for silencing telomeric variant surface glycoprotein genes in *Trypanosoma brucei*. Cell. 2009;137:99–109.

29. Alsford S, Kawahara T, Isamah C, Horn D. A sirtuin in the African trypanosome is involved in both DNA repair and telomeric gene silencing but is not required for antigenic variation. Mol Microbiol. 2007;63:724–36.

30. Rudenko G. Epigenetics and transcriptional control in African trypanosomes. Essays Biochem. 2010;48:201–19.

31. Stanne TM, Kushwaha M, Wand M, Taylor JE, Rudenko G. TbISWI regulates multiple polymerase I (Pol I)-transcribed loci and is present at Pol II transcription boundaries in *Trypanosoma brucei*. Eukaryot Cell. 2011;10:964–76.

32. Hughes K, Wand M, Foulston L, Young R, Harley K, Terry S, et al. A novel ISWI is involved in VSG expression site downregulation in African trypanosomes. EMBO J. 2007;26:2400–10.

33. Benmerzouga I, Concepción-Acevedo J, Kim HS, Vandoros AV, Cross GAM, Klingbeil MM, et al. *Trypanosoma brucei* Orc1 is essential for nuclear DNA replication and affects both VSG silencing and VSG switching. Mol Microbiol. 2013;87:196–210.

34. Denninger V, Fullbrook A, Bessat M, Ersfeld K, Rudenko G. The FACT subunit TbSpt16 is involved in cell cycle specific control of VSG expression sites in *Trypanosoma brucei*. Mol Microbiol. 2010;78:459–74.

35. Wang QP, Kawahara T, Horn D. Histone deacetylases play distinct roles in telomeric VSG expression site silencing in African trypanosomes. Mol Microbiol. 2010;77:1237–45.

36. Povelones ML, Gluenz E, Dembek M, Gull K, Rudenko G. Histone H1 plays a role in heterochromatin formation and VSG expression site silencing in *Trypanosoma brucei*. In: Ullu E, editor. PLoS Pathog. 2012;8:e1003010.

37. Pena AC, Pimentel MR, Manso H, Vaz-Drago R, Pinto-Neves D, Aresta-Branco F, et al. T rypanosoma brucei histone H1 inhibits RNA polymerase I transcription and is important for parasite fitness in vivo. Mol Microbiol. 2014;93:645–63.

38. Günzl A, Bruderer T, Laufer G, Schimanski B, Tu L, Chung H, et al. RNA polymerase I transcribes procyclin genes and variant surface glycoprotein gene expression sites in *Trypanosoma brucei*. Eukaryot Cell. 2003;2:542–51.

39. Berriman M, Ghedin E, Hertz-Fowler C, Blandin G, Renauld H, Bartholomeu DC, et al. The genome of the African trypanosome *Trypanosoma brucei*. Science. 2005;309:416–22.

40. Clayton CE. Networks of gene expression regulation in *Trypanosoma brucei*. Mol Biochem Parasitol. 2014;195:96–106.

41. Kolev NG, Franklin JB, Carmi S, Shi H, Michaeli S, Tschudi C. The transcriptome of the human pathogen *Trypanosoma brucei* at single-nucleotide resolution. In: Beverley SM, editor. PLoS Pathog. 2010;6:e1001090.

42. Siegel TN, Hekstra DR, Kemp LE, Figueiredo LM, Lowell JE, Fenyo D, et al. Four histone variants mark the boundaries of polycistronic transcription units in *Trypanosoma brucei*. Genes Dev. 2009;23:1063–76.

43. Wright JR, Siegel TN, Cross GAM. Histone H3 trimethylated at lysine 4 is enriched at probable transcription start sites in *Trypanosoma brucei*. Mol Biochem Parasitol. 2010;172:141–4.

44. Mandava V, Fernandez JP, Deng H, Janzen CJ, Hake SB, Cross GAM. Histone modifications in *Trypanosoma brucei*. Mol Biochem Parasitol. 2007;156:41–50.

45. Schulz D, Zaringhalam M, Papavasiliou FN, Kim H-S. Base J and H3.V regulate transcriptional termination in *Trypanosoma brucei*. PLoS Genet. 2016;12:e1005762.

46. Reynolds D, Hofmeister BT, Cliffe L, Alabady M, Siegel TN, Schmitz RJ, et al. Histone H3 variant regulates RNA polymerase II transcription termination and dual strand transcription of siRNA loci in *Trypanosoma brucei*. In: Figueiredo L, editor. PLoS Genet. 2016;12:e1005758.

47. Van Luenen HGAM, Farris C, Jan S, Genest PA, Tripathi P, Velds A, et al. Glucosylated hydroxymethyluracil, DNA base J, prevents transcriptional readthrough in Leishmania. Cell. 2012;150:909–21.

48. Nagarajavel V, Iben JR, Howard BH, Maraia RJ, Clark DJ. Global "bootprinting" reveals the elastic architecture of the yeast TFIIIB–TFIIIC transcription complex in vivo. Nucleic Acids Res. 2013;41:8135–43.

49. Schimanski B, Nguyen TN, Gunzl A. Characterization of a multisubunit transcription factor complex essential for spliced-leader RNA gene transcription in *Trypanosoma brucei*. Mol Cell Biol. 2005;25:7303–13.

50. Birch JL, Zomerdijk JCBM. Structure and function of ribosomal RNA gene chromatin. Biochem Soc Trans. 2008;36:619–24.

51. Siegel TN, Hekstra DR, Wang X, Dewell S, Cross GAM. Genome-wide analysis of mRNA abundance in two life-cycle stages of *Trypanosoma brucei* and identification of splicing and polyadenylation sites. Nucleic Acids Res. 2010;38:4946–57.

52. Giaever G, Chu AM, Ni L, Connelly C, Riles L, Véronneau S, et al. Functional profiling of the *Saccharomyces cerevisiae* genome. Nature. 2002;418:387–91.

53. Palenchar JB, Bellofatto V. Gene transcription in trypanosomes. Mol Biochem Parasitol. 2006;146:135–41.

54. Kogan S, Trifonov EN. Gene splice sites correlate with nucleosome positions. Gene. 2005;352:57–62.

55. Brogaard K, Xi L, Wang J-P, Widom J. A map of nucleosome positions in yeast at base-pair resolution. Nature. 2012;486:496–501.

56. Mavrich TN, Jiang C, Ioshikhes IP, Li X, Venters BJ, Zanton SJ, et al. Nucleosome organization in the Drosophila genome. Nature. 2008;453:358–62.

57. Lombraña R, Álvarez A, Fernández-Justel JM, Almeida R, Poza-Carrión C, Gomes F, et al. Transcriptionally driven DNA replication program of the human Parasite *Leishmania major*. Cell Rep. 2016;16:1774–86.

58. Jin C, Felsenfeld G. Nucleosome stability mediated by histone variants H3.3 and H2A.Z. Genes Dev. 2007;21:1519–29.

59. Martínez-Calvillo S, Vizuet-De-Rueda JC, Florencio-Martínez LE, Manning-Cela RG, Figueroa-Angulo EE. Gene expression in trypanosomatid parasites. J Biomed Biotechnol. 2010;2010:525241.

60. Ling X, Harkness TAA, Schultz MC, Fisher-Adams G, Grunstein M. Yeast histone H3 and H4 amino termini are important for nucleosome assembly in vivo and in vitro: redundant and position-independent functions in assembly but not in gene regulation. Genes Dev. 1996;10:686–99.

61. Huang H, Liu H, Sun X. Nucleosome distribution near the 3′ ends of genes in the human genome. Biosci Biotechnol Biochem. 2013;77:2051–5.

62. Mavrich TN, Ioshikhes IP, Venters BJ, Jiang C, Tomsho LP, Qi J, et al. A barrier nucleosome model for statistical positioning of nucleosomes throughout the yeast genome. Genome Res. 2008;18:1073–83.

63. Satchwell SC, Drew HR, Travers AA. Sequence periodicities in chicken nucleosome core DNA. J Mol Biol. 1986;191:659–75.

64. Nelson HC, Finch JT, Luisi BF, Klug A. The structure of an oligo(dA). oligo(dT) tract and its biological implications. Nature. 1987;330:221–6.

65. Vazquez MP, Mualem D, Bercovich N, Stern MZ, Nyambega B, Barda O, et al. Functional characterization and protein–protein interactions of trypanosome splicing factors U2AF35, U2AF65 and SF1. Mol Biochem Parasitol. 2009;164:137–46.

66. Lee T-Y, Huang H-D, Hung J-H, Huang H-Y, Yang Y-S, Wang T-H. dbPTM: an information repository of protein post-translational modification. Nucleic Acids Res. 2006;34:D622–7.

67. Liang X, Haritan A, Uliel S. *Trans* and *cis* splicing in trypanosomatids: mechanism, factors, and regulation. Eukaryot Cell. 2003;2:830–40.

68. Struhl K, Segal E. Determinants of nucleosome positioning. Nat Struct Mol Biol. 2013;20:267–73.

69. Clayton CE. Life without transcriptional control? From fly to man and back again. EMBO J. 2002;21:1881–8.

70. Kramer S. Developmental regulation of gene expression in the absence of transcriptional control: the case of kinetoplastids. Mol Biochem Parasitol. 2012;181:61–72.

71. Stanne TM, Rudenko G. Active VSG expression sites in *Trypanosoma brucei* are depleted of nucleosomes. Eukaryot Cell. 2010;9:136–47.

72. Figueiredo LM, Cross GAM. Nucleosomes are depleted at the VSG expression site transcribed by RNA polymerase I in African trypanosomes. Eukaryot Cell. 2010;9:148–54.

73. Haber JE. Mating-type genes and MAT switching in *Saccharomyces cerevisiae*. Genetics. 2012;191:33–64.

74. Tiengwe C, Marcello L, Farr H, Dickens N, Kelly S, Swiderski M, et al. Genome-wide analysis reveals extensive functional interaction between DNA replication initiation and transcription in the genome of *Trypanosoma brucei*. Cell Rep. 2012;2:185–97.

75. Narayanan MS, Rudenko G. TDP1 is an HMG chromatin protein facilitating RNA polymerase I transcription in African trypanosomes. Nucleic Acids Res. 2013;41:2981–92.

76. Pandya UM, Sandhu R, Li B. Silencing subtelomeric VSGs by *Trypanosoma brucei* RAP1 at the insect stage involves chromatin structure changes. Nucleic Acids Res. 2013;41:7673–82.

77. McCall M, Brown T, Kennard O. The crystal structure of d(G-G-G-G-C-C-C-C). A model for poly(dG).poly(dC). J Mol Biol. 1985;183:385–96.

78. Ruan J-P, Arhin GK, Ullu E, Tschudi C. Functional characterization of a *Trypanosoma brucei* TATA-binding protein-related factor points to a universal regulator of transcription in trypanosomes. Mol Cell Biol. 2004;24:9610–8.

79. Schwartz S, Meshorer E, Ast G. Chromatin organization marks exon–intron structure. Nat Struct Mol Biol. 2009;16:990–5.

80. Adl SM, Simpson AGB, Lane CE, Lukeš J, Bass D, Bowser SS, et al. The revised classification of eukaryotes. J Eukaryot Microbiol. 2012;59:429–514.

81. Ullu E, Matthews KR, Tschudi C. Temporal order of RNA-processing reactions in trypanosomes: rapid *trans* splicing precedes polyadenylation of newly synthesized tubulin transcripts. Mol Cell Biol. 1993;13:720–5.

82. Matthews KR, Tschudi C, Ullu E. A common pyrimidine-rich motif governs *trans*-splicing and polyadenylation of tubulin polycistronic pre-mRNA in trypanosomes. Genes Dev. 1994;8:491–501.

83. Nyambega B, Helbig C, Masiga DK, Clayton C, Levin MJ. Proteins associated with SF3a60 in *T. brucei*. PLoS ONE. 2014;9:e91956.

84. Naguleswaran A, Gunasekera K, Schimanski B, Heller M, Hemphill A, Ochsenreiter T, et al. *Trypanosoma brucei* RRM1 is a nuclear RNA-binding protein and modulator of chromatin structure. MBio. 2015;6:e00114–5.

85. Baumann M, Mamais A, McBlane F, Xiao H, Boyes J. Regulation of V(D)J recombination by nucleosome positioning at recombination signal sequences. EMBO J. 2003;22:5197–207.

86. Getun IV, Wu ZK, Bois PRJ. Organization and roles of nucleosomes at mouse meiotic recombination hotspots. Nucleus. 2012;3:244–50.

87. Chen X, Chen Z, Chen H, Su Z, Yang J, Lin F, et al. Nucleosomes suppress spontaneous mutations base-specifically in eukaryotes. Science. 2012;335:1235–8.

88. Hickman MA, Rusche LN. Transcriptional silencing functions of the yeast protein Orc1/Sir3 subfunctionalized after gene duplication. Proc Natl Acad Sci USA. 2010;107:19384–9.

89. Hirumi H, Hirumi K. Continuous cultivation of *Trypanosoma brucei* blood stream forms in a medium containing a low concentration of serum protein without feeder cell layers. J Parasitol. 1989;75:985–9.

90. Brun R, Schönenberger. Cultivation and in vitro cloning or procyclic culture forms of *Trypanosoma brucei* in a semi-defined medium. Short communication. Acta Trop. 1979;36:289–92.

91. Rudenko G, Chaves I, Dirks-Mulder A, Borst P. Selection for activation of a new variant surface glycoprotein gene expression site in *Trypanosoma brucei* can result in deletion of the old one. Mol Biochem Parasitol. 1998;95:97–109.

92. Cole HA, Howard BH, Clark DJ. Genome-wide mapping of nucleosomes in yeast using paired-end sequencing. Methods Enzymol. 2012;513:145–68.

93. Langmead B, Salzberg SL. Fast gapped-read alignment with Bowtie 2. Nat Methods. 2012;9:357–9.

94. Benson G. Tandem repeats finder: a program to analyze DNA sequences. Nucleic Acids Res. 1999;27:573–80.

# 11

# Silencing markers are retained on pericentric heterochromatin during murine primordial germ cell development

Aristea Magaraki[1], Godfried van der Heijden[2], Esther Sleddens-Linkels[1], Leonidas Magarakis[5], Wiggert A. van Cappellen[6], Antoine H. F. M. Peters[3,4], Joost Gribnau[1], Willy M. Baarends[1*†] ⓘ and Maureen Eijpe[1†]

## Abstract

**Background:** In the nuclei of most mammalian cells, pericentric heterochromatin is characterized by DNA methylation, histone modifications such as H3K9me3 and H4K20me3, and specific binding proteins like heterochromatin-binding protein 1 isoforms (HP1 isoforms). Maintenance of this specialized chromatin structure is of great importance for genome integrity and for the controlled repression of the repetitive elements within the pericentric DNA sequence. Here we have studied histone modifications at pericentric heterochromatin during primordial germ cell (PGC) development using different fixation conditions and fluorescent immunohistochemical and immunocytochemical protocols.

**Results:** We observed that pericentric heterochromatin marks, such as H3K9me3, H4K20me3, and HP1 isoforms, were retained on pericentric heterochromatin throughout PGC development. However, the observed immunostaining patterns varied, depending on the fixation method, explaining previous findings of a general loss of pericentric heterochromatic features in PGCs. Also, in contrast to the general clustering of multiple pericentric regions and associated centromeres in DAPI-dense regions in somatic cells, the pericentric regions of PGCs were more frequently organized as individual entities. We also observed a transient enrichment of the chromatin remodeler ATRX in pericentric regions in embryonic day 11.5 (E11.5) PGCs. At this stage, a similar and low level of major satellite repeat RNA transcription was detected in both PGCs and somatic cells.

**Conclusions:** These results indicate that in pericentric heterochromatin of mouse PGCs, only minor reductions in levels of some chromatin-associated proteins occur, in association with a transient increase in ATRX, between E11.5 and E13.5. These pericentric heterochromatin regions more frequently contain only a single centromere in PGCs compared to the surrounding soma, indicating a difference in overall organization, but there is no de-repression of major satellite transcription.

**Keywords:** Pericentric heterochromatin, Primordial germ cell, Centromere, Histone modifications, H3K9me3, H4K20me3, ATRX, HP1, Immunochemistry, Major satellites

## Background

Chromatin is composed of DNA, histones, and other tightly associated proteins. Modifications of the DNA and of histones directly or indirectly control the regulation of DNA-related processes like transcription. Globally, the chromatin in a nucleus can be functionally divided into active and accessible euchromatin and inactive and condensed heterochromatin. Heterochromatin exists in two forms: facultative and constitutive heterochromatin. Facultative heterochromatin is a flexible form of heterochromatin and can be found in various chromosomal regions,

*Correspondence: w.baarends@erasmusmc.nl
†Willy M. Baarends and Maureen Eijpe contributed equally to this work
[1] Department of Developmental Biology, Erasmus MC, University Medical Center, Rotterdam, The Netherlands
Full list of author information is available at the end of the article

when gene-coding regions need to be repressed. Its size varies from gene clusters to an entire chromosome (the inactive $X$ in female cells). Facultative heterochromatin is frequently marked by specific histone modifications such as H2AK119Ub and H3K27me3, mediated by the polycomb repressor complexes (PRC) 1 and 2, respectively. Constitutive heterochromatin forms at specific regions of the genome, which are characterized by arrays of tandem DNA repeats: at the centromeres (minor satellite repeats), telomeres (telomeric repeats), and pericentric regions (major satellite repeats). Here we focus on the pericentric heterochromatin. A known hallmark of this chromatin type is the lack of histone modifications that generally mark active chromatin, such as histone acetylation. Conversely, there is an accumulation of repressive histone marks such as H3K9me3 and H4K20me3 [1–5]. The presence of H3K9me3 results in recruitment of different heterochromatin protein (HP) isoforms that contribute to heterochromatin establishment and maintenance of this chromatin state [6, 7]. The basic unit of the major satellites in the mouse is an A/T-rich ~230-bp-long monomer, which can be repeated many times, leading to regions of up to several megabases in size. In an interphase mouse nucleus, pericentric constitutive heterochromatin can be visualized as 4′,6-diamidino-2-phenylindole (DAPI)-dense regions, termed chromocenters, with each chromocenter consisting of multiple pericentric regions from different chromosomes. The periphery of each chromocenter contains the centromeres of the chromosomes as individual entities [8].

Maintenance of the heterochromatic nature of pericentric DNA is important for proper cell functions; failure impairs cell viability, induces chromosomal instabilities, and increases the risk of tumorigenesis [2]. Therefore, pericentric heterochromatin has for a long time been considered as an inert, highly condensed, and inaccessible domain. In recent years, however, it has become clear that the biology of pericentric heterochromatin is more complicated. Emerging evidence indicates that some well-controlled dynamical changes of pericentric heterochromatin structure may occur, which are associated in some cases with brief bursts of major satellite transcription. Transcription of major satellites has been shown to occur during canonical cell processes, e.g. during the normal cell cycle [9, 10], cell differentiation [11, 12], and during early [13, 14] and late [15] embryonic development. For example, in pre-implantation mouse embryos, the paternal pericentric domains initially lack heterochromatin marks, such as H3K9me3 and HP1 proteins. This likely relates to the fact that the paternal genome enters the oocyte as a protamine-packaged compact structure, largely devoid of nucleosomes. After fertilization, the DNA rapidly decondenses as protamines

are removed and replaced by maternal histones that lack pericentric heterochromatin histone modifications [16–19]. Concomitantly, active DNA demethylation occurs [16, 20]. In contrast, maternal pericentric heterochromatin displays the typical somatic histone posttranslational modification marks. Interestingly, major satellites are transcribed (in forward direction) solely from the paternal pronucleus at the 2-cell stage, which might reflect the above-described specific epigenetic status of the paternal genome [21]. Then, a burst in transcription of the major satellites (in reversed direction) from both parental genomes facilitates the reorganization of pericentric heterochromatin from nuclear precursor bodies to the typical somatic like chromocenters in the developing embryo. This is completed by the 4- to 8-cell stage after which pericentric heterochromatin displays its specific H3K9me3–HP1 chromatin state [14, 22].

Developing mouse primordial germ cells (PGCs) also undergo genome-wide epigenetic reprogramming, and this occurs between E8.0 and E13.5. It includes changes in histone modifications (e.g. global loss of H3K9me2 and relative enrichment of H3K27me3 compared to somatic cells as assessed by immunofluorescence experiments), reactivation of the inactive $X$ chromosome in the female embryos, and global loss of DNA methylation, the last reaching its lowest levels at E13.5, both in male and female embryos [23, 24].

Initiation of imprint erasure in PGCs takes place between E10.5 and E11.5 [25, 26], and concomitantly, it has been reported that PGCs lose the DAPI-dense chromocenters [25]. These events are accompanied by a transient apparent loss of H3K9me3, HP1 proteins, and other heterochromatin marks [27]. In this study, we focus specifically on the pericentric heterochromatin in germ cells between E10.5 and E13.5 of mouse embryo development. Since we experienced difficulties to reproduce the previously reported transient loss of pericentric heterochromatin marks [27], we decided to revisit the possible loss and re-establishment of pericentric heterochromatin marks and of chromocenters during PGC development, by testing different preparation methods and fixation conditions. It is well known that different fixation and preparation methods may lead to variations in immunostaining results, and these should thus be interpreted with caution. In particular, the inability to detect a protein does not always result from its absence, but could be caused, for example, by epitope masking. Using a method that is known as "drying-down" or "spreading" of (meiotic) nuclei [28], we observed persistence of H3K9me3, HP1 isoforms, and H4K20me3 on pericentric heterochromatin of PGCs. Based on these results, we conclude that the reported loss and re-establishment of pericentric heterochromatin signature [27] may reflect a structural change in pericentric heterochromatin, affecting epitope availability, rather than the actual loss of the markers. In addition, we found ATRX, a chromatin remodeler known to associate with constitutive heterochromatin [29, 30], to be highly enriched at pericentric heterochromatin in PGCs at E11.5 compared to the somatic cells of the same developmental stage. Lastly, immunofluorescent analysis of centromere and pericentromere (adjacent to the centromeres) staining showed that pericentromeres do not cluster together in the same fashion as in the surrounding somatic cells, and this may explain the weak DAPI staining of pericentric heterochromatin in developing PGCs. Still, consistent with the overall persistence of histone modifications and the enrichment of ATRX, no increased transcription of major satellite repeats was detected in isolated E11.5 PGCs. Together, our data indicate that although the pericentric heterochromatin in E11.5 mouse PGCs may exist in a different chromatin conformation and is organized more frequently as small regions containing a single centromere compared to somatic cells, this phenomenon is neither associated with a complete loss of heterochromatin hallmarks nor with a burst in transcription of major satellite repeats.

## Results

### H3K9me3 remains present in pericentric heterochromatin throughout germ cell development

We first reanalysed the reported dynamics of H3K9me3 [27] in PGCs of E10.5–E13.5 mouse embryos. For this, we used a fluorescent immunohistochemical approach. Since fixation conditions may influence epitope availability, we fixed and embedded embryos using different protocols. OCT4 (E10.5 and E11.5) or TRA98 (E13.5) was used as germ cell markers. For H3K9me3 staining, we did not observe any robust and reproducible staining pattern using paraffin-embedded tissue sections. In contrast, cryosectioning of paraformaldehyde-fixed samples did produce the typical pattern of H3K9me3 enrichment in heterochromatin areas of somatic cells. Interestingly, using two different fixation protocols, one involving fixation only prior to freezing and embedding (regular fixation), and another protocol that included a postfixation step after sectioning (extended fixation, see Methods for details), two different staining patterns were observed. Using regular or extended fixation, two embryos or gonads were analysed for each studied protein or histone modification per stage, and qualitative analyses for the patterns in PGC nuclei compared to surrounding somatic cells were recorded for at least 20 PGCs. Using both methods, the pattern of H3K9me3 immunostaining in PGCs was similar to that of surrounding somatic cells at E10.5, despite the overall DAPI weak appearance of the PGC chromocenters (Fig. 1, panel a and b). Using the

regular fixation procedure, we observed an overall reduction of H3K9me3 signal solely from E11.5 germ cells, in accordance with previous observations [27] (Fig. 1, panel a). In contrast, when using extended fixation, H3K9me3 signal was retained on the pericentric heterochromatin as the PGCs developed between E10.5 and E13.5 (Fig. 1, panel b). As an alternative approach, and to further ensure epitope availability, we used a drying-down alias meiotic spread method that is commonly used to study the localization of chromatin modifications and associated proteins in nuclei of meiotic prophase cells [28]. It involves mixing of a cell suspension on a glass slide covered with a Triton X100-containing fixative, followed by gradual drying, whereby the nuclei spread on glass. The spreading results in loss of most cytoplasmic and loosely DNA-associated proteins and flattening of the nuclear chromatin. Therefore, we will further refer to this type of preparation as nuclear spreads. We prepared slides containing nuclei from E10.5 to E11.5 gonadal regions and from E13.5 male and female gonads. Since these preparations are relatively flat, and a single plane image at optimal focus provides a reproducible estimate for the amount of signal, we decided to include a quantitative analysis with this approach (see Methods for details concerning the quantification method). Similar to the results obtained with the extended fixation protocol, H3K9me3 signal was retained in the pericentric heterochromatin of PGCs from E10.5 to E13.5, although the signals are reduced in pericentric heterochromatin of E11.5 PGCs compared to pericentric heterochromatin of the soma (Fig. 2a, b). This indicates that there may be some difference between the pericentric heterochromatin structure of PGCs and somatic cells, but there is no major loss of H3K9me3 from the pericentric regions in E11.5 PGCs. In accordance with Kagiwada et al. [26], we did not observe complete loss of the DAPI-dense regions at any of the examined stages in all protocols tested, but the regions appeared less DAPI intense and at the same time smaller. Importantly, our results indicate that the previously reported absence of pericentric heterochromatin marks in E11.5 PGCs might be a consequence of the chosen experimental methodology.

### HP1 isoforms are stably recruited to pericentric heterochromatin of developing germ cells

Specific histone modifications recruit certain proteins. Members of the heterochromatin protein 1 (HP1) protein family bind H3K9me3 and mark constitutive heterochromatin. In mammals, there are three different HP1 isoforms: HP1α, HP1β, and HP1γ also known as CBX5, CBX1, and CBX3, respectively. We examined the localization of these three isoforms during PGC development. In the regular fixation protocol and using paraffin

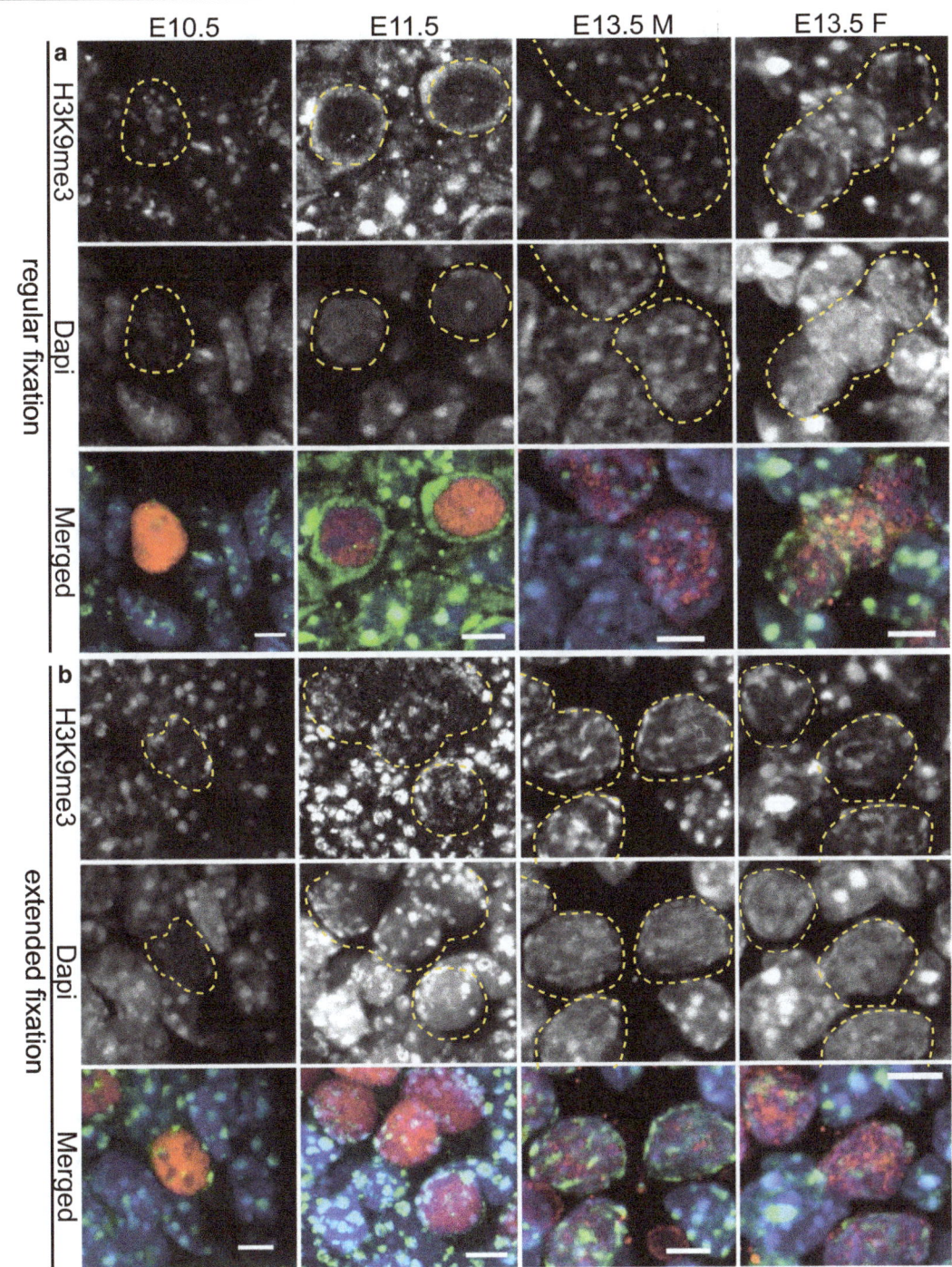

**Fig. 1** H3K9me3 signal enrichment persists on pericentric heterochromatin throughout germ cell development upon extended tissue fixation. **a** Immunofluorescent analysis of H3K9me3 (*green*) in cryosections of E10.5 and E11.5 trunks containing germ cells and of E13.5 male and female gonads using the regular fixation protocol. H3K9me3 staining is present in DAPI (*blue*)-dense regions of E10.5 PGCs and somatic cells. At E11.5 H3K9me3 transiently disappears from pericentric heterochromatin of E11.5 PGCs only. Thereafter, H3K9me3 returns in E13.5 DAPI-dense regions. **b** Using extended fixation, H3K9me3 is retained in DAPI-dense regions of PGCs and somatic cells in all embryonic stages examined. E10.5 and E11.5 germ cells are marked with OCT4 (*red*), while E13.5 germ cells are marked with TRA98 (*red*). Using regular or extended fixation, two embryos or gonads were analysed per stage, and at least 20 PGC nuclei were recorded. Representative germ cells are marked with *yellow dashed circles*. *Scale bars* represent 5 µm

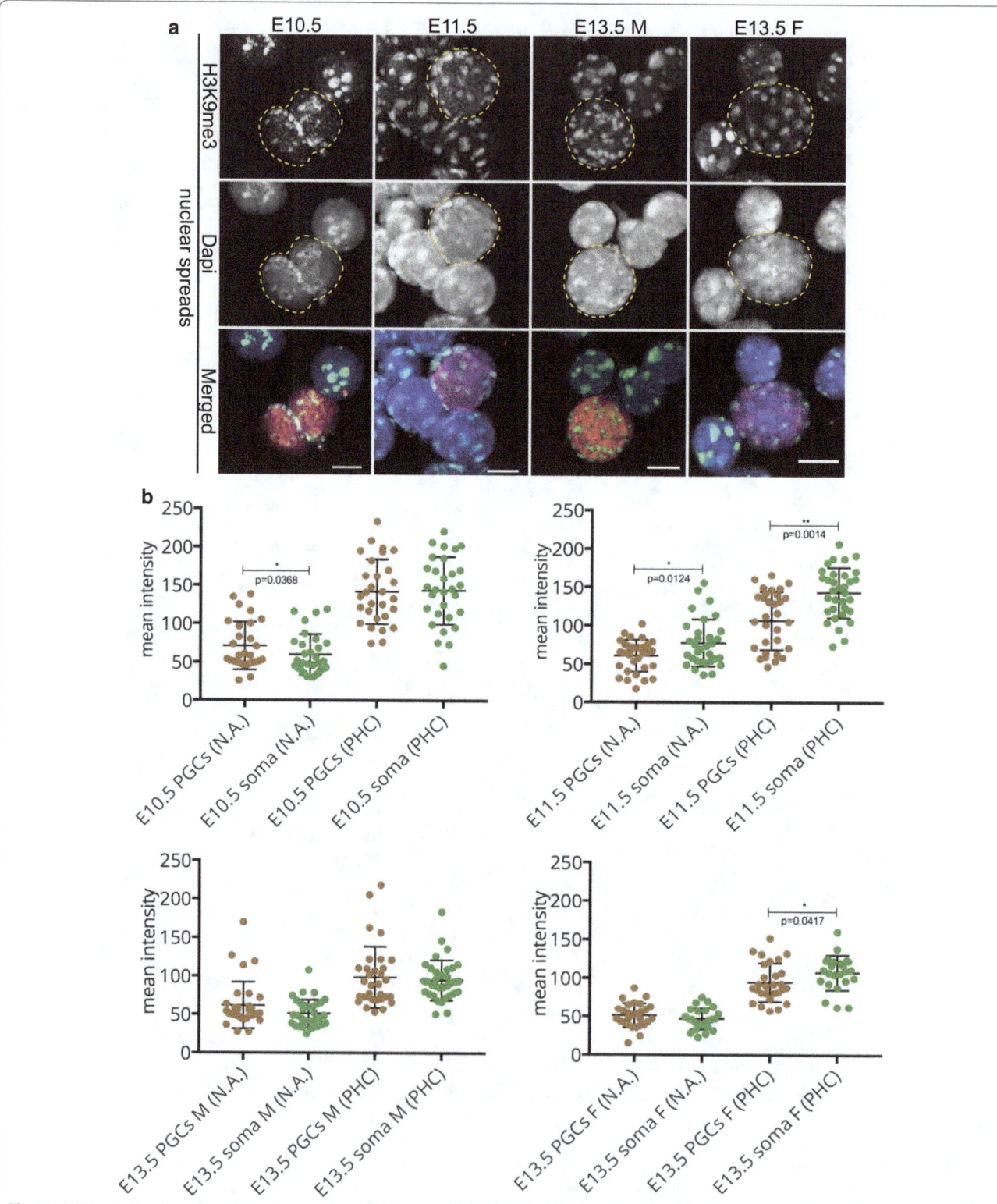

**Fig. 2** H3K9me3 signal enrichment persists on pericentric heterochromatin throughout germ cell development in spread preparations. **a** Spread preparations immunostained for H3K9me3 (*green*). E10.5 and E11.5 germ cells are marked with OCT4 (*red*), while E13.5 germ cells are marked with TRA98 (*red*). Representative germ cells are marked with *yellow dashed circles*. *Scale bars* represent 10 μm. **b** Quantification analysis of mean H3K9me3 levels in the whole nuclear area (N.A.) and pericentric heterochromatin regions (PHC) in the different embryonic stages examined. Four or more embryo trunks or gonads were pooled for the nuclear spread preparations and 20–30 PGC, and somatic nuclei were recorded. *Asterisks* (* or **) indicate significant differences

sections, HP1α immunostaining marked pericentric heterochromatin in E10.5 PGCs. Already at this stage, the signal for HP1α appeared lower in PGCs compared to surrounding somatic cells. Thereafter, HP1α was undetectable in developing germ cells (Additional file 1, panel A), which is in accordance with previous studies [27]. Using extended fixation conditions, HP1α signal was reduced (E10.5, E11.5 some PGCs, E13.5 female germ cells) or absent (E11.5 some PGCs, E13.5 male germ cells). It should be noted that in E13.5 male gonad sections, we could not reproducibly detect HP1α even in the surrounding somatic cells of sections, using either regular or extended fixation protocols (Additional file 1, panel A and B). This may be due to the different consistency of the male versus the female gonad at this age, causing differential and variable effects of the fixation protocols. Specifically, we observe that at E13.5 the ovarian tissue is softer and more vulnerable to dissociation compared to the developing testis, which seems more compact. When nuclear spreads from genital ridges or embryonic gonads were examined, HP1α immunostaining was readily detectable and enriched in DAPI-dense regions of all cells, but this enrichment was more clear in the pericentric heterochromatin areas of the soma compared to those of the corresponding PGCs in all developmental stages examined, but most clearly at E13.5 (Fig. 3a, b).

Like HP1α, HP1β is also known to predominantly localize to heterochromatin. When we use our regular fixation protocol, we observed accumulation of HP1β signal in DAPI-dense regions of both somatic and germ cell nuclei at E10.5 and E11.5 stages. However, at E13.5, hardly any enrichment was observed in the DAPI-dense regions of both male and female germ cells (Additional file 2, panel A). Following extended fixation, HP1β signal was preserved on pericentric heterochromatin in all developmental germ cell stages examined (Additional file 2, panel B), similar to what was observed in nuclear spread preparations (Fig. 4a). The quantitative measurements showed (similar to HP1α) that HP1β is present at a higher level in the pericentric heterochromatin of the soma compared to that of the PGCs (Fig. 4b).

The last HP1 isoform, HP1γ, is known to interact with both constitutive heterochromatin and euchromatin [31–33]. Examination of HP1γ in nuclear spreads revealed a clear immunostaining signal for this HP1 isoform in DAPI-dense regions in the nuclei of germ cells throughout development. Interestingly, HP1γ levels were significantly higher in the nuclei (and pericentric heterochromatin areas) of E10.5 and E11.5 PGCs, compared to the surrounding soma. At E13.5, HP1γ signals were similar in PGCs and somatic nuclei (Fig. 5a, b). Comparable

results were obtained upon regular or extended fixation in paraffin sections of E10.5, E11.5 genital ridges, and E13.5 female gonads (Additional file 3). However, at E13.5 of male development, detection of HP1γ in the surrounding somatic cells of paraffin sections was difficult and variable, using either our regular or extended fixation protocol. This was similar to our HP1α results. We could detect accumulation of HP1γ in DAPI-dense regions in some E13.5 male germ cells (Additional file 3, panel A and B, arrowhead), but not in all.

Taken together, the persistent detection of the H3K9me3 on pericentric heterochromatin during PGC development is consistent with the patterns observed for the HP1 proteins, whereby decreases in HP1α and HP1β may be at least partially compensated by an increase in HP1γ.

## H4K20me3 is retained at pericentric heterochromatin of E11.5 PGCs

An additional histone mark that participates in the establishment of pericentric heterochromatin is H4K20me3 [4, 5]. This histone modification is mediated by the histone methyltransferase SUV4-20H2 in a SUV39H and HP1-dependent manner [4, 34]. SUV39H is the enzyme responsible for establishing trimethylation of H3K9 [2]. Similar to H3K9me3, H4K20me3 is strongly enriched at DAPI-dense regions [4, 5]. Again we performed comparative immunofluorescence in sections processed with regular and extended fixation, and the more quantitative analyses in nuclear spreads. When using the regular fixation procedure on paraffin-embedded embryo sections, H4K20me3 signal intensity was similar in developing PGCs and surrounding somatic cells at E10.5 (Fig. 6a). However, at E11.5, the immunostaining signal for this histone modification transiently disappeared from the DAPI-dense chromocenters of the PGCs only, while it was strongly retained in the surrounding somatic cells (Fig. 6a). Two days later, at ~E13.5, H4K20me3 signal reappeared, albeit at low levels compared to the surrounding soma and only in some TRA98 (red)-positive cells, regardless of the embryo sex. When using the extended fixation protocol, H4K20me3 signal was retained on pericentric heterochromatin throughout PGC development, but clearly reduced in the PGCs compared to the surrounding somatic cells at E11.5–E13.5 (Fig. 6b). Lastly, upon analysis of nuclear spread preparations, we indeed observed that H4K20me3 is significantly reduced in pericentric heterochromatin regions of PGCs compared to the soma in all stages examined (Fig. 7a, b). Importantly, similar to H3K9me3, H4K20me3 did not fully disappear from the DAPI-dense regions of E11.5 PGCs.

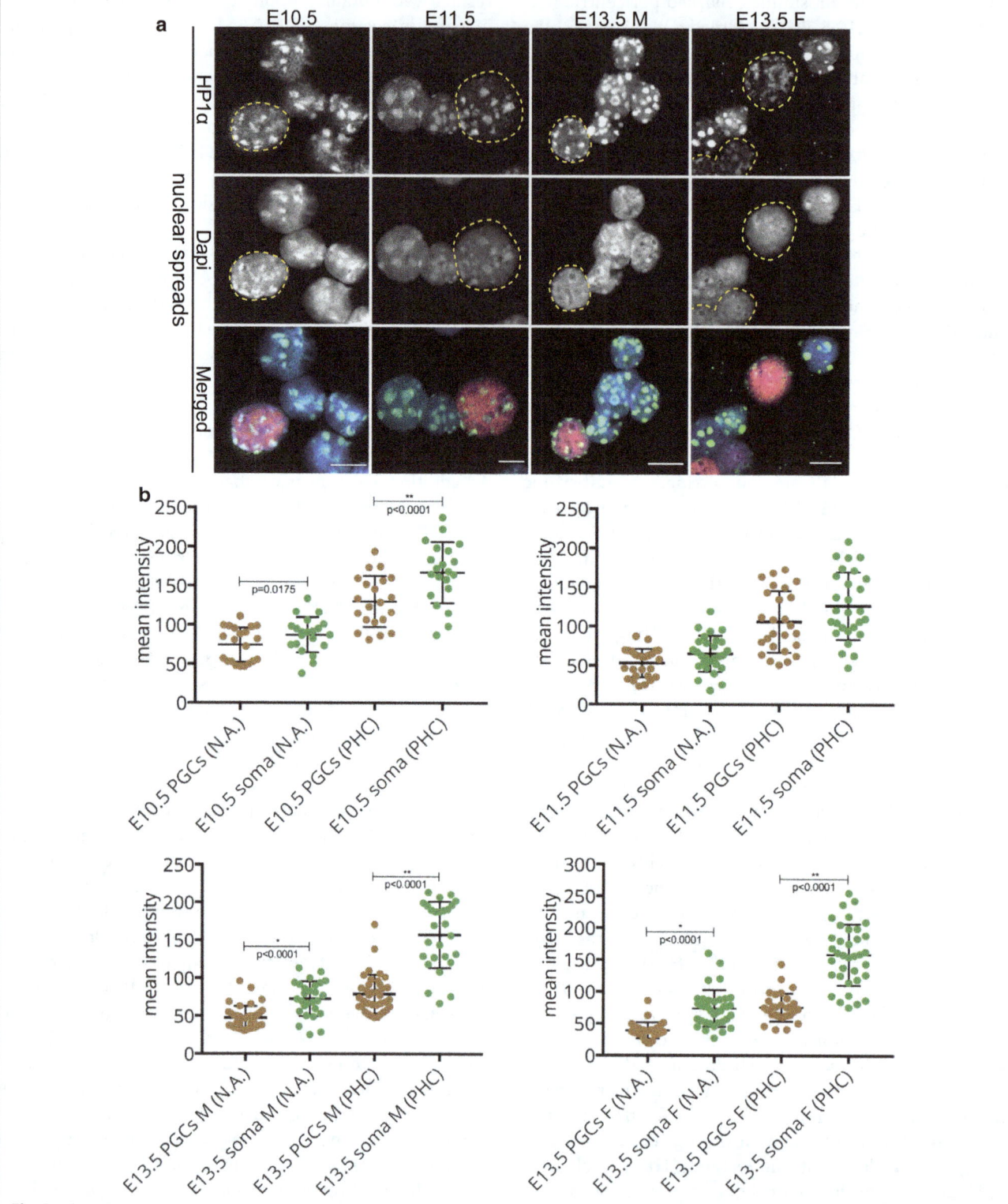

**Fig. 3** HP1α is recruited to pericentric heterochromatin in E10.5–E13.5 germ cells. **a** Nuclear spread preparations from E10.5 to E11.5 embryos and from E13.5 male and female gonads were stained for HP1α (*green*). The pericentric heterochromatin of germ cells was always decorated with HP1α. E10.5 and E11.5 germ cells are marked with OCT4 (*red*), while E13.5 germ cells are marked with TRA98 (*red*). Four or more embryo trunks or gonads were pooled for the nuclear spread preparations and 20–30 PGC, and somatic nuclei were recorded. Representative germ cells are marked with *yellow dashed circles*. *Scale bars* represent 10 µm. **b** Quantification analysis of HP1α levels in the whole nuclear area (N.A.) and pericentric heterochromatin regions (PHC) in the different embryonic stages examined. *Asterisks* (* or **) indicate significant differences

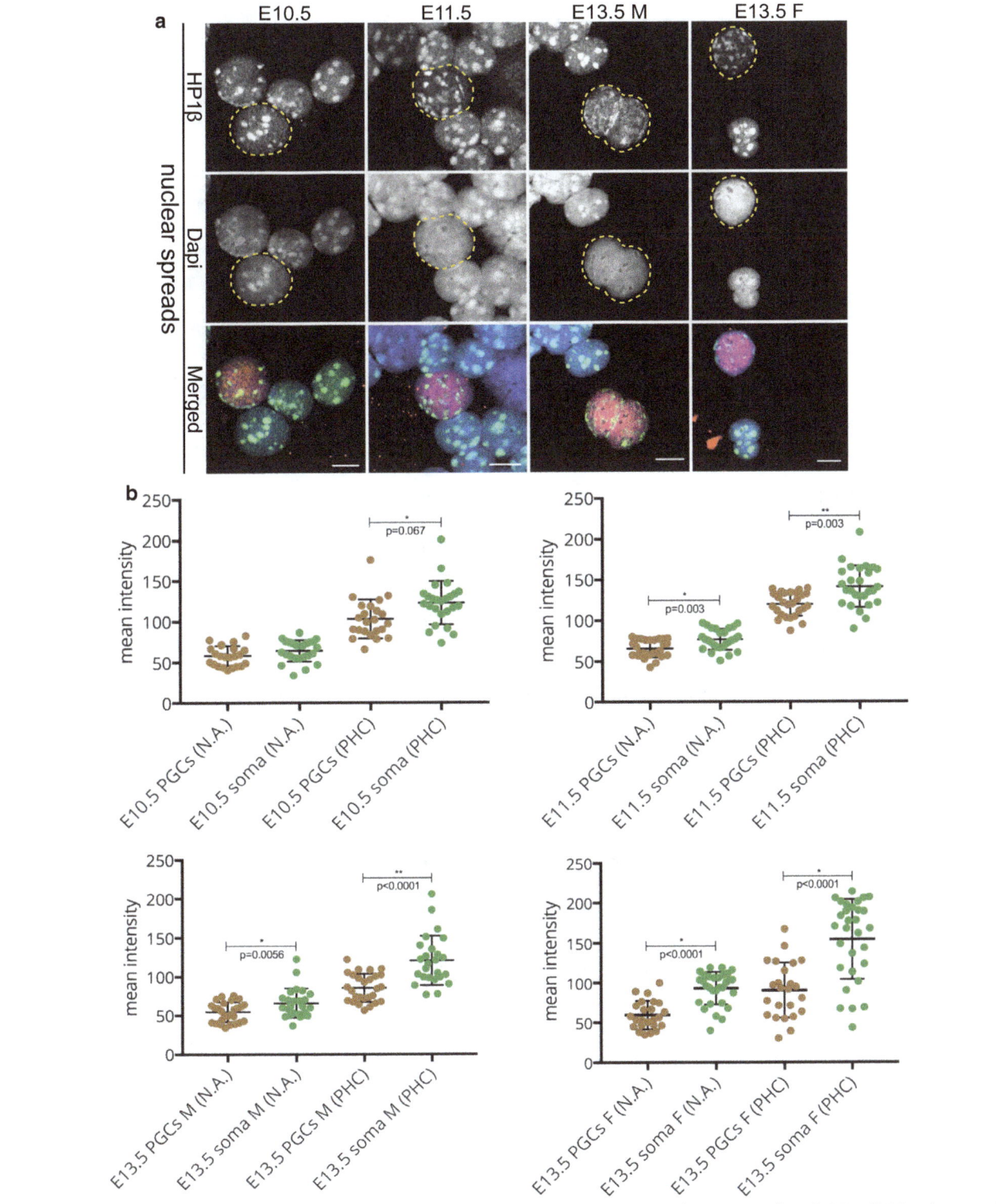

**Fig. 4** HP1β is recruited to pericentric heterochromatin in E10.5–E13.5 germ cells. **a** Nuclear spread preparations from E10.5 to E11.5 embryos and from E13.5 male and female gonads were stained for HP1β (*green*). This protein was always present in DAPI (*blue*)-dense regions of germ cells in all stages examined. E10.5 and E11.5 germ cells are marked with OCT4 (*red*), while E13.5 germ cells are marked with TRA98 (*red*). Four or more embryo trunks or gonads were pooled for the nuclear spread preparations and 20–30 PGC, and somatic nuclei were recorded. Representative germ cells are marked with *yellow dashed circles*. *Scale bars* represent 10 µm. **b** Quantification analysis of HP1β levels in the whole nuclear area (N.A.) and pericentric heterochromatin regions (PHC) in the different embryonic stages examined. *Asterisks* (* or **) indicate significant differences

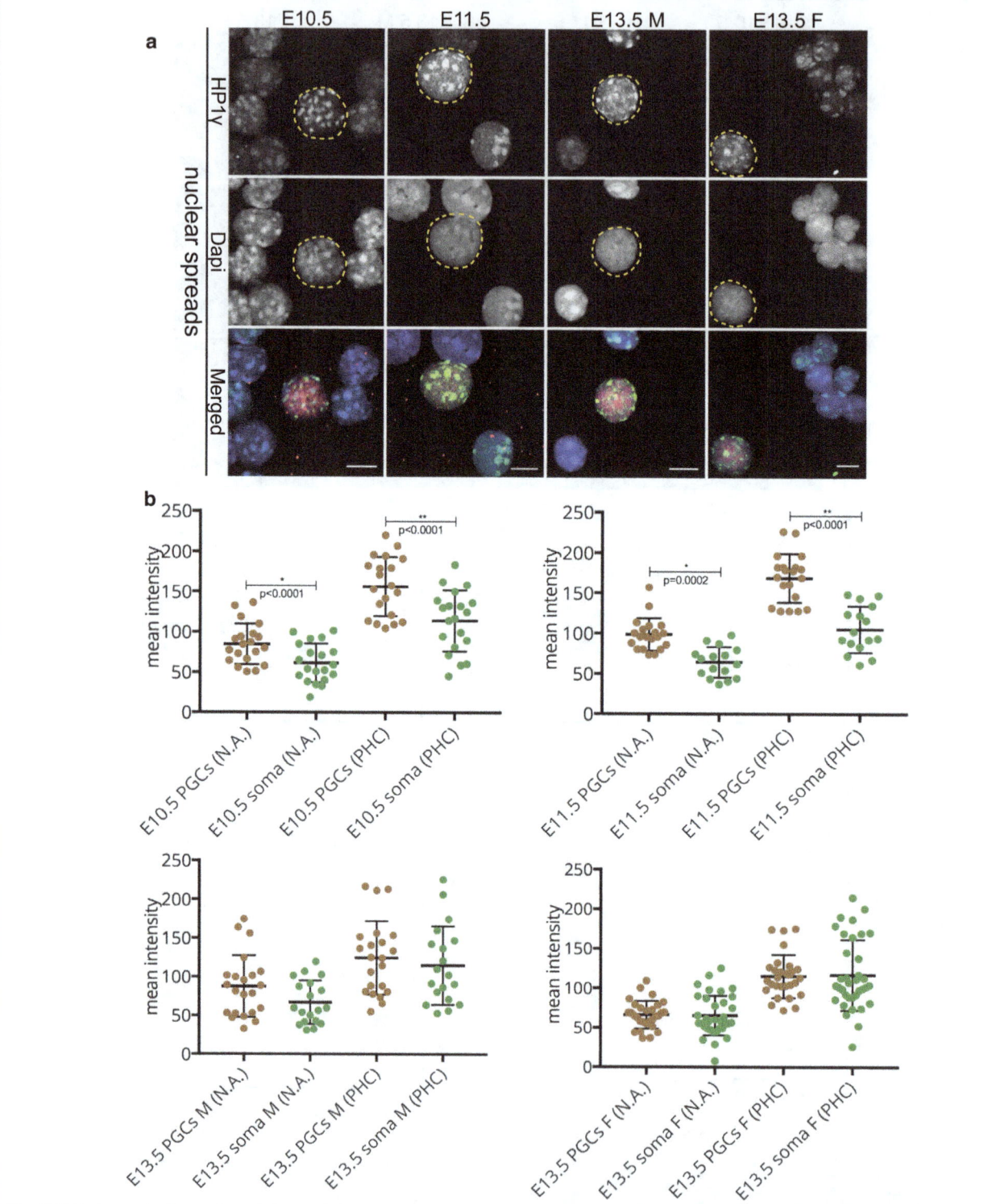

**Fig. 5** HP1γ is recruited to pericentric heterochromatin in E10.5–E13.5 germ cells. **a** Nuclear spread preparations from E10.5 to E11.5 embryos and from E13.5 male and female gonads were stained for HP1γ (*green*). This protein is also stably recruited to pericentric heterochromatin of PGCs, in the three stages examined. E10.5 and E11.5 germ cells are marked with OCT4 (*red*), while E13.5 germ cells are marked with TRA98 (*red*). Four or more embryo trunks or gonads were pooled for the nuclear spread preparations and 20–30 PGC, and somatic nuclei were recorded. Representative germ cells are marked with *yellow dashed circles*. *Scale bars* represent 10 μm. **b** Quantification analysis of HP1γ levels on the whole nuclear area (N.A.) and pericentric heterochromatin regions (PHC) in the different embryonic stages examined. The levels of HP1γ isoform at E10.5 and E11.5 in the pericentric heterochromatin regions are notably more enriched compared to those of the corresponding somatic cells. *Asterisks* (* or **) indicate significant differences

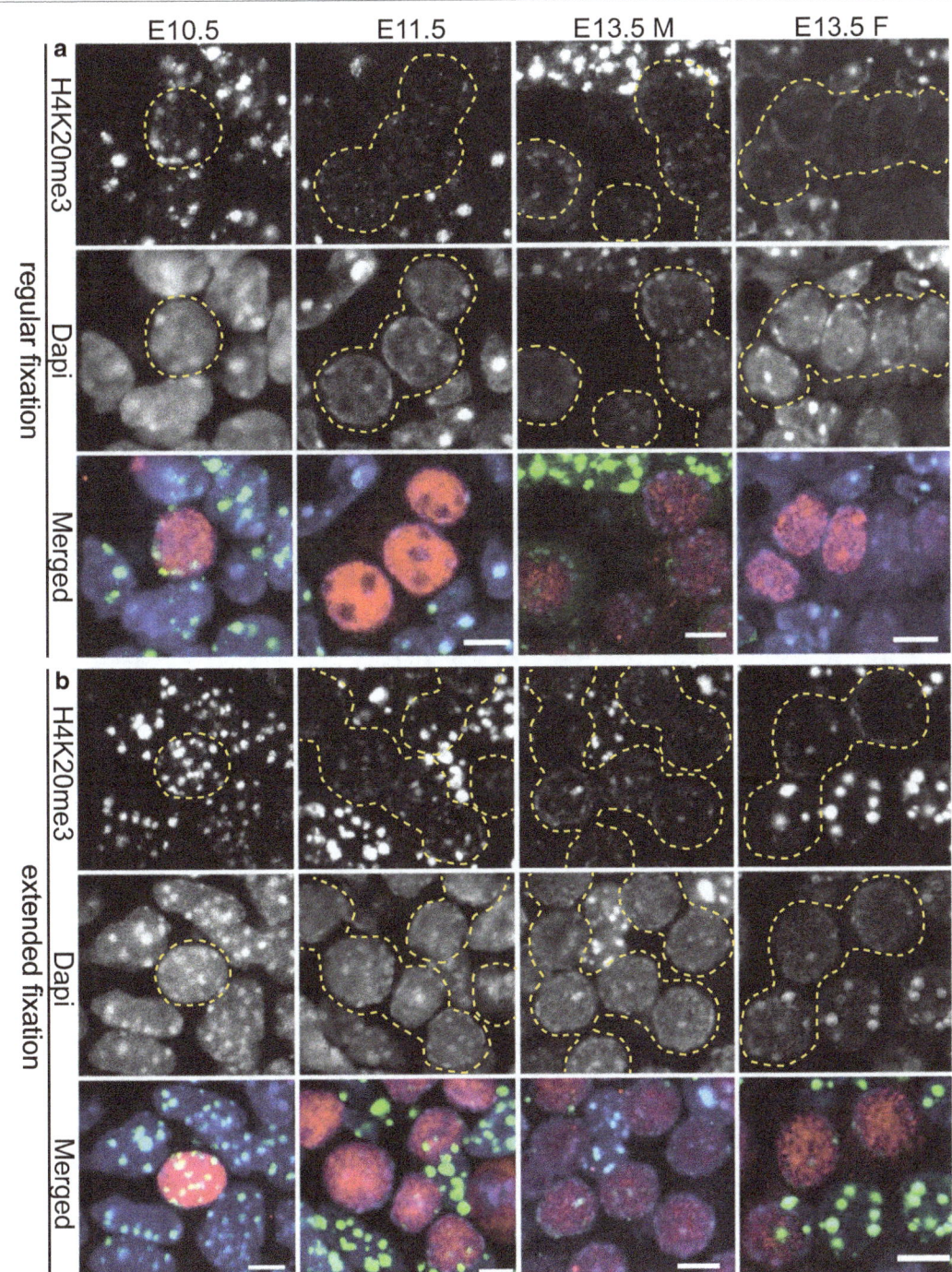

**Fig. 6** H4K20me3 enrichment is detected in pericentric heterochromatin of developing germ cells. **a** Paraffin sections of E10.5 and E11.5 embryo trunks and of E13.5 male and female gonads were immunostained using anti-H4K20me3 (*green*) after applying the regular fixation protocol. At E10.5 H4K20me3 is present in pericentric heterochromatin of PGCs and surrounding soma. At E11.5, H4K20me3 is lost from DAPI (*blue*)-dense regions of PGCs, while it is maintained in the somatic cells. At E13.5 H4K20me3 reappears in germ cells, but in substantially reduced levels compared to the surrounding gonadal somatic cells. **b** When applying the extended fixation protocol, H4K20me3 is retained at pericentric heterochromatin in PGC nuclei from E10.5 to E13.5. However, when compared to the H4K20me3 pattern in the surrounding somatic cells, the levels are reduced. E10.5 and E11.5 germ cells are marked with OCT4 (*red*), while E13.5 germ cells are marked with TRA98 (*red*). Using regular or extended fixation, two embryos or gonads were analysed per stage, and at least 20 PGC nuclei were recorded. Representative germ cells are marked with *yellow dashed circles*. *Scale bars* represent 5 μm

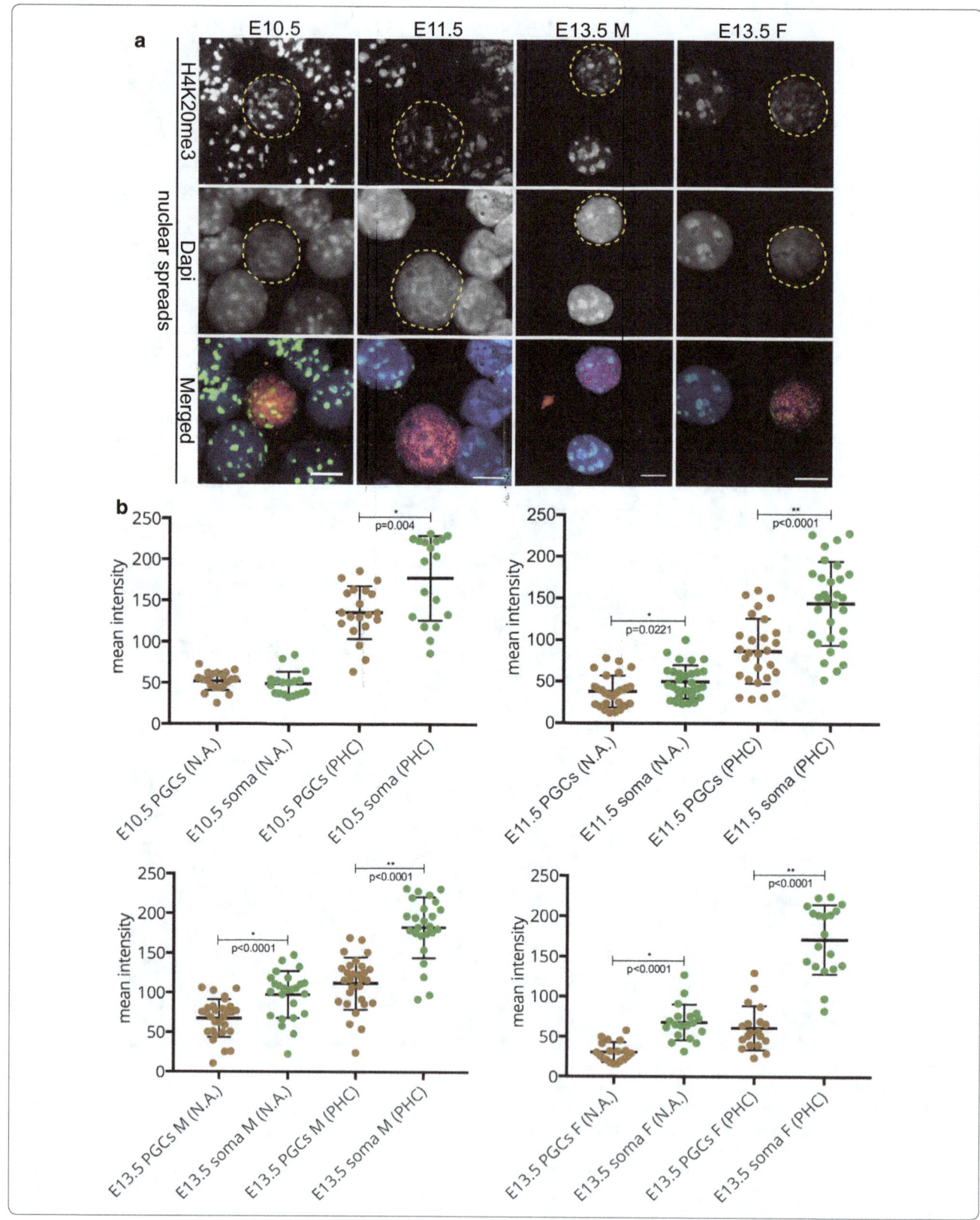

(See figure on previous page.)
**Fig. 7** H4K20me3 enrichment is detected in pericentric heterochromatin of developing germ cells in spread preparations. **a** Nuclear spread preparations from E10.5 to E11.5 embryos and from E13.5 male and female gonads were stained for H4K20me3 (*green*). This modification is enriched at pericentric heterochromatin in PGCs throughout germ cell development when analysed. E10.5 and E11.5 germ cells were identified by OCT4 (*red*), and E13.5 germ cells with TRA98 (*red*). Three or more embryo trunks or gonads were pooled for the nuclear spread preparations and 20–30 PGC, and somatic nuclei were recorded. Representative germ cells are marked with *yellow dashed circles*. *Scale bars* represent 10 μm. **b** Quantification analysis of H4K20me3 levels in the whole nuclear area (N.A.) and pericentric heterochromatin regions (PHC) in the different embryonic stages examined. Overall, the levels of H4K20me3 are significantly reduced in PGC pericentric heterochromatin areas compared to those of the surrounding somatic cells. *Asterisks* (* or **) indicate significant differences

## ATRX is enriched in pericentric heterochromatin of primordial germ cells

The α-thalassaemia mental retardation X-linked protein ATRX is a chromatin remodeler and a prominent marker of pericentric heterochromatin in somatic cells, in the mouse zygote, and in neonatal spermatogonia [29, 30, 35, 36]. We explored the presence of ATRX in developing germ cells, only in nuclear spread preparations, since we observed that this protocol yielded the most reproducible and quantifiable results. At E10.5, the levels of ATRX in germ cells and somatic cells were similar (Fig. 8). Interestingly, in the germ cells of E11.5, ATRX immunostaining was increased in pericentric heterochromatin compared to that of the soma. At E13.5, ATRX levels were again comparable in gonadal somatic cells and PGCs (Fig. 8). We did not observe any relocalization of ATRX to the nuclear periphery in any of the E11.5 PGCs examined, in contrast to what was previously reported [27].

## Spatial organization of constitutive heterochromatin in germ cells

In order to examine the organization of the chromocenters during germ cell development more globally, we stained nuclear spreads of all developmental stages examined within this study with CREST antisera, a marker of chromosome centromeres. Additionally, we used H4K20me3 to visualize the pericentric heterochromatin. In somatic cells, the chromocenters consisted of more than one pericentric domain, as indicated by the multiple CREST signals within the DAPI-dense (or H4K20me3 enriched) regions (Fig. 9a). This organization reflects the clustering of groups of pericentromeres, which is

a common hallmark of chromocenters [8]. However, already at E10.5, we observed a large number of individual pericentric regions, containing only a single CREST signal, within the nucleus of germ cells (Fig. 9a, examples indicated by arrowheads). Counting the number of CREST signals detected per H4K20me3-positive pericentric heterochromatin area, and subsequent analyses of the frequency distribution of pericentric heterochromatin areas containing 1, 2, 3, 4, or more CREST signals revealed that pericentric heterochromatin areas with a single CREST signal are more frequently observed in PGCs compared to somatic nuclei. Conversely, pericentric heterochromatin areas with 4 or more CREST signals are more frequent in the somatic nuclei, at all stages examined (Fig. 9b). This observation may explain the size, number, and intensity differences between the DAPI-dense regions in germ cell versus somatic cell nuclei. Thus, we demonstrate that the organization of the chromocenters in E10.5–E13.5 PGCs is different compared to that of the somatic surrounding cells. A summary of all immunostainings of pericentric heterochromatin markers is presented in Additional file 4.

## Major satellites are not transcribed in E11.5 murine PGCs

In order to examine whether the altered organization of the chromocenters during PGC development is associated with reduced transcriptional repression of major satellites, we FACS-sorted PGCs and somatic cells from developing gonads isolated from E11.5 embryos carrying a transgene-encoding GFP under the control of the *Oct4* promoter. We isolated RNA and analysed mRNA expression in the FACs-sorted cell fractions, heart tissue, and NIH3T3 cells

(See figure on next page.)
**Fig. 8** ATRX is more enriched at pericentric heterochromatin of E11.5 PGCs compared to that of the surrounding soma. **a** Analysis of ATRX (*green*) localization patterns in nuclear spread preparations of E10.5 and E11.5 embryo trunks and E13.5 male and female gonads. ATRX is enriched at pericentric heterochromatin of germ and somatic cells in all stages examined. At E11.5, ATRX levels seem to be higher in PGCs compared to the surrounding somatic cells, while at E10.5 and E13.5, the levels of ATRX are comparable between germ cell and somatic nuclei. E10.5 and E11.5 germ cells were identified by OCT4 (*red*) and E13.5 germ cells with TRA98 (*red*). Three or more embryo trunks or gonads were pooled for the nuclear spread preparations and 20–30 PGC, and somatic nuclei were recorded. Representative germ cells are marked with *yellow dashed circles*. *Scale bars* represent 10 μm. **b** Quantification analysis of ATRX levels in the whole nuclear area (N.A.) and pericentric heterochromatin regions (PHC) in the different embryonic stages examined. The levels of ATRX at E11.5 pericentric heterochromatin are significantly increased compared to the soma. *Asterisks* (* or **) indicate significant differences

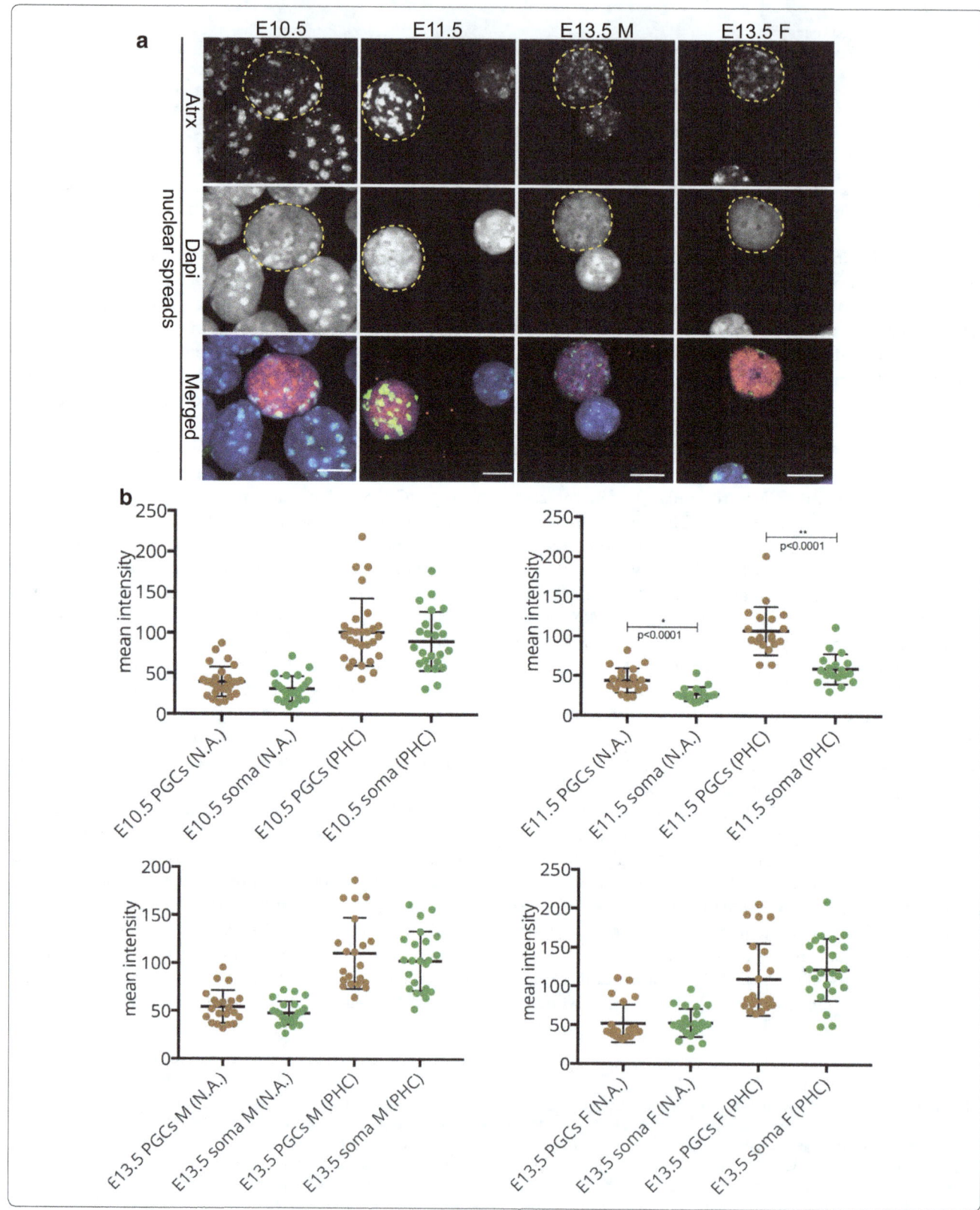

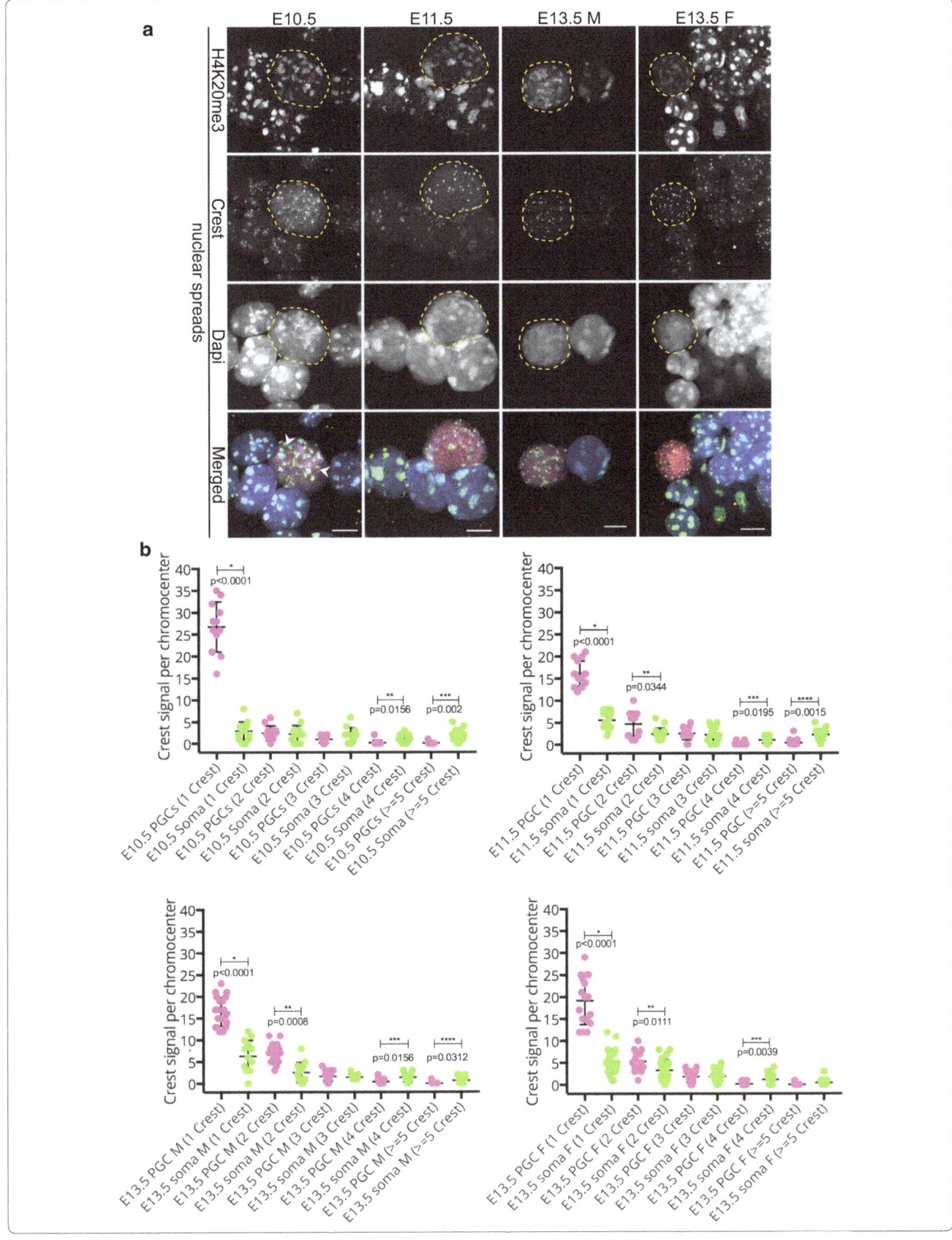

(See figure on previous page.)

**Fig. 9** Pericentromeres are more frequently found as single units in PGCs compared to the somatic cells. **a** Nuclear spread preparations were immunostained for CREST (*yellow*) and H4K20me3 (*green*). At E10.5, many small-sized chromocenters [identified by DAPI (*blue*) density and enrichment of H4K20me3) and their corresponding CREST signals can be observed in the PGCs (E10.5 and E11.5 germ cells were identified by OCT4 (*red*) and E13.5 germ cells with TRA98 (*red*)]. Examples of pericentric regions containing a single CREST focus are indicated by *arrowheads*. This pattern of dispersed pericentric heterochromatin organization in germ cells, as opposed to the clustering of pericentric heterochromatin regions and associated centromeres in somatic cells, is observed at least until E13.5. Representative germ cells are marked with *yellow dashed circles*. Three or more embryo trunks or gonads were pooled for the nuclear spread preparations and 20–30 PGC, and somatic nuclei were recorded. *Scale bars* represent 10 μm. **b** Frequency distribution of the number of pericentric heterochromatin regions (defined by delineation of the H4K20me4 areas) containing a specific number of CREST signals as indicated on the *X* axis. *Asterisks* (\*, \*\*, \*\*\* or \*\*\*\*) indicate significant differences

(Fig. 10). From the results of this analysis (Fig. 10a), we conclude that there is a clear enrichment of *Oct4* and *Atrx* mRNA in the E11.5 PGC fraction, in accordance with our immunocytochemical observations. A low level of major satellite transcription is detected in the E11.5 PGCs as well as in the soma, as can be inferred from a similar difference between the + and − RT samples of both fractions. In heart and NIH3T3 cells, the level of major satellite transcription is somewhat higher. This result indicates that there is no increase in transcription of major satellites in PGCs compared to the somatic fraction (Fig. 10b).

## Discussion

At the time of their specification, PGCs are epigenetically identical to the surrounding epiblast and therefore primed towards a somatic fate [37, 38]. In order to activate their germ cell transcriptional network, and at the same time repress their somatic fate, PGCs go through a series of extensive reprogramming events, which have been thoroughly characterized. The reprogramming encompasses DNA demethylation at several genomic loci, including the imprinted genes, but also involves changes in histone modifications [25, 26, 37, 39]. An additional reprogramming cycle has been reported to take place specifically at E11.5, when many histone modifications are transiently lost, including those marking constitutive heterochromatin and its readers [27]. In our study, we carefully re-evaluated epigenetic remodelling targeting specifically constitutive heterochromatin, from the period when PGCs enter the genital ridges (E10.5) [40], until E13.5, when female germ cells enter meiosis, while male germ cells continue to be mitotically active

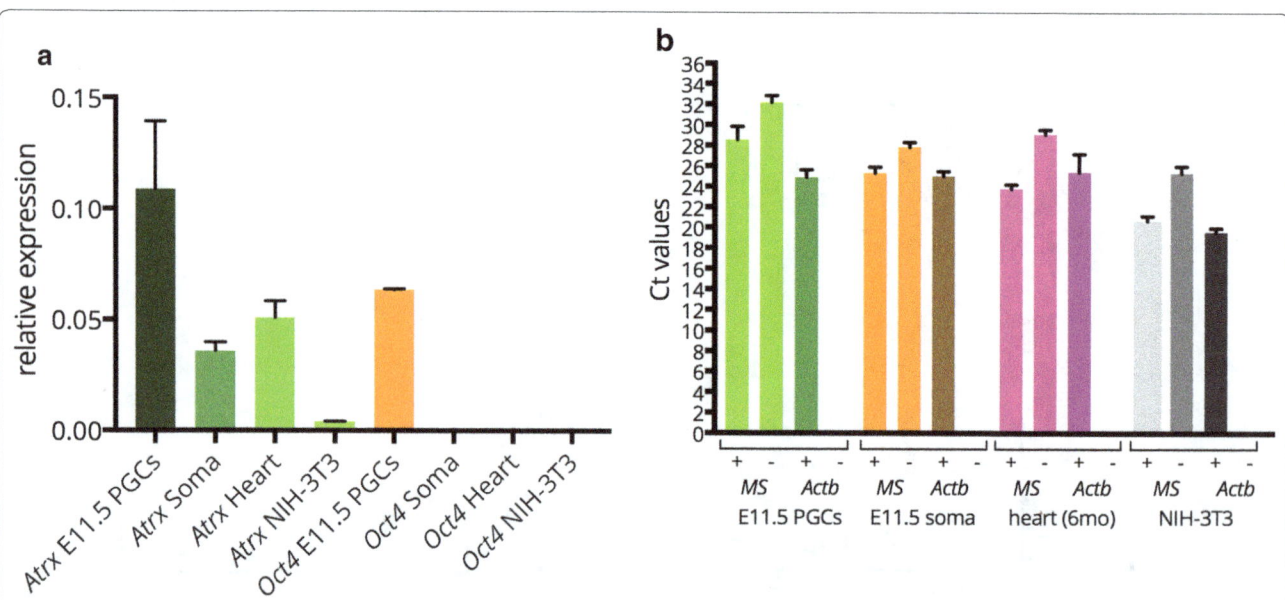

**Fig. 10** *Atrx* is more highly expressed in E11.5 PGCs compared to the soma, and low major satellite transcription is detected in both cell fractions. **a** Normalized expression values derived from qRT-PCR of E11.5 sorted OCT4-GFP-positive (E11.5 PGCs) and negative (E11.5 Soma) cells, show that expression of *Atrx* and *Oct4* is higher in PGCs compared to the soma, heart tissue, or NIH-3T3 cells. **b** Ct values obtained after qRT-PCR [Ct values of + or − RT reactions are shown, whereby no Ct value was obtained for the −RT performed with the *Actin (Actb)* primers] of major satellites (MS) and *Actb* in OCT4-GFP-positive (E11.5 PGCs) and negative (E11.5 Soma) cells, heart tissue, and NIH-3T3 cells. The difference between + and − RT reaction is a bit smaller for the E11.5 fractions compared to heart and NIH3T3 cells. *Error bars* represent standard deviations of two experiments

(until E15.5, when they enter a mitotically quiescent phase) [41]. Taking into account that epitope availability can be compromised under certain fixation conditions, we decided to test different preparation and fixation protocols. Indeed, when using our regular fixation protocol in sections, we observed loss of constitutive heterochromatin marks such as H3K9me3, H4K20me3, and HP1α exclusively in the germ cell nuclei, but not in the somatic nuclei, at E11.5. In striking contrast, upon extended fixation in sections, and in nuclear spread preparations, loss of these marks could not be reproduced. We obtained the most consistent results using nuclear spread preparations. This type of single cell methodology may in this case be superior to the former two, due to a better penetrance of the fixative and/or of the antibodies into the spread chromatin [19, 42]. In addition, loss of proteins that localize to the nucleoplasm and cytoplasm, and loss of proteins that are loosely associated with chromatin, may reduce the background signals, when histone modifications are studied. Previous studies used cytospin preparations to examine reprogramming taking place in germ cells [27]. The discrepancy between our results using nuclear spreads and these results using whole fixed cells may thus be attributed to higher background signals or reduced epitope accessibility in a more three-dimensional environment, whereby structural cellular and nuclear components such as membrane and matrix are still present. In support of our results, previous studies [26] could also not reproduce chromatin changes of H1 linker histone or loss of H3K27me3 at E11.5 reported elsewhere [27]. This again illustrates that testing different experimental methodologies is important in order to correctly understand and characterize epigenetic phenomena during different developmental states. In addition, somite counting at these early stages of development is a prerequisite for consistent developmental staging of the different embryos examined, to reduce inter-individual variability and thus improve reproducibility. From our quantification analysis, we can conclude that the level of H3K9me3 at pericentric heterochromatin is transiently reduced in PGCs at E11.5, when ATRX is most increased in these areas, compared to the patterns in surrounding somatic cells. HP1α and HP1β are even more clearly reduced in the pericentric heterochromatin of germ cells compared to somatic cells at (almost) all stages that were studied, but this appears to be compensated, at least in part, by a relative increase of HP1γ. Finally, H4K20me3 is also clearly reduced in pericentric heterochromatin of germ cells compared to surrounding somatic cells between E10.5 and E13.5. It should be taken into account that the differences in level intensities for HP1 isoforms and ATRX between PGCs and somatic cells may be somewhat over- or underestimated, since differences in

chromatin structure may also result in differential binding of these proteins to chromatin, which may affect the degree of retention during the nuclear spreading procedure. This issue does not apply to the histone modifications we examined, as they are tightly associated with the DNA. For ATRX, we could confirm the enrichment in PGCs by quantitative RT-PCR. However, for HP1γ in particular, the relative enrichment of this protein in the pericentric heterochromatin of PGCs is more outspoken in the nuclear spreads compared to the sections. Still, most importantly, none of the analysed markers were completely lost at any of the examined stages of development in mouse PGCs.

In light of our observations, it would also be interesting to re-examine whether H3K64me3, a newly identified histone modification marking constitutive heterochromatin, is truly absent from E12.0 to E13.5 germ cells as has been reported [43]. For this immunolocalization study, cryosections of embryo trunks and gonads were stained, using a protocol very similar to our own regular fixation protocol [43]. As our results suggest, such a protocol may not be suitable for answering constitutive heterochromatin localization questions, since epitopes may be masked.

Interestingly, our results show that ATRX, a chromatin remodeler and crucial factor for heterochromatin formation [29, 44], is maintained on pericentric heterochromatin throughout germ cell development. In addition, ATRX is enhanced in these locations of E11.5 PGCs compared to the surrounding somatic cells. Importantly, ATRX has been reported to transcriptionally block expression of major satellites from the maternal genome in the mouse zygote [36]. At this stage, in the early zygote, the maternal pericentromeres are labelled with the classical somatic histone modifications, while these marks are absent from the paternal genome, where transcription of major satellites has been recorded [13]. In addition, studies in embryonic stem cells report that ATRX, together with the histone chaperone DAXX, safeguards the genome against expression of tandem repeats, even when DNA methylation levels are absent from those regions [45]. It would be interesting to examine whether ATRX also performs such a repressive function in PGCs. An additional repressive mechanism against expression of those repeats could involve the formation of 5-hydroxyl-methylated DNA at pericentromeres in PGCs, gradually replacing 5-methyl-cytosine during PGC reprogramming [46]. Therefore, it is possible that more than one mechanism exists for silencing major satellite transcription in PGCs. In our study, we observed a similar low level of major satellite transcription in E11.5 PGCs and somatic cells. Thus, the differences that we observed in pericentric heterochromatin chromatin modification between PGCs and surrounding somatic cells, are not accompanied by a burst of major satellite transcription in PGCs as has been observed in the mouse pre-implantation embryos [22]. This is consistent with the fact that we observed that the vital players of constitutive heterochromatin are continuously present and that ATRX is enriched at pericentric repeats of E11.5 PGCs.

Nevertheless, analyses of the general distribution pattern of centromeres and adjacent pericentric heterochromatin revealed that there is a different organization of constitutive heterochromatin in germ cells compared to the surrounding somatic cells. Specifically, the somatic pericentromere organization into large chromocenters is much reduced in germ cells, where pericentromeres are mainly found as individual units or organized into small chromocenters. We have not identified the cause or the consequence of such an altered pericentromere organization, but this organization may be a natural consequence of germ cell development as they move from a somatic fate towards the more stem cell-like fate of a primordial germ cell and eventually towards the gonocyte. A similar phenomenon of a more dispersed constitutive heterochromatin has been described to take place upon reprogramming of mouse embryonic fibroblasts towards induced pluripotent stem cells, but also in the Nanog-positive cells of the inner cell mass of developing blastocysts [47]. In addition, DAPI-rich regions appear to spread upon induction of embryonic stem cells towards 2-cell stage-like cells [48]. Conversely, when cells differentiate, chromocenters appear to cluster. For example, when male germ cells reach their ultimate differentiated state in mouse adult testes, all chromocenters fuse into a single chromocenter in the nucleus of round, elongating, and condensed spermatids [12]. In addition, differentiation of myoblasts towards myocytes is also accompanied by centromere clustering and chromocenter formation, as well as further enrichment of H3K9me3 and H4K20me3. This differentiation is accompanied with transcriptional activation of major and minor satellite repeats [11].

## Conclusions

The present study reveals that pericentric heterochromatin organization in the embryonic PGC nucleus has changed dramatically from a clustered pattern into individual distribution, but the known hallmarks of heterochromatin are still present. In addition, ATRX, in combination with other mechanisms, may provide an extra level of protection against expression of major satellite transcripts. The observed changes in pericentric chromatin organization could be related to the transition of the germ cells from a somatic fate towards a stem cell-like one.

## Methods

### Collection of mouse embryos for immunofluorescence and immunocytochemistry

Female DBA2 mice were mated with C57BL/6 males to produce F1 fetuses. Mating was confirmed the next morning by the presence of a vaginal plug and recorded as E0.5. At E10.5, E11.5, and E13.5, embryos were dissected out of the uteri and were assessed for somite counting. We scored embryos with 34–36 somites as E10.5 and 44–47 somites as E11.5. We could not determine with precision the somite number at E13.5 (60–62 somites), due to the advanced developmental stage of the embryo. Embryos were kept in ice-cold PBS at all times, before any further processing.

### Tissue processing for immunofluorescence and immunocytochemistry

After embryo isolation from the uteri, embryo regions containing the developing germ cells were dissected from E10.5 to E11.5 embryos. Gonads were isolated from the E13.5 embryos, and the sex was determined by morphology. E10.5 and E11.5 gonadal regions were fixed in ice-cold 4% PFA for 2 h and 3 h, respectively, followed by consecutive washes in PBS. Gonads were fixed for 1.5 h in ice-cold 4% PFA. Tissues were then processed for OCT or paraffin embedding using standard histology procedures. Cryo- and paraffin sections were 10 and 5 μm, respectively.

For the regular and extended fixation, sections were fixed for an additional 10 min at room temperature or for 30 min at 37 °C, respectively, followed by brief PBS washes. The fixation step was performed after the OCT or paraffin was removed from the sections.

### Drying-down or nuclear spread preparations of germ cells

Embryo trunks containing the germ cells from E10.5, E11.5 and gonads from E13.5 embryos were dissected, pooled as indicated in figure legends, and incubated in 500 μl TrypLE™ Express (Thermo Fisher Scientific) for 6 min at 37 °C. Dissociation was followed by two washes with 5% FBS in PBS. Spreads of nuclei for immunocytochemistry were obtained as described by [28].

### Quantification analysis of immunocytochemistry

Single plane images at optimal focus were acquired with a Zeiss LSM 700 microscope (Carl Zeiss, Jena) with the same exposure time for each nucleus of the same stage. A homemade ImageJ macro was used to measure the mean fluorescence intensity of each pericentric heterochromatin marker used for immunofluorescence, in the whole nucleus (defined by the DAPI-positive area), or in pericentric heterochromatin regions, defined by the area that contained signal above the set background threshold for the corresponding marker, and corresponding to regions more strongly stained with DAPI.

## Immunohistochemistry and immunocytochemistry

Heat-mediated (900 W in a microwave for 20 min) epitope retrieval in citrate buffer pH = 6 was performed on paraffin sections. The following staining protocol was performed in all samples. Sections and nuclear spreads were blocked with 2% BSA, 5% donkey serum in PBS (blocking solution) for 30 min at room temperature, followed by primary antibody incubation, diluted in blocking solution, at 4 °C overnight in a humid chamber. The next day, slides were washed in PBS (3 × 5 min) and blocked with secondary antibodies, diluted in blocking buffer, for 1 h at room temperature, in a humid chamber. Slides were then washed in PBS (3 × 5 min) and mounted with ProLong® Gold Antifade Mountant with DAPI (Thermo Fisher Scientific). Confocal imaging was performed on a Zeiss LSM700 microscope (Carl Zeiss, Jena). In this study, the following primary antibodies were used: goat anti-OCT3/4 (N-19) by Santa Cruz (sc-8628) diluted 1:800 for sections and 1:50 for spread preparations, rabbit anti-OCT4 by Abcam (ab19857) diluted 1:250 for sections and 1:50 for spread preparations, rat anti-TRA98 by Abcam (ab82527, 1:500), rabbit anti-DDX4/MVH by Abcam (ab13840, 1:300), anti-rabbit H3K9me3 by Abcam (ab8898, 1:300), rabbit anti-H4K20me3 diluted 1:300 [49], goat anti-HP1α by Abcam (ab77256) diluted 1:200 for sections and 1:400 for spread preparations, mouse anti-HP1β by Abcam (ab10478, 1:200), rabbit anti-HP1γ by Abcam (ab10480, 1:200), and rabbit anti-ATRX (H-300) by Santa Cruz (sc-15408) 1:250, human anti-CREST (CS-1058) by Cortex Biochem 1:1000. The following Alexa Fluor secondary antibodies were used: donkey anti-goat 555/488, donkey anti-rat 555, donkey anti-mouse 488, and donkey anti-rabbit 488 by Thermo Fisher Scientific. All the Alexa Fluor 555 antibodies were used at a dilution of 1:400, while the Alexa Fluor 488 antibodies were diluted 1:250. To detect CREST, we used donkey anti-human 488 DyLight 488 (SA5-10126) by Thermo Fisher Scientific at a 1:250 dilution.

## FACS sorting

Female DBA2 mice were mated with OCT4-GOF18/GFP C57BL/6 males to produce F1 fetuses carrying the OCT4-GFP transgene [50]. Staging of the embryos and dissociation of the tissue were performed as described above (*Drying-down or nuclear spread preparations of germ cells* section). Equal numbers of PGCs and somatic cells were isolated using the SORP-FACSAria II flow cytometer (BD). In more detail, 600 cells per cell population were sorted at 4 °C in 40ul lysis buffer (AM1722, Cells-to-cDNA™ II Kit, Thermo Fisher Scientific) in a 96-well plate containing additionally 2U/µl RNAseOUT (10777-019, Invitrogen). Thereafter, the lysis buffer containing the cells was split into four tubes (two for +RT and two for −RT experiments) and cDNA reactions were performed as described below (RT-qPCR). For NIH-3T3 and heart tissue, RNA was isolated with TRIzol reagent (15596026, Thermo Fisher Scientific). Thereafter, RNA was treated with Turbo DNAse (AM2238, Thermo Fisher Scientific) according to the manufacturer's instructions.

## RT-qPCR

For quantitative RT-PCR (RT-qPCR), the lysis buffer (Cells-to-cDNA™ II kit, Thermo Fisher Scientific; AM1722) containing the cells was processed for cDNA according to the manufacturer's instructions. cDNA from NIH-3T3 cells and heart tissue was made with Superscript III (18080093, Thermo Fisher Scientific) according to the manufacturer's instructions.

All samples were analysed in triplicate in a 15-µl final reaction volume using the BioRad CFX 384 Real-time System. Each reaction contained LightCycler® 480 SYBR Green I Master (04887352001; Roche), primers to a final concentration of 0,2 µM and 1 µl of cDNA. The following primers were used: *Oct4* Fw CCCCAATGCCGTGAAGTTG and Rv TCAGCAGCTTGGCAAACTGTT, *major satellites* Fw GGCGAGAAAACTGAAAATCACG and Rv AGGTCCTTCAGTGTGCATTTC [51], *Atrx* Fw GAGCTTGACGTGAAACGAAGAG and Rv TTGTTG CTGTTGCTGCTGAG, *Actin* Fw ACTATTGGCAAC GAGCGGTTC and Rv AGAGGTCTTTACGGATGTC AACG.

After an initial hold at 94 °C for 4 min, reaction mixtures underwent 40 cycles of 30 s at 94 °C, 30 s at 60 °C, and 30 s at 72 °C. Gene expression levels were normalized over *Actin* expression according to the 2-ΔCt method. For the major satellites, the Ct values were compared.

## Additional files

**Additional file 1.** Immunofluorescent analysis of HP1α in paraffin sections using regular and extended fixation protocols. **A** In E10.5 embryos, HP1α (*green*) is enriched at pericentric heterochromatin, but its levels are lower in PGCs compared to somatic cells. From E11.5 onwards, HP1α signal is depleted from DAPI (*blue*)-dense regions in PGCs. **B** Results were similar to **A** when using extended fixation conditions. However, at E11.5 some PGCs could be detected with some signal of HP1α still present at the pericentric heterochromatin (marked by *arrowhead*). Note that in E13.5 male gonads HP1α could not be reproducibly detected in somatic cells, in both **A** and **B**. For each stage, two embryos were analysed per fixation protocol and at least 20 nuclei were recorded. E10.5 and E11.5 PGCs were marked with OCT4 (*red*). E13.5 male and female germ cells were identified by the presence of DDX4/MVH (*red*). Representative images are shown with germ cells highlighted by *dashed yellow circles*, and *scale bars* represent 5 µm.

**Additional file 2.** Immunofluorescent analysis of HP1β in paraffin sections using regular and extended fixation protocols. **A** Using the regular fixation protocol, HP1β signal is enriched at DAPI (*blue*)-dense regions of E10.5 and E11.5 PGCs and somatic cells. HP1β is then substantially reduced in E13.5 female and male germ cells. **B** With the extended fixation protocol, HP1β signal is retained in pericentric heterochromatin of PGCs throughout development. For each stage, two embryos were analysed per fixation protocol and at least 20 nuclei were recorded. E10.5 and E11.5 PGCs were marked with OCT4 (*red*). E13.5 male and female germ cells were identified by the presence of TRA98 (*red*). Representative images are shown with germ cells highlighted by *dashed yellow circles*, and *scale bars* represent 5 μm.

**Additional file 3.** Immunofluorescent analysis of HP1γ in paraffin sections using regular and extended fixation protocols. **A** HP1γ (*green*) signal is enriched at DAPI (*blue*)-dense regions of E10.5 and E11.5 PGCs and somatic cells using the regular fixation protocol. Thereafter, at E13.5, HP1γ could not be detected in male germ cells, while it was still present in E13.5 female germ cell nuclei. **B** Upon application of the extended fixation protocol, enrichment of HP1γ signal was observed in pericentric heterochromatin of E10.5 and E11.5 PGCs. Similar to **A**, HP1γ could not be detected at pericentric heterochromatin of male E13.5 germ cells, while it was still present in E13.5 female germ cells. Note that in both protocols (**A**, **B**) HP1γ could not reproducibly be detected in pericentric heterochromatin of E13.5 somatic cells. For each stage, two embryos were analysed per fixation protocol and at least 20 PGC nuclei were recorded. E10.5 and E11.5 PGCs were marked with OCT4 (*red*). E13.5 male and female germ cells were identified by the presence of TRA98 (*red*). Representative images are shown with germ cells highlighted by *dashed yellow circles*, and *scale bars* represent 5 μm.

**Additional file 4.** Summary of the immunosignals at pericentric heterochromatin of PGCs. The table displays whether a certain histone modification or chromatin-binding protein at the pericentric heterochromatin is detected (+), not detected (−) or detected in some but not all (*) nuclei of PGCs at the embryonic stages indicated. Differences in the degree of enrichment between the pericentric heterochromatin of PGCs and somatic cells are only taken into account in the last column, whereby less enrichment or more enrichment at pericentromeric heterochromatin in PGCs compared to the soma is indicated as <or>, respectively. n.d.: not determined.

## Abbreviations

DAPI: 4′,6-diamidino-2-phenylindole; E11.5: embryonic day 11.5; HP1: heterochromatin protein 1; PGC: primordial germ cell; PRC1 and PRC2: polycomb repressor complexes 1 and 2; RT-qPCR: reverse transcriptase-quantitative polymerase chain reaction.

## Authors' contributions

AM and ME conceived the work and designed the experiments, GvdH, JG, AHFMP, and WMB contributed to the design of the experiments, AM and ESL acquired and analysed the data, LM and WAC analysed part of the data, AM, ME, and WMB interpreted the data, AM and WMB wrote the draft of the manuscript, and GvdH, JG, AHFMP, and ME critically revised the manuscript. All authors read and approved the final manuscript.

## Author details

[1] Department of Developmental Biology, Erasmus MC, University Medical Center, Rotterdam, The Netherlands. [2] Division of Reproductive Medicine, Department of Obstetrics and Gynecology, Erasmus MC, Rotterdam, The Netherlands. [3] Friedrich Miescher Institute for Biomedical Research (FMI), Basel, Switzerland. [4] Faculty of Sciences, University of Basel, Basel, Switzerland. [5] Division of Reproductive Medicine, Department of Obstetrics and Gynecology, Central Hospital of Karlstad, Karlstad, Värmland, Sweden. [6] Erasmus Optical Imaging Center, Erasmus MC, Rotterdam, The Netherlands.

## Acknowledgements

We thank Dr. Sarra Merzouk for scientific discussions and Tsung Wai Kan for the sorting experiments.

## Competing interests

The authors declare that they have no competing interests.

## Funding

This work was supported by a NWO-VIDI grant 917.10.367 to ME.

## References

1. Rea S, Eisenhaber F, O'Carroll D, Strahl BD, Sun ZW, Schmid M, Opravil S, Mechtler K, Ponting CP, Allis CD, Jenuwein T. Regulation of chromatin structure by site-specific histone H3 methyltransferases. Nature. 2000;406:593–9.
2. Peters AHFM, O'Carroll D, Scherthan H, Mechtler K, Sauer S, Schöfer C, Weipoltshammer K, Pagani M, Lachner M, Kohlmaier A, Opravil S, Doyle M, Sibilia M, Jenuwein T. Loss of the Suv39 h histone methyltransferases impairs mammalian heterochromatin and genome stability. Cell. 2001;107:323–37.
3. Lehnertz B, Ueda Y, Derijck AAHA, Braunschweig U, Perez-Burgos L, Kubicek S, Chen T, Li E, Jenuwein T, Peters AHFM. Suv39 h-mediated histone H3 lysine 9 methylation directs DNA methylation to major satellite repeats at pericentric heterochromatin. Curr Biol. 2003;13:1192–200.
4. Schotta G, Lachner M, Sarma K, Ebert A, Sengupta R, Reuter G, Reinberg D, Jenuwein T. A silencing pathway to induce H3-K9 and H4-K20 trimethylation at constitutive heterochromatin. Genes Dev. 2004;18:1251–62.
5. Kourmouli N, Jeppesen P, Mahadevhaiah S, Burgoyne P, Wu R, Gilbert DM, Bongiorni S, Prantera G, Fanti L, Pimpinelli S, Shi W, Fundele R, Singh PB. Heterochromatin and tri-methylated lysine 20 of histone H4 in animals. J Cell Sci. 2004;117(Pt 12):2491–501.
6. Bannister aJ, Zegerman P, Partridge JF, Miska Ea, Thomas JO, Allshire RC, Kouzarides T. Selective recognition of methylated lysine 9 on histone H3 by the HP1 chromo domain. Nature. 2001;410:120–4.
7. Lachner M, O'Sullivan RJ, Jenuwein T. An epigenetic road map for histone lysine methylation. J Cell Sci. 2003;116(Pt 11):2117–24.
8. Guenatri M, Bailly D, Maison C, Almouzni G. Mouse centric and pericentric satellite repeats form distinct functional heterochromatin. J Cell Biol. 2004;166:493–505.
9. Lu J, Gilbert DM. Proliferation-dependent and cell cycle-regulated transcription of mouse pericentric heterochromatin. J Cell Biol. 2007;179:411–21.
10. Boyarchuk E, Filipescu D, Vassias I, Cantaloube S, Almouzni G: Pericentric heterochromatin state during the cell cycle controls the histone variant composition of centromeres. J Cell Sci. 2014;127:3347–59.
11. Terranova R, Sauer S, Merkenschlager M, Fisher AG. The reorganisation of constitutive heterochromatin in differentiating muscle requires HDAC activity. Exp Cell Res. 2005;310:344–56.
12. Govin J, Escoffier E, Rousseaux S, Kuhn L, Ferro M, Thévenon J, Catena R, Davidson I, Garin J, Khochbin S, Caron C. Pericentric heterochromatin reprogramming by new histone variants during mouse spermiogenesis. J Cell Biol. 2007;176:283–94.
13. Puschendorf M, Terranova R, Boutsma E, Mao X, Isono K, Brykczynska U, Kolb C, Otte AP, Koseki H, Orkin SH, van Lohuizen M, Peters AHFM. PRC1 and SUV39H specify parental asymmetry at constitutive heterochromatin in early mouse embryos. Nat Genet. 2008;40:411–20.
14. Casanova M, Pasternak M, El Marjou F, Le Baccon P, Probst AV, Almouzni G. Heterochromatin reorganization during early mouse development requires a single-stranded noncoding transcript. Cell Rep. 2013;4:1156–67.
15. Rudert F, Bronner S, Garnier JM, Dollé P. Transcripts from opposite strands of gamma satellite DNA are differentially expressed during mouse development. Mamm Genome. 1995;6:76–83.
16. Santos F, Hendrich B, Reik W, Dean W. Dynamic reprogramming of DNA methylation in the early mouse embryo. Dev Biol. 2002;241:172–82.

17. Santos F, Peters AH, Otte AP, Reik W, Dean W. Dynamic chromatin modifications characterise the first cell cycle in mouse embryos. Dev Biol. 2005;280:225–36.

18. Torres-Padilla M-E, Bannister AJ, Hurd PJ, Kouzarides T, Zernicka-Goetz M. Dynamic distribution of the replacement histone variant H3.3 in the mouse oocyte and preimplantation embryos. Int J Dev Biol. 2006;50:455–61.

19. van der Heijden GW, Dieker JW, Derijck AAHA, Muller S, Berden JHM, Braat DDM, van der Vlag J, de Boer P. Asymmetry in histone H3 variants and lysine methylation between paternal and maternal chromatin of the early mouse zygote. Mech Dev. 2005;122:1008–22.

20. Santos F, Dean W. Epigenetic reprogramming during early development in mammals. Reproduction. 2004;127:643–51.

21. Albert M, Peters AHFM. Genetic and epigenetic control of early mouse development. Curr Opin Genet Dev. 2009;19:113–21.

22. Probst AV, Okamoto I, Casanova M, El Marjou F, Le Baccon P, Almouzni G. A strand-specific burst in transcription of pericentric satellites is required for chromocenter formation and early mouse development. Dev Cell. 2010;19:625–38.

23. Hackett JA, Zylicz JJ, Surani MA. Parallel mechanisms of epigenetic reprogramming in the germline. Trends Genet. 2012;28:164–74.

24. Hill PWS, Amouroux R, Hajkova P. DNA demethylation, Tet proteins and 5-hydroxymethylcytosine in epigenetic reprogramming: an emerging complex story. Genomics. 2014;104:324–33.

25. Hajkova P, Erhardt S, Lane N, Haaf T, El-Maarri O, Reik W, Walter J, Surani MA. Epigenetic reprogramming in mouse primordial germ cells. Mech Dev. 2002;117:15–23.

26. Kagiwada S, Kurimoto K, Hirota T, Yamaji M, Saitou M. Replication-coupled passive DNA demethylation for the erasure of genome imprints in mice. EMBO J. 2013;32:340–53.

27. Hajkova P, Ancelin K, Waldmann T, Lacoste N, Lange UC, Cesari F, Lee C, Almouzni G, Schneider R, Surani MA. Chromatin dynamics during epigenetic reprogramming in the mouse germ line. Nature. 2008;452:877–81.

28. Peters AH, Plug AW, van Vugt MJ, de Boer P. A drying-down technique for the spreading of mammalian meiocytes from the male and female germline. Chromosome Res. 1997;5:66–8.

29. McDowell TL, Gibbons RJ, Sutherland H, O'Rourke DM, Bickmore WA, Pombo A, Turley H, Gatter K, Picketts DJ, Buckle VJ, Chapman L, Rhodes D, Higgs DR. Localization of a putative transcriptional regulator (ATRX) at pericentromeric heterochromatin and the short arms of acrocentric chromosomes. Proc Natl Acad Sci USA. 1999;96:13983–8.

30. Baumann C, Schmidtmann A, Muegge K, De La Fuente R. Association of ATRX with pericentric heterochromatin and the Y chromosome of neonatal mouse spermatogonia. BMC Mol Biol. 2008;9:29.

31. Takada Y, Naruse C, Costa Y, Shirakawa T, Tachibana M, Sharif J, Kezuka-Shiotani F, Kakiuchi D, Masumoto H, Shinkai Y, Ohbo K, Peters AHFM, Turner JMA, Asano M, Koseki H. HP1γ links histone methylation marks to meiotic synapsis in mice. Development. 2011;138:4207–17.

32. Smallwood A, Hon GC, Jin F, Henry RE, Espinosa JM, Ren B. CBX3 regulates efficient RNA processing genome-wide. Genome Res. 2012;22:1426–36.

33. Vakoc CR, Mandat SA, Olenchock BA, Blobel GA. Histone H3 lysine 9 methylation and HP1gamma are associated with transcription elongation through mammalian chromatin. Mol Cell. 2005;19:381–91.

34. Schotta G, Sengupta R, Kubicek S, Malin S, Kauer M, Callén E, Celeste A, Pagani M, Opravil S, De La Rosa-Velazquez IA, Espejo A, Bedford MT, Nussenzweig A, Busslinger M, Jenuwein T. A chromatin-wide transition to H4K20 monomethylation impairs genome integrity and programmed DNA rearrangements in the mouse. Genes Dev. 2008;22:2048–61.

35. Ishov AM, Vladimirova OV, Maul GG. Heterochromatin and ND10 are cell-cycle regulated and phosphorylation-dependent alternate nuclear sites of the transcription repressor Daxx and SWI/SNF protein ATRX. J Cell Sci. 2004;117(Pt 17):3807–20.

36. De La Fuente R, Baumann C, Viveiros MM: ATRX contributes to epigenetic asymmetry and silencing of major satellite transcripts in the maternal genome of the mouse embryo. Development. 2015;142:1806–17.

37. Seki Y, Hayashi K, Itoh K, Mizugaki M, Saitou M, Matsui Y. Extensive and orderly reprogramming of genome-wide chromatin modifications associated with specification and early development of germ cells in mice. Dev Biol. 2005;278:440–58.

38. Ohinata Y, Ohta H, Shigeta M, Yamanaka K, Wakayama T, Saitou M. A signaling principle for the specification of the germ cell lineage in mice. Cell. 2009;137:571–84.

39. Seki Y, Yamaji M, Yabuta Y, Sano M, Shigeta M, Matsui Y, Saga Y, Tachibana M, Shinkai Y, Saitou M. Cellular dynamics associated with the genome-wide epigenetic reprogramming in migrating primordial germ cells in mice. Development. 2007;134:2627–38.

40. Molyneaux KA, Stallock J, Schaible K, Wylie C. Time-lapse analysis of living mouse germ cell migration. Dev Biol. 2001;240:488–98.

41. Yoshioka H, McCarrey JR, Yamazaki Y. Dynamic nuclear organization of constitutive heterochromatin during fetal male germ cell development in mice. Biol Reprod. 2009;80:804–12.

42. Baarends WM, Wassenaar E, van der Laan R, Hoogerbrugge J, Sleddens-Linkels E, Hoeijmakers JHJ, de Boer P, Grootegoed JA. Silencing of unpaired chromatin and histone H2A ubiquitination in mammalian meiosis. Mol Cell Biol. 2005;25:1041–53.

43. Daujat S, Weiss T, Mohn F, Lange UC, Ziegler-Birling C, Zeissler U, Lappe M, Schübeler D, Torres-Padilla M-E, Schneider R. H3K64 trimethylation marks heterochromatin and is dynamically remodeled during developmental reprogramming. Nat Struct Mol Biol. 2009;16:777–81.

44. Sadic D, Schmidt K, Groh S, Kondofersky I, Ellwart J, Fuchs C, Theis FJ, Schotta G. Atrx promotes heterochromatin formation at retrotransposons. EMBO Rep. 2015;16:836–50.

45. He Q, Kim H, Huang R, Lu W, Tang M, Shi F, Yang D, Zhang X, Huang J, Liu D, Songyang Z. The Daxx/Atrx complex protects tandem repetitive elements during DNA hypomethylation by promoting H3K9 trimethylation. Cell Stem Cell. 2015;17:273–86.

46. Yamaguchi S, Hong K, Liu R, Inoue A, Shen L, Zhang K, Zhang Y. Dynamics of 5-methylcytosine and 5-hydroxymethylcytosine during germ cell reprogramming. Cell Res. 2013;23:329–39.

47. Fussner E, Djuric U, Strauss M, Hotta A, Perez-Iratxeta C, Lanner F, Dilworth FJ, Ellis J, Bazett-Jones DP. Constitutive heterochromatin reorganization during somatic cell reprogramming. EMBO J. 2011;30:1778–89.

48. Ishiuchi T, Enriquez-Gasca R, Mizutani E, Bošković A, Ziegler-Birling C, Rodriguez-Terrones D, Wakayama T, Vaquerizas JM, Torres-Padilla M-E. Early embryonic-like cells are induced by downregulating replication-dependent chromatin assembly. Nat Struct Mol Biol. 2015;22:662–71.

49. Peters AHFM, Kubicek S, Mechtler K, O'Sullivan RJ, Derijck AAHA, Perez-Burgos L, Kohlmaier A, Opravil S, Tachibana M, Shinkai Y, Martens JHA, Jenuwein T. Partitioning and plasticity of repressive histone methylation states in mammalian chromatin. Mol Cell. 2003;12:1577–89.

50. Yoshimizu T, Sugiyama N, De Felice M, Yeom YI, Ohbo K, Masuko K, Obinata M, Abe K, Scholer HR, Matsui Y. Germline-specific expression of the Oct-4/green fluorescent protein (GFP) transgene in mice. Dev Growth Differ. 1999;41:675–84.

51. Maze I, Feng J, Wilkinson MB, Sun H, Shen L, Nestler EJ. Cocaine dynamically regulates heterochromatin and repetitive element unsilencing in nucleus accumbens. Proc Natl Acad Sci USA. 2011;108:3035–40.

# 12

# PRC2 is required for extensive reorganization of H3K27me3 during epigenetic reprogramming in mouse fetal germ cells

Lexie Prokopuk[1], Jessica M. Stringer[1], Kirsten Hogg[1], Kirstin D. Elgass[2] and Patrick S. Western[1*] (iD)

## Abstract

**Background:** Defining how epigenetic information is established in the germline during fetal development is key to understanding how epigenetic information is inherited and impacts on evolution and human health and disease.

**Results:** Here, we show that Polycomb Repressive Complex 2 is transiently localized in the nucleus of mouse fetal germ cells, while DNA methylation is removed from the germline. This coincides with significant enrichment of tri-methylated lysine 27 on histone 3 near the nuclear lamina that is dependent on activity of the essential PRC2 catalytic proteins, Enhancer of Zeste 1 and/or 2.

**Conclusions:** Combined, these data reveal a role for Polycomb Repressive Complex 2 and trimethylated lysine 27 on histone 3 during germline epigenetic programming that we speculate is required to repress target sequences while DNA methylation is removed.

**Keywords:** Epigenetics, Germ cells, Epigenetic reprogramming, Histone modifications, H3K27me3

## Background

The sperm and oocyte convey epigenetic information to the offspring that is essential for development and post-natal life. To ensure that appropriate epigenetic information is transmitted, germ cells undergo extensive reprogramming during fetal development. This allows existing parent-specific information to be reset and potential epigenetic errors to be removed from the germline. The most evident change during germline epigenetic reprogramming is the extensive removal of DNA methylation between embryonic day (E) 8.0 and E13.5, after which new DNA methylation is established in the male and female germlines in a sex-specific manner. Removal of germline DNA methylation occurs during two developmental stages. The first occurs during proliferation and migration of germ cells in the early embryo and involves reduction of DNA methylation levels from

70 to 30% [1]. This is followed by further reductions in DNA methylation as the germ cells enter the developing gonads and initiate sex-specific development. Combining this extensive epigenetic reprogramming results in global DNA methylation levels of ~14 and 7% in male and female germ cells at E13.5, respectively [1].

During this later period of reprogramming, DNA methylation is removed from imprint control regions (ICRs) of maternally and paternally imprinted genes, non-imprinted genes, intergenic and intronic sequences, and from the inactive X-chromosome in XX germ cells [2]. DNA methylation is also lost from many endogenous retroviral elements (ERVs), although relatively new ERVs are protected and retain relatively high levels of DNA methylation [1–6].

In addition to DNA demethylation, germline reprogramming involves extensive reorganization of histone modifications [7, 8]. At E8.5, dimethylated lysine 9 on histone 3 (H3K9me2) is exchanged for trimethylated lysine 27 on histone 3 (H3K27me3), which is maintained in the germ cell nucleus until at least E12.5 [9], although

---
*Correspondence: patrick.western@hudson.org.au
[1] Department of Molecular and Translational Science, Centre for Genetic Diseases, Hudson Institute of Medical Research, Monash University, Clayton, VIC 3168, Australia
Full list of author information is available at the end of the article

transient loss of H3K27me3 has been reported at E11.5 [7].

H3K27me3 is catalyzed by Polycomb Repressive Complex 2 (PRC2) and leads to epigenetic repression of target sequences. PRC2 is comprised of three core components: Enhancer of Zeste 1 or 2 (EZH1 and EZH2); Suppressor of Zeste 12 (SUZ12); and Embryonic Ectoderm Development (EED). While EZH2 is the primary catalytic component of PRC2 [10–12], EED is critical for physically binding H3K27me3 via five tandemly repeated WD motifs, thereby playing an essential role in PRC2 assembly [13, 14]. Inactivation of any one of the three core PRC2 protein subunits severely compromises the functional activity of the complex and results in the loss of H3K27me3. Complete loss of PRC2 function in mice is embryonic lethal [15–17], while low function (hypomorphic) mutations [18, 19] or tissue-specific deletion leads to growth and skeletal malformations [20], heart and immune cell defects [21] and increased susceptibility to tumorigenesis [22]. Conditional deletion of *Eed* in the male germline from around birth results in complete male infertility demonstrating an essential role for PRC2 in male germline development [23]. Although deletion of *Ezh2* in growing oocytes is compatible with normal fertility, offspring are born underweight, indicating that PRC2 acts as a maternal factor in oocytes [24]. However, PRC2 is required for germline development in *Drosophila* oocytes, where it regulates cell cycle progression during oocyte specification [25]. Moreover, PRC2 is required in the *Caenorhabditis elegans* germline to regulate the balance between H3K27me3 patterning catalyzed by MES 2, 3 and 6 (*C. elegans* PRC2), and MES4 which catalyzes H3K36me3 [26]. Despite these observations, the function of PRC2 in male and female in mammalian fetal germ cells remains unknown.

H3K27me3 is enriched at the promoter regions of many PRC2 target genes and plays an essential role in cell differentiation [27, 28]. In fetal germ cells, H3K27me3 is enriched at developmental genes [29–31] and is also enriched on nucleosomes that are retained at the promoters of developmental genes in mature sperm, indicating that PRC2 may regulate epigenetic information that is transmitted to offspring [32–34].

In this study, we identify a key period of transient PRC2 enrichment in gonadal germ cells as they undergo epigenetic reprogramming. Moreover, we demonstrate that PRC2 is required for significant transient enrichment of H3K27me3 near the nuclear lamina, specifically during the developmental period in which germline DNA methylation levels are at their lowest. We propose that PRC2 and H3K27me3 are required for epigenetic reprogramming in fetal germ cells and may provide a transient mechanism that protects certain sequences from

aberrant expression during the period of reduced DNA methylation in the developing germline.

## Results

### H3K27me3 is highly enriched in fetal germ cells undergoing epigenetic reprogramming

Initially, we used quantitative flow cytometric analyses to determine the overall cellular levels of H3K27me3 in E11.5, E13.5 and E15.5 male and female germ cells as they undergo epigenetic reprogramming. Germ cells were identified based on germ cell-specific expression of an *Oct4*-eGFP transgene [35–38] (Fig. 1A). At E11.5, XX and XY germ cells contained high levels of H3K27me3 (Fig. 1A, B), with no evidence in either sex of a germ cell population in which H3K27me3 was low or negative (Fig. 1B). Relative to E11.5 germ cells, H3K27me3 levels were moderately reduced in E13.5 germ cells of both sexes and then maintained until E15.5 (Fig. 1A). At all stages, germ cells contained substantially higher H3K27me3 levels than the gonadal somatic cells, which maintained relatively constant, lower levels of H3K27me3 (Fig. 1A). No H3K27me3 staining was detected in oocytes in which *Eed* had been conditionally deleted through expression of *Zp3-Cre*, confirming that the H3K27me3 antibody used specifically detects H3K27me3 (Additional file 1: Fig. S1).

### PRC2 is transiently expressed in fetal germ cells undergoing epigenetic reprogramming

Next, we determined the temporal and spatial profile of the core PRC2 protein components EED, EZH2, SUZ12 and H3K27me3 in XX and XY fetal germ cells. Immunofluorescence (IF) was carried out on E10.5, E11.5, E12.5, E13.5 and E15.5 male and female gonads, coinciding with entry of germ cells into the developing gonad, epigenetic reprogramming in gonadal germ cells and early sex-specific germline development (Figs. 2, 3, 4). At E10.5, EED was not detected in either XX or XY germ cell (Fig. 2). EZH2 was detected with low staining intensity in the nucleus of some E10.5 XY germ cells but not XX germ cells (Fig. 3), and SUZ12 was detected with low staining intensity in the nucleus of both E10.5 XX and XY germ cells (Fig. 4). However, EED, EZH2 and SUZ12 were all readily detected in the nucleus of E11.5 and E12.5 XX and XY germ cells (Figs. 2, 3, 4). This pattern continued in E13.5 XX and XY germ cells (Figs. 2, 3, 4), with the exception of EED, which was not detected in E13.5 or E15.5 XX germ cells (Fig. 2). At E15.5, EED was weakly detected in the cytoplasm and nucleus of XY germ cells (Fig. 2). EZH2 was readily detected in E15.5 XX and XY germ cells (Fig. 3), while SUZ12 was only detected in XY germ cells at E15.5 (Fig. 4). We have summarized the relative staining intensity for each protein in E10.5–E15.5 XX and XY germ cells (Table 1).

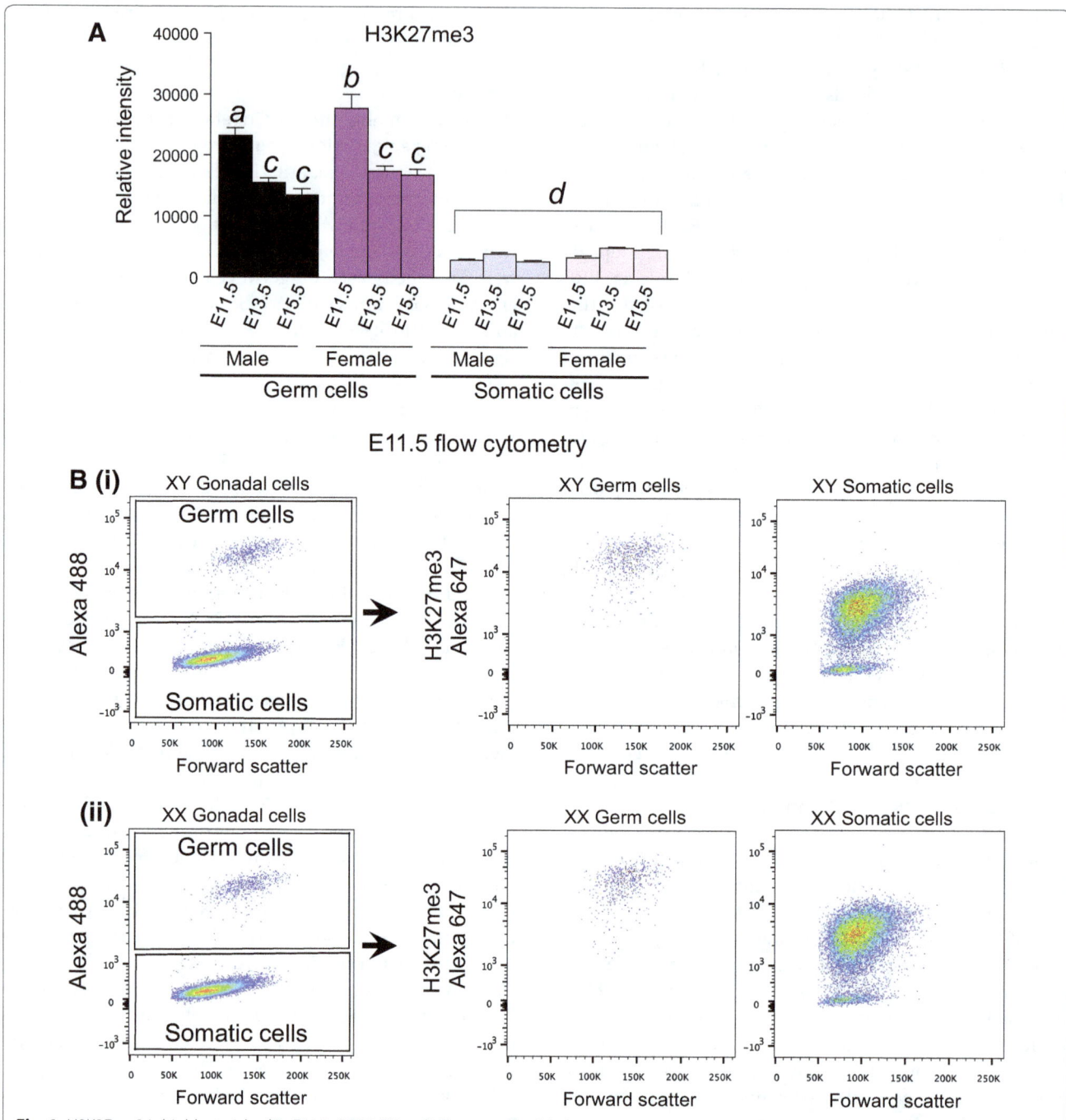

**Fig. 1** H3K27me3 is highly enriched in E11.5–E15.5 XX and XY germ cells. **A** Relative H3K27me3 antibody fluorescence intensities were measured using flow cytometry of dissociated and stained male (XY) and female (XX) fetal gonads at E11.5, E13.5 and E15.5. Germ cells were separated from somatic cells based on *Oct4*-eGFP expression (shown in **B**) and average ± SEM H3K27me3 levels measured in each population (*n* = 4). **B** Flow cyto-metric scatter plots showing gating of *Oct4*-eGFP positive germ cells and *Oct4*-eGFP negative somatic cells in XY (*i*) and XX (*ii*) E11.5 fetal gonads. Distribution of H3K27me3 intensities is shown for the germ cell (*middle plots*) and somatic cell populations (*right-hand plots*)

## H3K27me3 is enriched near the nuclear lamina in fetal germ cells undergoing epigenetic reprogramming

Consistent with previous studies [7, 9, 39], H3K27me3 was detected relatively uniformly throughout the nucleus of E10.5 XX and XY germ cells (Fig. 5a). Although

H3K27me3 was readily detected in E11.5 XX and XY germ cells, it was enriched close to the nuclear lamina, with little or no staining detected in the center of the nucleus (Fig. 5a). By E12.5, H3K27me3 staining was more obviously localized near the nuclear lamina in both XX

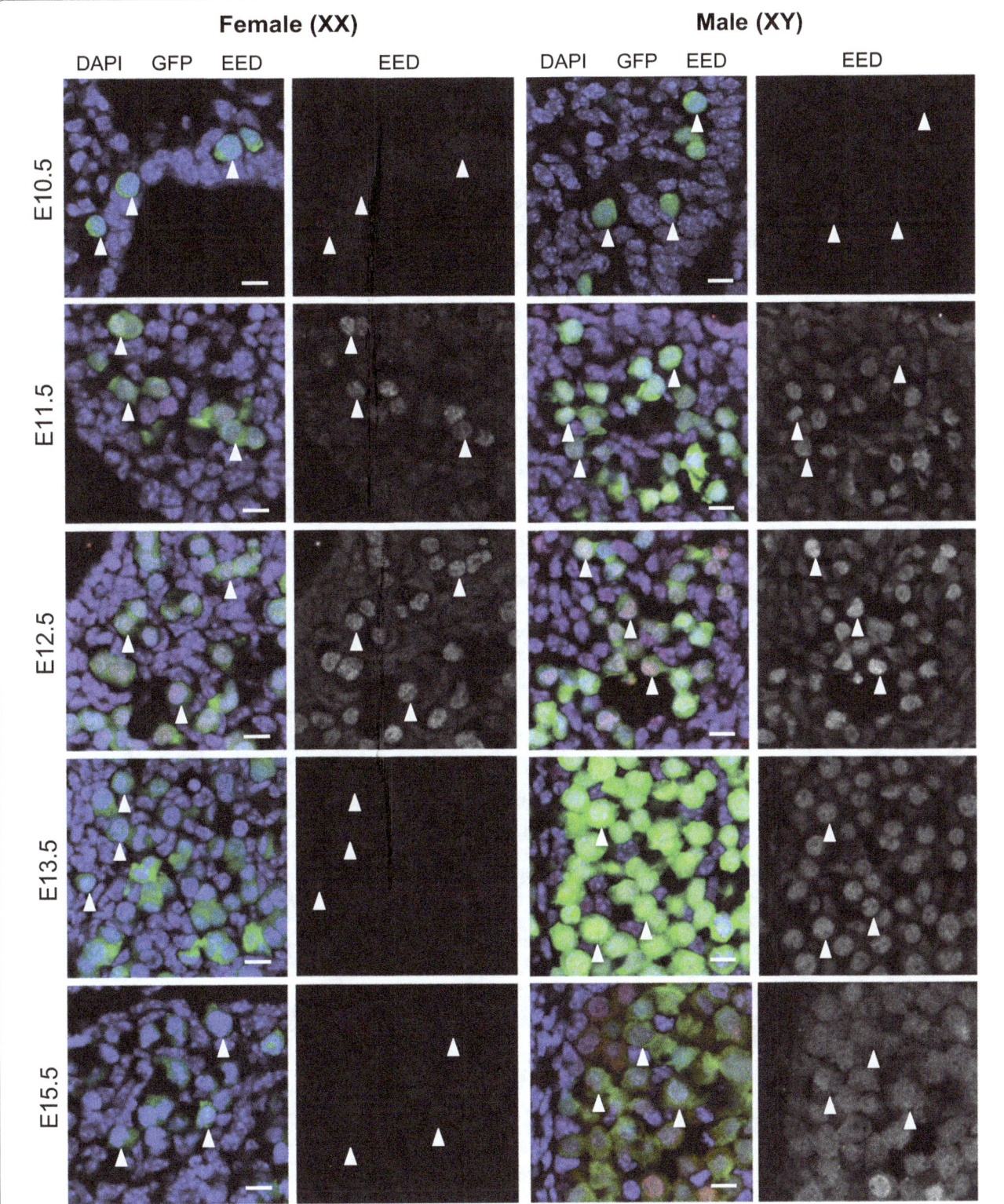

**Fig. 2** EED is enriched in E11.5–E12.5 male and female germ cells. Confocal images of EED immunofluorescence in sections of XX and XY E10.5–E11.5 bipotential gonad and E12.5–E15.5 developing ovaries and testes. *Left panels* are merged images: eGFP marking germ cells (*green*), EED (*red*) and DAPI nuclear stain (DNA; *blue*). *Right panels* are single-channel grayscale images showing EED staining. *White arrowheads* indicate *Oct4*-eGFP-expressing germ cells. Representative images chosen from 3 to 4 biological replicates for each time point. 10 µm *scale bars*

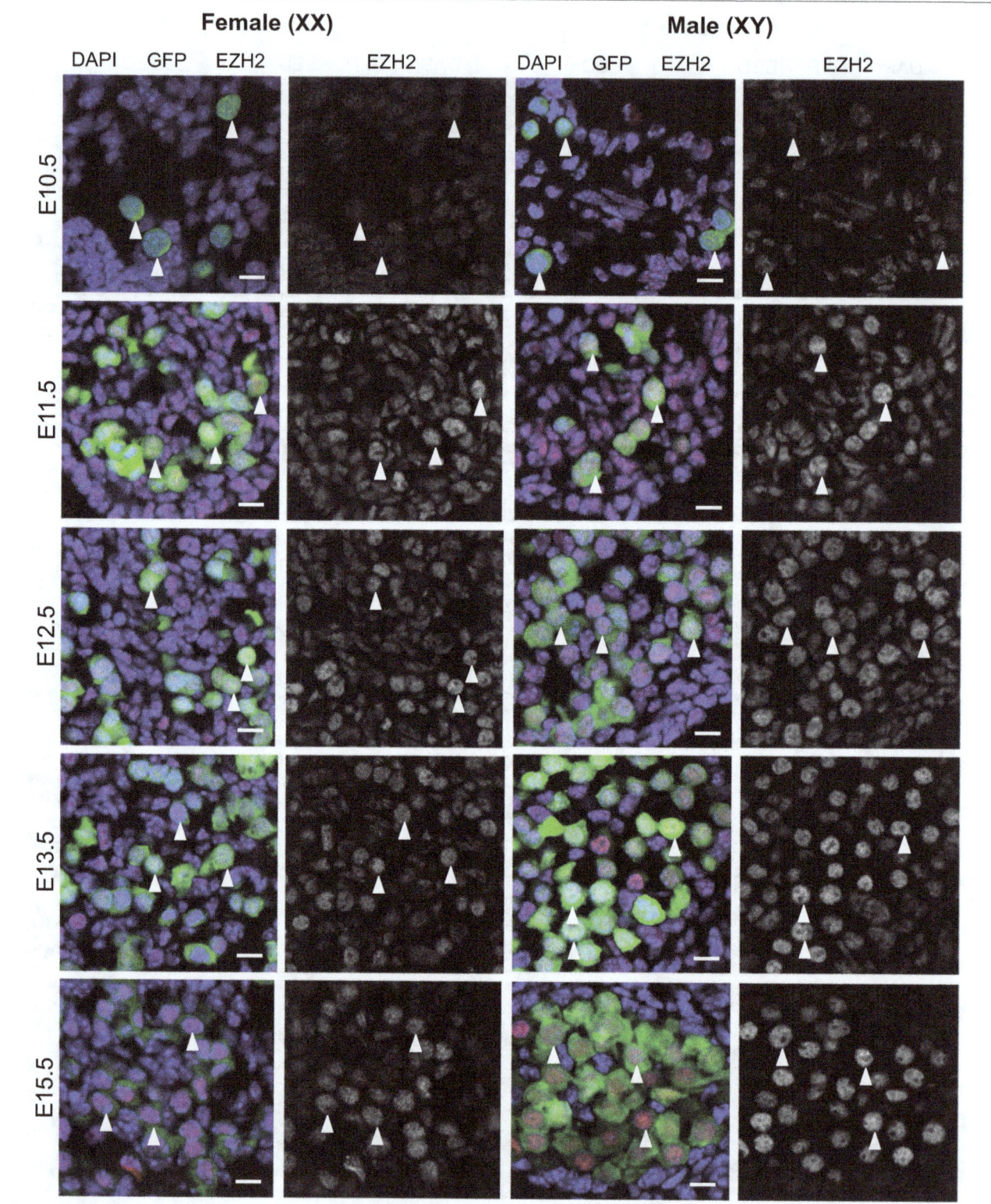

**Fig. 3** EZH2 protein is enriched in the nucleus of E11.5–E15.5 male and female germ cells. Confocal images of EZH2 immunofluorescence in sections of XX and XY E10.5–E11.5 bipotential gonad and E12.5–E15.5 developing ovaries and testes. *Left panels* are merged images: eGFP marking germ cells (*green*), EZH2 (*red*) and DAPI (DNA; *blue*). *Right panels* are single-channel grayscale images showing EZH2 staining. *White arrowheads* indicate *Oct4*-eGFP-expressing germ cells. Representative images chosen from 3 to 4 biological replicates for each time point. 10 µm *scale bars*

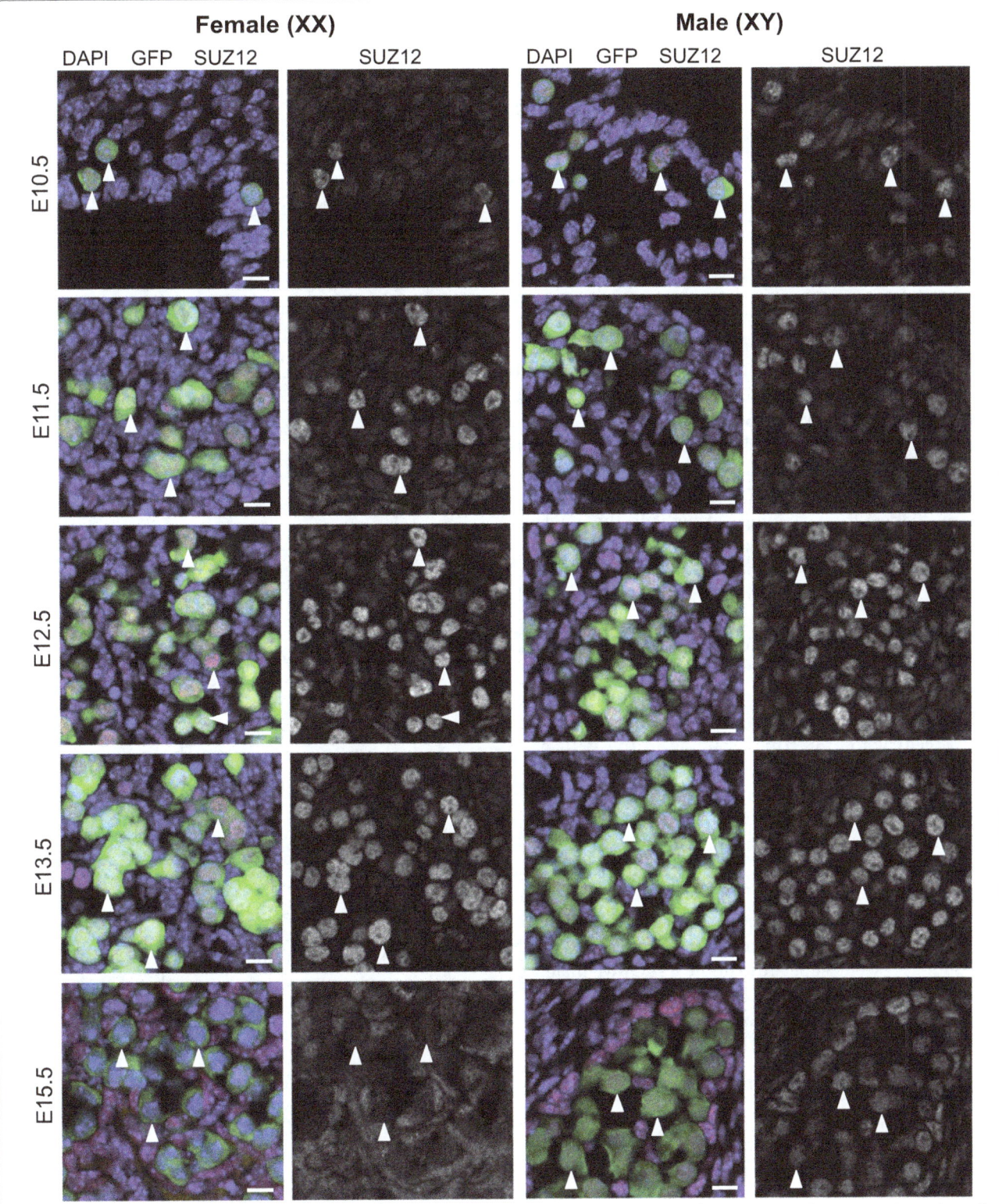

**Fig. 4** SUZ12 is enriched in the nucleus of E10.5–E13.5 male and female germ cells. Confocal images of SUZ12 immunofluorescence in sections of XX and XY E10.5–E11.5 bipotential gonad and E12.5–E15.5 developing ovaries and testes. *Left panels* are merged images: eGFP marking germ cells in *green*, SUZ12 (*red*) and DAPI (DNA; *blue*). *Right panels* are single-channel grayscale images showing SUZ12 staining. *White arrowheads* indicate *Oct4*-eGFP-expressing germ cells. Representative images chosen from 3 to 4 biological replicates for each time point. 10 µm *scale bars*

**Table 1 Relative staining intensities for PRC2 proteins in E10.5–E15.5 germ cells**

|            | E10.5 | E11.5 | E12.5 | E13.5 | E15.5    |
|------------|-------|-------|-------|-------|----------|
| Female XX  |       |       |       |       |          |
| EED        | –     | ++    | +++   | –     | –        |
| EZH2       | –     | ++    | ++    | ++    | ++       |
| SUZ12      | +     | ++    | +++   | +++   | –        |
| Male XY    |       |       |       |       |          |
| EED        | –     | ++    | +++   | ++    | –/+ Cyto |
| EZH2       | –/+   | ++    | +++   | +++   | +++      |
| SUZ12      | +     | ++    | +++   | +++   | ++       |

and XY germ cells. H3K27me3 nuclear staining remained enriched close to the nuclear lamina in XX and XY germ cells at E13.5, but by E15.5 H3K27me3 was distributed throughout the nucleus in a pattern similar to that observed in E10.5 germ cells (Fig. 5a).

To quantify the proportion of germ cells that underwent a change from staining throughout the nucleus (uniform nuclear localization; UNL) to staining located toward the nuclear lamina (peripheral nuclear localization; PNL), we assessed the proportions of cells that had UNL and PNL at E10.5, E11.5, E12.5 and E15.5 in male and female gonads based on confocal images. At E10.5, H3K27me3 staining conformed to the UNL pattern in 75/90% of XX and XY germ cells (Fig. 5b). However, at E11.5 this pattern was significantly different, with 85/95% cells with H3K27me3 staining in the PNL pattern in XX and XY germ cells, and this pattern was maintained in E12.5 XX and XY germ cells. However, by E13.5 H3K27me3 staining returned to the UNL pattern in around 40% of XY germ cells but not in XX germ cells. By E15.5, 90/94% of XX and XY germ cells contained UNL, rather than PNL localization (Fig. 5b).

To obtain greater resolution mapping of H3K27me3 localization in developing germ cells, we performed immunofluorescent super-resolution dSTORM imaging of H3K27me3 in E10.5, E11.5, E12.5 and E15.5 germ cells. Consistent with our observations using confocal imaging, H3K27me3 was distributed throughout the nucleus in both XX and XY germ cells at E10.5 (Fig. 6). This pattern changed dramatically at E11.5, with H3K27me3 located in a broadband around the periphery of the nucleus and by E12.5 H3K27me3 was typically detected in a relatively tight band around the nuclear periphery (Fig. 6). Remarkably, by E15.5 this pattern had completely reverted, and H3K27me3 was again localized throughout the nucleus in both XX and XY germ cells (Fig. 6). At this stage, XX and XY germ cells either have entered meiotic prophase or are mitotically arrested, respectively [36–38, 40–42].

To quantify the distribution of H3K27me3 located from the center to the outer edge of the nuclei of E10.5, E11.5, E12.5 and E15.5 germ cells, we developed an expanding circular quantification algorithm to measure blink intensity in the super-resolution images. Briefly, H3K27me3-derived pixels were counted in concentric circles increasing in 0.1 μm increments from the center of each nucleus to the nuclear lamina. At E10.5, H3K27me3 staining detected by dSTORM imaging was evenly distributed from the center of the nucleus to the nuclear lamina (Fig. 7). However, at E11.5 and E12.5 this pattern was strikingly different, with almost all enrichment for H3K27me3 staining located within 0–2 μm of the nuclear lamina in both sexes (Fig. 7). By E15.5, H3K27me3 was evenly distributed from the center of the nucleus to the nuclear lamina (Fig. 7). This confirmed a highly significant change in H3K27me3 from the UNL pattern at E10.5 to the PNL pattern during E11.5–E13.5 and again to the UNL pattern at E15.5.

### EZH1/2 is required for H3K27me3 enrichment near the nuclear lamina in fetal germ cells undergoing epigenetic reprogramming

Our examination of EED, EZH2 and SUZ12 protein expression revealed that all three core PRC2 proteins were present in XX and XY germ cells between E11.5 and E12.5, coincident with the enrichment of H3K27me3 near the nuclear lamina (Figs. 2, 3, 4; Table 1). To determine whether PRC2 function was required for H3K27me3 enrichment near the nuclear lamina during this period of germ cell development, we blocked EZH1/2 function using the highly specific EZH1/2 inhibitor GSK126 [43]. Initially, we validated the ability of GSK126 to block enrichment of H3K27me3 in germ cells. E11.5 whole gonad–mesonephros complexes were isolated from developing XX and XY embryos and cultured for 48 h on an organ culture membrane in medium containing increasing doses of GSK126 or vehicle control (DMSO). Flow cytometric analysis of GSK126 and control-treated gonads confirmed that GSK126 dose-dependently reduced H3K27me3 levels in germ cells in a 48-h period (Additional file 1: Fig. S2A). GSK126 was very well tolerated in these cultures, with even the highest dose having no effect on germ cell viability (Additional file 1: Fig. S2B). Moreover, we observed no impact on proliferation of either germ cells or somatic cells after 48-h treatment from E11.5 (Additional file 1: Fig. S2C, males shown). Similarly, in E12.5 fetal gonads cultured with 10 μm GSK126 for 72 h, germ cells entered mitotic arrest normally and somatic cell proliferation was normal, both in the Sertoli cell compartment and in the interstitial compartment (Additional file 1: Fig. S2D). This provided high confidence that GSK126 was well

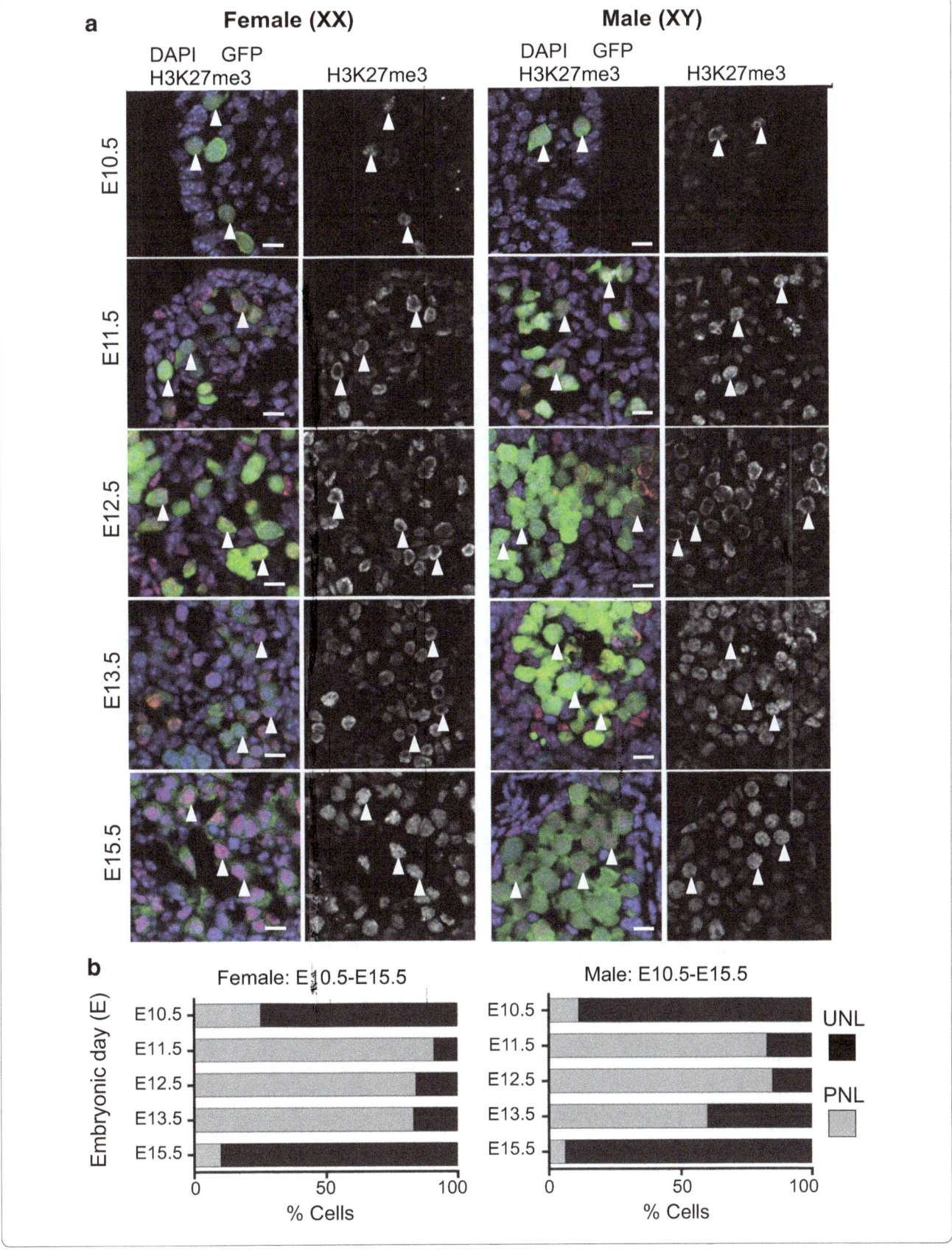

(See figure on previous page.)

**Fig. 5** H3K27me3 is enriched in the nucleus of E10.5–E15.5 male and female germ cells. **a** Confocal images of H3K27me3 immunofluorescence in sections of XX and XY E10.5–E11.5 bipotential gonad and E12.5–E15.5 developing ovaries and testes. *Left panels* are merged images: eGFP marking germ cells in *green*, H3K27me3 (*red*) and DAPI (DNA; *blue*). *Right panels* are single-channel grayscale images showing H3K27me3 staining. *White arrowheads* indicate *Oct4*-eGFP-expressing germ cells. Representative images chosen from 3 to 4 biological replicates for each time point. 10 µm *scale bars*. **b** Quantification of H3K27me3 localization in E10.5–E15.5 male and female germ cells from wild-type fetal testis and ovary sections. Percentages of cells with uniform nuclear localization (UNL; *black bars*) and peripheral nuclear localization (PNL; *gray bars*) analyzed using ImageJ Cell Counter are shown in the stacked histogram. Data represent 3–6 biological replicates for each sex, at each stage (XX/XY cells counted: $n = 16/19$ at E10.5; $n = 44/59$ at E11.5; $n = 196/214$ at E12.5; $n = 235/213$ at E13.5 and $n = 176/204$ at E15.5)

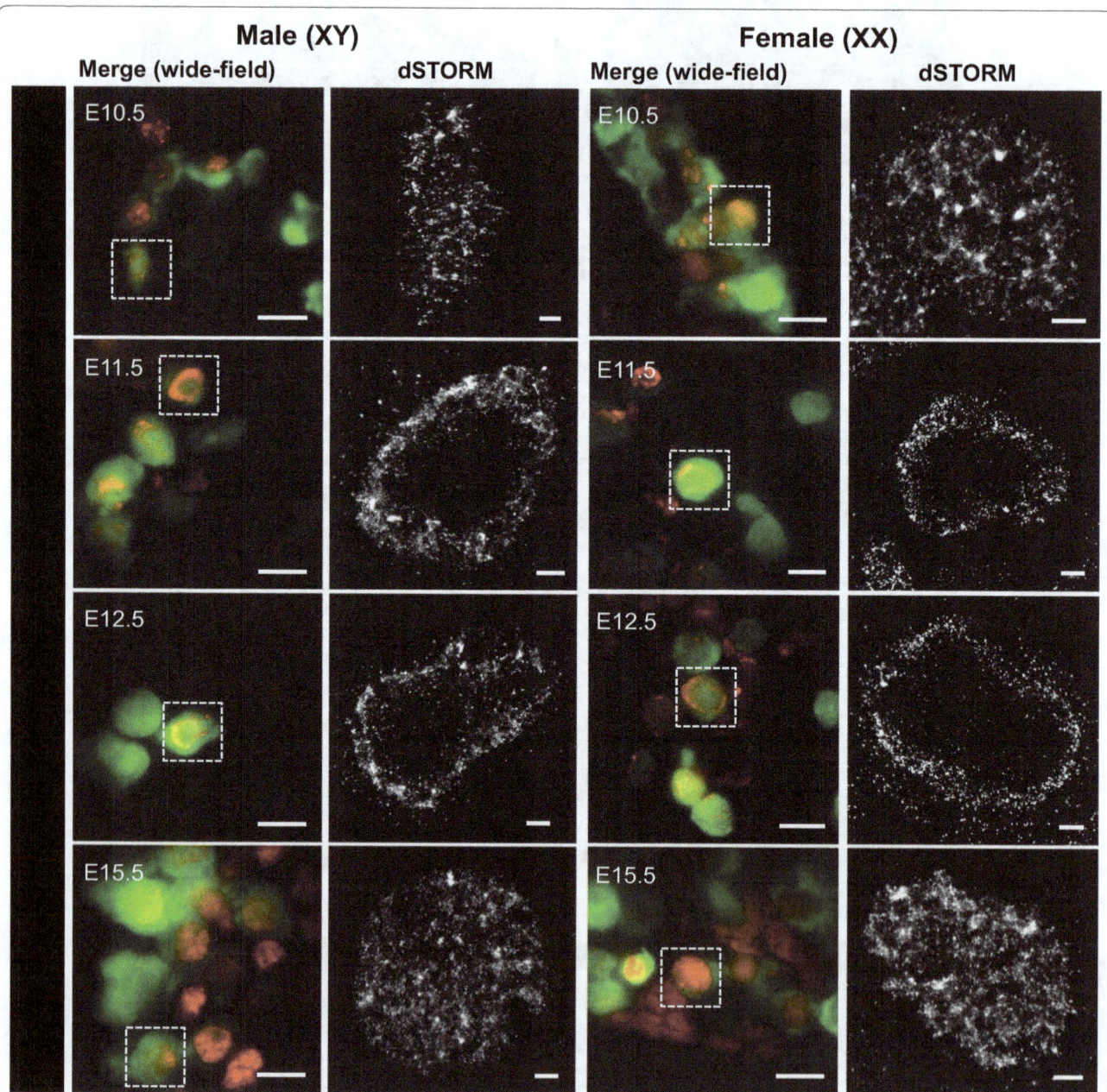

**Fig. 6** H3K27me3 is transiently relocated to the germ cell nuclear periphery in E11.5–E13.5 male and female germ cells. dSTORM super-resolution images in sections of XX and XY E10.5–E11.5 bipotential gonad and E12.5–E15.5 developing ovaries and testes. *Left panels* are merged wide-field (×160) images: eGFP marking germ cells (*green*), and H3K27me3 (*red*). *White dotted boxes* indicate super-resolved germ cell. 10 µm *scale bars*. *Right panels* dSTORM super-resolution of H3K27me3 antibody (*grayscale*). 1 µm *scale bars*. Representative images chosen from 3 to 8 super-resolved images in three biological replicates for each time point

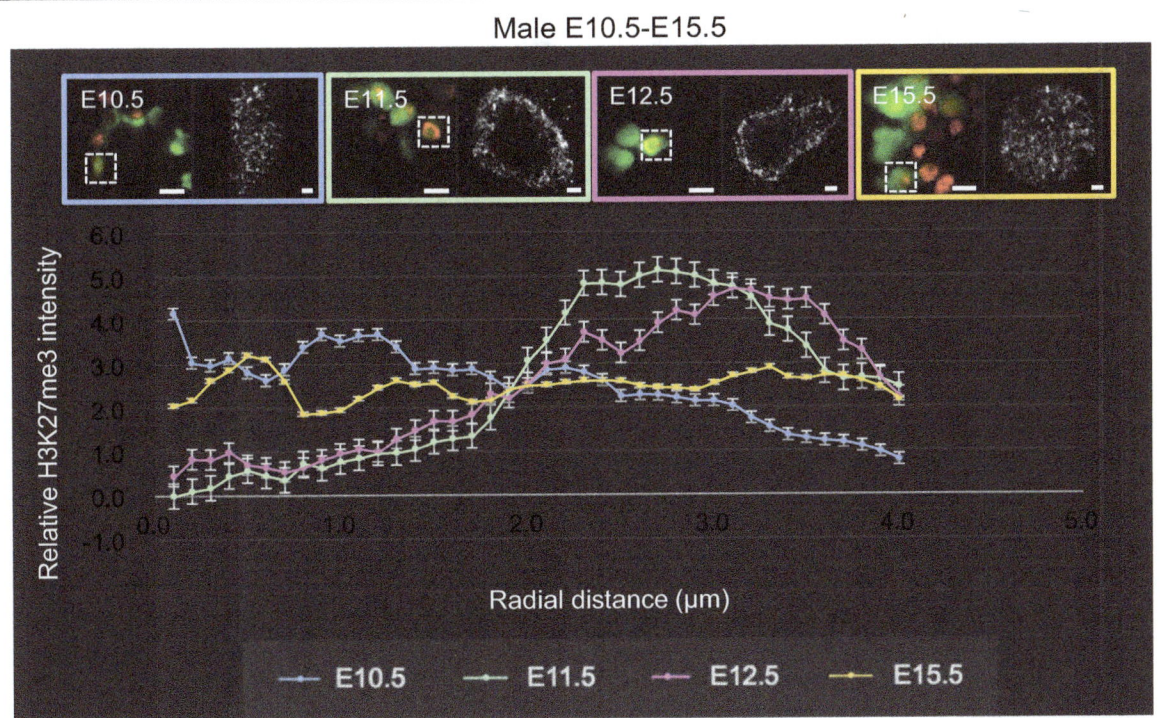

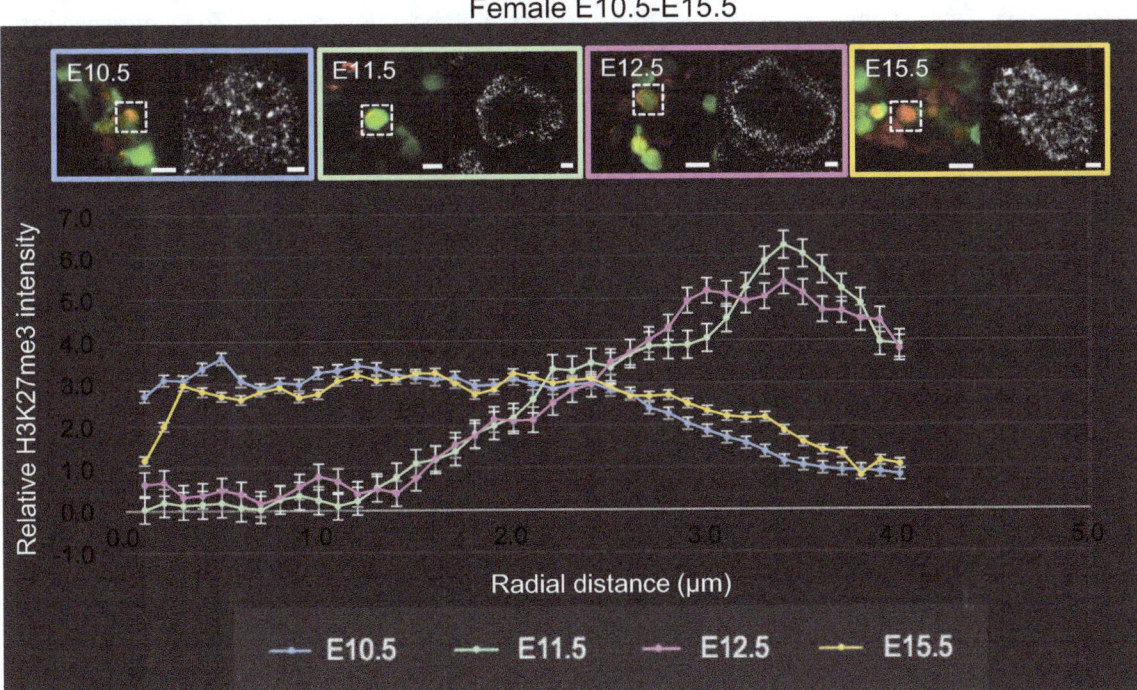

**Fig. 7** H3K27me3 is significantly enriched near the nuclear lamina in fetal germ cells undergoing epigenetic reprogramming. Radial histogram quantification of H3K27me3 localization of dSTORM super-resolution immunofluorescent images in wild-type male and female germ cells E10.5 (*blue*), E11.5 (*green*), E12.5 (*pink*) and E15.5 (*yellow*). *Left image* of each colored panel (×160) represents merged channels, *green* marking germ cells (eGFP), and H3K27me3 (*red*). 10 μm *scale bars*. *Right-hand* image of each *panel* shows dSTORM super-resolution images of germ cells (H3K27me3 in grayscale). 1 μm *scale bars*. Data represent 3–8 super-resolved images from three biological replicates. The *Y*-axis shows relative H3K27me3 intensity, and the *X*-axis shows radial distance from nucleus center (μm). Error bars ± SEM

tolerated in fetal gonad cultures and demonstrated highly efficient, GSK126-dependent, depletion of H3K27me3 in germ cells within the 48- and 72-h culture periods.

We next blocked EZH1/2 activity in E11.5 fetal gonads cultured with GSK126 for 48 h. Confocal imaging and analysis of H3K27me3 staining intensity using ImageJ demonstrated that H3K27me3 was reduced by 75 and 73% in germ cells of GSK126-treated male and female gonads compared to the germ cells of male and female control-treated gonads, respectively ($P < 0.0001$; Additional file 1: Fig. S3, Fig. 8a). Moreover, while H3K27me3 was detected in the periphery of the germ cell nucleus in control gonads, it was detected throughout the germ cell nucleus in GSK126-treated gonads (Fig. 8a), indicating a failure of this epigenetic mark to be enriched near the nuclear periphery when EZH1/2 function was blocked. Quantification of UNL and PNL staining revealed PNL staining in 89/75% of XX and XY germ cells in gonads treated with vehicle control, but only 15/4% in XX and XY germ cells of gonads treated with GSK126 for 48 h (Fig. 8c). To obtain a higher-resolution analysis of H3K27me3 localization in GSK126-treated germ cells, we performed super-resolution imaging on the same GSK126-treated and control gonads. In control-treated gonads, H3K27me3 staining was restricted to the periphery of the germ cell nucleus (Fig. 8b) in a pattern that was very similar to that observed in vivo in wild-type E12.5 germ cells (Fig. 6). However, in GSK126-treated gonads, remaining H3K27me3 was distributed throughout the germ cell nucleus (Fig. 8b) in a pattern that was indistinguishable from that observed in vivo at E10.5 and E15.5 (Fig. 6). Quantitative analysis of the localization of this staining demonstrated that the remaining H3K27me3 in germ cells of GSK126-treated gonads was evenly distributed throughout the nucleus and was not different from the pattern detected in untreated E10.5 and E15.5 germ cells. H3K27me3 was not enriched near the nuclear periphery of germ cells in GSK126-treated gonads, and its distribution was significantly different from the H3K27me3 enrichment observed near the

nuclear lamina in normally developing E11.5 and E12.5 germ cells (Fig. 9). Interestingly in XY control gonads cultured for 48 h, H3K27me3 staining extended to ~3 µm from the nuclear periphery but was absent or low in the center of the nucleus. This pattern was comparable to the staining pattern observed by confocal microscopy in many E13.5 XY germ cells (Fig. 5b), indicating that H3K27me3 had already begun to relocate throughout the nucleus of control-treated gonads. We therefore examined XY gonads cultured for 24 h with control and GSK126 medium (Additional file 1: Fig. S4A). H3K27me3 levels were reduced by 57% in samples treated with GSK126 for 24 h compared to controls ($P < 0.0001$; Additional file 1: Fig. S4B). Quantification of UNL and PNL staining in confocal images revealed PNL staining in 85% of XY germ cells in gonads treated with vehicle control, but only 20% in XY germ cells of gonads treated with GSK126 for 24 h (Additional file 1: Fig. S4C). Super-resolution imaging of germ cells in these gonads revealed peripheral staining of H3K27me3 in control gonads and complete loss of peripheral staining in GSK126-treated gonads (Additional file 1: Fig. S5).

H3K27me3 enrichment is associated with the repression of many developmentally important genes, perhaps most notably the *Hox* genes. To identify genes for which depletion of H3K27me3 enrichment might alter expression, we treated male E12.5 fetal gonads for 72 h with GSK126. This covers a period during which many developmental genes are highly regulated in germ cells as they undergo early male germline differentiation. Initially, we confirmed efficient depletion of H3K27me3 using flow cytometric analysis of 10% of the gonadal cells collected for the GSK126 and vehicle control-treated cultures (Fig. 10a). This revealed an 80% decrease in H3K27me3 in germ cells isolated from GSK126-treated gonads compared to gonads treated with vehicle control. The remaining 90% of cells from the cultured gonads were subject to fluorescence-activated cell sorting (FACS), allowing purification of the germ cells and global expression analysis using microarrays. Surprisingly, analysis of

---

(See figure on next page.)

**Fig. 8** Transient enrichment of H3K27me3 near the nuclear lamina is lost when EZH1/2 is blocked. Confocal and dSTORM super-resolution images of immunofluorescence staining at E11.5 XX and XY gonads cultured for 48 h with either DMSO (control) or 10 µm GSK126. **a** Confocal images (×80) showing efficacy of H3K27me3 depletion by GSK126, eGFP (germ cells; *green*) and H3K27me3 (*red*). 10 µm *scale bars*. The reduction in H3K27me3 in GSK126-treated samples is quantified in Supp. Figure 3. **b** Super-resolution dSTORM images. *Left panels* show merged wide-field (×160) images: eGFP (germ cells: *green*) and H3K27me3 (*red*). *White dotted boxes* indicate super-resolved germ cells. 10 µm *scale bars*. *Right panels* show dSTORM super-resolution images of H3K27me3 (grayscale) control and GSK126-treated germ cells. 1 µm *scale bars*. Representative images chosen from 3 to 5 super-resolved images in three biological replicates for each time point. **c** Quantification of H3K27me3 localization in E11.5 XX and XY gonads cultured for 48 h with either DMSO (control) or 10 µm GSK126. ×80 confocal immunofluorescent images were analyzed by ImageJ Cell Counter. Percentages of cells with uniform nuclear localization (UNL; *black bars*) and peripheral nuclear localization (PNL; *gray bars*) are shown in the stacked histogram. Data represent 3–6 biological replicates for each sex, for each treatment group. (XX/XY cells counted: vehicle control $n = 159/99$; GSK126 $n = 180/115$)

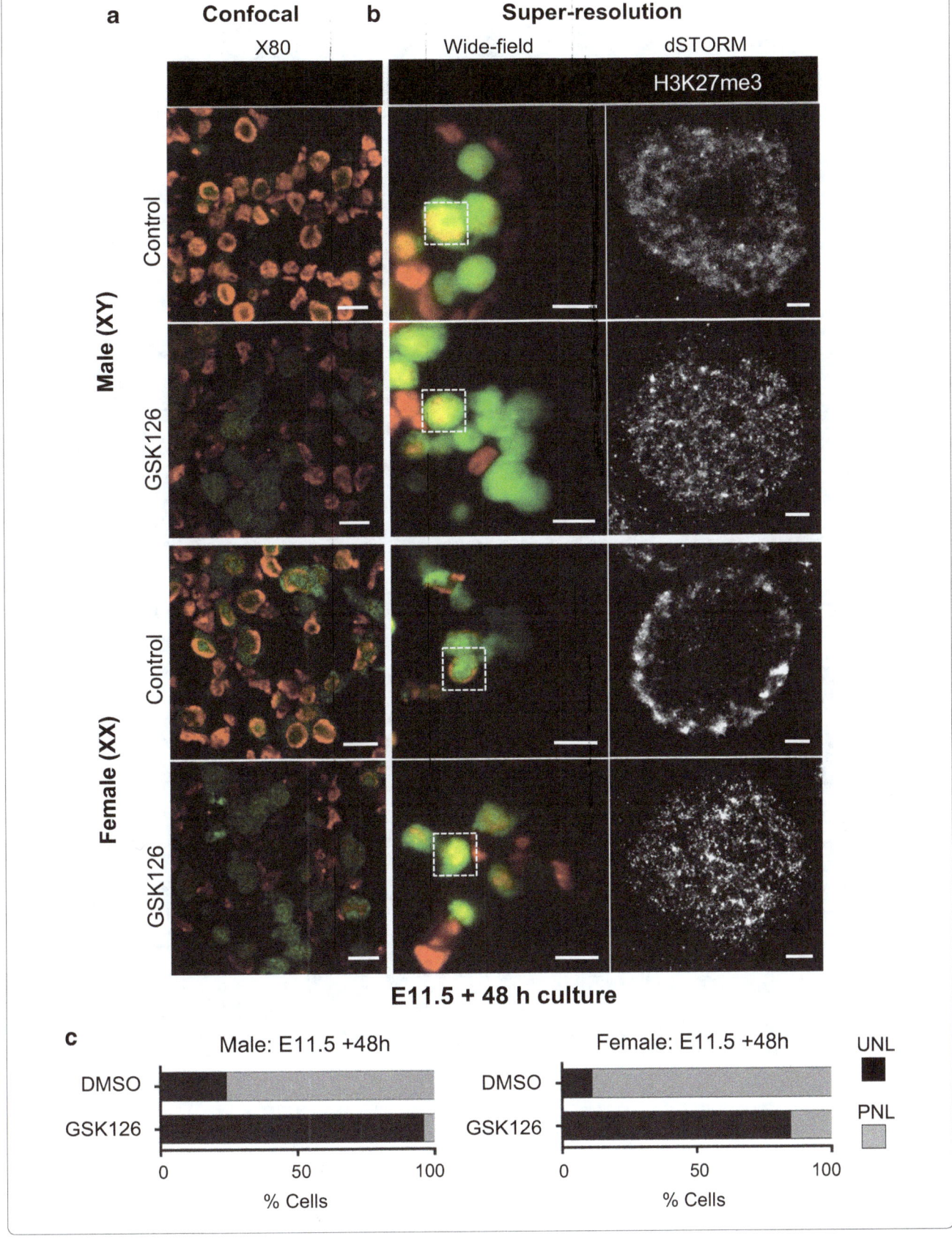

**a** Confocal   X80

**b** Super-resolution   Wide-field   dSTORM   H3K27me3

Male (XY) — Control / GSK126

Female (XX) — Control / GSK126

**E11.5 + 48 h culture**

**c**

Male: E11.5 +48h
- DMSO
- GSK126
0   50   100
% Cells

Female: E11.5 +48h
- DMSO
- GSK126
0   50   100
% Cells

UNL
PNL

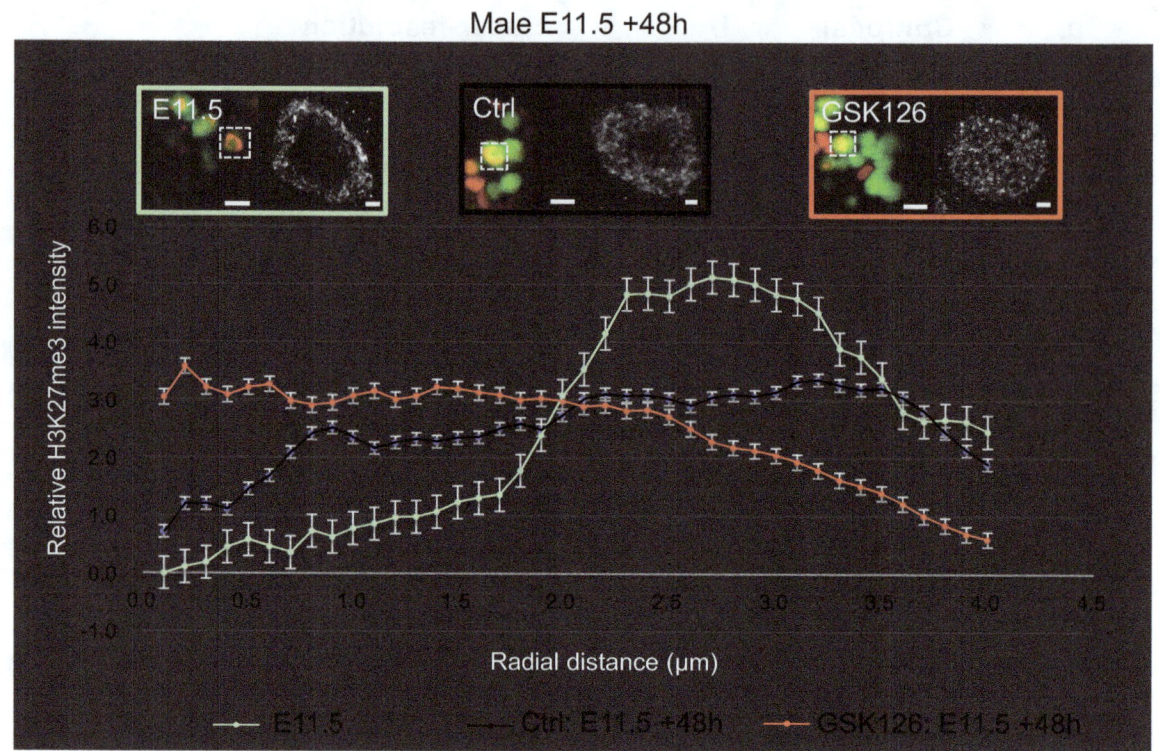

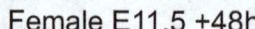

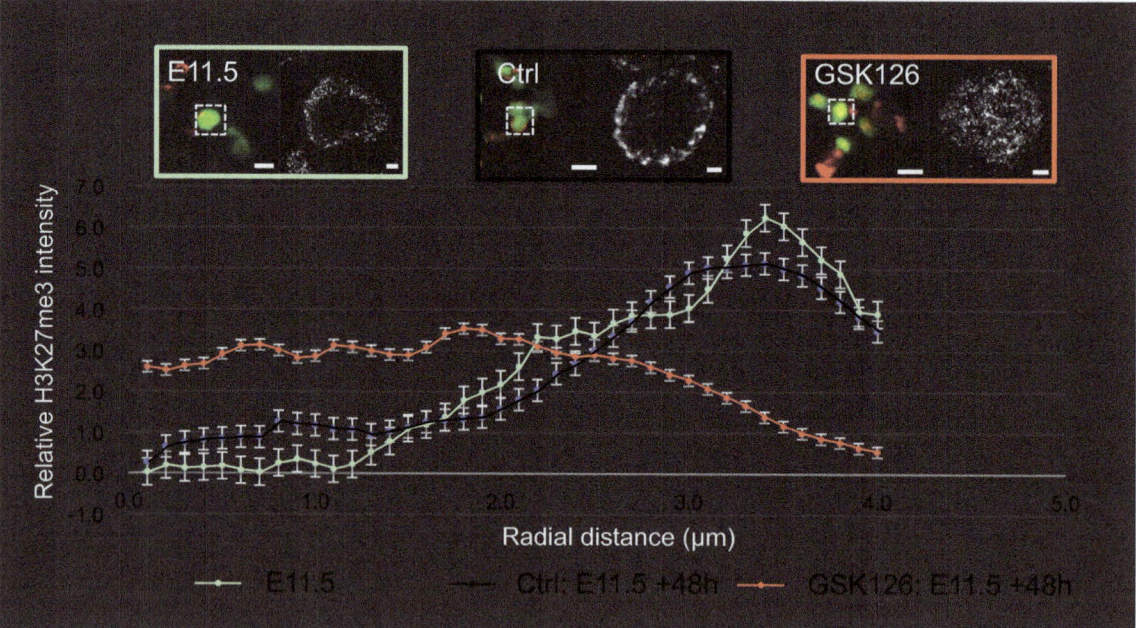

**Fig. 9** Blockage of EZH1/2 significantly reduces H3K27me3 enrichment near the nuclear lamina in fetal germ cells undergoing epigenetic repro-gramming. Radial histogram quantification of dSTORM super-resolution immunofluorescent images of sections from E11.5 fetal testes and ovaries cultured for 48 h. E11.5 wild-type control is shown in *green*, E11.5 + 48 h vehicle control (DMSO) is shown in *blue*, and E11.5 + 48 h of GSK126 treatment (10 μm) shown in *red*. *Left image* of each *colored panel* (×160) represents merged channels, with *green* marking germ cells (eGFP) and H3K27me3 shown in the *red* channel. 10 μm *scale bars*. The *right-hand* image of each *panel* shows dSTORM super-resolution images of germ cells (H3K27me3 in grayscale). 1 μm *scale bars*. 3–8 super-resolved images from three biological replicates. *Y*-axis shows relative H3K27me3 intensity, and *X*-axis shows radial distance (μm) from the center of the nucleus. *Error bars* ± SEM

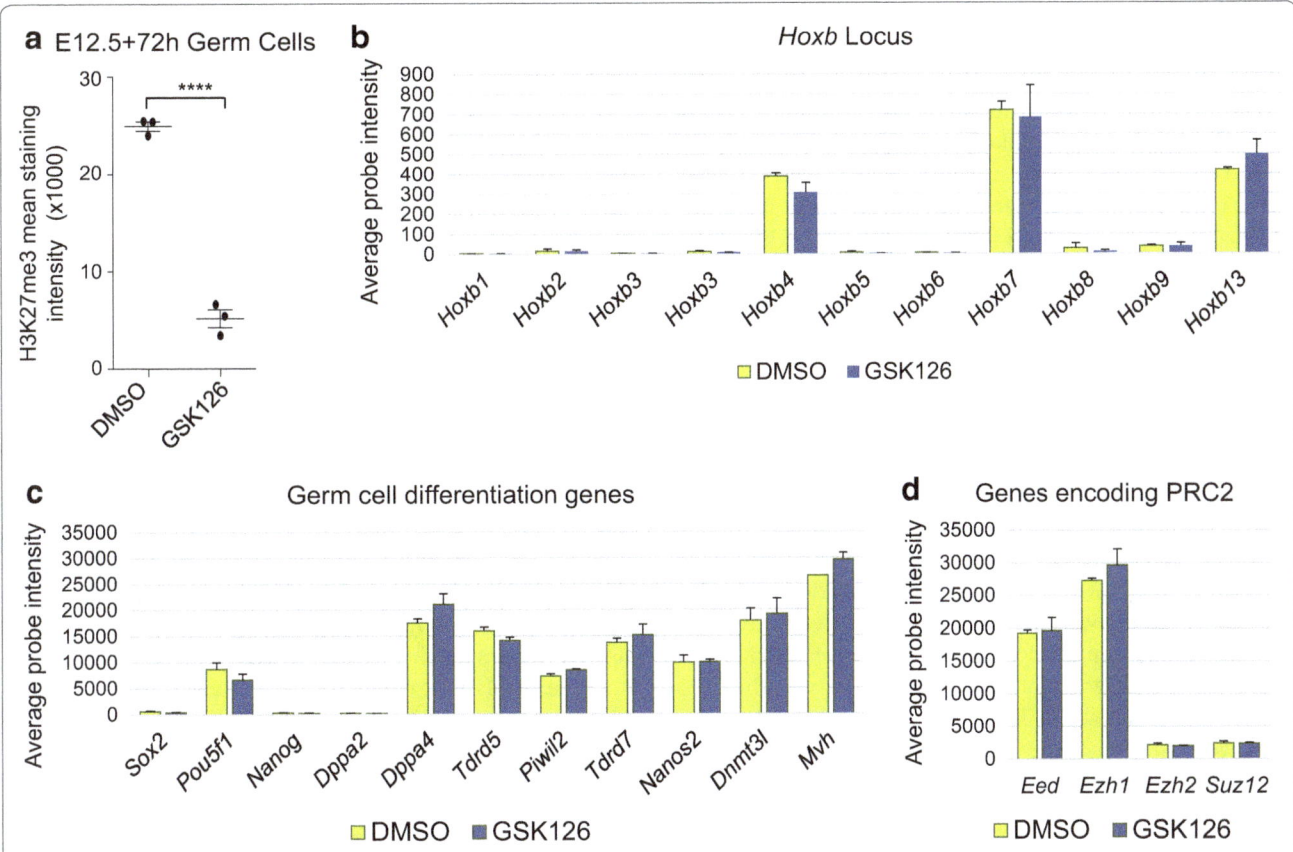

**Fig. 10** Gene expression profiles remained normal after H3K27me3 enrichment was reduced in fetal male germ cells. **a** Average H3K27me3 intensities were measured using flow analysis of FACS sorted E12.5 germ cells cultured for 72 h with a vehicle control (DMSO; *green*) or 10 µm GSK126 (*blue*). **b–d** Expression microarray relative probe signal intensities normalized to *Sdha*, *Canx* and *Mapk1* for genes expressed in E12.5 germ cells cultured for 72 h with DMSO or 10 µm GSK126: **b** *Hoxb* locus. **c** A selection of male germ cell development genes; *Sox2*, *Pou5f1*, *Nanog*, *Dppa2*, *Dppa4*, *Tdrd5*, *Piwil2*, *Tdrd7*, *Nanos2*, *Dnmt3l*, *Mvh*. **d** PRC2 genes; *Ezh1*, *Ezh2*, *Eed* and *Suz12*. Values ± SEM; $n = 3$ for each treatment

global gene expression patterns using GeneSpring GX software revealed no significant differences in germ cell gene expression in gonads treated with GSK126 and vehicle control. In addition to global GeneSpring analysis, we used validated reference genes *Sdha*, *MapK1* and *Canx* [44] to normalize the raw array hybridization signals of known germ cell development, PRC2-regulated genes and genes encoding PRC2 proteins, and compared their relative expression in GSK126 and vehicle-treated samples. Consistent with the GeneSpring analysis, homeobox genes of the A, B, C and D clusters (Fig. 10b; *Hoxb* cluster shown), known male germ cell development genes (Fig. 10c), pluripotency genes (Fig. 10c), and PRC2 genes *Ezh1*, *Ezh2*, *Eed* and *Suz12* (Fig. 10d) were not significantly different in germ cells isolated from gonads treated with GSK126 or vehicle control for 72 h. In all cases, the expression of these genes was similar to what is expected for during normal germ cell development, with generally very low signals for the *Hox* genes and pluripotency genes *Nanog* and *Sox2*, moderate expression of *Oct4*, which is

maintained in male germ cells at E15.5, high expression of *Dppa4*, *Piwil2*, *Tdrd5*, *Tdrd7*, *Nanos2*, *Dnmt3l* and *Mvh* which are upregulated in male germ cells by E15.5 and high expression of *Eed* and *Ezh1*, but relatively low expression of *Ezh2* and *Suz12*.

## Discussion

With the exception of DNA methylation, reprogramming of epigenetic information in the developing germline is poorly understood. Here, we demonstrate that the epigenetic modifier complex PRC2, which catalyzes H3K27 methylation, is transiently expressed in germ cells soon after they enter the gonad and undergo the final phase of DNA demethylation [3]. Coincident with increased PRC2 protein localization to the nucleus of germ cells, we observed substantial enrichment of H3K27me3 near the nuclear lamina in the gonadal germ cell population. This peripheral nuclear enrichment of H3K27me3 occurred during E11.5–E13.5, coinciding with the final phase of DNA demethylation and the lowest levels of DNA methylation in germ cells. By E15.5, H3K27me3 was again uniformly localized throughout nucleus in XX and XY germ cells. These findings demonstrate significantly three-dimensional reorganization of chromatin in the germ cell nucleus that is likely to be important for epigenetic reprogramming. In addition, we provide evidence that H3K27me3 enrichment near the nuclear lamina in E11.5–E13.5 germ cells is dependent on EZH1/2 activity, which comprise the essential catalytic proteins that mediate PRC2 activity. Combined, these data indicate that H3K27me3 mediates global reorganization of chromatin toward the nuclear lamina during the critical period during which germ cell DNA methylation levels are at their lowest. However, despite depletion of 80% of the H3K27me3 in male fetal germ cells, development of these cells was not altered within the 72-h culture window, and neither genes that mark male germline differentiation, nor known H3K27me3 targets, were altered.

A previous study reported reorganization of germ cell chromatin soon after germ cells enter the gonad [7]. This led to the proposal that H3K27me3 is removed and replaced in gonadal germ cells to ensure that epigenetic information is reset in the fetal germline [7]. However, another study concluded that H3K27me3 is maintained at relatively constant levels in gonadal germ cells, but fluctuates with cell cycle progression [9]. In the present study, we used flow cytometry to quantify H3K27me3 levels per cell and directly assess the variation in H3K27me3 levels in gonadal germ cells at E11.5, E13.5 and E15.5. In addition, we examined H3K27me3 in sections of gonads using both standard confocal imaging and super-resolution imaging at E10.5, E11.5, E12.5, E13.5 and E15.5. Using flow cytometry, we detected high levels of H3K27me3 in E11.5 germ cells, which moderately declined in E13.5 and E15.5 germ cells. H3K27me3 levels in germ cells were fourfold to sevenfold higher than those detected in the gonadal somatic cells at all stages analyzed. Consistent with this, confocal and super-resolution imaging revealed higher levels of H3K27me3 in germ cells than in somatic cells at all stages analyzed. Using these combined approaches, we found no evidence of extensive loss of H3K27me3 in either XX or XY germ cell after they entered the gonad. This is consistent with the conclusions of Kagiwada et al. [9], who also observed relatively constant levels of H3K27me3 in gonadal germ cells, although levels fluctuated with cell cycle. It remains possible that H3K27me3 is removed and replaced in a very short temporal window. However, based on our findings we conclude that widespread loss of H3K27me3 is unlikely to occur as part of epigenetic reprogramming in mouse germ cells in the few days after they enter the developing gonads.

Rather than loss of H3K27me3 from germ cells at E11.5, we observed substantial transient enrichment of H3K27me3 near the germ cell nuclear lamina. This persisted until E13.5, but by E15.5 H3K27me3 was again localized throughout the germ cell nucleus in both XX and XY germ cells. In male germ cells, the redistribution of H3K27me3 back through the nucleus at E15.5 coincided with the initiation of de novo DNA methylation in mitotically arrested male germ cells. However, H3K27me3 was also distributed throughout the nucleus by the time female germ cells exit the cell cycle and have entered meiotic prophase by E15.5 [37].

The redistribution of H3K27me3 toward the nuclear lamina in E11.5–E13.5 male and female germ cells is both striking and intriguing. One possibility is that H3K27me3 regulates specific gene expression programs in the male and female germlines during this time. However, our analysis of global gene expression patterns in which H3K27me3 was depleted by 80% revealed no differences in gene expression in the 72-h window analyzed. Markers of male germline development were expressed normally, and there was no evidence that known targets of PRC2/H3K27me3 were de-repressed. This was surprising as previous studies have identified genes that are enriched for H3K27me3 in male and female germ cells during this period, suggesting that PRC2-mediated H3K27me3 enrichment might substantially regulate sex-specific germ cell development [29–31]. One caveat for the current study is that around 20% of the global H3K27me3 levels remained in germ cells after drug treatment, and this may be sufficient to maintain normal regulation of gene expression. However, dSTORM imaging clearly demonstrated that the H3K27me3 that remained in GSK126-treated cells was no longer localized toward the nuclear lamina, demonstrating that at least this function was compromised in the drug-treated gonads. Moreover, as we observed similar enrichment of H3K27me3 near the nuclear lamina in both sexes, and inhibition of EZH1/2 and depletion of H3K27me3 affected this equally in both sexes, it is unlikely that the global reorganization of H3K27me3 we observed can explain the regulation of sex-specific programs. It is more likely that H3K27me3 mediates a repressive activity that affects sequences common to both male and female germ cells.

Based on the current data, the most plausible explanation for the enrichment of H3K27me3 near the nuclear periphery is that H3K27me3-enriched chromatin is relocalized to the nuclear lamina in order to achieve appropriate transcriptional repression. Although it remains possible that H3K27me3 is removed from the center of the nucleus between E10.5 and E11.5 and re-established in the center of the nucleus between E13.5 and E15.5, we consider this unlikely for two reasons. Firstly, blocking EZH1/2 which catalyzes H3K27me3 prevented enrichment of H3K27me3 at the germ cell nuclear periphery in gonads cultured from E11.5 for 24 or 48 h. Secondly, studies in other cell systems provide substantial evidence that localization of chromatin to the nuclear periphery is associated with transcriptional repression [45, 46].

Indeed, it is highly plausible that transcriptional repression plays an essential role in maintaining genomic integrity during epigenetic reprogramming. DNA methylation is at its lowest level soon after germ cells enter the gonad. This is considered to involve a period during which germ cells may be vulnerable to aberrant expression of sequences such as ERVs that would otherwise contain DNA methylation and be repressed [2, 6, 47–49]. Previous studies have demonstrated enrichment of H3K27me3 at ERVs including young LTR and LINE elements, and at intergenic regions, introns and ICRs in male and female gonadal germ cells undergoing epigenetic reprogramming [6, 31]. Although repression of ERVs is in part achieved by SETDB1 through its catalysis of H3K9me3 in E13.5 fetal germ cells, this activity does not account for the repression of all ERVs, indicating that other mechanisms also repress ERVs during this period [6]. Moreover, recent work has demonstrated a role for H3K27me3 in ERV repression in embryonic stem cells in which DNA methylation levels are reduced by treatment with 5-azacytidine and vitamin C [50].

Interestingly, although H3K27me3 was enriched near the nuclear lamina in germ cells undergoing reprogramming, EZH2, EED and SUZ12 staining was consistently detected throughout the nucleus. Yet blocking EZH2 activity prevented enrichment of the remaining H3K27me3 near the nuclear lamina. This implies that although PRC2 is required for catalyzing H3K27me3 in the nucleus at this stage, the complex is not relocated toward the nuclear lamina with H3K27me3-enriched chromatin. One possible explanation is that relocalization of H3K27me3 may depend on an additional mechanism(s) in the germ cell nucleus.

In this study, we demonstrate PRC2-dependent enrichment of H3K27me3 near the nuclear lamina when DNA methylation is at its lowest level in germ cells, but no change in expression of protein coding genes, with the caveat that EZH1/2 was blocked and H3K27me3 depleted in a limited 72-h window. However, EZH1/2-dependent reorganization of H3K27me3 in both sexes suggests that H3K27me3 plays roles other than repressing coding genes during this developmental period. One possibility may be that PRC2 and H3K27me3 play a transient role in repressing ERVs during epigenetic reprogramming of the germline through a mechanism that involves their relocalization to the nuclear periphery and sequestration into repressive chromatin structures. However, further work is required to determine whether this is the case.

## Conclusions

We have identified substantial nuclear reorganization of the repressive histone modification H3K27me3 in germ cells as they undergo epigenetic reprogramming and reach their lowest DNA methylation levels. This occurs during a transient period of PRC2 enrichment in germ cells of both sexes and is dependent on EZH2 activity, the enzyme that catalyzes H3K27 methylation. Determining the repressive mechanisms that regulate germline epigenetic state during germline reprogramming is essential for understanding how altered epigenetic function in the germline contributes to inherited disease. It is therefore imperative that further work seeks to understand the mechanisms that regulate epigenetic and genetic integrity in the germline prior to fertilization.

## Methods

### Mouse strains, animal housing, breeding and ethics

Mice were housed at Monash Medical Centre Animal Facility using a 12-h light–dark cycle. Food and water were available ad libitum, and room temperature was 21–23 °C with controlled humidity. With the exception of breeding pairs and neonatal mice (up to three weeks old), males and females were weaned at 21 days and kept in cages of up to five individuals. All animal work was undertaken in accordance with Monash University Animal Ethics Committee (AEC) approvals. Temporal and spatial profiling of PRC2 components and H3K27me3 was carried out on embryos collected from *Oct4* [*Pou5f1*]-eGFP on 129svJ background males crossed with Swiss females (obtained from Monash Animal Services). Females were checked for vaginal plugs daily, and detection of a plug was noted as E0.5. Embryos E12.5 or older were sexed by the presence (male) or absence (female) of testis cords in gonads. Embryos younger than E12.5 were sexed using PCR as described [51].

### Flow cytometry

Gonad collection, dissociation, fixation, antibody staining and flow cytometry are as previously described [52]. Gonad samples stained with rabbit IgG control antibody were used as negative controls to set flow cytometry gates for H3K27me3 intensity, and mesonephros samples were used as a germ cell negative control to gate for eGFP. Cell proliferation was assessed by dosing cultured gonads with 20 μm 5-ethynyl-2 deoxyuridine (EdU) for the final 2 h of culture prior to tissue collection, dissociation and fixation. Cell cycle analysis was carried out as described

[53] with germ cells identified by their expression of mouse vasa homologue (MVH) and somatic cells identified by anti-Mullerian hormone (AMH) staining. Cells were stained with 20 µg/ml propidium iodide, allowing quantitation of cellular DNA content. Proliferation was measured by gating EdU-positive cells against propidium iodide to identify cells actively in S-phase. Cells in G1 and G2/M were identified by their DNA content and the absence of S-phase activity. All flow cytometry was performed on a FACS Canto instrument and data analyzed in FlowJo and GraphPad Prism. For all analyses, 3–9 biological replicates were analyzed and statistical significance determined using one-way ANOVA with Tukey's multiple comparison or $t$ test as appropriate. $P$ values <0.05 were considered significant.

### Tissue fixation and embedding

Gonads were fixed in 4% paraformaldehyde (PFA) in PBS for IF and confocal imaging, or 2% PFA plus 0.02% glutaraldehyde (GA) in PBS (dSTORM super-resolution microscopy) for 20–90 min according to embryo age (Additional file 1: Table S1). Samples were then washed twice in PBS and left in 30% sucrose in PBS overnight at 4 °C. Samples were then placed in disposable cryostat molds (Sakura Finetek, 4565) filled with OCT (Sakura Finetek, 4583) and frozen in dry ice. Blocks were and stored at −80 °C.

### Immunofluorescence

Eight-micron sections were cut from OCT-embedded gonads fixed in 4% PFA, mounted on SuperFrost Plus slides and dried for 5 min before immersing in 1× PBS. Sections were then permeabilized by incubation in 1% Triton × 100 (Sigma, T8787) in PBS for 10 min at room temperature (RT). Slides were washed three times for 5 min each in PBS. Sections were blocked in PBS containing 5% BSA (Sigma, A9647) and 10% donkey serum (Sigma, D9663) and incubated for 45 min at RT. Blocking solution was replaced by PBS containing 1%BSA and appropriately diluted primary antibodies (Additional file 1: Table S2) and incubated for 1 h at RT. Slides were washed three times for 5 min in PBS and secondary antibodies diluted in 1% BSA in 1 x PBS according to antibody dilutions outlined in Additional file 1: Table S3. Secondary antibody incubation was carried out in a dark box for 1 h at RT. Slides were washed three times in PBS (5 min each wash) and mounted in ProLong Gold® containing DAPI (Life Technologies, P36931) and left in a dark box over night to dry. For control slides, only a secondary antibody was applied. Confocal images were taken as single optical sections using a Nikon® C1 inverted confocal microscope. All pictures were taken at 80×, using a 40× oil immersion lens. Nuclear localization of H3K27me3

was assessed in 3–6 biological replicates per embryonic stage. Each germ cell was visually assessed for UNL or PNL staining based on the presence of staining throughout the nucleus or only around the nuclear periphery, respectively. Cell counts were recorded using ImageJ Plugin Cell Counter. Data were expressed as the percentage of cells with UNL and PNL staining and analyzed for statistically significant differences between stages using one-way ANOVA with Tukey's multiple comparison test or differences between treatments using Student's $t$ test with $P \leq 0.05$ considered significant. H3K27me3 mean intensity was measured in three biological replicates for each culture period (E11.5 +48 h male and females, >119 cells analyzed; E11.5 +24 h males, >68 cells analyzed). ImageJ ROI manager was used to calculate the mean intensity for H3K27me3, and unpaired $t$ tests were used to statistically analyze treatments groups, with $P \leq 0.05$ considered significant.

### dSTORM super-resolution imaging

Two-micron sections were cut from OCT-embedded gonads fixed in 2% PFA and 0.02% GA (Additional file 1: Table S1) and mounted on poly-L-lysine (Sigma) coated #1.5 Coverslips (Grale HDS) enclosed within 8-well sticky chambers (Ibidi, 80828) and stained.

Super-resolution images were recorded using a custom-built (Monash Micro Imaging) dSTORM instrument based on an Olympus IX-71 microscope equipped with Plan ApoChromat oil immersion 100× 1.4NA objective (Olympus part #N1480900), 1.6× magnification changer, Toptica 488 nm laser (200 mW), Gem 561 nm laser (500 mW), and Oxxius 637 nm laser (150 mW), suitable Olympus fluorescence filter cubes and Andor iXon Ultra 897 High Speed EMCCD camera with single-photon sensitivity for single-molecule detection. The final excitation steering mirror and beam expansion lenses were mounted on a translation stage for free adjustment of the total internal reflection fluorescence (TIRF) angle. The system was operated at a TIRF angle appropriate for the respective sample to concentrate excitation power and reduce background fluorescence. Samples were mounted on a manual $x,y$ translation stage to minimize sample drift.

Alexa Fluor® 647 super-resolution imaging was performed in imaging buffer at pH8.0 containing 50 mM MEA and a glucose scavenging system (10% Glucose, 40 µg/ml catalase, 0.5 mg/ml glucose oxidase) in PBS. The sample was illuminated continuously with 70 mW 637 nm laser power at the appropriate TIRF angle. After an initial pumping period of <30 s to drive dyes into the dark state, single-molecule blinking time series were acquired for 10,000 frames at a camera EM gain of 50 and exposure time of 20 ms.

The acquired data were reconstructed to super-resolved images of 10-nm pixel size using the open-source software rapidSTORM version 3.3.1. Blinks with a local signal-to-noise ratio (SNR) <80 were discarded. Images were first color coded for temporal appearance of blinks to detect sample drift, then (if necessary) corrected for drift using the linear drift correction available in rapidSTORM and exported as 8-bit grayscale images.

## Analysis of H3K27me3 staining in dSTORM super-resolution images

Radial histogram analysis was carried out for all representative super-resolution images for each embryonic time point and treatment group. This involved the use of an algorithm (macro was installed and analyzed on ImageJ) to measure staining intensity of H3K27me3, beginning at the center of the cell nucleus out to the nuclear periphery. H3K27me3 intensity was measured in ever-increasing concentric circles (radial diameter) expanding in at 0.1 μm increments from the center of the nucleus until the nuclear lamina was reached. Values on each radial diameter line were collated and averaged to provide values ± standard error of the mean across 3–6 biological replicates. Data were processed and graphed in Microsoft Excel.

## Organ culture

All culture reagents were purchased from Life Technologies unless otherwise stated. Embryos were collected at E11.5 from Swiss females mated to 129T2svJ Oct4-eGFP transgenic males. Gonad plus mesonephros was cultured on 30-mm organotypic cell culture inserts (Merck Millipore; PICM03050) in 1200-μl culture media (250 μm sodium pyruvate, 15 mM Hepes, 1X nonessential amino acids (Life Technologies, 11140), 1 mg/ml N-acetylcysteine (Sigma, A9165), 55 μm β-mercaptoethanol (Life Technologies, 21985) and 10% FCS in DMEM/F12 with Glutamax (Life Technologies, 10565) containing either DMSO (vehicle control) or 10 μm GSK126 (EZH1/2 inhibitor; SelleckChem, S7061). Media preparations also contained 1× penicillin/streptomycin (Life Technologies, 15070). Gonads were randomly allocated to each culture treatment condition and cultured for 48 h in 37 °C/5% $CO_2$ conditions. Culture media was refreshed daily. Gonads were processed for flow cytometry, FACS, IF and dSTORM super-resolution imaging.

## FACS and cell viability

Four to six cultured gonads were collected, pooled together and dissociated with 0.25% trypsin as previously described. The reaction was stopped by adding 500 μl of culture media containing 10% fetal calf serum, and cells were collected by centrifugation. Cells were resuspended in 300 μl culture media containing 10% fetal calf serum and propidium iodide added to a final concentration of 2 μg/ml. Cells were run on an Influx 2 cell sorter. Non-viable cells were quantified and excluded as a percentage of the whole single cell population based on propidium iodide staining. Oct4-eGFP positive (germ) and negative (somatic) cells separated based on eGFP fluorescence and collected in separate tubes. Cells were collected by centrifugation and snap frozen for RNA extraction.

## RNA preparation and expression microarray analysis

RNA was extracted from FACS-purified cells (~20,000 cells per sample, $n = 3$ for vehicle and GSK126 treatments, respectively) using TRIzol reagent, treated with Turbo DNAse 1 (Ambion) according to the manufacturer's instructions and resuspended in RNAase-free water. RNA quantity and quality were assessed using a Qubit instrument and a Bioanalyzer. Total RNA (100 ng/sample) was labeled using Agilent One Colour Low Input Quick Amp Labelling Kit v.6.6, and dye incorporation assessed using a NanoDrop ND-1000 Spectrophotometer. In total, 600 ng of Cy3-labeled cRNA (specific activity >6 pmol Cy3/μg) was fragmented at 60 °C according to the Agilent protocol and hybridized to Agilent SurePrint G3 Mouse Gene Expression 8 × 60K Arrays for 17 h at 67 °C according to manufacturer's instructions. Arrays were washed and immediately scanned on an Agilent C, DNA microarray scanner using one-color scan settings for 8 × 60K arrays. The scanned images were analyzed with Feature Extraction Software 11.0.1.1 (Agilent) using default parameters (protocol GE1-1100_Jul11 and Grid: 028005_D_F_20120201) to obtain background subtracted and spatially detrended Processed Signal intensities. Data were analyzed using GeneSpring GX software using the proprietary analysis pathway. In addition, raw signal intensities for a selection of genes were independently normalized against validated reference genes for fetal germ cells using intensities across all probes for Sdha, Canx and MapK1 and the data expressed as normalized signal intensities.

### Abbreviations

AMH: anti-Mullerian hormone (also known as Mullerian inhibitory substance); Cyto: cytoplasm; EED: Embryonic Ectoderm Development; E: embryonic day; eGFP: enhanced GFP; ERVs: endogenous retroviral elements; ERV1: Class I; ERVK: Class II; EZH1: Enhancer of Zeste 1; EZH2: Enhancer of Zeste 2; FACS: fluorescence-activated cell sorting; GSK126: EZH1/2 inhibitor; H3K9me2: dimethylated lysine 9 on histone 2; H3K9me3: trimethylated lysine 9 on histone 3; H3K27me3: trimethylated lysine 27 on histone 3; IAPs: intracisternal-A-particles; LINE: long interspersed nuclear elements; LINE1: LINE subgroup 1; MVH: mouse vasa homologue (also known as DDX4); PNL: peripheral nuclear localization; PRC2: Polycomb Repressive Complex 2; SETDB1: SET domain, bifurcated 1; SINE: short interspersed elements; SUZ12: Suppressor of Zeste 12; TEs: transposable elements; UNL: uniform nuclear localization.

## Authors' contributions

PW conceived the project. PW, LP, JS and KH designed the experiments. LP, KH, KE and JS performed the experiments. LP, KE, KH, JS and PW analyzed the data. LP, JS and PW wrote the paper. All authors read and approved the final manuscript.

## Author details

[1] Department of Molecular and Translational Science, Centre for Genetic Diseases, Hudson Institute of Medical Research, Monash University, Clayton, VIC 3168, Australia. [2] Monash Micro Imaging, Monash University, Clayton, VIC 3800, Australia.

## Acknowledgements

This work was supported by National Health and Medical Research Grants GNT1043939 and GNT1051223 awarded to PSW, funding from the Monash University Faculty of Medicine, Nursing and Health Sciences funding granted to PW and the Victorian Government's Operational Infrastructure Support Program. LP was supported by an Australian Postgraduate Award. We thank Monash Animal Research Platform staff for assistance with mouse care, Camden Lo (Monash Micro Imaging) for imaging advice and Jodee Gould (Monash Health Translational Precinct Medical Genomics Facility) for assistance with micro-array studies.

## Competing interests

The authors declare that they have no competing interests.

## References

1. Seisenberger S, Andrews S, Krueger F, et al. The dynamics of genome-wide DNA methylation reprogramming in mouse primordial germ cells. Mol Cell. 2012;48(6):849–62.
2. Hackett JA, Sengupta R, Zylicz JJ, et al. Germline DNA demethylation dynamics and imprint erasure through 5-hydroxymethylcytosine. Science. 2013;339(6118):448–52.
3. Hajkova P, Erhardt S, Lane N, et al. Epigenetic reprogramming in mouse primordial germ cells. Mech Dev. 2002;117(1–2):15–23.
4. Guibert S, Forne T, Weber M. Global profiling of DNA methylation erasure in mouse primordial germ cells. Genome Res. 2012;22(4):633–41.
5. Kobayashi H, Sakurai T, Miura F, et al. High-resolution DNA methylome analysis of primordial germ cells identifies gender-specific reprogramming in mice. Genome Res. 2013;23(4):616–27.
6. Liu S, Brind'amour J, Karimi MM, et al. Setdb1 is required for germline development and silencing of H3K9me3-marked endogenous retroviruses in primordial germ cells. Genes Dev. 2014;28(18):2041–55.
7. Hajkova P, Ancelin K, Waldmann T, et al. Chromatin dynamics during epigenetic reprogramming in the mouse germ line. Nature. 2008;452(7189):877–81.
8. Seki Y, Yamaji M, Yabuta Y, et al. Cellular dynamics associated with the genome-wide epigenetic reprogramming in migrating primordial germ cells in mice. Development. 2007;134(14):2627–38.
9. Kagiwada S, Kurimoto K, Hirota T, Yamaji M, Saitou M. Replication-coupled passive DNA demethylation for the erasure of genome imprints in mice. EMBO J. 2013;32(3):340–53.
10. Cao R, Wang L, Wang H, et al. Role of histone H3 lysine 27 methylation in Polycomb-group silencing. Science. 2002;298(5595):1039–43.
11. Czermin B, Melfi R, Mccabe D, Seitz V, Imhof A, Pirrotta V. Drosophila enhancer of Zeste/ESC complexes have a histone H3 methyltransferase activity that marks chromosomal Polycomb sites. Cell. 2002;111(2):185–96.
12. Muller J, Hart CM, Francis NJ, et al. Histone methyltransferase activity of a Drosophila Polycomb group repressor complex. Cell. 2002;111(2):197–208.
13. Denisenko O, Shnyreva M, Suzuki H, Bomsztyk K. Point mutations in the WD40 domain of Eed block its interaction with Ezh2. Mol Cell Biol. 1998;18(10):5634–42.
14. Margueron R, Li G, Sarma K, et al. Ezh1 and Ezh2 maintain repressive chromatin through different mechanisms. Mol Cell. 2008;32(4):503–18.
15. Pasini D, Bracken AP, Jensen MR, Lazzerini Denchi E, Helin K. Suz12 is essential for mouse development and for EZH2 histone methyltransferase activity. EMBO J. 2004;23(20):4061–71.
16. Faust C, Schumacher A, Holdener B, Magnuson T. The eed mutation disrupts anterior mesoderm production in mice. Development. 1995;121(2):273–85.
17. O'carroll D, Erhardt S, Pagani M, Barton SC, Surani MA, Jenuwein T. The polycomb-group gene Ezh2 is required for early mouse development. Mol Cell Biol. 2001;21(13):4330–6.
18. Rinchik EM, Carpenter DA. N-ethyl-N-nitrosourea-induced prenatally lethal mutations define at least two complementation groups within the Embryonic Ectoderm Development (EED) locus in mouse chromosome 7. Mamm Genome. 1993;4(7):349–53.
19. Rinchik EM, Carpenter DA, Selby PB. A strategy for fine-structure functional analysis of a 6- to 11-centimorgan region of mouse chromosome 7 by high-efficiency mutagenesis. Proc Natl Acad Sci USA. 1990;87(3):896–900.
20. Schumacher A, Faust C, Magnuson T. Positional cloning of a global regulator of anterior-posterior patterning in mice. Nature. 1996;384(6610):648.
21. Sauvageau M, Sauvageau G. Polycomb group genes: keeping stem cell activity in balance. PLoS Biol. 2008;6(4):e113.
22. Lessard J, Schumacher A, Thorsteinsdottir U, Van Lohuizen M, Magnuson T, Sauvageau G. Functional antagonism of the Polycomb-Group genes eed and Bmi1 in hemopoietic cell proliferation. Genes Dev. 1999;13(20):2691–703.
23. Mu W, Starmer J, Fedoriw AM, Yee D, Magnuson T. Repression of the soma-specific transcriptome by Polycomb-repressive complex 2 promotes male germ cell development. Genes Dev. 2014;28(18):2056–69.
24. Erhardt S, Su IH, Schneider R, et al. Consequences of the depletion of zygotic and embryonic enhancer of zeste 2 during preimplantation mouse development. Development. 2003;130(18):4235–48.
25. Iovino N, Ciabrelli F, Cavalli G. PRC2 controls Drosophila oocyte cell fate by repressing cell cycle genes. Dev Cell. 2013;26(4):431–9.
26. Gaydos LJ, Rechtsteiner A, Egelhofer TA, Carroll CR, Strome S. Antagonism between MES-4 and Polycomb repressive complex 2 promotes appropriate gene expression in C. elegans germ cells. Cell Rep. 2012;2(5):1169–77.
27. Bracken AP, Dietrich N, Pasini D, Hansen KH, Helin K. Genome-wide mapping of Polycomb target genes unravels their roles in cell fate transitions. Genes Dev. 2006;20(9):1123–36.
28. Bernstein BE, Mikkelsen TS, Xie X, et al. A bivalent chromatin structure marks key developmental genes in embryonic stem cells. Cell. 2006;125(2):315–26.
29. Lesch BJ, Dokshin GA, Young RA, Mccarrey JR, Page DC. A set of genes critical to development is epigenetically poised in mouse germ cells from fetal stages through completion of meiosis. Proc Natl Acad Sci USA. 2013;110(40):16061–6.
30. Sachs M, Onodera C, Blaschke K, Ebata KT, Song JS, Ramalho-Santos M. Bivalent chromatin marks developmental regulatory genes in the mouse embryonic germline in vivo. Cell Rep. 2013;3(6):1777–84.
31. Ng JH, Kumar V, Muratani M, et al. In vivo epigenomic profiling of germ cells reveals germ cell molecular signatures. Dev Cell. 2013;24(3):324–33.

32. Erkek S, Hisano M, Liang CY, et al. Molecular determinants of nucleosome retention at CpG-rich sequences in mouse spermatozoa. Nat Struct Mol Biol. 2013;20(7):868–75.

33. Brykczynska U, Hisano M, Erkek S, et al. Repressive and active histone methylation mark distinct promoters in human and mouse spermatozoa. Nat Struct Mol Biol. 2010;17(6):679–87.

34. Hammoud SS, Nix DA, Zhang H, Purwar J, Carrell DT, Cairns BR. Distinctive chromatin in human sperm packages genes for embryo development. Nature. 2009;460(7254):473–8.

35. Szabo PE, Hubner K, Scholer H, Mann JR. Allele-specific expression of imprinted genes in mouse migratory primordial germ cells. Mech Dev. 2002;115(1–2):157–60.

36. Western PS, Miles DC, Van Den Bergen JA, Burton M, Sinclair AH. Dynamic regulation of mitotic arrest in fetal male germ cells. Stem Cells. 2008;26(2):339–47.

37. Miles DC, Van Den Bergen JA, Sinclair AH, Western PS. Regulation of the female mouse germ cell cycle during entry into meiosis. Cell Cycle. 2010;9(2):408–18.

38. Western PS, Ralli RA, Wakeling SI, et al. Mitotic arrest in teratoma susceptible fetal male germ cells. PLoS ONE. 2011;6(6):e20736.

39. Seki Y, Hayashi K, Itoh K, Mizugaki M, Saitou M, Matsui Y. Extensive and orderly reprogramming of genome-wide chromatin modifications associated with specification and early development of germ cells in mice. Dev Biol. 2005;278(2):440–58.

40. Hilscher B, Hilscher W, Bulthoff-Ohnolz B, et al. Kinetics of gametogenesis. I. Comparative histological and autoradiographic studies of oocytes and transitional prospermatogonia during oogenesis and prespermatogenesis. Cell Tissue Res. 1974;154(4):443–70.

41. Adams IR, Mclaren A. Sexually dimorphic development of mouse primordial germ cells: switching from oogenesis to spermatogenesis. Development. 2002;129(5):1155–64.

42. Bullejos M, Koopman P. Germ cells enter meiosis in a rostro-caudal wave during development of the mouse ovary. Mol Reprod Dev. 2004;68(4):422–8.

43. Mccabe MT, Ott HM, Ganji G, et al. EZH2 inhibition as a therapeutic strategy for lymphoma with EZH2-activating mutations. Nature. 2012;492(7427):108–12.

44. Van Den Bergen JA, Miles DC, Sinclair AH, Western PS. Normalizing gene expression levels in mouse fetal germ cells. Biol Reprod. 2009;81:362–70.

45. Bickmore WA, Van Steensel B. Genome architecture: domain organization of interphase chromosomes. Cell. 2013;152(6):1270–84.

46. Croft JA, Bridger JM, Boyle S, Perry P, Teague P, Bickmore WA. Differences in the localization and morphology of chromosomes in the human nucleus. J Cell Biol. 1999;145(6):1119–31.

47. Walsh CP, Chaillet JR, Bestor TH. Transcription of IAP endogenous retroviruses is constrained by cytosine methylation. Nat Genet. 1998;20(2):116–7.

48. Kato Y, Kaneda M, Hata K, et al. Role of the Dnmt3 family in de novo methylation of imprinted and repetitive sequences during male germ cell development in the mouse. Hum Mol Genet. 2007;16(19):2272–80.

49. Bourc'his D, Bestor TH. Meiotic catastrophe and retrotransposon reactivation in male germ cells lacking Dnmt3L. Nature. 2004;431(7004):96–9.

50. Walter M, Teissandier A, Pérez-Palacios R, Bourc'his D. An epigenetic switch ensures transposon repression upon dynamic loss of DNA methylation in embryonic stem cells. eLife. 2016;5:e11418.

51. Mcfarlane L, Truong V, Palmer J, Wilhelm D. Novel PCR assay for determining the genetic sex of mice. Sex Dev. 2013;7(4):207–11.

52. Wakeling SI, Miles DC, Western PS. Identifying disruptors of male germ cell development by small molecule screening in ex vivo gonad cultures. BMC Res Notes. 2013;6(1):168.

53. Hogg K, Western PS. Differentiation of fetal male germline and gonadal progenitor cells is disrupted in organ cultures containing knockout serum replacement. Stem Cells Dev. 2015;24(24):2899–911.

# PERMISSIONS

The contributors of this book come from diverse backgrounds, making this book a truly international effort. This book will bring forth new frontiers with its revolutionizing research information and detailed analysis of the nascent developments around the world.

We would like to thank all the contributing authors for lending their expertise to make the book truly unique. They have played a crucial role in the development of this book. Without their invaluable contributions this book wouldn't have been possible. They have made vital efforts to compile up to date information on the varied aspects of this subject to make this book a valuable addition to the collection of many professionals and students.

This book was conceptualized with the vision of imparting up-to-date information and advanced data in this field. To ensure the same, a matchless editorial board was set up. Every individual on the board went through rigorous rounds of assessment to prove their worth. After which they invested a large part of their time researching and compiling the most relevant data for our readers.

The editorial board has been involved in producing this book since its inception. They have spent rigorous hours researching and exploring the diverse topics which have resulted in the successful publishing of this book. They have passed on their knowledge of decades through this book. To expedite this challenging task, the publisher supported the team at every step. A small team of assistant editors was also appointed to further simplify the editing procedure and attain best results for the readers.

Apart from the editorial board, the designing team has also invested a significant amount of their time in understanding the subject and creating the most relevant covers. They scrutinized every image to scout for the most suitable representation of the subject and create an appropriate cover for the book.

The publishing team has been an ardent support to the editorial, designing and production team. Their endless efforts to recruit the best for this project, has resulted in the accomplishment of this book. They are a veteran in the field of academics and their pool of knowledge is as vast as their experience in printing. Their expertise and guidance has proved useful at every step. Their uncompromising quality standards have made this book an exceptional effort. Their encouragement from time to time has been an inspiration for everyone.

The publisher and the editorial board hope that this book will prove to be a valuable piece of knowledge for researchers, students, practitioners and scholars across the globe.

# LIST OF CONTRIBUTORS

**Jie Lan**
FR 8.3, Biological Sciences, Genetics/Epigenetics, University of Saarland, Campus A2.4, 66123 Saarbrücken, Germany
Present Address: Faculty of Medicine, Free University of Brussels, C.P. 614, Building GE, 5th floor, 808 Route de Lennik, 1070 Brussels, Belgium

**Konstantin Lepikhov, Pascal Giehr and Joern Walter**
FR 8.3, Biological Sciences, Genetics/Epigenetics, University of Saarland, Campus A2.4, 66123 Saarbrücken, Germany.

**Arlette Rwigemera, Fabien Joao and Geraldine Delbes**
Institut National de la Recherche Scientifique, Centre INRS – Institut Armand-Frappier, 531, boulevard des Prairies, Laval, QC H7V 1B7, Canada

**Masahiro Okada, Mitsuhiro Kanamori, Kazue Someya, Hiroko Nakatsukasa and Akihiko Yoshimura**
Department of Microbiology and Immunology, Keio University School of Medicine, 35 Shinanomachi, Shinjuku-ku, Tokyo 160-8582, Japan

**Lenka Gahurova**
Epigenetics Programme, Babraham Institute, Cambridge CB22 3AT, UK.
Present Address: Laboratory of Developmental Biology and Genetics, Department of Molecular Biology, University of South Bohemia, 37005 Ceske Budejovice, Czech Republic.

**Shin-ichi Tomizawa**
Department of Histology and Cell Biology, School of Medicine, Yokohama City University, Yokohama 236-0004, Japan.

**Sébastien A. Smallwood**
Epigenetics Programme, Babraham Institute, Cambridge CB22 3AT, UK.
Friedrich Miescher Institute for Biomedical Research, 4058 Basel, Switzerland.

**Kathleen R. Stewart-Morgan**
Epigenetics Programme, Babraham Institute, Cambridge CB22 3AT, UK.
Present Address: Biotech Research and Innovation Centre (BRIC), University of Copenhagen, 2200 Copenhagen, Denmark.

**Heba Saadeh**
Epigenetics Programme, Babraham Institute, Cambridge CB22 3AT, UK.
Computer Science Department, KASIT, University of Jordan, Amman, Jordan.

**Jeesun Kim**
Department of Epigenetics and Molecular Carcinogenesis, The University of Texas M.D. Anderson Cancer Center, Smithville, TX 77030, USA.

**Simon R. Andrews**
Bioinformatics Group, Babraham Institute, Cambridge CB22 3AT, UK.

**Taiping Chen**
Department of Epigenetics and Molecular Carcinogenesis, The University of Texas M.D. Anderson Cancer Center, Smithville, TX 77030, USA.

**Gavin Kelsey**
Centre for Trophoblast Research, University of Cambridge, Cambridge CB2 3EG, UK.

**Srihari Radhakrishnan**
Bioinformatics and Computational Biology Program, Iowa State University, Ames, IA 50011, USA.
Department of Ecology, Evolution and Organismal Biology, Iowa State University, 251 Bessey Hall, Ames, IA 50011, USA.

**Robert Literman**
Ecology and Evolutionary Biology Program, Iowa State University, Ames, IA 50011, USA.

**Beatriz Mizoguchi**
Interdepartmental Genetics and Genomics Program, Iowa State University, Ames, IA 50011, USA.
Department of Ecology, Evolution and Organismal Biology, Iowa State University, 251 Bessey Hall, Ames, IA 50011, USA.

**Nicole Valenzuela**
Department of Ecology, Evolution and Organismal Biology, Iowa State University, 251 Bessey Hall, Ames, IA 50011, USA.

**Gautier Richard, Nathalie Prunier-Leterme , Denis Tagu and Gaël Le Trionnaire**
EGI, UMR 1349, INRA, Institut de Génétique, Environnement et Protection des Plantes (IGEPP), Domaine de la Motte, BP 35327, Le Rheu, France.

**Fabrice Legeai and Anthony Bretaudeau**
BIPAA, UMR 2 BIPAA, UMR 1349, INRA, Institut de Génétique, Environnement et Protection des Plantes (IGEPP), Campus Beaulieu, Rennes, France.
Genscale, INRIA, IRISA, Campus Beaulieu, Rennes, France.
Genouest, INRIA, IRISA, Campus Beaulieu, Rennes, France.

**Gaël Le Trionnaire**
CNRS, UMR 6553, EcoBio, University of Rennes 1, 35042 Rennes, France.

**Allwyn Pereira  and Renato Paro**
Department of Biosystems Science and Engineering, ETH Zurich, 4058 Basel, Switzerland.
Faculty of Sciences, University of Basel, 4056 Basel, Switzerland.

**Laia Ribas and Francesc Piferrer**
Institut de Ciències del Mar, Consejo Superior de Investigaciones Científicas (CSIC), Passeig Marítim, 37–45, 08003 Barcelona, Spain.

**Konstantinos Vanezis**
Imperial Centre for Translational and Experimental Medicine, Hammersmith Hospital, Du Cane Road, London W12 0NN, UK.

**Marco Antonio Imués**
Departamento de Recursos Hidrobiológicos, Universidad de Nariño, Torobajo, Pasto, Colombia.

**Jorke H. Kamstra and Peter Aleström**
Faculty of Veterinary Medicine, Department of Basic Sciences and Aquatic Medicine, CoE CERAD, Norwegian University of Life Sciences, Dep., 0033 Oslo, Norway

**Liana Bastos Sales**
Institute for Environmental Studies, VU University Amsterdam, Amsterdam, The Netherlands.

**Juliette Legler**
Institute for Environmental Studies, VU University Amsterdam, Amsterdam, The Netherlands.
Institute for Environment, Health and Societies, College of Health and Life Sciences, Brunel University London, Uxbridge, UK.

**Johannes Petrus Maree and Hugh-George Patterton**
Department of Biochemistry, Stellenbosch University, Matieland 7602, South Africa.

**Megan Lindsay Povelones**
Department of Biology, Pennsylvania State University (Brandywine Campus), Media, PA 19063, USA.

**David Johannes Clark**
Division of Developmental Biology, Eunice Kennedy Shriver National Institute for Child Health and Human Development, National Institutes of Health, Bethesda, MD, USA.

**Gloria Rudenko**
Department of Life Sciences, Imperial College London, South Kensington, London SW7 2AZ, UK.

**Aristea Magaraki, Esther Sleddens-Linkels, Joost Gribnau, Willy M. Baarends and Maureen Eijpe**
Department of Developmental Biology, Erasmus MC, University Medical Center, Rotterdam, The Netherlands.

**Godfried van der Heijden**
Division of Reproductive Medicine, Department of Obstetrics and Gynecology, Erasmus MC, Rotterdam, The Netherlands.

**Anto. ine HF. M. Peters**
Friedrich Miescher Institute for Biomedical Research (FMI), Basel, Switzerland.
Faculty of Sciences, University of Basel, Basel, Switzerland.

**Leonidas Magarakis**
Division of Reproductive Medicine, Department of Obstetrics and Gynecology, Central Hospital of Karlstad, Karlstad, Värmland, Sweden.

**Wiggert A. van Cappellen**
Erasmus Optical Imaging Center, Erasmus MC, Rotterdam, The Netherlands.

**Lexie Prokopuk, Jessica M. Stringer, Kirsten Hogg and Patrick S. Western**
Department of Molecular and Translational Science, Centre for Genetic Diseases, Hudson Institute of Medical Research, Monash University, Clayton, VIC 3168, Australia.

**Kirstin D. Elgass**
Monash Micro Imaging, Monash University, Clayton, VIC 3800, Australia.

# Index

www.ingramcontent.com/pod-product-compliance
Lightning Source LLC
Chambersburg PA
CBHW080405190526
45161CB00003B/136